LEVELS 2 & 3
DIPLOMA

TERRY GRIMWOOD | ANDY JEFFERY

ELECTRICAL INSTALLATIONS

BUILDINGS AND STRUCTURES

Published by Pearson Education Limited, Edinburgh Gate, Harlow, Essex, CM20 2JE.

www.pearsonschoolsandfecolleges.co.uk

Text © Pearson Education Limited 2013
Edited by Liz Cartmell
Index by Indexing Specialists (UK) Ltd
Typeset by Tek-Art, Crawley Down, West Sussex
Original illustrations © Pearson Education 2013
Illustrated by Tek-Art, Crawley Down, West Sussex
Cover photo/illustration © SuperStock Science Photo Library

The rights of Andy Jeffery, Terry Grimwood and Damian McGeary to be identified as authors of this work have been asserted by them in accordance with the Copyright, Designs and Patents Act 1988.

W 621.31924 GRI

First published 2013

16 15 14 13

10 9 8 7 6 5 4 3 2 1

British Library Cataloguing in Publication Data
A catalogue record for this book is available from the British Library

ISBN 978 1 447 94025 8

Copyright notice
All rights reserved. No part of this publication may be reproduced in any form or by any means (including photocopying or storing it in any medium by electronic means and whether or not transiently or incidentally to some other use of this publication) without the written permission of the copyright owner, except in accordance with the provisions of the Copyright, Designs and Patents Act 1988 or under the terms of a licence issued by the Copyright Licensing Agency, Saffron House, 6–10 Kirby Street, London EC1N 8TS (www.cla.co.uk). Applications for the copyright owner's written permission should be addressed to the publisher.

Printed in Slovakia by Neografia

Acknowledgements
The authors and publisher would like to thank the following for their kind permission to reproduce their photographs:

(Key: b-bottom; c-centre; l-left; r-right; t-top)

Alamy Images: Adrian Sherratt 68, Art Directors and TRIP 317b, Catchlight Visual Service 231, David J. Green 312, 323c, 325, DWImages 241, Geoff du Feu 305t, Imagebroker 485, Ingram Publishing 202, Jean Schweitzer 480, Jim West 102, John Boud 174cr, Kris Bailey 35, Mark Richardson 110, Mint Photography 474, Pat Tuson 475t, Paul Glendell 306, Phil Degginger 30, Pixel Shepherd 216t, Richard Heyes 169 (Scutch hammer), 363, Robert Wilkinson 316b, Roberto Orecchia 269c, stu49 322tl, 330, Wayne Hutchinson 482, Wiskerke 471, ZUMA Wire Service 485b; **Construction Photography:** BuildPix 476, Jean-Francois Cordella 196; **Corbis:** 97; **CSCS:** / g_studio / Photos.com 522, / Jupiter Images / Photos.com 523; **Digital Vision:** 443; **DK Images:** Peter Anderson 86; **Fotolia.com:** knee0 348t, Maksym Dykha 195, seraphine5 348b, Silvano Rebai 161, tr3gi 341, xalanx 510; **Sid**
Frisby: 305bl; **Getty Images:** 216b; **Imagemore Co., Ltd:** 91, 265r, 453; **Imagestate Media:** John Foxx Collection 197, 204, 379, 457; **Masterfile UK Ltd:** 512, 514; **Pearson Education Ltd:** Clark Wiseman / Studio 8 74, 83, 164t, 166 (Pad saw), 167, 169 (Claw hammer), 169 (Soft face mallet), 170c, 172tr, 172cr, 172bl, 173, 177t, 192, 280l, 280r, 319, 322br, 425t, 426t, 502, Coleman Yuen 270, David Sanderson 76l, 76r, 95, Gareth Boden 84t, 84b, 88c, 88bl, 88br, 105, 130b, 134 (FP Gold), 134 (PVC multicore), 134 (Single), 138b, 144, 145, 162, 164b, 165, 166 (Jig saw), 166 (Wood saw), 169 (Ball pein hammer), 169 (Lump hammer), 170t, 170b, 174tr, 174br, 175b, 180t, 180c, 184, 186, 311, 316tl, 328, 404, 408t, 445, 455c, 455b, 458c, 458b, HL Studios 92, 96t, 96b, 97b, 97bl, Joey Chan 266b, 456, Jules Selmes 70, 93, 116, 131, 138t, 139, 141, 164c, 171 (Flat file), 171 (Half round file), 171 (Hand file), 171 (Knife file), 171 (Square file), 171b, 176bl, 177c, 178, 179, 181, 188, 190, 314, 317t, 331, 333, 409bl, 416, 422, 423, 425b, 426b, 427, Ken Vail Graphic Design 118, 125, Naki Photography 34, 72, 134 (Armoured), 134 (Flexible), 134 (Mineral insulated), 137, 143, 149, 174tl, 301, 408c, 408b, 409tl, 409tc, 409bc, 454t, 454c, Stuart Cox 71, Trevor Clifford 166 (Hacksaw), 171 (Round file), 171 (Three square file), 176c, Tsz-shan Kwok 418; **PhotoDisc:** 127, 130cr, 172br, 201, 264l, 264r, 315, 452; **Photos.com:** 215, Comstock 352, Marco Hegner 134 (Tri-rated); **Rex Features:** SIPA / Niviere-Chamussy 199; **Science Photo Library Ltd:** GIPhotoStock 130cl, 323t, 323b, Simon Fraser 472; **Shutterstock.com:** 7505811966 269t, Alaettin YILDIRIM 305br, Deymos 175c, Flegere 107, Igorsky 316tc, Ivaschenko Roman 265l, Jouke van Keulen 266t, kriangkrai wangjai 439, Martin Fischer 237, Nomad_Soul 393, Patrick Power 65, Piorr Wardynski 269b, pokchu 26, R-O-M-A PAGE DESIGN, Skyline 305c; **Sozaijiten:** 111; **SuperStock:** Glow Images 337, imagebroker.net 467; **Veer/Corbis:** Alexander Yurinsky 419, Bluewren 100, Darren 479, hoch2wo 499, Martin33 1, Tatyana Alekiseva-Sabeva 475b; **www.imagesource.com:** 66

All other images © Pearson Education

In some instances we have been unable to trace the owners of copyright material, and we would appreciate any information that would enable us to do so.

Websites
Pearson Education Limited is not responsible for the content of any external internet sites. It is essential for tutors to preview each website before using it in class so as to ensure that the URL is still accurate, relevant and appropriate. We suggest that tutors bookmark useful websites and consider enabling students to access them through the school/college intranet.

Acknowledgements

Every effort has been made to contact copyright holders of material reproduced in this book. Any omissions will be rectified in subsequent printings if notice is given to the publishers. The following materials have been reproduced with kind permission from the following organisations:

Renewables First for the table of minimum head and flow rates, page 489.

Contents

Introduction

This book supports the Level 2 and Level 3 Diploma in Electrical Installations (Buildings and Structures) currently offered by City & Guilds and EAL, although at the time of writing other awarding organisations may be developing similar qualifications. The Diploma has been prepared by Summit Skills, who have been working with the awarding organisations to provide a qualification for those seeking a career in the electrotechnical industry.

The Diploma has been approved on the Qualifications and Credit Framework (QCF); the government framework which regulates all vocational qualifications. The QCF ensures that qualifications are structured and titled consistently and that they are quality assured.

The book has been written by vocational lecturers with many years of experience in both the electrical and other associated industries as well as in further education, where they currently teach electrical qualifications.

Who the qualification is aimed at

The standard industry electrical course, the Level 3 NVQ Diploma is funded only for those who are working as apprentices in the electrical industry. The Diploma in Electrical Installations would be suitable for people who are:

- not currently working in the sector but ware looking for a qualification to support their efforts to find employment as electrician improvers and trainees
- working in the industry but do not have the academic qualifications, or breadth of work, to undertake the Level 3 NVQ Diploma.

In essence:

- The Level 2 qualification is designed for people who are new to the industry and provides the basic skills and experience they need.
- The Level 3 qualification is aimed at those who have already completed the Level 2 qualification or have some relevant experience and knowledge of the industry.

This qualification tests both practical and knowledge-based skills. It will not qualify you as an electrician. For this, you will need to meet certain performance criteria. These can be found in the National Occupational Standards, which were created and structured by Summit Skills. Successful completion of Level 3 NVQ Diploma will meet the requirements necessary for entry into the profession. It is possible to map the qualifications gained in the Level 3 Diploma across to the Level 3 NVQ Diploma and be left to complete only the outstanding performance units.

Your tutor or assessor will be able to explain how you may progress onto the Level 3 NVQ Diploma. However, you should be aware that the relevant performance units need to be carried out in industry.

About this book

This book supports both Level 2 and Level 3 Diplomas in Electrical Installations in a single volume, enabling you to progress upwards from Level 2, but at the same time to use the resource as revision if you are taking the Level 3 qualification.

The chapters match the qualification units as follows (numbering refers to the City & Guild 2365 qualification units).

LEVEL 2

Unit 201	Health and safety in building services engineering
Unit 202	Principles of electrical science
Unit 203	Electrical installations technology
Unit 204	Installation of wiring systems and enclosures
Unit 210	Understand how to communicate with others within building services engineering

LEVEL 3

Unit 301	Understand the fundamental principles and requirements of environmental technology systems
Unit 302	Principles of electrical science
Unit 303	Electrical installations: fault diagnosis and rectification
Unit 304	Electrical installations: inspection, testing and commissioning
Unit 305	Electrical systems design
Unit 308	Career awareness in building services engineering

EAL Electrical Installation qualifications relating to content in this book include:

- Level 2 600/6724/X
- Level 3 501/1605/8.

Each unit consists of a set of outcomes which, in turn, form the structure of each chapter. As far as possible the order in which the outcomes run through each unit is followed. However, there are occasions when the order has been changed to create a logical path through the chapter. In all cases the entire content of each unit is explored.

Installation of wiring systems and enclosures is a predominantly practical unit. The associated chapter, therefore, includes some practical exercises such as cutting, threading and bending conduit. It concentrates on the basic skills, tools and tasks needed in the electrical industry rather than providing a detailed handbook of workshop projects.

There are progress and knowledge checks throughout the book, which will enable you to assess your own level of knowledge and understanding at various stages of the course.

Using this book

This book is to be used as part of your training and also as a reference work to support your career within the electrical industry. It is not intended to be a handbook to an actual installation. It is not a code of practice or guidance note. British Standards, manufacturer's data or HSE and IET documents are among the material to be used for actual installation work.

Features of this book

This book has been fully illustrated with artworks and photographs. These will help to give you more information about a concept or a procedure, as well as helping you to follow a step-by-step procedure or identify a particular tool or method.

This book as contains a number of different features to help your learning and development.

Safety tip

This feature, colour-coded to match industry practice, gives you guidance for working safely on the tasks in this book.

Progress check

A series of short questions, usually appearing at the end of each learning outcome, give you the opportunity to check and revise your knowledge.

Key terms

These are new or difficult words. They are picked out in **bold** in the text and defined in the margin.

Knowledge check

This feature is a series of multiple choice questions at the end of each chapter.

Working practice

These features show professional practice case studies that focus on employability skills and problem solving.

ACKNOWLEDGEMENTS

Pearson would like to thank all those who have contributed to the development of this book, making sure that standards and quality remained high through to the final product.

Particular thanks go to Richard Swann who has reviewed the book and made many invaluable comments to ensure accuracy throughout.

Thanks also go to Glen Lambert, Carl Weymouth and Thomas Hendriks at Oaklands College for their patience, assistance, advice and support during the photo shoot.

Andy and Terry would both like to thank their colleagues in the electrical department at Oaklands College, and Pen Gresford for her enthusiasm and encouragement. Terry would like to thank Jessica for her endless support and Andy would like to thank his wife, Andrea, and sons, Matthew and Sam, for their support, advice and proofreading. He would also like to thank Olivia Medcalf for her ideas on the communications unit.

Candidate handbook answers

The answers to all the questions in this book can be found on the Training Resource Disk. They can also be downloaded from Pearson's website at the following URL: www.pearsonfe.co.uk/ElectricalInstDiploma

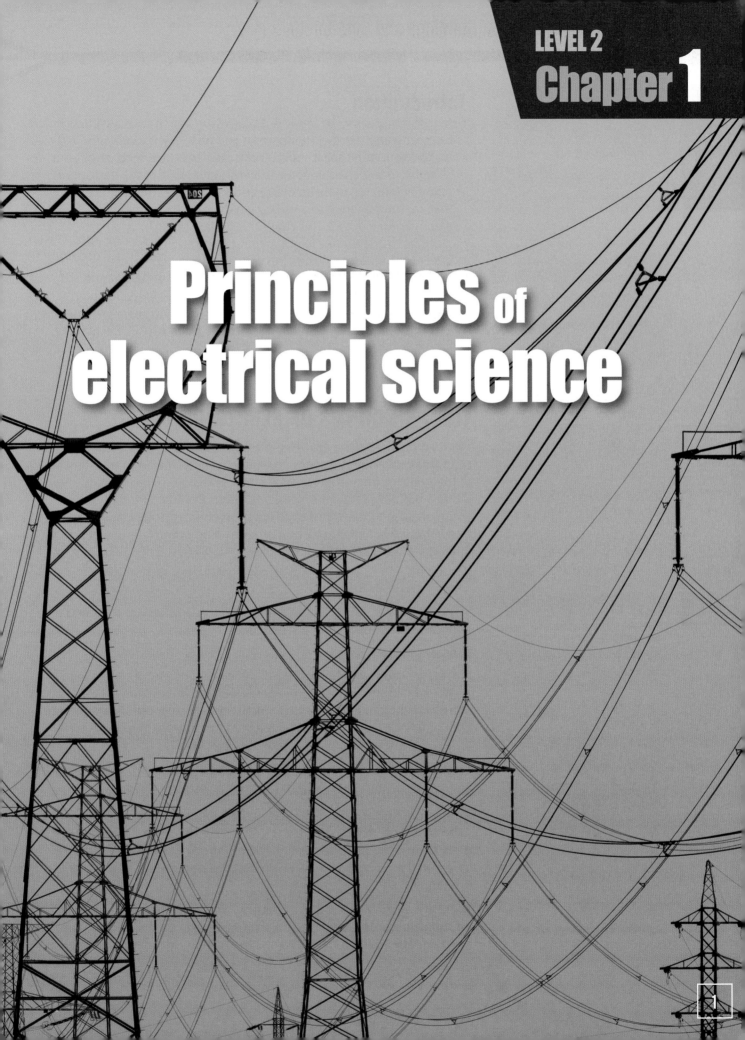

Principles of electrical science

Chapter 1

This chapter covers:

- the principles of electricity
- the principles of basic electrical circuits
- the principles of magnetism and electromagnetism
- the operating principles of a range of electrical equipment
- the principles of basic mechanics
- the principles of a.c. theory
- electrical quantities in Star Delta configurations.

Introduction

Electricity and the science behind it can be tricky, especially because you can't see it – all you can do is see the effects of it or feel it. In this chapter you will investigate some of the basic scientific principles of electricity, and the applications and uses that are fundamental to a technician's role in the building services engineering industry. This chapter is broken down into logical sections to enable you to grasp the science and progress on to the Level 3 science syllabus. You will be introduced to the basic concepts followed by worked examples, backed up by a range of practice questions at different levels. The order of the learning outcomes has been changed slightly from the C&G specification in some instances: this is to allow a logical build-up of knowledge before you progress to the next subject.

THE PRINCIPLES OF ELECTRICITY

Electricity is simply the movement of electrons carrying energy. As electrons move along a conductor they have the ability to transfer some of their energy into other forms such as heat, light or movement. Your job as a building services engineer is to install and maintain systems that make use of this transferred electron energy. Electricians need to understand what electricity is in order to predict how it will react under different conditions.

LO1

The structure of matter

In order to understand electricity, we need to also consider some basic scientific principles.

Matter

Matter can come in different states and can change from one state to another. An example is ice turning to water and then steam. All three states – the ice, water and steam – have the same molecular chemical structure; they simply change state when subjected to a temperature change.

The atom

An atom is made up of subatomic particles and it is the properties of an atom that are of particular interest to an electrician.

Atoms consist of a very small nucleus carrying a positive charge surrounded by orbiting electrons that are negatively charged. The nucleus of an atom is made up of small particles called neutrons and protons (imagine the nucleus to be the Sun and the electrons to be the planets orbiting it).

Normally, atoms are electrically neutral, that is, the negative charge of the orbiting electrons is equal to the positive charge on the nucleus. One of the general rules that applies to electricity (and magnetism) is that like

Key term

Matter – something that has weight and takes up space.

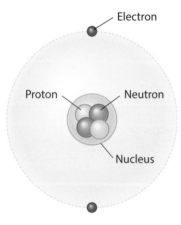

Figure 1.1: Atom showing nucleus with electrons, neutrons and protons

charges repel and unlike charges attract, so within an atom the electrons in the outer shell are attracted to the positive protons. Neutrons are neutral in charge but act as the glue holding the atom together. It is the electrons in the outer shell that determine if a material is going to be a good insulator (like glass) or a good conductor (like copper). In insulators the electrons are firmly bound to the atom taking a lot of force to break them free. In good conductors such as metals, the electrons appear to be relatively free and with a little encouragement they can be broken off and set free to find something more positive. It is this movement of 'free' electrons that is known as current and hence electricity. A free-moving electron is said to be a 'charge carrier' because it is carrying a negative charge of energy as it moves through a conductor.

To get a feel of the strength of an electric current, imagine the flow of negatively charged electrons as a flow of water down a river. A strong current would be a large amount of water passing a point in a certain time. So a strong or large electric current would be a large volume of electric **charge** (free electrons) passing a point in a certain time. It is important for you to understand charge because without it there would be no electricity. Charge is a quantity of electricity and it is the movement of charge that gives rise to current flow in a circuit.

Charge, Q, is measured in **coulombs**; time, t, is measured in seconds and current, I, is measured in **amperes**.

The rate of flow of electric charge is therefore calculated using the following formula:

$$I = \frac{Q}{t}$$

> **Key terms**
>
> *Charge* – sometimes called electric charge or electrostatic charge, it is a quantity of electricity. Charge can be positive (+ve) or negative (–ve).
>
> *Coulomb* – one coulomb is 6.24×10^{18} electrons.
>
> *Ampere* – one ampere equals one coulomb of electrons passing by every second.
>
> *Luminaire* – a term used in the electrical industry to describe the whole light fitting, including the housing, reflector, lamp and any internal control gear.

Worked example

If a charge of 180 coulombs (180 C) flows through a **luminaire** every minute, what is the electric current in the luminaire?

Q = 180 C

t = 1 minute (or 60 seconds)

$I = \frac{Q}{t}$

Putting the values in place of the symbols gives:

$I = \frac{180}{60}$

I = 3 A

Or, 3 amps will flow through the lamp if a charge of 180 coulombs passed through it in 1 minute.

> **Link**
>
> For another worked example, go to www.pearsonfe.co.uk/ElectricalInstDiploma.

Activity 1.1

1 If a charge of 180 C flows in a luminaire every 2 minutes, what is the current?
2 What is the charge in coulombs if a current of 3.2 A flows for 2 seconds?

Potential difference and voltage

A current will flow if 'free' electrons are encouraged to break away from an atom. What encourages this flow to happen? Using the example of water, imagine water in a pond and also water in a river. Water in a pond is still, so there is no flow or current. Water in a river, however, does have the potential to flow. The strength of the current flow depends on the difference in height between the top of the river and the bottom – the steeper the drop, the faster the current. In an electrical circuit this is the same basic principle.

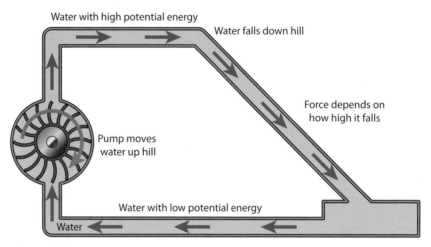

Figure 1.2: Flow of water and potential energy

An electric circuit has a high 'charge' point and a lower 'charge' point. These are the +ve and −ve terminals of a battery of supply. The height in the circuit is known as the voltage and the difference between the high point and the low point is called the 'potential difference' (otherwise known as p.d.). The battery or electrical source acts as a pump, moving the charge to the top of the hill where it has potential energy and then it can fall down the hill giving up potential energy in the form of heat or light (for example, going through a heat element or a luminaire). Potential difference, often referred to as 'voltage across' or simply 'voltage', is measured in volts and the greater the p.d., the greater potential for current flow in a circuit.

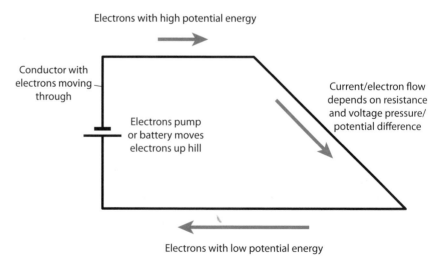

Figure 1.3: Flow of electrons and potential difference

Sources of an electromotive force (emf)

There are a number of ways in which electromotive force, or emf, can be generated. Electricity can be the output of a chemical, magnetic or thermal reaction. The way electricity is generated depends on a number of factors and circumstances.

Magnetic emf source – if electricity is required at a location where there happens to be a large volume of moving water, then it makes sense, from an ecological point of view, to use this force of nature to turn a turbine. By moving a conductor within a magnetic field, a current can be generated in that conductor as long as there is a circuit. This effect is called electromagnetic induction and this is how a basic turbine works.

Magnetic field + conductive circuit + movement = electromotive force and current flow

Thermal emf source – if two metals with different chemical properties are connected together, an emf will be generated across the junction if the two sides of the junction are at slightly different temperatures. Effectively there is a potential difference between the two sides that is enough to cause electrons to want to move. The principle of converting heat energy directly to electrical energy is known as the Seebeck or thermoelectric effect. A use for this effect is the thermocouple where the device is used to measure temperature.

Chemical emf source – a battery cell works on the principle of creating an emf by chemical reaction. Within the cell there is a chemical called an acid solution or 'electrolyte'. If two different types of metal strips are put into the acid, an electromotive force appears between the two plates. If the circuit is completed by joining the metal plates at the top with a conductor, a path is now available for the free electrons to move through.

The −ve plate is the zinc and the +ve plate is made of copper. As the connected plates are put into the acid electrolyte solution, the zinc will start to dissolve. The zinc atoms then leave the zinc plate and enter the solution. As each zinc atom leaves the plate and enters the acid, it leaves behind two electrons. This has the effect of making the zinc plate more negative and the atom more positive (due to losing two of its electrons). The +ve zinc atoms are not attracted to the +ve hydrogen atoms in the acid solution and they are pushed to the −ve charged copper plate where the hydrogen grabs a copper electron. When the reaction occurs between the copper electron and the hydrogen atom, the copper plate becomes more +ve due to the loss of an electron. This reaction causes a potential difference to appear between the two plates, creating what is effectively a 'chemical pump' or an emf source.

Key term

Electromotive force (emf) – a source of energy that can cause current to flow.

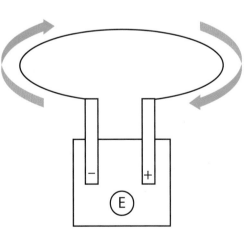

Figure 1.4: Electron flow through an electrolyte

Effects of electric current

Chemical effect – as described with a battery cell, by passing a current through an electrolyte a chemical reaction will take place. The process that happens can be useful. One such process is called 'electrolysis' and is used to electroplate metals. An example of electroplating is chrome plating to give a soft metal a harder-wearing finish. The chrome coating is useful for tools or car parts to stop corrosion or rust.

Magnetic effect – if you connect a battery to a circuit consisting of a lamp, conductor and a switch, and turn the circuit on, several things happen. Firstly, the lamp goes on. If you put an ammeter into the circuit you would notice the current growing from '0' to a value. If you had a meter that could measure magnetism, you would notice that as the current grows, a magnetic field would grow. The magnetic field would grow to a maximum at the same time as the current – they are directly related and proportional.

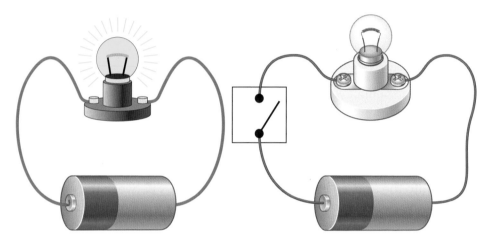

Figure 1.5: Simple battery circuit and broken circuit

Thermal effect – electrons passing through a conductor of any type will produce a third effect: heat. As an electron is given energy to move by the emf, some energy is converted into other forms. This might be light and heat. Remember, energy is never lost, it is only ever converted into another form. Sometimes this heating effect is useful, as in a heating element of a fire or an infrared light. At other times the heating effect of current flow is not useful – cables overheating in a domestic installation causing a house fire.

Electrical quantities and SI units

You will come across many terms, symbols and abbreviations when dealing with the science around electricity. It is very important to get a good understanding of how they are used and applied.

The language of science has been standardised into the SI system of units so that it can be understood all over the world. Most countries now use SI units.

There are seven base units, which are shown in Table 1.1. Notice the columns: base quantity (the quantity we are actually using), base unit (the full measurement name) and the symbol (the symbol is a shorter version of the base unit and comes after the number to show what the number represents, for example the temperature is 20 kelvin or 20 K). You will need to be familiar with these units.

Base quantity	Base unit	Symbol
Length	metre	m
Mass	kilogram	kg
Time	second	s
Electric current	ampere	A
Temperature	kelvin	K
Amount of substance	mole	mol
Luminous intensity	candela	cd

Table 1.1: SI base units

As well as the base units you will need to be familiar with others called derived units. Some base units and derived units can be seen in Table 1.2 below with their meanings, descriptions and some applications.

Base quantity	Base unit	Symbol	Description
Area	metre squared	m^2	Used to measure a surface such as a floor area for numbers and types of circuits required (2 × A1 circuits if the floor area is 200 m^2 – BS 7671:2008 Onsite Guide Appendix H, standard circuits for households).
Length	metre	m	Used to measure the length of a cable run.
Volume	metres cubed	m^3	Used to measure the capacity of a heating system cylinder.
Mass	kilogram	kg	The amount of material. This must not be confused with weight. Building services (BS) engineers need to know how much energy is required to change mass from one state to another.
Weight	newton	N	Weight takes into account the mass of an object and the effect of gravity on it. Mass is constant but gravity and hence weight can vary (compare Moon and Earth gravity!).
Temperature (t)	kelvin	K	Note 0°K = −275°C or 'absolute zero' Also, 0°C = 275°K and a change of 1°K = 1°C
Energy (W)	joule	J	The ability to do work.
Time	second	s	60 s = 1 minute and 60 mins = 1 hour. BS engineers need to know this to work out speeds for motors (revs/min vs revs/s).
Force (F)	newton	N	You need to know how much force is applied on a conductor in a magnetic field to work out motor torque.

Table 1.2: SI derived units

Continued ▼

Base quantity	Base unit	Symbol	Description
Electric current (I)	ampere	A	The symbol 'I' actually comes from the original French name l'intensité du courant, as comparisons were made against the intensity of flow of water (still a very good analogy).
Luminous intensity	candela	Cd	You may have heard of the term candle power. This is the light power at a source of illumination.
Magnetic flux (Ø)	weber	Wb	If you can, imagine the flux lines coming out of a magnet.
Magnetic flux density (B)	weber per metre squared or tesla	Wb/m^2 or T	An example of how you can actually work out the formula from the units – webers/metres squared or flux divided by area.
Frequency	herz	Hz	The number of full cycles of a sine wave that occur in one second: 50 Hz in the UK and 60 Hz in the USA.
Resistance	ohm	Ω	The level of resistance a conductor will put up against current flow. It is affected by the material, heat, length and cross-sectional area of the conductor.
Voltage pressure or potential difference or simply voltage	volt	V	Voltage is a measure of the pressure available to force electrons to break free from their atom and start flowing. If there is a different level of voltage between two points on a circuit, there is said to be a voltage drop or potential difference.
Resistivity (ρ)	ohm metre square	Ωm^2	A unique constant number given to individual materials that can be used to work out what a material's resistance is likely to be. All materials have different resistivity values and hence resistance (or opposition to current flow).

Table 1.2: SI derived units

Multiples and sub-multiples

These units sometimes lead to very small or very large numbers that are not always easy to deal with in calculations. For this reason, multiples and sub-multiples exist and are given special names and prefix symbols such as those shown in Table 1.3.

Multiplier	Name	Symbol prefix	As a power of 10
1 000 000 000 000	tera	T	1×10^{12}
1 000 000 000	giga	G	1×10^{9}
1 000 000	mega	M	1×10^{6}
1 000	kilo	k	1×10^{3}
1	unit		
0.001	milli	m	1×10^{-3}
0.000 001	micro	μ	1×10^{-6}
0.000 000 001	nano	n	1×10^{-9}
0.000 000 000 001	pico	p	1×10^{-12}

Table 1.3: Common names and prefixes for multipliers

Table 1.3 shows that multipliers are arranged into groups of thousands, millions, billions and thousandths, millionths, billionths, etc. By dividing a number by 1 000 and replacing the 0s with a prefix k, the number looks simpler and, providing you have good calculator skills, it becomes much easier to handle.

Worked example

Convert 34 000 000 J into a number with a prefix.

Divided by 1 000 it would become 34 000 kJ.

If you divide 34 000 000 by 1 000 000 it would become 34 MJ.

Link

For another worked example, go to www.pearsonfe.co.uk/ ElectricalInstDiploma.

Progress check 1.1

1 What does the symbol prefix 'm' mean?
2 What does the prefix 'μ' mean?
3 What is the base unit for weight?

Activity 1.2

1 Convert 0.0048 F into a number with a prefix.
2 Convert 4.5 mF into a number without a prefix.
3 Show 1.2 TN as MN.
4 Change 64 μf into mf.
5 Represent 0.000 000 135 A as mA.

Electrical maths for building services engineers

You will be aware by now that a certain amount of maths is required in the building services sector. Amongst others, you will need to be able to work out:

- values of current a circuit can carry
- the size of a cable based on what it is going to power
- the force on a conductor
- the speed of a motor
- the missing sides or angles on power triangles
- the torque of a turning force.

The list goes on – in fact, most aspects of a professional engineer's work require maths. Some of the most important maths skills which are going to be covered in this section include:

- basic rules of numbers
- transposition of formulae
- trigonometry, triangles and Pythagoras.

You will use a wide range of formulae by the end of Level 3 training and the formulae will rarely be ready to use – you will have to change it around to make what you want to find the subject. The rules are fairly straightforward but the key is lots of practice until it becomes second nature. Once you have practised and mastered transposition you will be able to change around the most complex formulae as needed. Understanding the actual number you have been given or which you have taken off a meter is also essential, because if you get it wrong you could put yourself or others at serious risk.

Basic mathematical concepts

For all engineers to get the same correct answers, everyone needs to be working to the same basic rules.

You may have seen the term BODMAS before. It is used to decide in what order you should complete a calculation. If you have a mathematical expression and need to find out what it is equal to, then you need to tackle it in the right order. If you do it in the wrong order you will get the wrong answer, as can be seen in the following example:

$5 \times (5 + 5) = 50$ (complete what is inside the brackets first to give $5 \times 10 = 50$)

If this was done in a different order you could get:

$5 \times 5 + 5 = 25 + 5 = 30$ (giving a wrong answer)

How does BODMAS work?

BODMAS stands for:

Brackets	If there are brackets around any part of the maths expression then this must be acted on first, ignoring all other parts.
Other operations	This refers to powers and roots being dealt with next.
Division	Division and multiplication must be completed next, starting from the left.
Multiplication	Multiplication and division both have the same priority but you must start from the left and work to the right. Note that sometimes a number might be next to a bracket – even though there is no '×' sign, it still means multiplication needs to happen.
Addition	Having completed all the other maths operations, there should only be addition and subtraction left in the expression to deal with.
Subtraction	Addition and subtraction have the same priority so start from the left and work to the right until all the terms have been completed.

Worked example

$4(7 - 3)^2 \times 5 + 3$

Brackets:

$(7 - 3) = 4$

Operations:

$(4)^2 = 16$

Division and multiplication:

There is no division so move on to multiplication (starting from left!)

$4(16) = 64$

$64 \times 5 = 320$

Addition and subtraction:

Final solution:

$320 + 3 = 323$

Link

For another worked example, go to www.pearsonfe.co.uk/ElectricalInstDiploma.

Activity 1.3

Resolve the following mathematical expressions:

1 $(9 + 7) \times 1 + 3$
2 $(4 + 4 - 3 + 1)^2 + 3 + 3$
3 $(7 + 8) \div 3 + 3$
4 $(7 + 7 \times 7) \times (3 + 3)$

Formulae transposition (rearranging formulae)

The ability to change formulae around is an important skill for engineers. If there are three things in a formula, you normally will know two of the values but need to work out the third. This is where the difficulty occurs, as the formula will not necessarily be arranged in such a way to give you the answer. You will need to rearrange the formula to find what you need.

There are several ways to approach formulae transposition. The best way is to practise long hand until you are completely confident.

Worked example

$5 = 3 + 2$

We can rearrange this simple formula in a number of different ways:

$5 - 3 = 2$

$5 - 2 = 3$

$-3 = -5 + 2$ (or $-3 = 2 - 5$)

$-2 = 3 - 5$ (or $-2 = -5 + 3$)

If the number moves to the other side of the '=' it changes sign from + to – or – to +.

Numbers can also be replaced with letters, as follows:

A = B + C

A – B = C

A – C = B

–C = –A + B (or –C = B – A)

–B = C – A (or –B = –A + C)

When a number or symbol moves from one side of the 'equals' to the other, the sign must change from +ve to –ve or vice versa.

Multiplication and division operations

Electrical equations are not all simply addition and subtraction: other functions such as multiplication, division, square roots and squares are common.

Worked example

Rearrange the following formula to make 'I' the subject.

Imagine you have numbers for V and R (V = 100 volts and R = 50 ohms) but you need to work out what I is equal to.

V = IR

It doesn't matter what side of the '=' the 'I' is but it needs to be alone. To do this you need to remove the 'R' by rearranging the formula.

One option is to divide by 'R'. If you divide something by itself the answer is 1 – regardless of whether it is a number or even a symbol!

This means if you divide R by R the answer will be 1. If you divide one side of the formula by something the rule states you must do the same to the other side as well. This keeps the formula balanced (like a see-saw). Remember this, as it is very important!

$$V = I \times R$$

$$\frac{V}{R} = \frac{I \times R}{R}$$

Both sides have been divided by R. Remember, if you have R divided by R this is the same as 1 so they must cancel out.

$$\frac{V}{R} = \frac{I \times \cancel{R}}{\cancel{R}}$$

$$\frac{V}{R} = I$$

Notice R has moved from one side to the other to leave the 'I' the subject, all by itself. Also note that in moving from one side to the other it has turned from multiplication to division (moved from above the line to below the line). This would be the same if it were to move in the reverse direction. This means that if something moves from one side of a formula to the other, the operation becomes the opposite.

Examples:

$\sqrt{A}$ becomes A^2 and vice versa

Ax becomes $\frac{1}{A}$

Summary of simple rules

Rule 1

Whatever you do to one side of a formula you do to the other side also.

Rule 2

If something is divided by itself it is equal to '1' and can be cancelled out/simplified.

Rule 3

To make something the subject of a formula, you need to get it by itself.

Rule 4

By moving a term from one side of the formula to the other, you reverse the operation so divide becomes multiply and vice versa.

Trigonometry

One of the most important maths tools that you will need to use is trigonometry. Trigonometry is used to find missing angles or missing sides of a right-angled triangle. This is necessary in many electrical applications such as:

- power factors
- impedance triangles
- a.c. single-phase theory
- a.c. three-phase theory
- power triangles.

The 'opposite' is always the side opposite the angle being considered.

If the angle being investigated changed to the top right-hand corner then the opposite and adjacent would swap – simply change around the sides to suit.

The hypotenuse is always the longest side. The adjacent is always the side that is left after you have worked out the others – simple!

Sine, cosine and tangent

Now you know how to work out hypotenuse, adjacent and opposite sides, the angles can be introduced. Sine, cosine and tangent functions on your calculator can be used to find the relationship between the sides of a triangle:

$$\text{Sin } \emptyset = \frac{\text{Opposite}}{\text{Hypotenuse}}$$

$$\text{Cos } \emptyset = \frac{\text{Adjacent}}{\text{Hypotenuse}}$$

$$\text{Tan } \emptyset = \frac{\text{Opposite}}{\text{Adjacent}}$$

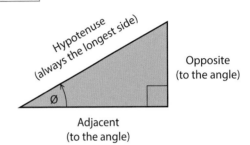

Figure 1.6: Pythagoras' theorem

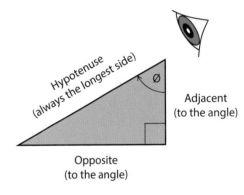

Figure 1.7: How to identify the sides and angles in Pythagoras' theorem

Link

For more information on Pythagoras' theorem, go to www.pearsonfe.co.uk/ElectricalInstDiploma.

To remember these formulae use the following name, **SOHCAHTOA**, which shows:

- **sine** (shortened to sin) of an angle equals **opposite** over **hypotenuse**
- **cosine** (shortened to cos) of an angle equals **adjacent** over **hypotenuse**
- **tangent** (shortened to tan) of an angle equals **opposite** over **adjacent**.

To find the angle when you know two sides you use the INV function (or SHIFT) depending on the two sides you have. If you have the opposite and the hypotenuse, divide them and this will give you a value for sine Ø. To find the actual angle Ø, you press the INV button on the calculator before pressing the sine button. This will convert a decimal value to an actual angle.

Link

For another worked example, go to www.pearsonfe.co.uk/ElectricalInstDiploma.

Worked example

The opposite is 4.9 cm and the hypotenuse is 6.9 cm – what is the angle?

$$\text{Sin } \varnothing = \frac{\text{Opposite}}{\text{Hypotenuse}}$$

$$\text{Sin } \varnothing = \frac{4.9}{6.9}$$

$$\text{Sin } \varnothing = 0.71$$

$$(\text{INV}) \text{ Sin}^{-1} 0.71 = 45.23°$$

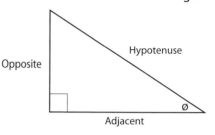

Activity 1.4

1 The opposite is 33 m and the hypotenuse is 80 m – what is the angle?

2 The adjacent is 49 mm and the angle is 60° – how long is the hypotenuse?

3 The hypotenuse is 1 km and the angle is 24° – how long is the adjacent?

LO2

THE PRINCIPLES OF BASIC ELECTRICAL CIRCUITS

Calculating the resistance of a conductor

Conductors and insulators

The atomic molecular structure of a material can determine whether it is going to make a good conductor, insulator or something in between. If the atomic structure has very few free electrons in its outer shell then it is likely to be a good conductor. If the outer shell of the atom has lots of electrons it is likely to be a good insulator.

Resistance and resistivity

A material's ability to resist the flow of electrons through it depends on the type of material it is. Each type of material is unique and will have a different effect on current flow. The unique property of a conductor that affects this is called resistivity, symbol ρ (Greek letter 'rho'). The resistance of a conductor is directly proportional to the resistivity of the material – if the resistivity is high, the resistance to current flow will be high.

Resistance also goes up when the distance that the current has to flow increases – the further you go, the more effort it is.

To return to the water analogy – if there is a large pipe and a small pipe, it is much easier for water to flow in the large pipe. This is true for electricity and conductors. If there is a very thin conductor it will be more difficult for electrons to flow through it, as there are fewer atoms with free electrons available. A smaller cross-sectional area will reduce current flow and create a higher resistance. This means resistance is inversely proportional to cross-sectional area.

Given the resistivity, cross-sectional area and the length of a conductor, it is therefore possible to work out the resistance of any conductor. Putting all of these factors together in a formula gives:

$$R = \frac{\rho l}{A}$$

Worked example

Calculate the resistance of an aluminium cable of length 10 km and diameter 4 mm.

Firstly, convert all the units to the correct dimensions – never mix dimensions!

$10 \text{ km} = 10 \times 10^3$ (1 000 m in a km)

$4 \text{ mm} = 4 \times 10^{-3}$ (1 000 mm in a m)

Resistivity, ρ, of aluminium is 28.2×10^{-9} (see Table 1.4 on page 16).

$A = \frac{1}{4} \pi d^2$

$A = \frac{1}{4} \pi (4 \times 10^{-3})^2$

$A = \frac{1}{4} \pi 4 \times 10^{-6} = 3.142 \times 10^{-6}$

(Notice what happens with the powers inside and outside the brackets when they are combined.)

Using the formula $R = \frac{\rho l}{A}$, and replacing the symbols with the correct numbers:

$R = \dfrac{(28.2 \times 10^{-9} \times 10 \times 10^3)}{(3.142 \times 10^{-6})}$

$R = 89.75 \ \Omega$

Link

For more worked examples, go to www.pearsonfe.co.uk/ElectricalInstDiploma.

Activity 1.5

Work out the resistance of the following conductors:

1 a 25 m length of 2.5 mm² copper
2 400 m of 1.5 mm² copper
3 50 m of 8 mm² aluminium
4 25 m of 6 mm² copper
5 100 m of 1 mm² copper earth conductor.

Material	Resistivity (Ωm)	Conductor or insulator?
Silver	15.9×10^{-9}	Conductor
Copper	17.5×10^{-9}	Conductor
Gold	24.4×10^{-9}	Conductor
Aluminium	28.2×10^{-9}	Conductor
Tungsten	56×10^{-9}	Conductor
Brass	75.0×10^{-9}	Conductor
Iron	100×10^{-9}	Conductor
Sea water	2×10^{-1}	Poor conductor
Silicon	6.2×10^{2}	Semiconductor
Glass	10×10^{10} to 10×10^{14}	Insulator
Wood	1×10^{8} to 1×10^{11}	Insulator
Hard rubber	1×10^{13}	Insulator
Air	1.3×10^{16} to 3.3×10^{16}	Insulator

Table 1.4: A table of material resistivity

The formula for resistance shows that resistivity, length and area directly affect resistance but there is a fourth thing that can affect resistance, temperature. Table 1.4 shows resistance specifically at room temperature, 20°C. This is not always the case as you will install in very hot locations such as boiler rooms or very cold locations such as industrial freezers. Heat excites electrons and makes them move around faster, bumping into each other on their travels. This movement causes more resistance. If a conductor is placed in a hot environment, there will be proportionally more resistance (think how much less gets done on a hot sunny day!). The amount the resistance changes with temperature is due to the temperature coefficient of the particular conductor and is given by the formula:

$R_f = R_o(1 + \alpha t)$

where t is the temperature change in °C, α is the temperature coefficient (measured in $\Omega/\Omega/°C$: assume copper to have a value of $0.004\ \Omega/\Omega/°C$), R_o is the resistance at 0°C (in Ω) and R_f is the final resistance (in Ω).

Worked example

A length of copper has a resistance of $R_o = 2\ \Omega$ at 0°C – what would its resistance change to if the temperature increased to 20°C?

$R_f = R_o(1 + \alpha t)$

$R_f = 2(1 + 0.004 \times 20)$

$R_f = 2.16\ \Omega$

Applying Ohm's law

A voltage source connected to a conductor to create a circuit will cause current to flow but conductors can be connected together in different ways.

Imagine two identical conductors are connected together end to end. Their length has now doubled. If they are identical, the overall resistance has also doubled because the free electrons travel twice as far and overcome twice the resistance. This is called a 'series circuit'.

If the identical conductors are placed next to each other, there are two identical paths for the free electrons to flow down. This is a 'parallel circuit'. Because there are two identical paths for the electrons to flow down, they split evenly and half will go down one route and the other half down the other route. The extra path means the overall resistance to electron flow in a parallel circuit such as this has halved. Consider the relationship between voltage pressure, current and the resistance of a conductor. All of these three points are reliant on each other. This relationship is described by Ohm's law.

The current through a metallic conductor, maintained at constant temperature, is directly proportional to the potential difference between its ends:

$I \, \alpha \, V$

This means conductors that have a constant value for the ratio of $\frac{V}{I}$ must follow Ohm's law:

$R = \frac{V}{I}$

Consider a simple series circuit. If 'R' is constant and the voltage pressure increases, this means the current must also increase for the balance to continue. This is just like a tank of water with a tap at the bottom. Imagine the height of the water in the tank being the voltage pressure and the rate of flow of the water out of the tap being the current. As the water in the tank reduces, the flow reduces.

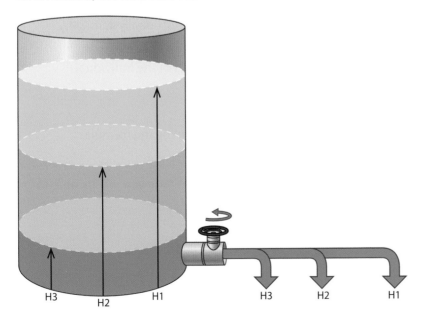

Figure 1.8: Water pressure in a tank – a water electricity analogy

Calculating power in basic electrical circuits

Resistance is mostly a constant value, and voltage and current can change. You have seen that the resistance of a conductor is based on conductor material, temperature, the cross-sectional area and length. All that needs to be discussed now is the effect on resistance, current and voltage when you connect resistive conductors together in different ways.

Imagine the resistance of a conductor is measured using an ohmmeter and its value noted down. Now take a further identical conductor and connect the two ends to make it longer – you would expect the new ohm reading to be double. This is correct as you have created a conductive path twice as long for the electrons to travel along, hence double the resistance.

Series circuits

For a series circuit to exist, resistors must be connected end to end (see Figure 1.9). It does not matter if the resistors are individual pieces of copper conductor or discrete resistor components – the fact is they have resistance and they are connected in a line. The current only has one path to flow down so the electrons cannot be tempted down any other paths.

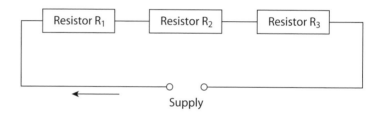

Figure 1.9: Series circuit

The voltage pressure required to force electrons to break free and flow is across the resistors. If, as per Figure 1.9, there are three resistors, there will be three voltage pressures, one across each resistor. The supply voltage is across all three but this will be divided up proportionally across each resistor, depending on the size of the resistor the current has to overcome.

The resistance in a series circuit must be the total of the individual resistances added together:

$R_t = R_1 + R_2 + R_3$

The relationship between R, V and I is also known by the formula:

$R = \dfrac{V}{I}$

Based on these two facts and a known supply voltage you will be able to work out what current is being forced to flow.

Worked example

The supply in the circuit is 12 V and the resistors are all 2 Ω each.

Step 1 – add the resistors up:

$R_t = R_1 + R_2 + R_3$

$R_t = 2 + 2 + 2 = 6\,\Omega$

Step 2 – now apply the relationship formula called Ohm's law (rearrange to make I the subject, as discussed earlier in the chapter, and remember current is measured in amps, A).

Step 3 – put the known values into the formula and find the value of current flowing in this circuit:

$$I = \frac{V}{R}$$

$$I = \frac{12}{6} = 2\,A$$

Because there is only one path for the current, the current will be the same wherever it is measured in a series circuit and is said to be constant. The only way the current will change is if the voltage supply is changed or the resistors are replaced with other values.

Let's look at this series circuit in a little more detail. If the current is a constant 2 A anywhere in this series circuit and the resistor values are all fixed at 2 Ω, it should be possible to work out the individual voltage pressure across a single resistor by applying Ohm's law again. Ohm's law can be applied to each part of the circuit.

The current in resistor 1 is 2 A, the resistor has a value of 2 Ω, so using Ohm's law:

$$I = \frac{V_1}{R_1}$$

$V_1 = I \times R_1$

$V_1 = 4\,V$

This calculation is carried out across all three resistors, giving 4 V across each resistor. The potential difference between one end of the resistor and the other is known as voltage drop.

If all the individual voltage drops across each resistor were added up, it would equal the supply voltage for this series circuit:

$V_t = V_1 + V_2 + V_3$

$V_t = 4 + 4 + 4 = 12\,V$

Voltage drop and the regulations

If there is a very long cable it is effectively made up of lots of series of connected resistances. If a **load** is connected to the very end of the cable you would need to ensure there was a high enough voltage available to run the device.

> **Key term**
>
> *Load* – refers to any component or device that requires power in a circuit, for example a light, motor or cooker.

Because of voltage drop in cables there is guidance given in BS 7671 (Electrical Installation Requirements, 17th edition).

BS 7671 states that lighting circuits should not exceed 3 per cent and power circuits should not exceed 5 per cent.

Safe working

All electrical circuits must be checked for voltage drop. If the voltage drop is greater than 3 per cent for lighting and 5 per cent for power, the circuit needs to be redesigned and will probably require a larger conductor.

Activity 1.6

For each of the five examples below, calculate:

- the total resistance
- the total circuit current
- the voltage drop across each resistor.

1 A series circuit has three resistors: $R_1 = 3\ \Omega$, $R_2 = 6\ \Omega$, $R_3 = 9\ \Omega$ and is connected across a 12 V supply.

2 A series circuit has four resistors: $R_1 = 1.2\ \Omega$, $R_2 = 2.6\ \Omega$, $R_3 = 9\ \Omega$, $R_4 = 9\ \Omega$ and is connected across a 24 V supply.

3 A series circuit has five resistors: $R_1 = 12\ \Omega$, $R_2 = 36\ \Omega$, $R_3 = 29\ \Omega$, $R_4 = 5\ \Omega$, $R_5 = 6\ \Omega$ and is connected across a 36 V supply.

4 A series circuit has four resistors: $R_1 = 2\ k\Omega$, $R_2 = 3\ k\Omega$, $R_3 = 6.7\ k\Omega$, $R_4 = 5\ k\Omega$ and is connected across a 230 V supply.

5 A series circuit has four resistors: $R_1 = 1.7\ M\Omega$, $R_2 = 2.3\ M\Omega$, $R_3 = 2.9\ M\Omega$, $R_4 = 5.7\ M\Omega$ and is connected across a 400 V supply.

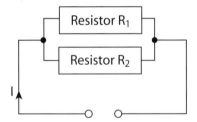

Figure 1.10: Parallel circuit

Parallel circuits

Instead of connecting the conductors end on to make one long resistive conductor, imagine connecting them side by side (see Figure 1.10).

In a parallel circuit the voltage source will push the electrons around until they reach a junction. At the junction the electrons have two choices – they now have two paths to travel down. Current is very lazy and will travel down the path of least resistance. If the paths were of equal resistance, an equal proportion of electrons would flow down each one – the resistance is effectively halved. If there were three equal resistors or conductors connected in parallel (next to each other), then the overall resistance effect would be one third, allowing the current to split into three equal parts. This is different to the series circuit where the current was constant. Now the voltage is a constant pressure as it is connected across the three resistances. The current is now the thing that varies as it leaves the source, splits into three at the junction, passes through each of the three resistors and then joins back together again at the other junction.

$$I_t = I_1 + I_2 + I_3$$

With resistors in series it is straightforward to find the overall effect – simply add them up. Resistors in parallel need to be added up as fractions as we have seen; two identical resistors in parallel equate to half of one resistor and three identical resistors in parallel equate to one third of one resistor.

Key fact

In a parallel circuit, the current varies and the voltage is a constant.

The relationship and calculation for parallel resistors is found by the following formula (three in this case):

$$\frac{1}{R_t} = \frac{1}{R_1} + \frac{1}{R_2}$$

$$\frac{1}{R_t} = \frac{R_2 + R_1}{R_1 \times R_2}$$

$$\frac{R_t}{1} = \frac{R_2 \times R_1}{R_1 + R_2}$$

Worked example

Now consider $R_1 = 5\ \Omega$ and $R_2 = 7\ \Omega$:

$$\frac{1}{R_t} = \frac{1}{R_1} + \frac{1}{R_2}$$

$$\frac{1}{R_t} = \frac{R_2 + R_1}{R_1 \times R_2} \quad \text{now, put in the values:}$$

$$\frac{R_t}{1} = \frac{5 \times 7}{7 + 5}$$

$$R_t = \frac{35}{12}$$

$$R_t = 2.92\ \Omega$$

If the supply voltage is 12 V, what is the total circuit current?

The total resistance is 2.92Ω and the total supply voltage is 12 V, so apply Ohm's law to find the total circuit current:

$$I = \frac{V}{R}$$

$$I = \frac{12}{2.92} = 4.1\ A$$

If the total circuit current is 4.1 A and this is the value that leaves the supply travelling towards the junction between the two resistors – what happens when it reaches this junction? It splits – but how much goes down each branch? Apply Ohm's law to each resistor to find out.

Step 1 – the total current is 4.1 A and the total voltage 'dropped across' both resistors in parallel is 12 V. If the voltage across R_1 (5 Ω) is 12 V, applying Ohm's law to resistor R_1 gives the following:

$$I_1 = \frac{V}{R_1}$$

$$I_1 = \frac{12}{5} = 2.4\ A$$

Step 2 – now applying Ohm's law to the other resistor, R_2, gives the following:

$$I_2 = \frac{V}{R_2}$$

$$I_2 = \frac{12}{7} = 1.7\ A$$

Step 3 – to prove the branch currents are correct, simply add them:

$I_t = 2.4 + 1.7 = 4.1$ A

Tip: always sketch out the circuit and label it as it will make the task quicker in the long run.

More than two parallel resistors

If multiple resistors are connected in parallel, fractions can still be used to add them up. However, a much easier method is to use a scientific calculator. The function is $\boxed{x^{-1}}$. This function effectively divides your resistor into 1, i.e. $\frac{1}{R}$. To add up three resistors, say 3 Ω each, follow the sequence:

$3 \boxed{x^{-1}} + 3 \boxed{x^{-1}} + 3 \boxed{x^{-1}} =$. All we have worked out so far is $\frac{1}{R_t}$, so to find R_t, press $\boxed{x^{-1}} =$ again and you have your answer, 1, as expected.

Activity 1.7

1 A parallel circuit consists of two resistors: $R_1 = 2.4$ Ω and $R_2 = 1.2$ Ω. If they are connected to a 12 V supply, calculate:

   (a) the total resistance
   (b) the total circuit current
   (c) the current in each resistor.

2 The circuit in Question 1 has a third resistor connected in parallel of 5.9 Ω. Carry out the same calculations and find:

   (a) the total resistance
   (b) the new total circuit current
   (c) the current in each of the three legs.

3 A parallel circuit consists of four resistors: $R_1 = 4$ Ω, $R_2 = 1.8$ Ω, $R_3 = 3.7$ Ω and $R_2 = 6.7$ Ω. If they are connected to a 24 V supply, calculate:

   (a) the total resistance
   (b) the total circuit current
   (c) the current in each resistor.

4 The circuit in Question 3 has a fifth resistor connected in parallel of 12.3 Ω and the voltage supply is changed to 40 V. Carry out the same calculations and find:

   (a) the total resistance
   (b) the new total circuit current
   (c) the current in each of the five legs.

5 The circuit consists of four resistors connected in parallel. If $R_1 = 12$ Ω, $R_2 = R_1 + 20\%$, $R_3 = R_2 + 20\%$, $R_4 = R_3 + 20\%$ and if the circuit is connected to a 36 V supply, calculate and find:

   (a) the individual resistor values
   (b) the total resistance
   (c) the total circuit current
   (d) the current in each of the four legs.

Combined circuits – series and parallel

An electrical circuit may not be a simple series or parallel circuit. Electrical circuits may consist of a combination. The rule is to simplify wherever you can and remember you can always apply Ohm's law to one single part of a circuit. A small group of parallel resistors can be turned into an equivalent single resistor. This process of simplification can be repeated until there is one equivalent resistor for the whole network. Once the total resistance has been calculated, the total current can be calculated and then individual voltages and currents can be found using Ohm's law. It is also worth remembering that voltages in a series circuit will always add up to the supply voltage, and currents in a parallel circuit will always add up to the total circuit current. This is a good way to check your calculations.

Worked example

Calculate the total resistance of this circuit and the current flowing through the circuit when the applied voltage is 110 V.

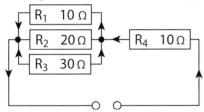

Step 1: Find the equivalent resistance of the parallel group (R_p):

$$\frac{1}{R_p} = \frac{1}{R_1} + \frac{1}{R_2} + \frac{1}{R_3}$$

$$\frac{1}{R_p} = \frac{1}{10} + \frac{1}{20} + \frac{1}{30}$$

$$\frac{1}{R_p} = \frac{6+3+2}{60} = \frac{11}{60}$$

Therefore: $R_p = \frac{60}{11} = 5.45\ \Omega$

Step 2: Add the equivalent resistor to the series resistor R_4:

$$R_t = R_p + R_4$$

$$R_t = 5.45 + 10 = 15.45\ \Omega$$

Step 3: Calculate the current:

$$I = \frac{V}{R_t} = \frac{110}{15.45} = 7.12\ A$$

Activity 1.8

For Questions 1 and 2 below, calculate:

- the total circuit resistance
- the total circuit current
- the voltage drop across each resistor in the network.

1 A parallel resistor network containing four equal 4 Ω resistors is connected to a fifth series resistor, R_5, with a resistance of 12 Ω. If the whole circuit is connected across a 12 V supply, carry out the calculations above.

2 The same circuit as Question 1 has all five resistors replaced with 3 Ω resistors and the supply is changed to 24 V. Carry out the calculations above.

Power in a circuit

Power is defined as the rate at which work is done. Clearly, when an electron is forced free from an atom and pushed around a circuit, work must have been done. Power is also the heating effect of passing a current through a conductor – you probably have a heater at home that has a power rating written on it somewhere. The power rating describes how much heat can be expected – the higher the power rating, the higher the heating effect.

The basic power formula for a circuit with a known voltage and current is given by the following:

$P = I \times V$

Ohm's law can be used to give another power formula:

$V = I \times R$

The Ohm's law formula for V can be put into the power formula:

$P = I \times (I \times R)$

$P = I^2 \times R$

By rearranging Ohm's law again to make 'I' the subject, you can put this into the power formula to get a third formula for power as follows:

$I = \dfrac{V}{R}$

$P = \left(\dfrac{V}{R}\right) \times V$

$P = \dfrac{V^2}{R}$

Link

For more worked examples, go to www.pearsonfe.co.uk/ ElectricalInstDiploma.

Worked example

A resistor has a current of 20 mA passing through it when a voltage supply of 200 V is connected across it. What is the power?

$P = V \times I$

$P = 200 \times 20 \times 10^{-3}$

$P = 4\,W$

Activity 1.9

1 Calculate the power if a 100 mΩ resistor is connected to a 36 V supply.

2 What is the resistor value if the power is 10.2 W when a current of 3.9 A passes through it?

3 If the voltage across a 1.2 kΩ resistor is 210 V, what is the power dissipated?

4 A voltmeter registers 125 V across a resistor and an ammeter reads 30 A. What is the power dissipated in the resistor?

5 What is the voltage drop across a 15 Ω resistor when the power measures 120 W?

How instruments are connected in circuits in order to measure electrical quantities

The majority of tests (except resistance tests) are all live. For this reason great care must be taken and all guidance followed. The use of test instruments is where science and practical skills come together.

Some very important electrical terms and principles have been established. Firstly, electrons move from one ion to another causing the effect 'current flow'. The statement 'current flows through a conductor' is therefore a fact worth remembering. Secondly, the potential difference that exists from one side of an electrical circuit to the other side is the pressure responsible for encouraging the current flow. The statement 'potential difference is measured across a circuit or a load' is also well worth remembering.

Before the meter is connected

There are a few things that need to be considered before connecting a meter to a circuit. Unfortunately electricians do get electric shocks and a number of accidents actually occur at the testing stage. You need to be sure of the following:

1 Am I qualified to use this meter?

2 Do I have the correct meter for the job?

3 Do the instruments, leads, probes and accessories meet the Health and Safety Guidance Note GS38 standards?

4 Is it damage-free and working correctly, and how can I prove this?

5 How do I know what to set it on and how does it work?

6 What am I trying to test and therefore how is it connected?

Once all of these points have been answered, you should be able to go to the next step and choose your meter for the test.

Measuring current – ammeter

As shown previously, current needs to flow through a load to create an effect (light, movement of a motor, heat from a resistor or heater element). For an ammeter to register current flow it needs the current to be directed through it. An ammeter must be connected in series. If you are measuring small currents this is not a great issue. However, there is always a risk when you disconnect a circuit to connect an ammeter in series with a load.

When measuring current in such a way it is a good idea to have an idea of the level of current expected, as your meter might not have the necessary range for the job. It is also a good idea to set the meter on the largest current range. The meter range can be turned down to fine tune the reading and get a more accurate reading (more decimal places). It is not always practical or safe to disconnect a circuit, as per Figure 1.12. By connecting an ammeter to a circuit, you are diverting all the current in the circuit through the meter. For this reason the ammeter must have a very small internal resistance or the circuit resistance and current will be affected.

⚠ Safe working

Current

Current flows through a conductor.

Current flows *in* a circuit – *never across* a circuit!

Voltage

Potential difference is the voltage difference between two points in a circuit.

Potential difference (voltage) is measured *across* a circuit or *across* a conductor – *never through* a circuit!

❗ Safe working

Test instrumentation must be made in accordance with BSEN 61243-3 (2 pole voltage detectors) and BSEN 61010 or BSEN 61557 for instruments.

Figure 1.11: Ammeter in circuit

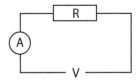

Figure 1.12: Multimeter in circuit

Figure 1.13: Clamp meter measuring a large current

Safe working

Never break into a live circuit to connect a series ammeter – the result could be fatal. Only in specific circumstances will highly trained electricians with special permission work live.

Measuring large currents

Breaking a circuit that is carrying 100 A to connect a series ammeter would be very unwise. If you are lucky enough to live, it will be an extremely frightening experience! If the circuit is safely isolated, the meter and connecting clips must still be able to take the largest current – this is still to be avoided. By using the principle of magnetic induction and clamp meters the circuit can stay connected without having to actually become part of the electrical circuit. It should be noted that it is still very dangerous with such high levels of current so all the normal precautions must be taken, as well as checking the meter for damage before and after use.

As current flows through the conductor, a magnetic field is generated. If a further coil is placed around the current-carrying conductor, this magnetic field can be picked up. The magnetic field will cut the conductors in the secondary coil and induce a further current that will travel through the meter. The value of current travelling through the meter can be stepped down so the value can be read easily. The main supply cable is acting like the primary side of a transformer and the meter clamp coils are acting like the secondary side of a transformer. This is why this type of arrangement is sometimes called a CT or current transformer. There is no actual electrical contact between the supply cable and the clamp meter – all the current being measured is induced by magnetism. The current and voltage levels in the clamp meter are still extremely high and dangerous, and great care should be taken to short the CT out before trying to move the meter. (The science of transformers will be covered later.) A CT meter may be used if you were measuring a 400 V 100 A supply with a clamp meter. Within the CT meter the current may be stepped down so it is in the 0–10 A range. This would mean you could have at the terminals of your clamp meter 4000 V at 10 A – enough to kill you several times over!

Measuring voltage – voltmeter

Voltage pressure supplies the driving force to cause current flow. A potential difference will exist across a load to make the electrons pass through the load. A voltmeter is therefore connected across a load in order to get a reading. See Figures 1.14 and 1.15 below for connections.

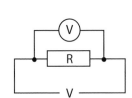

Figure 1.14: Voltmeter in circuit

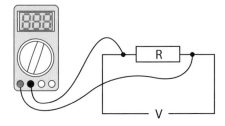

Figure 1.15: Multimeter measuring voltage

Whereas the ammeter has a very low internal resistance to encourage current to flow through it, the voltmeter has a very high resistance. To measure voltage as accurately as possible, most of the current should continue flowing through the load and only a very small proportion of current should pass through the voltmeter. There will always be a small error introduced by using a voltmeter or ammeter as you are adding components to the circuit you are measuring, but these can be minimised by careful calibration.

Measuring large voltages

Large voltages exist on the national grid and in large industrial sites and buildings. It is not wise to try and take live readings using a standard meter, even if it does have the range. Large installations often have voltage and current meters built in to the control panels so they can be monitored constantly or at a glance. Using the same magnetic transformer principles as the CT meter, large voltages can be measured safely. A large supply voltage cable will have a secondary coil on a transformer to step the voltage down to an acceptable level for metering equipment. As you will find out in the next section, stepping a voltage down will have the opposite effect on current and step it up.

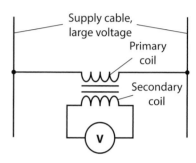

Figure 1.16: Voltage transformer to measure large supply voltages

Measuring resistance – ohmmeter

To measure the resistance of a circuit, it must be dead and safely isolated. Resistance is found by applying Ohm's law. The ohmmeter is essentially a combined ammeter and voltmeter.

$$R = \frac{V}{I}$$

If the voltage is known, the internal battery supply of the meter and the current can be measured by the ammeter, then Ohm's law can be applied. The ohmmeter connects the meter battery across the load to be measured. The current that flows around the circuit is dependent on the resistor. The meter divides the battery voltage by the current flowing and gives the resultant value for resistance. Care has to be taken to make sure any extra resistance introduced by the test leads is taken into account. The lead resistance can be measured by touching the probes together and then making a note that this needs to be taken off the final reading. Most modern meters have a built 'null' facility. By pushing the null button while holding the probes/crocodile clips together, the meter display will return to zero and is ready to be reconnected to the circuit to be tested.

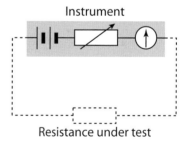

Figure 1.17: Ohmmeter

Measuring power quantities – wattmeter

In the same way that an ohmmeter uses the combination of voltage meter and ammeter, the wattmeter can follow the same principle to measure power in watts. For a d.c. circuit, power can be measured by taking a voltmeter and ammeter reading and then multiplying together:

$$P = V \times I$$

In the a.c. world this is not as straightforward as there are other things to consider such as capacitance, inductance and impedance – these are reactions to a.c. current that will be covered later in the chapter. These reactive components have a slightly different effect on the current if compared to just resistance, and need to be taken into account. For this reason a.c. single-phase power is measured using the same formula as above but is also multiplied by a 'power factor' as follows:

$P = V \times I \times$ (Power factor)

$P = VI \cos \varnothing$

A full description of the 'types of power' will be expanded on in a later chapter but at this stage it can be considered as a decimal multiplier that is a value less than 1 (1 being perfect). Power factor at this stage can also be considered as a power efficiency multiplier that needs to be considered in all a.c. circuits that contain a capacitor or coil (power factor will be covered fully later on in this chapter). You will need to measure the true power being consumed in a circuit which is why a specially connected wattmeter is required. The wattmeter usually has four terminals – two for the voltage coil and two for the current coil, as shown in Figure 1.18.

Measuring three-phase power quantities

If a single-phase supply can be measured in the way just described, then so can one phase on a three-phase supply, in theory. This can only happen if all the phases are balanced – drawing the same current and at the same voltage level. If the three phases are balanced, then the reading for one phase can be simply multiplied by three to give the total power. The wattmeter would be connected, as shown in Figure 1.19 below. You would only take a power reading like this if the three phases were balanced, otherwise there could be errors.

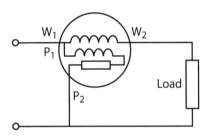

Figure 1.18: Wattmeter connected to load

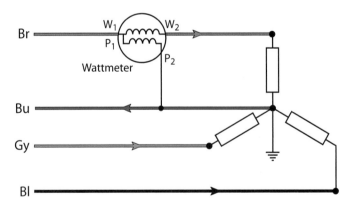

Figure 1.19: One-meter method

True power in a balanced three-phase supply = Voltage × current × power factor

The following equation shows the balanced power/current across the supply:

$P = \sqrt{3} \, V_L \, I_L \cos \varnothing$

Measuring three-phase power (unbalanced supply)

In most cases you will not be able to guarantee a balanced load in a three-phase supply. This is due to the large variations caused by different load requirements – simply, people on each phase change their mind about how much electricity they need so it constantly varies.

To take a measurement on an unbalanced supply requires three wattmeters to be connected, readings taken and then added up. The meters would be connected as shown in Figure 1.20.

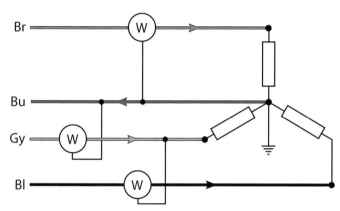

Figure 1.20: Three-phase circuit

Each of the wattmeter's readings would then be taken and added to give an overall value:

True three-phase unbalanced power = power W_1 + power W_2 + power W_3

There are lots of other methods to measure power but they are not covered at this level of study.

THE PRINCIPLES OF MAGNETISM AND ELECTROMAGNETISM

LO3

The physics involved in magnetism is very complex, but even without going into too much detail it is useful to remember that it is a fundamental force that attracts or repels certain types of materials, and that it occurs because of two atomic causes: the spin and orbital motions of electrons. Therefore, the magnetic characteristics of a material can change when alloyed with other elements. The unique class of materials that are strongly affected by magnetism are called ferromagnetic and they will attract anything that contains iron.

A permanent magnet is a material that, when placed into a strong magnetic field, will exhibit a magnetic field of its own, and continue to exhibit a magnetic field once it has been removed from the original field. This magnetic field is continuous without losing strength, as long as it is not subject to changes in environment such as temperature, a de-magnetising field or being hit. Magnetism is a difficult concept as all you can see are the effects of magnetism. The invisible lines, called flux lines, can only be seen by sprinkling iron filings on to paper that is close to a magnet.

Ferromagnetic metals that are attracted by magnets include iron, steel and other iron alloys, but also nickel and cobalt.

Magnetic materials and applications

If you moved a ferromagnetic material such as a pin close to a permanent magnet, the pin would be strongly attracted to the magnet.

- So magnetic effect can happen at a distance.

The pin is not initially magnetised but magnetic properties are induced into the pin.

- Magnetic induction never causes repulsion.

The material that the pin is made out of will determine if the pin stays magnetised long after it is removed from the permanent magnet. Some materials will stay magnetised, while others will lose it fairly quickly:

- Soft magnetic materials like stalloy, a soft iron alloy, can be magnetised and demagnetised easily. Typically, they are used in transformer cores and electromagnets.

- Hard magnetic materials like alnico and alcomax are very hard and can be made into very strong permanent magnets.

Rules of magnetism

There are several features that magnets display:

- Magnets have two poles, North and South.
- The magnetic lines of force (flux) never cross.
- Flux lines always form a closed loop.
- If the flux lines distort when brought close to another magnetic field they will always return to their original shape when moved away again.
- Outside of the magnet, flux lines run North to South.
- A magnet placed in a magnetic field experiences a force on it.
- The higher the concentration of flux lines, the stronger the magnet.
- Like poles repel.
- Unlike poles attract.

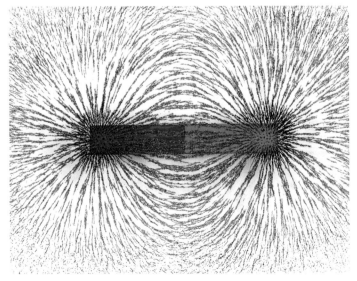

Figure 1.21: Iron filings around a bar magnet

Strength of a magnet

From Figure 1.21 you can see the lines of flux. The concentration of flux lines determines how strong the magnet will be. Magnetic flux, with the symbol φ, is measured in webers, Wb. The concentration of flux lines or magnetic flux density, B, is how many flux lines are in a specific area of the magnet. Hence, it is measured in webers/m² and found by the formula:

$$B = \frac{\varphi}{A}$$

where A is the cross-sectional area measured in m² and φ, the magnetic flux, is measured in Wb. It is worth noting that magnetic flux density is also measured in tesla, T. It has two units that are both perfectly acceptable.

Worked example 1

Calculate the magnetic flux density of a magnet with a flux 2 Wb and cross-sectional area of 0.13 m^2.

$$B = \frac{\varphi}{A}$$

$$B = \frac{2}{0.13} = 15.38$$

Link

For more worked examples, go to www.pearsonfe.co.uk/ElectricalInstDiploma.

Activity 1.10

1 What cross-sectional area does a magnet need to have to produce a magnetic flux density of 5 T when the magnetic flux is 5 Wb?

2 What is the magnetic flux density of a magnet with a cross-sectional area of 0.3 m^2 and a flux of 4 Wb?

3 What is the magnetic flux of a 160 mT magnet with a cross-sectional area of 200 mm^2?

4 A magnet of flux 120 µWb and cross-sectional area 200 mm^2 is required to have a flux density of 1 T to be used as a fire door magnet. Is this magnet strong enough to hold the fire door open?

5 A motor requires a magnet with a flux density of 0.65 T. If the flux is 200 mWb, what is the cross-sectional area of the magnet?

Electromagnetism

When current passes through a conductor, a magnetic field is induced around that conductor. The strength of the magnetic field is proportional to the amount of current passing through the conductor. It can only exist while the current is flowing. Control of an electromagnet can be achieved by simply putting a switch into the circuit.

It is important for lots of applications to know the direction in which the magnetic field is going. Motor movement is caused by the interaction of magnetic fields, so it is a good idea to know which way the motor will start spinning. The concentric circles of magnetic flux lines stretch along the whole length of a current-carrying conductor and the flux direction is relative to the direction of current, as can be seen in Figure 1.22.

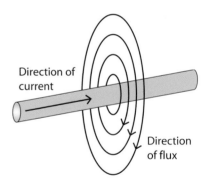

Direction of current

Direction of flux

Figure 1.22: Lines of magnetic force set up around a conductor

Maxwell's screw rule

An easy way to work out which way the flux lines are running is to imagine putting a screw into a piece of wood. As the screw is turning clockwise it is going into the wood. The direction of the screw is the direction of the current flow (away from you and into the wood) and the clockwise rotation represents the rotation of flux lines, as per Figure 1.23 below. This is known as Maxwell's screw rule and is named after the scientist who discovered it.

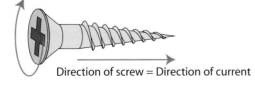

Rotation of screw = Rotation of magnetic field

Direction of screw = Direction of current

Figure 1.23: The screw rule

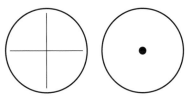

Figure 1.24: Direction of magnetic flux in current-carrying conductors

Another way of working out the direction of an induced magnetic field is to imagine the flight of an arrow or dart. The arrow is flying away from you and all you can see is the cross of feathers at the back of the arrow. The direction of the magnetic field is clockwise. If you were looking at the arrow coming towards you, all you would see is the point. The direction of current can be simplified by a circle with a cross or dot in it (see Figure 1.24).

Magnetism, applications and machines

If two current-carrying conductors were brought close to each other and the direction of the current was different (one current going into the paper, the other current coming towards you), the flux lines at the point closest between them would be travelling in the same direction. Remember the rules of magnetism: opposites attract, alike repel. This would mean the conductors would repel each other. If they were free to move, they would jump apart.

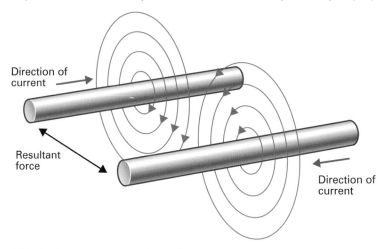

Figure 1.25: Force between current-carrying conductors

If the current is made to travel in the same direction, at the point between the conductors, the lines of flux would be going in opposite directions – opposites attract – so the conductors would move closer together.

This principle is how motor movement is created. Now consider putting a current-carrying conductor into a permanent magnetic field. With no current flowing the conductor will not have any magnetic properties but when the current is switched on, invisible concentric flux lines will appear along the length of the conductor. What exists now are two magnetic fields: one from the permanent magnets going from north to south and one from the conductor with induced electromagnetic field lines. These will interact and movement will occur – the trick is to work out which way.

The solenoid

A solenoid is a number of turns of insulated conductor wire closely wound in the same direction to form a coil. The coil is held in place by a core called a former. If the coils are connected to a circuit with a current supply and a switch, a solenoid is created. The direction of the current dictates the polarity of the electromagnet. Imagine taking your right hand and wrapping your fingers around the solenoid in the direction of the current flow; your thumb will point to the North Pole.

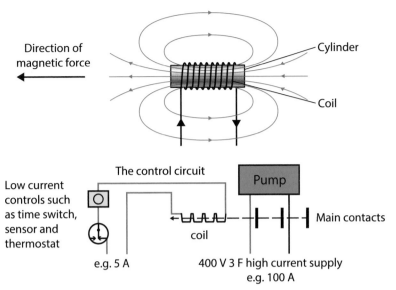

Direction of
magnetic force

Cylinder

Coil

The control circuit

Pump

Low current
controls such
as time switch,
sensor and
thermostat

coil

Main contacts

e.g. 5 A

400 V 3 F high current supply
e.g. 100 A

Control circuits often use one supply to switch another, e.g. a low
current supply to switch a high current load. An example of this
is a heavy-duty pump operated by a time switch and thermostat.

Figure 1.26: The principle of the solenoid coil and simple control circuit

The solenoid is a temporary magnet that acts like a permanent bar magnet
when switched on. Applications include relays, contactor controls, fire
doors, bells and buzzers, residual current devices (RCDs) and miniature
circuit breakers (MCBs).

Consider replacing the battery source with an ammeter and then pushing a
magnet into the coil at speed – what will happen? The ammeter will show a
current is flowing.

Lenz's law for induced current

As shown with the simple generator, for an electromagnetic field (emf) and
current to be generated, there must be movement. With the example shown
in Figure 1.27, you could either move the coil or the magnet. As long as there
is movement, an emf will be generated and subsequently an induced current.
Lenz's law gives a clue as to what direction the current will flow.

Lenz's law states:

'The direction of an induced current is such as to oppose the change
causing it.'

This can be seen in Figure 1.27: as the magnet moves into the coil, the
ammeter shows a negative deflection. But if the magnet is pulled out of the
coil, the current deflection will be positive.

There are also other factors that can affect the induced emf and current.

- The speed or velocity at which the magnet or coil is moved will
  have a direct effect on the induced emf and current. The quicker the
  movement, the larger the measured current.
- A larger number of coils will increase the current generated.
- A larger magnet will mean larger current.

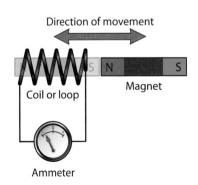

Direction of movement

Coil or loop

Magnet

Ammeter

Figure 1.27: Induced emf and current

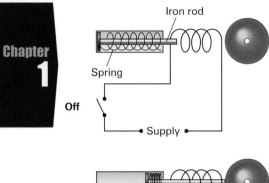

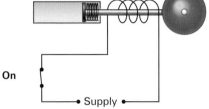

Off

On

Figure 1.28: Operation of a switch in the off and on positions on a solenoid in a door bell

Door bell

An energised solenoid is the same as a bar magnet. Like a bar magnet, if an iron bar was brought close to the solenoid, it would be attracted and move towards it. This principle is used for a door bell with a spring-loaded iron plunger sitting close to the de-energised solenoid coil. When the switch is closed the iron plunger is drawn into the solenoid coil and hits the bell, as shown in Figure 1.28.

Residual current devices (RCDs)

RCDs, miniature circuit breakers (MCBs) and RCBOs make use of electromagnetism in their operation. In the heart of an RCD is the toroid core. Under normal operating conditions this core is not magnetised. Wound around the magnetic core are two main current-carrying coils: one for the line conductor and one for the neutral conductor.

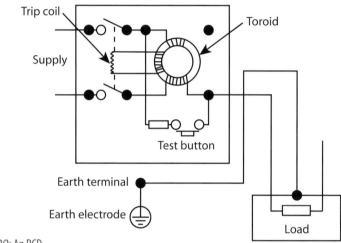

Figure 1.29: An RCD

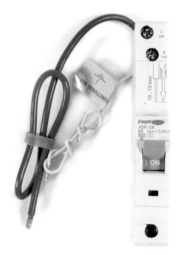

Figure 1.30: An RCBO

Key terms

Toroid – a small transformer designed to detect imbalance generated by earth fault currents.

Search coil – also known as the trip coil.

Contactor – a switch that uses small currents to control large currents. It is generally found in industrial applications such as motor control.

As long as the current on the line matches the current on the neutral, the induced magnetism in the **toroid** core cancels each other out. If, however, there is a slight leakage to earth (a contact between the live and earth) there will be an imbalance in the two coils. This imbalance will mean the induced magnetic flux will not be the same on both sides of the core and a third coil that sits between the line and neutral coil will detect a new generated magnetic flux. This **search coil** will trip a relay and cut the circuit off instantly, making it safe.

Contactor control and relays

Contactor controls use the electromagnetic principles of a solenoid. In a domestic situation, to turn a light on and off all that is required is a light switch. This is because the voltage and current levels involved are accordingly low. If the requirement was to switch much bigger currents, a different method might be better. A contactor control will use a small current to energise a relay which in turn will pull over a much bigger contactor. As the contactor is pulled over, a contact bridge is made for the larger load current to flow through. In this way the small control current is kept very separate from the load current.

In Figure 1.32 it can be seen under normal conditions the switch is open and is termed 'normally open'. When the coil in the relay is energised the light is switched on. Contactor relays can be configured in many different ways but remember contactors are still only simple switches and if you can follow a wiring diagram you will be able to wire a contactor.

Force on a conductor

A current-carrying conductor will move with a certain force when placed in a magnetic field. The force on a conductor is directly related to three things:

- the strength of the magnet (magnetic flux density or concentration of flux lines)
- the amount of current you pass through the conductor
- the length of the conductor that you put between the permanent magnetic poles.

Figure 1.31: A relay

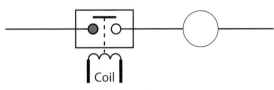

The formula for force on a conductor is as follows:

Force, $F = B \times I \times L$

where F is force in newtons, B is magnetic flux density in Wb/m² or tesla, I is the current in amps and L is length in metres.

Figure 1.32: One-way switch – off position

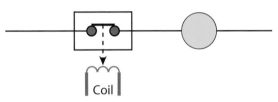

Current is often in mA and magnetic flux density is often in mWb/m², so be careful and convert first before trying calculations.

To try and remember this equation think of 'ForcefulBill' or 'Bill was not a nice man, he was very forceful'.

Figure 1.33: One-way switch – on position

Worked example

A 3 m conductor is placed in a magnetic field of 3 T when a current of 3 A is turned on. What force is exerted on the conductor?

$F = B \times I \times L$

$F = 3 \times 3 \times 3 = 27$ N

Link

For more worked examples, go to www.pearsonfe.co.uk/ ElectricalInstDiploma.

Activity 1.11

1 Calculate how long a conductor is when placed inside a 4 T magnetic pole pair. A force of 6 N is experienced when the current is 20.9 mA.

2 What force will a 0.6 m conductor experience if it has a 3 mA current flowing through it and is placed in a 5 T magnetic flux?

3 Calculate the magnetic flux density of a 22 m current-carrying conductor that is carrying a current of 2 A and experiences a force of 3 N.

4 What current is flowing in a conductive armature of 0.3 m length when it experiences a force of 0.68 N when moving through a 2 T magnetic field?

5 How long does a conductor need to be to experience a force of 3 N when a current of 250 mA passes through it? The magnetic flux density is 350 mT.

Fleming's right-hand rule

A motor and a generator are essentially the same pieces of machinery. With a motor, you put electrical energy in and get mechanical energy out. With a generator, you put mechanical energy in and get electrical energy out. The equipment is the same.

For a motor (and movement) to exist, three things must be present: a current, a conductor and a magnet. If there is no current source but instead you move the conductor through the magnetic field you will induce an emf, which in turn creates a current. This is the generator rule.

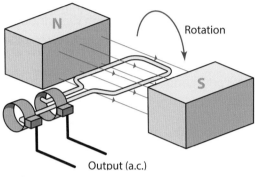

Figure 1.34: An a.c. generator

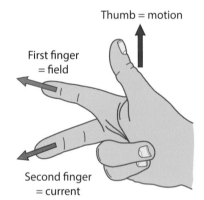

Figure 1.35: Fleming's right-hand rule

The generated voltage and current can be found using Fleming's right-hand or 'generator' rule, as shown in Figure 1.35.

To work out the direction of the current flow generated, line up the first finger in the direction of the permanent magnet field. Now align the thumb in the direction the conductor is being moved. The second finger will show the direction of the current generated.

Calculating magnitudes of a generated emf

As before with the motor rule, it is possible to work out the amount of emf (voltage) that will be generated by moving a conductor within the magnetic field using the following formula:

$E = B \times I \times V$

where E is the induced emf, B is the magnetic flux density, I is the length of the conductor and V is the velocity or speed at which the magnet is moving (moving either the coil or the magnet will have exactly the same effect).

Link

For more worked examples, go to www.Pearsonfe.co.uk/ ElectricalInstDiploma.

> **Worked example**
>
> If a conductor with a length of 0.4 m is moving at right angles to a magnetic field density of 1.5 T and is moving with a velocity of 3 m per second, what is the induced emf in the conductor?
>
> $E = B \times I \times V$
>
> $E = 1.5 \times 0.4 \times 3 = 1.8 \, V$

Activity 1.12

1 A 0.2 m long conductor moves through a magnetic field of density 12.3 T at a velocity of 1800 cm/minute. Calculate the emf induced in the conductor.

2 A conductor is moved through a magnetic field of flux density 200 mT at a velocity of 18 000 mm/minute. The conductor is twice the length of the conductor in the previous question. What emf is generated?

3 What is the velocity of a 15 m conductor that generates 230 V when moved through a 15 T magnetic field?

4 What speed is required to generate a voltage of 210 V when the generator components consist of an armature of 12 m and a magnet of flux density 2.1 T?

5 A voltage of 120 V is required from a generator with an armature of length 13 m and that has a maximum velocity capability of 10 rev/s. What size magnet is required?

Sinusoidal waveform

Instantaneous value

As discussed earlier, an alternator will produce a sine wave, otherwise known as an alternating current (a.c.). In the UK the a.c. supply is 50 Hz. As the loop conductor passes through and cuts the largest concentration of magnetic field lines, the largest current is made. As the conductor rotates to the vertical point, no flux lines are cut and no emf or current is induced. At any point in time of the cycle the instantaneous value of voltage or current can be taken. In electrical terms, this value is called the instantaneous value.

Average value

If the sine wave was split up into a number of time samples, an average value could be taken. Using the standard method of averages in maths:

$$\text{Average value (mean)} = \frac{\text{The sum of the instantaneous values}}{\text{The number of samples}}$$

If more samples are taken the average value (or mean) could become much more accurate. Over an entire 360° sine wave, the average value would be '0', as in a symmetrical sine wave the negative part of the wave is identical and opposite to the positive section so they cancel each other out. Over half a cycle, say the positive part, the average would work out as follows:

$$\text{Average value (mean of } \tfrac{1}{2} \text{ cycle)} = \text{Maximum peak value} \times 0.637$$

Peak value

When the loop conductor is horizontally between the magnet poles it is cutting the highest concentration of magnetic flux lines. Following the theory that this is when the largest emf will be generated, then this means this will be the peak value. The positive part of the cycle will have a peak value and so will the negative part of the sine wave.

If a measurement is taken from the peak of the positive cycle to the trough of the negative part, this is the peak to peak value, as can be seen in Figure 1.36.

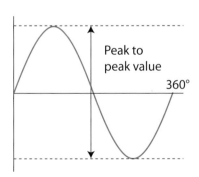

Figure 1.36: Peak to peak value

Safe working

Meters measure rms values. The peak value will actually be larger by a factor of 1.414.

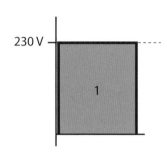

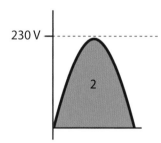

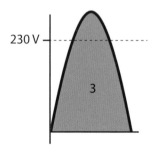

Figure 1.37: Peak value diagrams

Link

For more worked examples, go to www.Pearsonfe.co.uk/ElectricalInstDiploma.

Root mean squared (rms) value

It is important to understand the different effect a.c. and d.c. supplies have in circuits. To do this there needs to be a way to compare them. When a d.c. current is passed through a conductor with resistance it has a heating effect which is constant because the current is a constant. When an alternating current is connected to the same conductor, the heating effect will be different because the current grows to a maximum and then shrinks to '0' before changing direction and growing to a maximum in the opposite direction. The best way to work out the effect of a.c. is to work out its equivalent if it were d.c. This is done by calculating the rms value.

If the heat produced by 1 A of d.c. is equal to 100°C then 1 A of a.c. current would produce 70.7°C of heat. This 'heating effect' gives the comparison required to find the rms value using the following formula:

$$\frac{\text{Heating effect of 1A maximum a.c.}}{\text{Heating effect of 1A maximum d.c.}} = \frac{70.7°C}{100°C} = 0.707$$

Therefore:

The effective rms value of an a.c. waveform = $0.707 \times I_{max}$

And if the formula is rearranged to find the maximum value of current, it becomes:

$I_{max} = I_{rms} \times 1.414$

In Figure 1.37 a comparison of a constant d.c. supply (1) is made with an a.c. supply with the same peak value (2). Waveform 3 shows what an a.c. supply would need to be to have the same heating effect as the constant d.c. supply. From this diagram it can also be seen that the domestic 230 V a.c. supply is actually a lot more than 230 V peak. A domestic 230 V supply has a peak of 325.22 V and an rms value of 230 V. Most a.c. values and readings from meters are rms values.

Sine wave – frequency and period

50 Hz is the standard frequency in Europe. This means 50 full cycles where a wave starts and returns to the same place in one second. If 50 cycles were drawn out in a line to represent 1 second of time, that would mean 1 cycle is one fiftieth of a second. This can be shown in the following formula:

$$\text{Frequency} = \frac{1}{\text{The time for one cycle, T}}$$

where T is also known as the period.

Worked example

What is the period for a 50 Hz supply?

$T = \dfrac{1}{f}$

$T = \dfrac{1}{50} = 0.02$ or 20 mS

Activity 1.13

1 If a wave form has a period of 50 μs, what is the frequency? (Don't forget to take great care with the units and convert to base values.)

2 If one third of a full sine wave cycle is 33 mS, what is the frequency?

3 A square wave reaches its maximum value of 3 V twice in 20 mS. What is the frequency of this square wave?

4 A 100 GHz signal is measured on an oscilloscope. What is the length of one cycle?

5 If a sine wave has a frequency range 10–12 kHz, what is the period range?

THE OPERATING PRINCIPLES OF A RANGE OF ELECTRICAL EQUIPMENT

LO 6

Electromagnetism is the basis for how a wide range of electrical equipment works. The principle is always to switch on a supply (a.c. or d.c.) and cause a magnetic reaction. Some elements of electrical equipment have already been introduced within the earlier section on electromagnetism.

Electrical equipment

The d.c. generator

A single conductor that is moved through a magnetic field produces an emf. Providing there is an electrical circuit, a current will flow. That current will tend to move in the opposite direction to the movement causing it. This is called Lenz's law for induced current (otherwise known as Lenz's generator law). If the conductor length, the magnet or the speed is increased, the induced emf and current will increase. Imagine taking the conductor and making it into a loop, as shown in Figure 1.38.

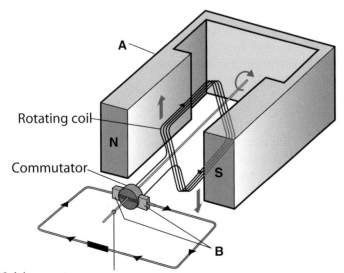

Figure 1.38: A d.c. generator

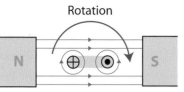

Figure 1.39: Wire loop within a permanent magnetic field

The coil can be moved around an axis by rotating it. As the conductor cuts the magnetic field lines, an emf and current will be induced into the coil. If you look at Figures 1.39 and 1.40, you will notice one side of the loop is moving upwards and the other side is moving downwards to give the rotation.

As the loop reaches the horizontal position it cuts the highest density of flux lines that exist directly between the magnet poles. However, as the loop reaches the vertical position, it is moving parallel to the magnetic flux lines – not cutting any at all! From this you can see that most emf is generated when the coil is horizontal and little or no emf is generated when the coil is vertical. To connect the turning loop up to a circuit, a commutator is used. For a single loop generator, a two copper segment commutator is required. The commutator is split and separated by an insulated material.

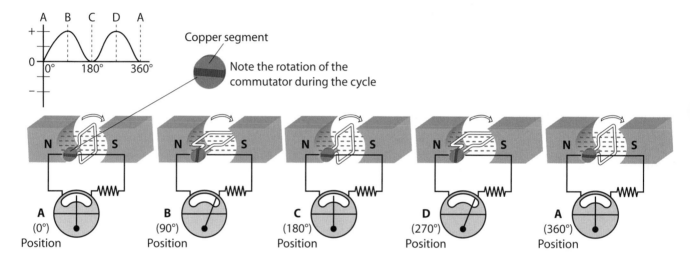

Figure 1.40: Voltage output for one complete revolution

The circuit is now complete and the loop is rotating within the magnetic field. The commutator allows the current to pass around the circuit in the same direction, as can be seen in Figure 1.41. As the loop rotates, the brushes keep contact with the moving commutator until the brushes reach the insulated divider. At this point the loop is about to start on the second half of its journey and the direction of the current changes. The commutator/brush arrangement allows the current to keep flowing in the same direction as the loop passes the vertical position.

To make a larger emf and current, more loops of wire are required. For each extra loop, a further connection is required and this is achieved by adding more segments to the commutator. In reality, lots of loops and segments will make a much smoother d.c. current. An armature is made up of lots of loops formed around a core for strength.

The brushes are made of carbon for several reasons. Carbon, as it heats up with friction, has a negative temperature coefficient. This means that unlike most metals, as it gets hotter its resistance gets less, making it a better conductor. Carbon also self-lubricates to minimise friction. Springs behind the carbon push the brushes on to the commutator segments to keep a constant pressure and good contact. Timely maintenance is important to make sure the carbon brushes are replaced when they have worn down.

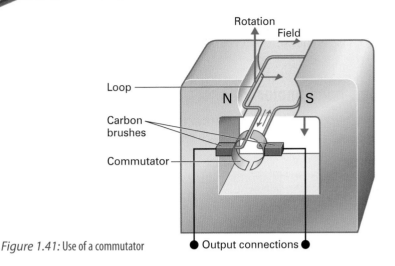

Figure 1.41: Use of a commutator

The a.c. generator

Up until now the discussion has been about direct current. This is current flow in one direction, as you would expect from a battery source. Alternating current is different as it flows in one direction and then turns around and flows in the opposite direction – it alternates. The d.c. generator keeps the current flowing in one direction via a commutator, segments and brushes. However, an a.c. generator keeps the current alternating via 'slip rings' and brushes, as shown in Figure 1.43.

The slip rings allow each end of the loop to be continuously connected to their side of the circuit, so as the loop enters and leaves the magnetic field, the current changes direction and this is fed to the outside circuit. As the loop makes one full rotation (360°), a full cycle of a.c. current is generated. One full cycle is called a sine wave and contains a positive and negative half cycle, as shown in Figure 1.42.

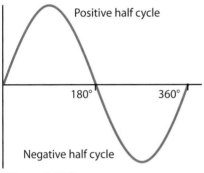

Figure 1.42: Sine wave

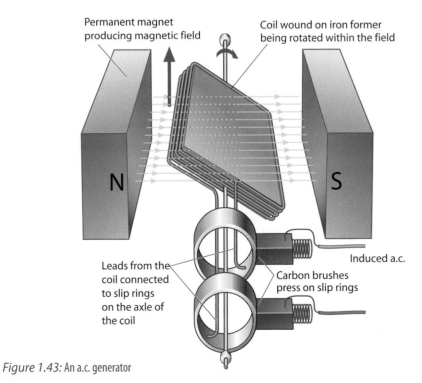

Figure 1.43: An a.c. generator

In the UK an a.c. generator will produce a sine wave that alternates 50 times a second. This is measured in hertz (Hz) and therefore the UK national supply is 50 Hz.

Fleming's left-hand 'motor' rule

The interaction that occurs is down to attracting and repulsion forces of magnetic flux lines.

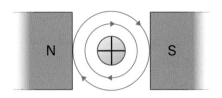

Figure 1.44: Current-carrying conductor between two magnetic poles

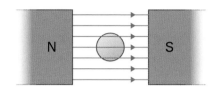

Figure 1.45: Field caused by two magnetic poles

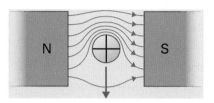

Figure 1.46: Current-carrying conductor in a magnetic field

As can be seen from these figures, the top of the conductor is experiencing a compression force as the flux lines are going in the same direction, hence repulsion. However, at the bottom of the conductor, the induced flux lines of the conductor and the permanent magnet are travelling in opposite directions. This causes attraction. At the top of the conductor there is a push and at the bottom a pulling force. This dual action will cause the conductor to be forced downwards – this is the d.c. motor principle.

A simple way to work out the direction of a current-carrying conductor in a magnetic field is to use Fleming's left-hand 'motor' rule, as seen in Figure 1.47.

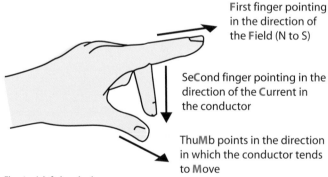

First finger pointing in the direction of the **F**ield (N to S)

Se**C**ond finger pointing in the direction of the **C**urrent in the conductor

Thu**M**b points in the direction in which the conductor tends to **M**ove

Figure 1.47: Fleming's left-hand rule

Line up the first finger with the permanent magnetic field in the direction of North to South. The second finger is lined up with the direction of current flow. The direction of movement will now be shown by the direction of your thumb.

The operating principle of transformers

Mutual inductance is the principle of how a transformer works. A changing magnetic flux caused by a changing current is required to set up a transformer effect. This is why transformers are used with a.c. supplies and not d.c.

In a transformer there is a second coil and a different type of inductance will happen. Imagine the switch is closed and a magnetic field builds up as the current grows to a stable high value. This growing magnetic field links the coils and creates the back emf. If a second separate coil is in range of the first coil, it will also be affected by the growing magnetic field. If this occurs a further back emf and opposing current will be developed in this second coil. This magnetic circuit will create a current in the second coil but with the difference that there is no supply current in this second coil to oppose.

Energy in a coil

If there is current flow in a coil, there will be power. If there is power there is work being done. Work done is the same as energy and energy can be stored in a coil. If an inductive circuit is switched off, arcing can be seen in the form of a spark. It is this spark that needs to be suppressed to avoid interference in other circuits when a fluorescent light is switched on. Suppression is achieved by placing an **RF suppression** capacitor across the starter switch terminals to soak up the spark energy. The stored energy within the coil can be calculated from the following formula:

$$W = \frac{L \times I^2}{2}$$

where W is the stored energy measured in joules, L is the inductance measured in henrys and I is the current in amps. The formula for coil energy is shown here but will mainly be used at Level 3.

Relationship between coils

The relationship between one coil and a second coil is dependent on:

- the number of coils on either side of the transformer
- the supply current
- the supply voltage.

If the number of coils on the supply side (primary coils) or the load side (secondary coils) are changed, the load voltage and current will also change.

There is no actual direct electrical contact between the two coils as they are insulated from each other and the magnet by a thin layer of special varnish or enamel. The coils are, however, magnetically linked and have a common magnetic circuit.

> **Key term**
>
> *RF suppression* – otherwise known as radio frequency suppression – it is used to suppress the high frequency signal given off by arcing at switch terminals as they operate.

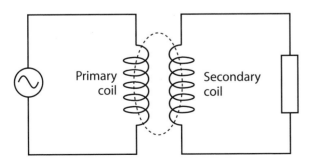

Figure 1.48: Mutual inductance

As mentioned previously, the material that a permanent magnet is made of is different to that of a temporary magnet. A transformer needs to be a temporary magnet as it needs to magnetise and demagnetise quickly.

The construction of a transformer has a direct impact on its efficiency. Losses can occur through different methods in a transformer.

Copper losses

Transformers heat up when they are used as current passes through the coils. This heating effect is energy being lost and makes the transformer less efficient. Large transformers are often cooled by oil passing around an outer skin. These losses are called copper losses and can be calculated by the power formula:

$P = I^2 \times R$

Eddy currents

As the magnetic flux rises and falls and currents are induced, current is also created in the transformer core. This current will pass around the magnet core and produce heat. This in turn causes energy loss. To prevent this, the core of the magnet is laminated, with each lamination being electrically separate from its neighbour by insulation. This stops the magnet being one big electrical conductor. The laminated separate conductors will still have eddy currents induced in them but they will be smaller and they will oppose and cancel each other out.

Hysteresis loss

Transformer cores are made from silicon steel alloys because they can be magnetised and demagnetised easily by the fast changing magnetic flux. These silicon alloys have low hysteresis loss compared to harder ferrous materials such as alnico or alcomax.

The difference between a fixed permanent magnet and a silicon steel transformer core can be seen in the hysteresis curves (Figure 1.49).

Transformer types

There are various types of transformers available that you will come across throughout a career in the building services industry. Each transformer type has a different purpose.

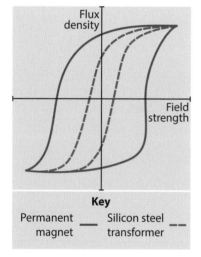

Key

Permanent magnet	——
Silicon steel transformer	– –

Figure 1.49: Hysteresis loop

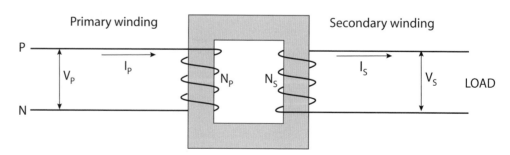

Figure 1.50: Double wound, core-type transformer

A standard a.c. core transformer is designed to step up or step down voltage and current. The core is made up of laminations. A core transformer has completely separate windings: one connected to the supply and one connected to the load, such as a motor or lighting. If the purpose of the transformer is to step down the supply voltage to a more useable level for a machine, then the supply side will have more coils. If the transformer is to be used as a step up transformer, the secondary or load side will have more coils than the primary. The construction of a core form transformer means that there are two legs to the transformer. The varying voltage creates a varying magnetic field in the core that links to the other coil in the magnetic circuit. The varying magnetic field linking the two coils induces a secondary voltage in the secondary coil with the same frequency as the supply side.

Although this type of transformer configuration is very common, it is not the most efficient at energy transfer. Some of the magnetic flux falls outside the core and is wasted.

Self-inductance transformers

The property of a coil to resist current flow is a measure of the coil's inductance, L, and is measured in henrys, H. If a coil is connected to a supply via a switch and the switch is closed suddenly, the current would build up in the coil and a magnetic field would build up, linking all the coils. This growing magnetic field would induce back emf and opposing current. The induced current will oppose the main current that produced it. As the current reaches a stable value, so will the magnetic field. This stability will mean no more back emf is induced. As the switch is turned off again the magnetic field collapses and this will again produce a back emf and opposing current. If this continues, as it would with a.c. (50 times a second), there would be a continuous process of current opposition. This is called self-induction. The induced emf therefore has a relationship between the change in current, the inductance of the coil and the time:

$$E = -L\frac{\Delta I}{\Delta t}$$

where E is the generated back emf (–ve represents 'back'), L is the inductance, measured in henrys, H. ΔI represents the change in current and Δt represents the change in time. This means the current changes from one value to another over a time period as one of the conditions for a back emf to be generated.

If the change in current produces the back emf, then so does the change in magnetic flux linking the coils together. The back emf can also be found by knowing the change in magnetic flux over time and the number of coils, as shown in the formula below:

$$E = -N\frac{\Delta \varphi}{\Delta t}$$

where E is the generated back emf (–ve represents 'back'), N is the number of coils, with no units. $\Delta \varphi$ represents the change in magnetic flux and Δt represents the change in time. This means the magnetic flux changes from one value to another over a time period as one of the conditions for a back emf to be generated.

Worked example

What is the back emf generated within a 10 coil inductor when the flux changes from 1.2 mWb to 5 mWb in 3 seconds?

$$E = -N \frac{\Delta \varphi}{\Delta t}$$

$$E = -N \frac{(\varphi_2 - \varphi_1)}{3}$$

$$E = -10 \times \frac{(5 \times 10^{-3} - 1.2 \times 10^{-3})}{3}$$

$$E = -\frac{3.8 \times 10^{-3}}{3} = -1.27 \text{ mV}$$

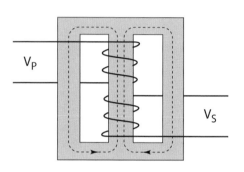

Figure 1.51: Shell-type transformer

Shell transformer

The shell form is a slightly more efficient configuration as there are three legs to the shape and the coils are around the central pillar. The windings can either be next to each other or wound on top of each other. As the varying voltage creates the varying magnetic flux, the outer pillars act as a parallel magnetic circuit, as can be seen in Figure 1.51.

Transformer calculations

Transformers are all about 'ratios'. If you can find the ratio values and you have one of the voltages or currents involved, you can work out the rest.

Points to consider:

1   A transformer can be used to step up or step down a current or a voltage.

2   A voltage transformer will step down if the secondary windings are less than the primary windings.

3   A voltage transformer will step up a voltage if the secondary windings are more than the primary windings.

4   A current transformer is completely the opposite to a voltage transformer.

5   A current transformer will step down a current if the secondary windings are more than the primary windings.

6   A current transformer will step up a current if the secondary windings are less than the primary windings.

7   Unless otherwise stated, the power rating on both sides of the transformer are the same – this means that the primary voltage × the primary current = secondary voltage × secondary current.

8   The formula for transformers is as follows:

$$\frac{V_p}{V_s} = \frac{N_p}{N_s} = \frac{I_s}{I_p}$$

Worked example

A transformer has a primary voltage of 230 V and primary windings with 100 turns. The secondary windings are 50 turns. What is the secondary voltage?

Step 1

What is this: a voltage or current transformer?

It is a voltage transformer.

Step 2

Is it a step up or step down voltage transformer?

It is a step down transformer as there are fewer turns on the secondary windings.

Step 3

What is the ratio?

The ratio is 100 : 50.

To simplify a ratio, simply divide both sides by the smallest number, in this case 50, to give:

$$\frac{100}{50} : \frac{50}{50} = 2 : 1$$

This is a 2 : 1 step down voltage transformer.

Step 4

Apply the ratio to the voltage to find what the secondary voltage should be.

$$\frac{230}{2} = 115 \text{ V}$$

Link

For more worked examples, go to www.pearsonfe.co.uk/ ElectricalInstDiploma.

It is a good idea to draw a diagram and label it to determine what information you have and what you need to find. The temptation with transformer questions is to do them in your head. This will come with time but the key is lots of practice – don't take chances with something that is this simple but also equally simple to slip up on.

Activity 1.14

1   A transformer has a primary voltage of 230 V and a secondary of 12 V. If the primary turns are 200, how many turns are there on the secondary side?

2   A core transformer has 3 A at the output and is supplied by 12 A, 400 V. What is the secondary voltage?

3   A shell transformer has a turns ratio of 20 : 1 and a secondary voltage of 23 V. What is the primary voltage?

4   A transformer is supplied with 200 V at 3 A. If the secondary current is 15 A, what is the secondary voltage?

Auto-transformers and isolating transformers

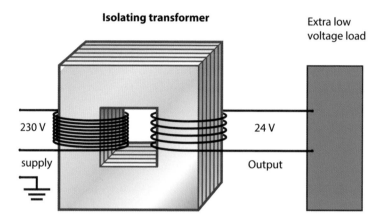

Figure 1.52: An auto-transformer and an isolating transformer

Auto-transformer

The main principle of transformers centres around the amount of coils on the primary and secondary sides. The output voltage is dependent on the turns ratio. If you were able to get access to the coils on the secondary side and decide exactly where you wanted to take your output from, you could effectively pick the amount of coils that made up your output. By varying the secondary coils, the output voltage also can be varied. An auto-transformer is a transformer having a part of its winding common to the primary and secondary circuits. The advantage of only having one copper winding to pay for can also be seen as a disadvantage, as there is one point of failure that could make the transformer very dangerous. There is a direct electrical circuit between the input and the output that could also lead to the input voltage appearing at the output under fault conditions.

Auto-transformers are suited to applications such as starting cage-type induction motors that require a variable voltage. Another type of transformer design is required where complete electrical separation must be maintained at all times.

Isolating transformer

Where water is concerned, electrical separation is particularly important. Transformers are still required to bring voltages down to extra low voltage levels but the possibility of completing an electric circuit and getting a shock must be minimised. Applications such as shaving sockets are supplied via isolating transformers that do not share a common earth. The two sides of the transformer are electrically separated from earth and from other systems. This means that a fault cannot give rise to the risk of electric shock.

THE PRINCIPLES OF BASIC MECHANICS

As an electrician, other scientific principles need to be understood if you are to make use of electricity efficiently. Basic mechanics will help determine the work, effort and power required to move an object.

Calculating quantities of mechanical loads

Mass and weight

Mass and weight are two terms that often get confused and misused. Mass is the amount of substance or 'stuff' that exists in an object. This amount of substance does not change unless you do something to the object (for example, break a bit off). The value of mass is therefore a constant and does not change wherever it is, even if it is on the Moon. The standard unit for mass is the kilogram or kg (1 kg = 1 000 gm).

Weight is one of the most misused terms in science. If someone asks you: 'How much do you weigh?', generally you will give an answer in stones or kilograms – this is not strictly correct! Weight is a combined effect of mass and **gravity**. Gravity or gravitational pull is an effect that the Earth has on all things that have a mass. The further away from sea level, the weaker the gravitation pull. Gravity is something you could observe if you dropped your drill from the top of a ladder and it accelerated to the ground. If you hold a drill you are holding the mass of the drill but also you are overcoming the gravitational acceleration that the Earth is exerting on it. What you can feel is the true weight of the drill, or the force. This force is measured in newtons, N, and is given by the formula:

Force = Mass × Acceleration

$F = ma$

Using acceleration due to gravity, 'a' can be replaced with the constant for gravity, 'g', giving:

$F = mg$

Worked example

A photovoltaic solar panel has a mass of 48 kg. What weight force would you need to consider when installing it on a roof?

$M = 48$ kg

$G = 9.81$ ms^{-2}

$F = mg$

$F = 48 \times 9.81 = 470.88$ N

> **Progress check 1.2**
>
> 1 How do you calculate the energy stored in a coil?
>
> 2 Define inductance in a coil.
>
> 3 What is the transformer formula?

LO4

> **Key term**
>
> *Gravity* – acceleration due to gravity can be taken as a constant value of 9.81 ms^{-2}. If you took your drill to the Moon it would weigh approximately one sixth in comparison to its weight on Earth.

> **Link**
>
> For more worked examples, go to www.pearsonfe.co.uk/ElectricalInstDiploma.

Activity 1.15

1 What is the force due to gravity on a control panel with a mass of 15 kg?
2 What is the mass of a cable drum with a weight force of 108.2 N?
3 How many cable drums of mass 27 kg each can a trolley rated at 1 kN carry in one trip?

Key term

Work done – this is the same as energy used. Both are measured in joules, J, and 1 joule of energy is said to have been used if a force of 1 N moves an object 1 m.

Work and energy

Every time an object is moved from one place to another, work has been done. The amount of **work done** depends on the force used and distance you have moved the object either up and down or side to side. For this instance we are just considering moving an object horizontally (vertically will come later).

Using common sense, the following statements must be true.

- More force used = more work done.
- Greater the distance moved = more work done.

The relationship between work, force and distance can be shown in the formula:

Work done (energy used) = Force × Distance

Worked example

How much work has been done if a force of 12 N moves a control panel 15 m across a room?

$F = 12$ N

$D = 15$ m

$W = Fd$

$W = 12 × 15 = 180$ J

Link

For more worked examples, go to www.pearsonfe.co.uk/ElectricalInstDiploma.

Activity 1.16

1 What energy do you use when moving a street light from a highways depot store to your van? It requires a force of 16 N and the distance is 25 m.
2 What energy is required when loading your van with three cable drums? Each drum takes 10 N of force and the distance is 10 m.
3 How far do you move if the total work done moving your site lockup box is 2500 J and the force used is 3 N?
4 What energy will be spent moving a 200 kg transformer core 12 m into position for installation?
5 What energy will be used moving 7 × 25 kg photovoltaic panels 20 m from the delivery pallet to winch position?

Potential energy (PE)

Energy comes in many forms. Energy does not get lost – it just converts to other forms. Energy put into a machine or process always equals energy that comes out of the machine or process. Some energy is not very useful, such as vibration, smoke, sparks or heat.

Up until now you have just considered moving an object across a floor – now imagine moving an object *up* a floor, vertically. Every object has potential energy or stored energy. An example might be a drill that is about to be picked up. This drill has potential energy – it also has a known mass (for example, a standard 36 V SDS cordless drill has a mass of around 3.5 kg). To be picked up, you will also have to overcome the gravitational pull of the Earth. The amount of potential energy is also related to how high the drill will be raised.

This can be shown by the formula:

$PE = m \times g \times h$

Worked example

Consider this – you have to drill a hole 1 500 mm above **fixed floor level (ffl)**. What potential energy (PE) will there be if the 3.5 kg SDS drill is to be lifted from your tool box on the floor to the marked spot on the wall?

$PE = mgh$

$m = 3.5$ kg

$g = 9.81$ ms^{-2}

$h = 1 500$ mm $= 1.5$ m (convert to base unit of length, m)

$PE = 3.5 \times 9.81 \times 1.5 = 51.5$ J

Key term

Fixed floor level (ffl) – the finished floor level that is the starting point for measurements. This term is often abbreviated to ffl.

Link

For more worked examples, go to www.pearsonfe.co.uk/ElectricalInstDiploma.

Activity 1.17

1 Calculate the potential energy of a cable drum that has a mass of 50 kg and which is to be lifted 0.75 m into the back of a truck.

2 Calculate the height a cable tray has been lifted if the potential energy is 1 kJ and the mass is 20 kg.

3 Calculate the mass of a tool chest that has a potential energy of 250 J and requires lifting 1.75 m back on to the store shelf.

4 Calculate the potential energy of a pallet of solar panels with a mass of 100 kg to be lifted on to a roof at a height of 6 m.

5 A control panel for a motor is to be lifted up a 30 m riser. The control panel is labelled up as 2000 N. What is the potential energy of the control panel?

Energy and power

If energy is expelled and work is done when moving an object a certain distance, what happens if the object is moved even quicker? Imagine you are moving two large boxes 20 m. You take your time to move the first box – you are not out of breath and you are not sweating. However, by the time you get to the second box, you are in a hurry and you move it in half the time. You are out of breath and you are now sweating. That is because you have used much more power! So the general rule is the shorter the time you take to move an object, the greater the power you use. Power is related to the work done in the amount of time it took and is measured in watts, W.

Chapter

1

Power can be expressed as a relationship between work done and the time taken, using the formula:

$$\text{Power, } P = \frac{\text{Work done (W)}}{\text{Time taken to do the work (t)}}$$

Worked example

A 25 kg drum of mineral-insulated cable has to be pushed into the storage lockup 30 m away. You have 2 minutes to complete the move as your supervisor wants to leave site – how much power will you use? (Keep this simple and ignore considerations such as slopes and friction – in reality these would need to be considered but not at this level.)

Step 1 – write down what you know/have been given:

m = 25 kg

d = 30 m

t = 2 mins = 120 s (convert to base unit of time, s)

F = mg

W = Fd

$P = \dfrac{W}{t}$

Step 2 – what can you work out from the facts you have?

F = mg

F = 25 × 9.81 = 245.25 N

and:

W = Fd

W = 245.25 × 30 = 7357.5 J (or 7.3575 kJ)

so:

$P = \dfrac{W}{t}$

$P = \dfrac{7357.5}{120} = 61.31$ W

Link

For another worked example, go to www.pearsonfe.co.uk/ ElectricalInstDiploma.

Activity 1.18

1   How much power will you use if you move a 50 kg crate of stripped copper 50 m to the recycling bin in 5 minutes?

2   You use 500 kJ of energy moving a fluorescent bulb crushing machine from the tailgate of a delivery truck to your electrical storage area which is 75 m away and it takes you half an hour. How much does it weigh?

3   How much power will you use if you move a conduit bending machine (mass of 20 kg) a distance of 30 m in 30 seconds?

4   How much power will you use moving a pallet of contactors with a mass of 78 kg from your van to the installation position 12.5 m away in 2 minutes?

5   You need to move a ground source heat pump 35 m from your secure lockup to the installation point. The pump weighs 75 kg and you have 20 minutes to do the task. How much power will you exert doing the move?

Calculating mechanical advantage gained by use of levers

As well as having to work out the true weight of an object, you might have to move it. If it is over a certain mass you will have to use science to gain some kind of mechanical advantage. This device could be a lever or a pulley of some kind.

Levers are classified into three types or orders.

First order lever

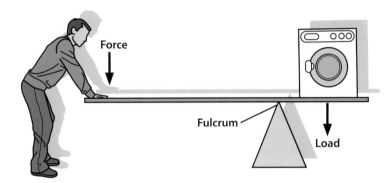

Figure 1.53: A first order lever

This lever system is similar to a see-saw where the fulcrum or pivot point is in the middle. The force (person pushing) is at one end and the other end of the lever is the load you are trying to lift.

Second order lever

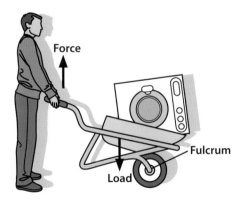

Figure 1.54: A second order lever

This type of lever system is found on a wheelbarrow. This time the fulcrum and the force (person lifting) are at either end, with the load in the middle.

Third order lever

This type of lever system is similar to a person lifting a dumb-bell. The load is at one end and the fulcrum is at the other end, with the force (muscle) in the middle.

Each one of these systems will give a mechanical advantage and enable more load to be moved. On a see-saw, for it to be balanced or in a state of equilibrium, both sides of the pivot have to be identical.

Figure 1.55: A third order lever

There is a rule that states the effort × the length to the fulcrum must equal the load × the length to the fulcrum for it to be balanced. This means that if you are trying to lift something really heavy, it can be done if you move the fulcrum closer to the load. This can be calculated using the formula:

$$F_1D_1 = F_2D_2$$

where F_1 is the effort, F_2 is the load force you are trying to lift, and D_1 and D_2 are the distances to the fulcrum from the person and the load respectively.

Worked example

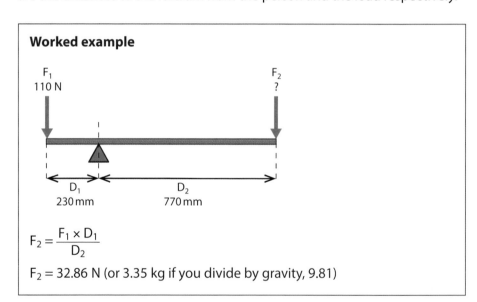

$$F_2 = \frac{F_1 \times D_1}{D_2}$$

F_2 = 32.86 N (or 3.35 kg if you divide by gravity, 9.81)

Link

For more worked examples, go to www.pearsonfe.co.uk/ElectricalInstDiploma.

Activity 1.19

1 A first order lever is used to raise a motor off the ground. The lever is 2.75 m long and you are 1.9 m from the fulcrum. The motor is 200 kg. What force will you exert?
2 The same scenario as Question 1, but this time the motor has increased to 300 kg. How much force will be required now?

Pulley systems

Levers are a good way of gaining mechanical advantage. Another way is to use a pulley system. There are several types of pulley system available.

Fixed single pulley system

In the single pulley system shown in Figure 1.55, a force of 2 kg × g (9.81) is required to overcome gravity. If the mass had to be moved 3 m, the person pulling the rope would need to pull in 3 m of rope. There is no real advantage apart from potentially convenience.

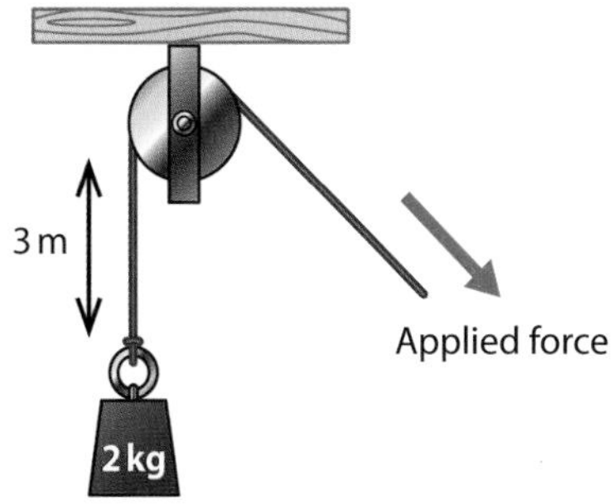

Moveable pulley system

With a moveable pulley system, as shown in Figure 1.56, the mass of 2 kg is shared between the two ropes suspending it. This means that the effort required to lift it is 1 kg × g = 9.81 N. In this system a mechanical advantage has been gained. This mechanical advantage (MA) can be calculated using the formula:

$$MA = \frac{Load}{Effort}$$

$$MA = \frac{2 \text{ kg} \times 9.81}{1 \text{ kg} \times 9.81} = 2$$

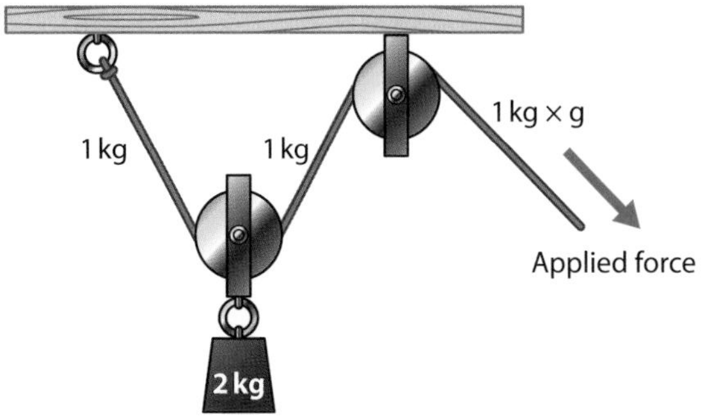

Multiple pulley systems

Where more pulleys and ropes are used, greater mechanical advantage can be gained. To calculate your effort, simply divide the load by the number of ropes that the mass is suspended by. In Figure 1.57, the mass is suspended by four ropes that share the 200 kg equally. Each rope is therefore subjected to 50 kg. Using this information, the applied force must be 50 kg × 9.81 = 490.5 N, which gives you an MA of 4.

Efficiency

Efficiency is defined as the ratio of useable output energy (or power) to the energy (or power) you have to put in. The symbol for efficiency is η (Greek, eta). Simply, no machine is perfect and no matter how much energy you put into it, not all of it will be converted to something useful like movement. Some energy or power will be lost in the form of heat, vibration or generation of noise or smoke.

Efficiency is extremely important to an electrician, as you will need to know how much electricity or effort to put into a machine to get the output you want. If there are too many losses or there is too much wasted energy, there may not be enough left to turn a motor or power a load. It is worth noting that it is impossible for a system to be 100 per cent efficient.

Efficiency is a ratio and has no units but is commonly shown as a percentage. Efficiency can be expressed with the formula:

$$\text{Efficiency, } \eta = \frac{\text{Useful output power (or energy)}}{\text{Input power (or energy)}} \times 100\%$$

Figure 1.55: A fixed single pulley system

Figure 1.56: A moveable pulley system

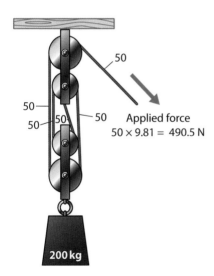

Figure 1.57: A multiple pulley system

Worked example

If the input power to a motor is 1100 W and the output power is 800 W, what is the efficiency of the motor?

$$\text{efficiency, } \eta = \frac{800}{1100} \times 100\% = 72.72\%$$

Link

For more worked examples, go to www.pearsonfe.co.uk/ElectricalInstDiploma.

Chapter

1

Activity 1.20

1 A machine has losses of 3.1 kW and it requires 56 kW to run. What is the required input power?

2 What is the efficiency of the above machine?

3 If the same machine is serviced and the overall efficiency improves by 1%, what is the new input power required?

4 A motor is 97.3% efficient and produces 16.7 kW of output power. What input power is required to achieve this?

5 A d.c. motor has a loss of 3.1 kW due to vibration and friction. If the input power is 22 kW, what is the efficiency of the motor?

LO7

THE PRINCIPLES OF A.C. THEORY

You have looked at the d.c. world, including Ohm's law and how current and voltage are related to resistance. In a d.c. circuit the analogy to water is very strong. If there is resistance then a larger pressure (voltage/potential difference) is required to force the current through the obstruction (resistance). This holds true for the a.c. single-phase circuits as well. Three-phase a.c. circuits are the next logical step, which is why they are covered last (and slightly out of order with the running order of outcomes in the specification).

The effects of components in a.c. circuits

An extra level of complexity comes into play with an a.c. circuit due to the way some components react to a varying supply. The reactions to an a.c. supply must be understood by an electrician when designing, testing or fault finding a circuit. An a.c. circuit is exactly that, an alternating circuit. This means that unlike a d.c. circuit, where current is pushed around a circuit in one direction, in an a.c. circuit the current changes direction 50 times a second (that is what 50 Hz means). In an a.c. circuit, the current starts at a value of '0' and builds up to a maximum before returning to '0' and repeating the process in a negative direction.

You have already seen that an a.c. supply acts in a slightly different way in comparison to a d.c. supply when connected to a load. In a d.c. circuit a capacitor acts as a total block to current due to the layer of insulation in the dielectric – all that can happen is the capacitor plates charge up to the same value as the supply battery. A coil of wire acts as a long piece of wire and hence has resistance when connected into a d.c. supply. In an a.c. circuit things are a little different.

Key term

Phasor – a phasor diagram is a slightly different representation of the sine wave. Imagine the phasor is rotating about a fixed point as per the generator coil spinning in a magnetic field. The direction of rotation is anticlockwise.

If an a.c. circuit was perfect and just had resistance in it, the voltage and current sine waves would be completely in phase, that is they would reach maximum peak values at exactly the same time as shown in Figure 1.58 and also represented by the **phasor** diagram in Figure 1.59.

This is not that easy to achieve if there is anything that looks like a coil or capacitor in the a.c. circuit. Remember, two cables lying next to each other can be considered a capacitor.

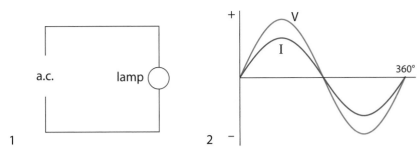

Figure 1.58: Circuit and sine wave diagrams

Figure 1.59: Phasor diagram for zero phase angle

Capacitors in an a.c. circuit

A battery is connected to a capacitor (two conductive plates separated by an insulating dielectric material) via a switch and a series ammeter. The capacitor will start to charge as soon as the switch is turned on. If a voltmeter is placed across the capacitor plates at the same time the initial voltage reading is '0', as there is no potential difference between the plates. For a potential difference to be present between the plates, one side must be positive compared to the other. The instant the switch goes on, the electrons on one side of the capacitor plates are heavily attracted to the +ve terminal of the battery source. The sudden rush of electrons or current flow will show on the ammeter. What has effectively happened is current flows before voltage is registering on the voltmeter. Current is leading the voltage! Strictly speaking, there is not even a circuit as the capacitor is non-conductive. If this battery source is replaced with an a.c. source the only difference is the situation is reversed 50 times a second. The overall effect of a capacitor in an a.c. circuit is the current appears to be leading the voltage by as much as 90°, as can be seen in Figure 1.60.

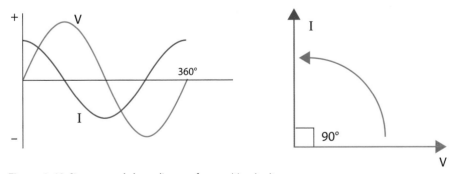

Figure 1.60: Sine wave and phasor diagrams for capacitive circuit

Inductors in an a.c. circuit

A current in a conductor will produce a magnetic field, as shown earlier in a solenoid. If the current is an alternating current, the magnetic field will be continuously changing. By winding the conductor into a coil

The sine wave used to represent this inductive circuit would look like this:

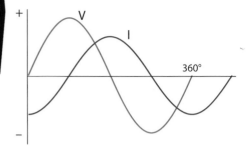

If we represented this as a phasor diagram, we end up with:

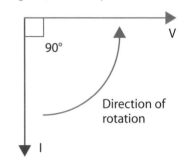

Figure 1.61: Sine wave and phasor diagrams for inductive circuit

Key term

Inductance – the ability of a coil to resist a changing current – measured in henrys, H.

there will be a series of conductors next to each other. A growing magnetic field in a coil will be cutting the coil next to it and hence inducing a second emf and current. This emf and current will be in direct opposition to the voltage and current causing it. A second emf created in this way is called back emf. By having a second emf and current in opposition, the circuit behaves like a big man pushing a small man. The small man has an effect on the big man but the big man does eventually win through (although a little weakened). The reality of an inductor is that the current lags behind the voltage as the coil resists any change in current flow. The property of a coil to resist a changing electric current by the production of an opposing back emf and current is known as **inductance** (L) and is measured in henrys (H). The back emf that occurs in an a.c. circuit with a coil can cause a current to lag behind the supply voltage by up to 90°, as shown in Figure 1.61.

Impedance

As you have now seen, the effects of a capacitor and an inductor in an a.c. circuit on current are opposite. In reality an a.c. circuit will contain capacitance and inductance, as well as resistance. The combined effect of all of these qualities of an a.c. circuit is known as impedance. In a d.c. circuit, Ohm's law states V = RI. For an a.c. circuit the same principle applies but the R is replaced with Z for impedance. Impedance is measured in ohms. More work on this subject follows in Chapter 6 but for now all you need to appreciate is the concept that impedance exists in an a.c. circuit and it is a combination of resistive and reactive components.

Characteristics of power quantities for an a.c. circuit

Power is defined as doing work or using energy in a set time. In a d.c. circuit you have already looked at the concept of the heating effect or power. Power is measured in watts and can be found by using the formula P = VI. This is not the whole story for an a.c. circuit. Reactive components (capacitors and inductors) make the current and voltage go out of phase with each other. This is particularly troublesome when you consider that a power reading is taken at an instant in time and at that instant the voltage could be 0 V and the current could be 2 A. This would mean that in this circuit at that particular instant in time, the power could be 0 W.

The power in an a.c. circuit is therefore made up of three parts:

- reactive power, KVAr (or VAr) – due to the capacitors and inductors
- true power, KW (or W) – due to the resistive part of the circuit
- apparent power, KVA (or VA) – due to the overall combined effect of impedance.

Power calculations are covered in Chapter 6.

Why power factor correction is required and how it may be achieved

An efficient a.c. circuit needs to draw the least amount of current possible for the job it is trying to do. The most efficient a.c. circuit only contains resistance. You have seen that a capacitor has the opposite effect to an inductor on voltage and current. If a circuit contains resistance and inductors, the overall effect on that circuit will be inductive. If a capacitor is added to this circuit, some of the inductive effect can be balanced out. If it is very carefully calculated, a capacitor can be put into an a.c. circuit and make it seem as though it just contains resistance. If this were the case the voltage and current would be completely in phase with each other. The angle between the voltage and current in an a.c. circuit is known as the power factor angle. **Power factor correction** is achieved by placing a capacitor in parallel with an inductive load. The capacitor used in this kind of supply correction is known as a power factor correction capacitor.

Industrial sites may contain very large equipment that runs on a.c. supplies. Large industrial sites mean a large number of inductors in the form of machine windings, transformer coils and fluorescent light chokes or ballasts. All of these 'inductors' will cause the site to have a large phase shift between the current and voltage (remember, a coil will cause current to lag behind voltage by up to 90°).

Power factor correction can be achieved by a number of different methods, including:

- capacitors
- load correction
- bulk correction
- **synchronous motor**.

A typical example of pf or power factor correction would be a large site with banks of **fluorescent** lights (discharge lighting). Each light contains an inductive coil or choke.

If the current and voltage sine waves were displayed using an oscilloscope, you would notice the voltage and current are out of phase. This leads to an overall effect of more current being drawn from the supply than is required to power the lights. A power factor correction capacitor is therefore required in each fluorescent tube to counter the effect of the coil. For industrial sites it may be practical to monitor this as a whole installation and then correct the site rather than the individual equipment. This is known as load correction or bulk correction and can be automated with correction added as the power factor angle drifts in to unacceptable levels.

On a much larger scale, this type of correction is carried out on the electricity supplier's network and again is automated.

Key terms
Power factor correction – also referred to as pf correction, it is the ability to bring the current in phase with the voltage.
Synchronous motor – power factor correction is used on large induction motors because it has a leading power factor.
Fluorescent – a fluorescent tube inductive coil is also referred to as a choke or ballast.

Progress check 1.3
1 Define what effect a capacitor and inductor will have on an a.c. circuit.
2 What is impedance?
3 What is power factor and why does it need correction?

LO5

ELECTRICAL QUANTITIES IN STAR DELTA CONFIGURATIONS

An introduction to three-phase networks

Three-phase supplies (or poly-phase supplies as they are sometimes referred to) are used in the UK to supply electricity to users. Electricity is generated in the power stations as a poly-phase supply. Poly-phase means multiple phases, in this case three phases, that have electricity induced in each of them in turn. There are many reasons for generating electricity in this way but the main benefits are cost, efficiency and ease of transportation to customers.

You have looked at direct current (d.c.) and alternating current (a.c.) but as an electrician you will also work with three-phase supplies – putting it simply, three single-phase supplies that are connected together. An a.c. single-phase supply consists of a live and a neutral wire but this is not quite the case for a three-phase supply. Each live wire does not necessarily need a separate neutral. A single-phase supply is generated by spinning a loop of wire within a magnetic field. A three-phase supply is no different apart from the fact that there are three loops of wire spinning. If the three loops of wire are equally spaced out there will be an angle of 120° between each of the lines.

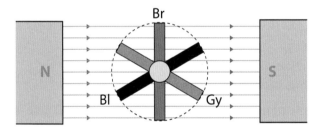

Figure 1.62: Three-phase generation

As a loop rotates past a magnetic pole the moving conductor will have a current induced in it – the greater the density of magnetic flux lines that are cut, the greater the induced current.

As the first loop passes the magnetic pole, the second loop will be approaching (120° behind). This loop will go through the same growth in current as it cuts across the most magnetic flux lines. This will occur again for the third loop. The rotation of all three loops of wire through the magnet poles produces three identical sinusoidal waveforms but each is separated by 120°. The way in which the three loops are connected together can be changed. One method is to connect the end of one armature to those of its two neighbour armatures. This type of connection is called a delta connection. The other method is to connect one end of each loop together at a central point and create a star connection.

Delta connection

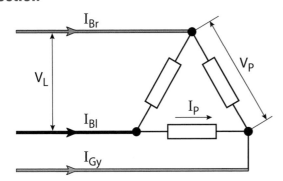

Figure 1.63: Delta connection

This type of connection has three lines only. Each end of the transformer phase is connected to another to form a triangle configuration. To get a supply voltage you connect between any two of the lines. There is only one level of voltage that can be found. You can see from Figure 1.63 that no earth or neutral are available on a delta configured system.

The current flowing through the transformer winding is called the phase current and the current flowing through the line is called the line current. As the current flows through the line and gets to the junction with the other two phases, it has two other possible routes to flow down. This means the current must split. On a delta system there are two different values of current flowing. The two different currents can be found by the formula $I_L = \sqrt{3} \times I_P$. On a delta system there is only one voltage so $V_L = V_P$.

Star connection

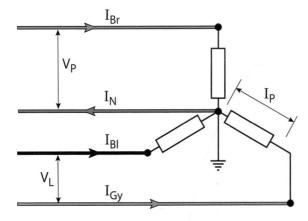

Figure 1.64: Star connection

A star configured supply is the typical supply that can be found in your street. It consists of three separate line conductors, each with a separate supply on it. A star system also has a neutral and an earth conductor. You can see from Figure 1.64 that the centre point of the star is connected down to earth and also onwards connected to give an extra conductor – this is the neutral.

A star connection has an advantage over the delta because not only can you go across two lines to get a particular voltage, you can also connect between any line and a neutral. Connecting between a line and a neutral will give a different voltage. This means two different voltage levels are available on the star connected supply. If you connect between two lines you will get a line voltage. If you connect between a line and a neutral (effectively across a single-phase) you will get the phase voltage. In a domestic street supply the line voltage would be 400 V and the phase voltage would be 230 V. These voltages can be found by the formula $V_L = \sqrt{3} \times V_P$.

The current flowing through the phase winding on a star system is also called the phase current and the current flowing through the line is called the line current. As the current flows along the line and reaches the junction with the phase winding, there is only one route so the line current and the phase current must be the same thing – it simply travels along the line, turns the corner and becomes the phase current. For this reason, $I_L = I_P$.

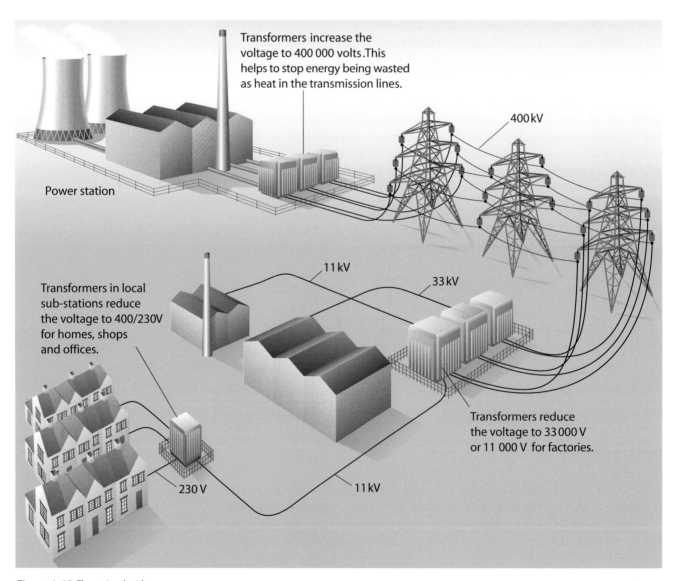

Transformers increase the voltage to 400 000 volts. This helps to stop energy being wasted as heat in the transmission lines.

400 kV

Power station

Transformers in local sub-stations reduce the voltage to 400/230V for homes, shops and offices.

11 kV

33 kV

Transformers reduce the voltage to 33 000 V or 11 000 V for factories.

230 V

11 kV

Figure 1.65: The national grid

The advantage of a star connected network is not only the different levels of voltage supply available but fewer neutrals are required. Effectively the three lines share the same neutral.

Three-phase supplies are the preferred way of local distribution to domestic streets and houses. If all the loads, i.e. the demands from houses split over a three-phase supply, are equal then the currents drawn on each of the lines are equal. If the currents are identical but 120° apart, the sum of these at any point in time is 'zero'.

This means that if the currents stay balanced (i.e. equal) the current that flows down the neutral is also 'zero' and the neutral can also be made smaller. Reducing the size of the conductor saves money for the supply company. However, if the currents become unbalanced, the out-of-balance current will flow down the neutral. If this out-of-balance current gets too big, the neutral can heat up (one of the effects of current flow in a conductor).

 **Safe working**

If the lines are not balanced in a star supply, the neutral can heat up and fail. This is a common problem on developments where planning has not taken into account growth in demand for electricity.

Case study

A college experienced such a rapid growth in learners that extra temporary buildings were built to house them. The extra electricity required for heating, computers and workshop machinery led to the supply becoming unbalanced. The neutral heated up because one line in particular was supplying most of the extra current. When it snowed and even more electricity was required, the path of the underground supply cable melted the snow with the excessive heat produced. Eventually the supply transformer fuse blew, putting the whole college into darkness for a few hours! The supply was redesigned, balanced and a new supply was installed.

Key facts

1   For a star connected network, a balanced load is preferred as no current flows down the neutral and means the neutral can be small.
2   A star connected supply (three line and neutral) has an advantage over delta because of the neutral and the two different available voltages.
3   Delta has one voltage and two currents, star has one current and two voltages.

Progress check 1.4

1   What happens if a three-phase supply is not balanced?
2   Draw and label a star network.
3   Draw and label a delta network.

Knowledge check

Attempt all 15 multiple choice questions, which are based on the basic science covered in this chapter. If you are not successful in getting all of the answers correct, go back over the chapter and attempt them again.

1 What is the unit for magnetic flux density?

   a Wb
   b T
   c Hz
   d Ω

2 34 000 000 J can also be written as what?

   a 34 MJ
   b 340 kJ
   c 3.4 MJ
   d 34,000 MJ

3 $7^2 \times (3 \times 3 + 3)$ can be simplified to what?

   a 444
   b 120
   c 588
   d 882

4 If copper has a resistivity of 17.5×10^{-9}, what is the resistance of a 25 m length of 2.5 mm^2 conductor?

   a 1.75 Ω
   b 0.175 Ω
   c 17.5 Ω
   d 175 Ω

5 What is the resistance of a series circuit with resistors of 3 Ω, 6 Ω, 9 Ω?

   a 1.64 Ω
   b 162 Ω
   c 27 Ω
   d 18 Ω

6 A parallel circuit consists of two resistors: $R_1 = 2.5$ Ω, $R_2 = 1.2$ Ω. If the circuit is connected to a 12 V supply, what is the total circuit current?

   a 10 A
   b 1.5 A
   c 15 mA
   d 15 A

7 A parallel circuit has four resistors of 1.7 MΩ, 2.3 MΩ, 2.9 MΩ, 5.7 MΩ. This parallel circuit is now connected to a 400 V supply. What is the total circuit current drawn by the loads?

   a 617.3 A
   b 6.3 A
   c 16 A
   d 0.617 mA

8 What is the magnetic flux density of a magnet with a cross-sectional area of 0.3 m^2 and a flux of 4 Wb?

   a 13.33 T
   b 1.2 T
   c 4.3 T
   d 43 mWb/m^2

9 A 280 kg mass is suspended by four ropes on a moveable pulley system. What is the mechanical advantage?

   a 5
   b 2
   c 4
   d 1

10 What force will a 0.6 m conductor experience if it has a 3 mA current flowing through it when placed in a 5 T magnetic field?

   a 90 T
   b 9.0 T
   c 0.009 T
   d 900 T

11 A sine wave reaches the same peak value twice in 50 mS. What is the frequency of the sine wave?

   a 50 Hz
   b 500 Hz
   c 200 Hz
   d 20 Hz

12 What energy has been used moving a 200 kg transformer core 12 m into position for installation?

   a 23.544 kJ
   b 2.44 kJ
   c 16.67 kJ
   d 16.67 J

13 A motor is 97.3% efficient and produces 16.7 kW of power at its output. What input power is required to achieve this?

   a 16.25 kW
   b 19.4 kW
   c 17.16 kW
   d 1.6 MW

14 What is the best power factor?

   a Unity
   b 0.99
   c 0.1
   d 0

15 A transformer has a turns ratio of 3 : 1 and a secondary current of 3 A. What is its primary current?

   a 9 A
   b 3 A
   c 0.1 A
   d 1 A

Health and Safety in building services engineering

Chapter 2

This chapter covers:

- health and safety legislation
- handling hazardous situations
- electrical safety requirements
- working with gases and heat-producing equipment
- using access equipment
- working in excavations and confined spaces
- applying safe working practices to manual handling.

Introduction

Whether it is an office or a construction site, the workplace is a dangerous environment. Equipment of all kinds is in constant use; there are electrical supplies, cables and other tripping hazards, dangerous tools, chemicals, many toxic or flammable, and people – all busy in whatever jobs and roles they are employed to do.

Accidents happen and can result in injury, disability and even death. While this is, of course, terrible for the victim, it is also traumatic for their families and loved ones. Accidents at work can lead to loss of earnings, personal problems and drastic changes to your life.

For the company, there may be a heavy financial cost resulting from lost hours, lost business, possible legal action and fines to pay. Added to this is the loss of credibility if your company is seen to be one where health and safety is not the main priority.

Because of this, successive governments are constantly introducing and updating regulations and processes to make sure that, as far as possible, working life, and the workplace itself, are safe and healthy.

A building services engineer is expected to have a general knowledge of health and safety laws and practices. Ignorance is no defence, so it will be pointless to tell a judge, 'I didn't know I wasn't supposed to work that way.' This chapter, then, is intended to help you gain the health and safety awareness you will need in the workplace.

LO1

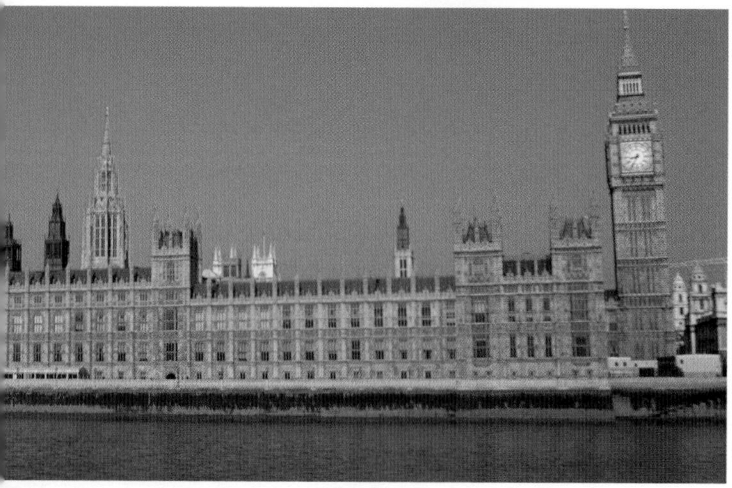

Figure 2.1: The Houses of Parliament

HEALTH AND SAFETY LEGISLATION

Aims of health and safety legislation

Legislation is a set of rules put in place by the government. It is debated and refined until it can be voted in and become an act of parliament. This legislation is intended to make sure that the workplace is safe and healthy, not only for the workers themselves, but for anyone in the vicinity or connected with the work. This includes everyone from site visitors to the general public.

Health and safety legislation is enforced by inspection, and prosecution if any serious breaches of the law are found. Serious accidents may be followed by an inquiry, both to investigate what went wrong and to establish how they can be prevented from happening again.

The responsibilities of individuals under health and safety legislation

Everyone who works on, visits or passes near to a work area is responsible for their own health and safety and for that of others.

Employers

Employers must make sure that the workplace and equipment they own (or are responsible for) is **fit for purpose** and safe. Employers are also responsible for providing safety equipment such as hard hats, ear defenders and safety features on tools and machinery. They must appoint people who are responsible for introducing and maintaining health and safety standards and processes, from managers who are responsible for the health and safety of the whole organisation, to first aiders who are there to help individual staff members if they are injured.

Employees

It is no use saying, 'Well, it isn't my job to worry about health and safety, I only work here'. The law does not see it that way and has placed responsibilities on the employee.

- Take care of yourself and any other person who may be affected by your actions.
- Report any equipment or other issues that pose risks to health and safety.
- Cooperate with your employer in relation to health and safety issues.
- Use any safety equipment provided – for example, you MUST wear a hard hat on a construction site even if you don't think it suits you!
- Do not interfere with or misuse anything provided in the interests of health and safety.

Contractors

A contractor is someone who has been taken on by an employer to do a specific job, but who is not an employee of that company. It is vital that:

- the right contractor is used for the job
- there is a clear understanding of what the contractor is expected to do and what is involved in the work, including any risks
- the contractor receives training before starting work, including health and safety training
- the contractor provides training to everyone who will use any specialist installation or equipment they install.

Visitors to site

A visitor to a workplace can be an inspector, a customer, a repair engineer, a delivery driver or anyone else who needs to enter the area on official business. Once on site visitors are exposed to many of the same hazards and risks as employees and contractors.

- Training – make sure visitors are aware of how to conduct themselves safely in your workplace. This may involve a short training talk, for example.
- Visitors must be provided with any personal protective equipment (PPE) needed in that area.
- Visitors are responsible for wearing the PPE and using any other safety equipment or procedures put in place.

> **Key term**
>
> *Fit for purpose* – using the right tool for the right job and the right environment for the work. For example, try assembling delicate electronic components in a dirty, tumbledown barn. The environment is simply not right for that job. It is not 'fit for purpose' and will lead to a faulty product.

Figure 2.2: A work area must be protected by barriers and warning signs

General public

Although the general public are banned from most construction sites, there are times when building services engineering work takes place in a public area, for example a shop, school, office or in the street. In these situations, the work area must be protected by barriers and warning signs making it clear that:

- the work area is hazardous
- only the engineer and other authorised people are allowed to cross the barriers.

Barriers and signs must force people to avoid the area, for example diverting people far enough around the base of a scaffold tower so that they will not be injured by falling objects, or bump into the tower and cause the engineer to fall off.

Members of the public have their own responsibilities to heed barriers and warning signs and not enter a work area.

Statutory health and safety materials

Workplace health and safety law is explained in a series of documents issued by the government. These are **statutory** documents, which means that they must be adhered to. If you don't, you are breaking the law and can be prosecuted.

There are two main types of regulations.

- General – these apply to everyone in the workplace, whether a construction site or an office.
- Specific – these deal with specific types of work, for example working at height or in a confined space.

This section will look at the general statutory documents used in the building services engineering workplace.

Progress check 2.1

1 What is legislation?
2 What are an employee's main health and safety responsibilities?
3 A customer wants to visit your site – how will you keep him or her safe?

Key term

Statutory – statutory documents are those which have been debated and issued by the government. The word 'statutory' comes from the fact that they are on the Statute Book, in other words, part of the law of the land.

The Health and Safety at Work Act 1974

The Health and Safety at Work Act 1974 is the central document on which all other statutory health and safety regulations are based. Its basic message is that:

- employers have a duty to keep their employees, contractors and visitors safe in the workplace
- employees must take care of their own health and safety and cooperate with their employers with regards to health and safety
- the Health and Safety Executive (HSE) has powers to enforce health and safety law on a company and can prosecute them if they do not cooperate.

Reporting Incidents, Diseases and Dangerous Occurrences Regulations (RIDDOR) 1995

RIDDOR is a regulation that makes employers, or anyone else in charge of a workplace, responsible for reporting:

- serious workplace accidents
- occupational diseases
- **dangerous occurrences**.

Accidents and other incidents that cause people harm in the workplace need to be reported as soon as possible. This is because an investigation may be required and also it will provide a record to help with treatment if there are any long-term effects from the incident. There is an online RIDDOR form on the HSE website. Types of reportable injury include:

- fracture (except to fingers, thumbs and toes)
- loss of a limb
- dislocated joint
- loss of sight (temporary or permanent)
- burns
- electric shock serious enough to cause unconsciousness, resuscitation or hospital admittance.

> **Key term**
>
> *Dangerous occurrence* – a near miss. This is when an accident occurs that could have caused serious injury but didn't, for example a scaffold collapses after working hours when the entire workforce are off site.

The Electricity at Work Regulations 1989 (revised in 2002)

Although the title of this document contains the word 'electricity', the Electricity at Work Regulations do not just apply to people working on electrical systems. The intention of the Electricity at Work Regulations is to make sure that workplace electrical installations and equipment are safe for all the people working on the premises. Here are some of the main points.

- Electrical systems must not cause a risk to anyone working near to them or working with them.
- Electrical systems must be regularly maintained.
- Electrical equipment will be strong and capable enough of doing its job without endangering anyone.
- There must be a means of cutting off the supply, both automatically if there is a fault, and manually if the system or equipment has to be worked on.

Personal Protective Equipment at Work Regulations 1992

Personal protective equipment, or PPE as it is usually called, is the safety gear used to protect your body while you are at work. PPE includes:

- hard hats
- protective boots
- high visibility (hi-vis) jackets or waistcoats
- ear defenders
- gloves
- masks and respirators.

The employer should provide necessary PPE to their employees free of charge. The employees must wear, and take good care of, the PPE provided. On most construction sites it is mandatory to wear a hard hat, high visibility clothing and protective footwear. The rule is: no PPE = no entry.

If any item of PPE is damaged or faulty, it should not be worn. Report the fault to your employer and dispose of the item.

Goggles protect the eyes from flying objects when using power tools such as drills, saws and cutters.

A hi-vis waistcoat makes sure you can be seen at all times. Compulsory on most construction sites.

Protective footwear protects feet from injury from heavy objects. Compulsory on most construction sites.

Figure 2.3: An employer must issue all employees with PPE

Control of Substances Hazardous to Health (COSHH) 2002

It would be easy to think that 'substances hazardous to health' means poisons or dangerous liquids such as acid, but these regulations apply to any substance used in the workplace, for example:

- paint
- varnish
- cleaning fluids
- adhesives.

Even the most everyday products contain ingredients that can be harmful, whether by contact with the skin or by inhaling fumes. Prolonged exposure to many substances, such as the glue used by an electrician to fit together parts of a plastic conduit system, can cause illness and injury.

COSHH is really a huge set of regulations covering a vast range of chemicals and substances, from dust to the components of nanotechnology. There are data sheets and guides for specific chemical types. These guides contain instructions and recommendations for their safe handling and use. The main requirements that apply to most substances are as follows.

- Use the substance according to the manufacturer's instructions.
- Be aware of the specific hazards before using a substance – this information should be provided by the manufacturer (on the tin or instruction leaflet, for example).
- Prepare a risk assessment (see page 77) before using any type of chemical substance.
- Provide good ventilation in the work area.
- Provide (and wear) suitable PPE, e.g. respirators, gloves and eye protection.

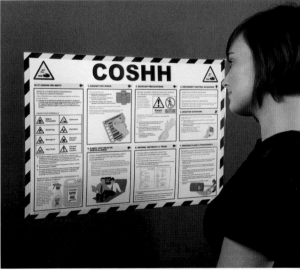

Figure 2.4: A COSHH notice

Provision and Use of Work Equipment Regulations (PUWER) 1998

Nearly every job, whether it is a practical trade or office based, involves the use of tools or work equipment. This can be anything from hammers and screwdrivers to computers. Tools get a lot of hard use in the workplace and once damaged, worn out or faulty, can become dangerous. For example, a loose hammer head might fly off while in use, and electrical equipment can heat up or cause an electric shock.

The Provision and Use of Work Equipment Regulations have been issued to give guidelines on the use and provision of tools and other work equipment. The main message of PUWER is that work equipment should be:

- the right equipment for the right job (do not use a pair of pliers to hammer in a nail, or a hammer to drive in a screw!)
- only used if it is safe
- maintained in a safe condition and, in some cases (e.g. electrical, compressed air and hydraulic equipment), inspected and tested
- accompanied by information, instructions and training
- fitted with safety devices, warning labels and markings.

Electrical equipment, which covers everything from a power tool to a television in a hotel room, must be tested regularly. This type of testing is called Portable Appliance Testing or PAT. You can read more about PAT on pages 80–82.

Control of Asbestos at Work Regulations 2012

Asbestos is a very dangerous substance and, while it comes under the COSHH regulations, it is so hazardous it has its own set of regulations. Asbestos must never be drilled, cut or removed by anyone who is:

- not trained to work on asbestos
- not wearing the appropriate PPE – which includes full body protection and respirator equipment.

Asbestos fibres cause a disease of the lungs called asbestosis. Not only is the worker at risk from asbestos dust and fibres, but so is anyone else in the vicinity. If asbestos has to be removed from a building or area, that area has to be shut own, sealed off and the asbestos itself cleared by fully trained, licensed and equipped professionals.

If you inadvertently drill or cut asbestos then it must be reported as a dangerous occurrence, covered by Reporting Incidents, Diseases and Dangerous Occurrences Regulations (RIDDOR) (see page 69).

The Control of Asbestos at Work Regulations also states the following.

- If existing asbestos is in good condition and not likely to be damaged, it can be left in place. However, its condition must be monitored and managed to make sure it is not disturbed.
- The presence of asbestos has to be made known to anyone who is going to carry out work in the area.
- Training is mandatory for anyone likely to be exposed to asbestos fibres while they are at work.

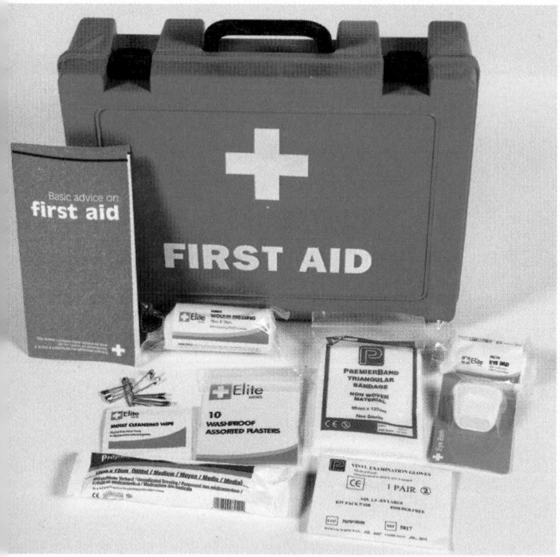

Health and Safety (First Aid) Regulations 1981

First aid is the initial treatment and help given to someone who has suffered an accident or been taken ill. In the workplace there should be trained staff who can carry out first aid immediately. The names of trained first-aiders and their contact details must be clearly displayed so they can be called as soon as an accident occurs. Any accident must be recorded in an accident book as soon as possible after the incident. It doesn't matter whether the injury or illness is work related or not.

An employer must provide first-aid equipment and facilities. The regulations require:

- a first-aid box
- an appointed person who is responsible for all first-aid arrangements and facilities
- information for employees about first-aid arrangements.

Figure 2.5: A first-aid box must be provided by employers

Workplace (Health, Safety and Welfare) Regulations 1992

Not only must people in the workplace be protected against accidents, it is also important to consider their welfare. This means their health and well-being. For example, proper hygienic toilet and washing facilities must be provided. There must be adequate lighting and working space. People in the workplace should, as far as possible, be comfortable enough to carry out their work without risk to their general health.

Work at Height Regulations 2005

Please refer to pages 95–100 for more information.

Confined Spaces Regulations 1997

Please refer to pages 102–103 for more information.

Manual Handling Operations Regulations 1992

Please refer to pages 105–105 for more information.

Progress check 2.2

1 What is RIDDOR?
2 List five items of PPE.
3 Who is authorised to work on asbestos?

Chapter 2

Activity 2.1

Research four more health and safety regulations relevant to the building services engineering sector and put together a table showing each regulation, its purpose and main points.

Non-statutory health and safety materials

Non-statutory documents are not law but guidelines, and what is called 'best practice'. They are, however, based on statutory regulations. Even though they are not, in themselves, mandatory, working according to non-statutory codes of practice and guidelines means that you will be applying safe and effective methods.

Most health and safety at work regulations are available from the HSE. The HSE also provides sets of non-statutory, easy-to-understand guidelines explaining how the regulations can be applied in everyday work situations.

The different roles enforcing health and safety legislation

In order to manage and maintain health and safety, both nationally and within a company, various organisations have been set up and roles created. Some of the main ones are outlined below.

Health and Safety Executive (HSE)

The HSE is a government agency which has the authority to enforce health and safety law. Its inspectors can carry out checks in the workplace and impose improvement notices on any business that does not comply with health and safety law. The HSE can also prosecute a business if it does not comply.

Intimidating as all this sounds, the HSE is also there to advise and help businesses make their workplaces safe for employees, visitors, contractors and the general public.

Safety manager

A safety manager is a member of the management team. He or she has responsibility for making sure that health and safety laws and practices are put in place throughout the company. The safety manager may be the head of the company or someone at senior level. They will not necessarily carry out the work themselves, but assign it to safety officers and representatives. They do, however, bear responsibility for health and safety compliance.

Safety officer

A more 'hands on' role, a safety officer is the person you are most likely to see with regards to keeping your workplace safe. The officer will carry out inspections and has the authority to order improvements. The safety officer may be full time in that role, as is the case in many large companies, or they may have other responsibilities, of which health and safety is part.

Safety representative

Safety representatives are usually members of the workforce who are responsible for making sure their particular part of the company or workplace is safe, and for bringing health and safety matters back to the management. So if the safety representative is told by other electricians that the ladders provided by the company are not safe, the representative will go to the management, usually via the safety officer, bring this to their attention and argue the case for the ladders to be replaced with safer ones.

First-aider

If someone has an accident in the workplace, the first person on the scene is a first-aider. First-aiders are employees who have been trained to deal with the immediate consequences of an accident, for example:

- to control heavy bleeding in the case of a cut
- to rig up support for an injured limb, such as a temporary sling or splint
- to attend to an unconscious patient by trying to rouse them or making sure they are safe and in the correct position.

The names and contact details of all first-aiders must be clearly displayed so that they can be called out quickly if there is an accident.

Many companies give all their staff some basic first-aid training on a regular basis, so that they can, at least, keep the victim of an accident safe and calm until a trained first-aider can get to the scene.

Figure 2.6: A first-aider contact notice

Fire warden

Fire is an ever present danger in all workplaces. Construction sites will contain many fire hazards such as gas bottles, chemicals, naked welding and brazing flames. There may also be paper and plastic packaging as well as wood and wood shavings everywhere. The fire warden is a designated employee who has been trained to get you to the correct assembly point if the fire alarm is sounded. Once at the assembly point, he or she will take a register to make sure everyone is present.

If someone is missing, the fire warden must not go back into the building to try to find them. Instead, he or she will report the missing person to the fire brigade and to the relevant safety officer.

The fire warden also has the authority to inspect the work area and order the removal of any fire hazards.

Progress check 2.3

1. Who is the HSE?
2. Who represents the workforce in matters of health and safety?
3. Who is the first person to call when someone is injured in the workplace?

HANDLING HAZARDOUS SITUATIONS

The workplace contains many **hazards**. Some, of course, are more immediately apparent and dangerous than others: a construction site, for example, is filled with hazards whereas a shop or office is relatively safe. Inevitably, accidents will happen and hazardous situations will arise. You need to know how to deal with those and this section will tell you both how to avoid, and how to handle, hazardous situations.

Hazardous situations on site

Table 2.1 shows some of the typical hazards found in the kind of workplace building services engineers such as electricians and plumbers will encounter. This is by no means a complete list, but shows the main ones. Some of these will be looked at in more detail later in the chapter.

LO2

Key term

Hazard – a situation that poses a threat. For example, a drill with a damaged power lead is a hazard because if you attempt to use it, the drill could give you an electric shock.

Chapter 2

Hazard	Possible result	Cause
Tripping hazards/trailing leads	• Falls resulting in: o sprains and broken bones o head injury	• Trailing leads and hoses • Untidy workplace
Slippery floors and surfaces	• Falls resulting in: o sprains and broken bones o head injury	• Wet floors • Newly cleaned and polished floors
Electric shock (see pages 88–89)	• Heart failure • Burns • Unconsciousness	• Faulty power tools and equipment • Damaged cables • Incorrect wiring and connection • Exposed cable joints and connections • Water around electrical fittings and supplies • No earthing arrangements
Fire (see pages 93–95)	• Burns • Explosion • Smoke inhalation	• Incorrect handling of gas bottles and other flammable substances • Smoking in a no smoking area • Presence of flammable materials such as paper and wood shavings • Carrying out welding or gas cutting operations near to flammable materials or in a flammable atmosphere
Dust and fumes	• Respiratory problems • Eye and skin problems	• Cutting and drilling masonry • Sweeping up • Chemical use – paints, cleaning fluid, etc. • Fibreglass insulation
Contaminants and irritants	• Respiratory problems • Eye and skin problems	• Chemical use – paints, cleaning fluid, etc.
Asbestos (see page 72)	• Major respiratory problems	• Cladding found in older buildings

Table 2.1: Common hazards on a construction site

Continued ▼

Hazard	Possible result	Cause
Working at height and in confined spaces (see pages 95–100 and 100–104)	• Falls resulting in: ○ sprains and broken bones ○ head injury ○ dropped materials and equipment causing head injuries to people below ○ breathing problems and even asphyxiation	• Ladders • Towers • Scaffolding • Working above trenches and other excavations • Working on a roof • Working in a confined place such as a duct or cramped plant room with insufficient ventilation
Gas bottles (see pages 90–92)	• Explosions • Fire • Asphyxiation • Low temperature burns	• Welding and cutting • Brazing • Laying bitumen on a flat roof • Paint stripping
Malfunctioning and damaged tools and equipment	• **Hazardous malfunction** resulting in: ○ eye injury ○ cuts, bruises and general injuries ○ electric shock ○ burns	• Damaged electric equipment and leads • Poorly maintained equipment • Using damaged or worn out tools • Using the wrong tool for the job
Manual handling and lifting (see pages 104–105)	• Back injury • Sprains and strain injury • Cuts to hand • Foot injury	• Incorrect lifting technique • Attempting to manually lift something that is too heavy or awkward • Not using lifting equipment • Failing to wear PPE such as gloves and protective footwear

Table 2.1: Common hazards on a construction site

Key term

Hazardous malfunction – this occurs when a tool or item of equipment goes wrong but does not injure anyone – but could have caused injury to the person using it, or to people in the vicinity.

Figure 2.7: Tripping hazards

Progress check 2.4

1 What is a hazard?

2 An electric shock can result in what?

3 What can cause manual handling and lifting injuries?

Safe systems of work

Risks can be reduced by using safe systems of work. These are procedures that allow you to think about, and prepare for, the possible hazards presented by a particular job.

Risk assessment

Most risk assessments are an automatic process. You look at a job activity, work out what tools you will need and how you will protect yourself from the hazards involved in the work. We do it all the time – before crossing the road, for example. The first thing we do is check that there is no traffic and that it is safe to cross. We are, in fact, assessing the risks involved in stepping out onto the road and trying to reduce them by crossing when there are no cars coming our way.

For larger, more complex work, a written risk assessment has to be produced. This will be a form with a set of headings. Typical headings are as follows.

- Tasks – break down the project into a series of tasks.
- Hazards – the hazards associated with each task.
- Control – how the risks are alleviated, for example what PPE is needed?
- Who is affected – who will be affected if an accident actually happens?
- Seriousness – this will be coded using letters or numbers. For example, if the hazard could end up in the loss of a limb or eye, then it will be graded as a 1 or an A. Something that results in a bruise but is not serious or life-threatening would be graded as a 3 or C.
- Likelihood – how likely is it that the hazard will actually cause an accident? Again, this is graded using a number or letter system.

Method statements

Method statements describe, in detail, the way a task is to be carried out. The method statement should outline the hazards involved and include a step-by-step guide as to how to do the job safely.

Method statements often form part of a **tendering** process and can give a potential customer an insight into your company and how it operates. Below is the type of information needed for a method statement.

> **Key term**
>
> *Tendering* – the tendering process is the part of a project when companies are invited to submit a price and a proposal to the customer. The customer will choose the most suitable offer and appoint that company to carry out the work.

Section	What it contains
Section 1	• A title • A brief description of the work to be carried out • Your company details • Start date and completion date for the work • Site address and contact details including emergency numbers, etc.
Section 2	• Summary of the main hazards that are present and the measures put in place to control them • Personal protective equipment • Environmental or quality procedures
Section 3	• Staff and the training they will need • Permits to work (see page 78) • Machinery shutdown and lock off procedures • Site access • Welfare and first aid
Section 4	• Step-by-step guide to how the work is to be carried out

Table 2.2: A guide to the sections in a method statement

Permit to work

If a particularly hazardous job is to be carried out, then a permit to work should be issued. The types of jobs and situations that require a permit to work are:

- working on, or near, live electrical equipment or systems
- working in a confined space
- working in deep trenches
- working with or near corrosive or toxic substances
- working near machinery in production areas such as assembly lines.

The permit to work must be authorised by a competent person. This is someone who fully understands the dangers associated with the job. The work must also be carried out by a trained and **competent person**. The work is detailed on the permit and the permit has a start and finish time.

The finish time is important because if the engineer has not returned by the time stated on the permit, it acts as a warning that there may be something wrong.

A permit to work covers one specific job and cannot be used again.

The layout of a typical permit to work is:

- description and scope of work to be carried out
- the location of the work
- who is responsible
- who will actually do the work
- precautions needed
- hazards identified
- PPE required
- authorisation to start work
- necessary checks to be carried out
- emergency arrangements
- what to do in unusual circumstances
- the time limit of the permit
- completion signatures
- cancellation signatures.

Safety signs and notices

Signs are an important part of workplace safety. They are not there simply as decoration. They must be heeded. If a sign instructs you to wear ear defenders then you must wear them. A standard colour scheme has been established for warning signs. Table 2.3 shows these standard colours and gives examples of signs.

Key term
Competent person – BS7671: 2008 and the associated guidance notes and codes of practice define a competent person as someone with the technical knowledge or experience to carry out electrical work without risk of injury to themselves or others.

Type of sign	Examples		Hazard	Symbol
Mandatory	Hearing protection must be worn	Protective footwear must be worn in this area	Toxic	
Hazard		RADIOACTIVE	Very toxic	T+
			Harmful	
Prohibition			Irritant	i
			Highly flammable	
Fire	Fire alarm		Extremely flammable	F+
Safe condition	Fire assembly point		Explosive	
			Dangerous to the environment	
			Oxidising	

Table 2.3: Safety signs

Hazardous substance signs and symbols

As well as warning signs and notices, symbols are used to denote harmful substances. One place you will see these is on the back of a tanker lorry. They give immediate information on whether the contents of the tanker are corrosive or flammable, etc. See Table 2.4 for examples of hazardous substance signs.

Progress check 2.5

1 State three headings found on a risk assessment.
2 What is the purpose of a method statement?
3 Why is the finish time on a permit to work so important?

Hazard	Symbol
Corrosive	

Table 2.4: Hazardous substance signs

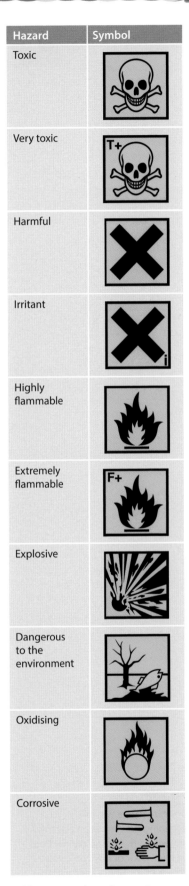

LO3

ELECTRICAL SAFETY REQUIREMENTS

Vital as it is to our working lives, electricity carries its own unique hazards:

- Shock – caused by direct contact with a live part such as an exposed conductor or terminal. Shock can result in a burn and, in some cases, cause unconsciousness and even heart failure.
- Fire – caused by:
  - an overheated cable
  - arcing due to loose connections
  - short circuit, which is an explosive increase in current caused by a direct contact between line and neutral, or line and earth.

Case study

A small manufacturer of agricultural equipment set up its operation in a former warehouse. Although production had started, there were still alterations to be completed in order to convert the premises into a fully functioning unit.

An electrician had installed a number of new circuits and the time had come to connect them into the main, three-phase distribution board. The board was mounted on a wall and was quite high up. The electrician brought a stepladder. As he climbed he reached out to grab the top of the distribution board to steady himself – it was an instinctive action and not normally dangerous.

What he did not know was that someone had removed a number of circuits and instead of disconnecting the cable from the distribution board and pulling it out, had simply snipped the cables and left them protruding from the top of the board, unseen from ground level. The electrician's hand came to rest on these cables.

The 230 V shock he received was painful, potentially lethal and made him feel ill for several hours afterwards.

Only trained and competent people should work on electrical systems – people who know that you must NEVER leave bare cables exposed to touch.

Common electrical dangers to be aware of on site

Construction sites are harsh environments and, because of this, electrical supplies, equipment and tools should be of a type designed to survive the rough treatment they are likely to receive. Some common electrical hazards in the workplace are looked at below.

Faulty and damaged equipment

All electrical work equipment should be regularly inspected and tested to make sure that there is no damage to:

- the casing
- the lead
- the plug.

It should also be tested to make sure that it works correctly and that there are no hidden electrical faults such as loose connections and poor earth **connectivity**. If any equipment appears to be damaged it must not be used, but replaced or sent away for repair.

The care and use of electrical work equipment and tools is covered by PUWER (see page 71) and the testing of electrical equipment is described in the *IEE Code of Practice for In-service Inspection and Testing of Electrical Equipment* (see pages 81 and 82 for the main requirements of the IEE Code of Practice).

Key term

Connectivity – a way of describing how good an electrical connection is. A loose connection will have poor or low connectivity and a secure connection will have good or high connectivity.

Classes of electrical tools and equipment

The classes of electrical equipment are among the key information contained in the code of practice:

- Class I – has an earth conductor connected to the frame of the equipment.
- Class II – relies on extra or toughened insulation to protect the user from an electric shock. There must be NO earth connection.
- Class III – protects the user from electrical shock by transforming the voltage down from the mains voltage of 230 V to an extra low voltage of, for example, 24 V. This extra low voltage cannot harm a human being, so even if there was a fault and the user received a shock it would not cause them injury. There must be NO earth connection.

Test intervals for electrical tools and equipment

The *IEE Code of Practice for In-service Inspection and Testing of Electrical Equipment* also contains a table with recommended intervals between testing. This depends on:

- the type of electrical equipment, e.g. hand-held, portable
- the environment the equipment is used in, for example electrical equipment used on a construction site will need to be tested more often than that used in a shop.

Types of electrical tools and equipment

As mentioned, the equipment-testing code of practice separates electrical equipment into a number of main types. These are shown in Table 2.5.

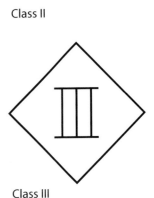

Class II

Class III

Figure 2.8: The Class II and Class III equipment labels

Electrical equipment	Description
Hand-held appliance or equipment	Held in the hand during its normal use, for example an electric drill or a soldering iron
Moveable equipment	Weighs up to 18 kg and can be unplugged and moved, for example a welding set
Portable appliance and equipment	Plugged in and can be moved while it is being used, for example a kettle
Stationary appliance or equipment	Fixed in place or weighing more than 18 kg and not intended to be moved regularly, for example a wall-mounted heater or a washing machine

Table 2.5: Main types of electrical tools and equipment

Activity 2.2

A collection of electrical tools and items of equipment are provided for you to examine. Using Table 2.5, decide whether each one is hand-held, moveable, portable or stationary.

Formal visual inspection

Unlike the user check, the formal visual inspection must be carried out by a competent person. For hand-held and portable equipment used by the public this has to be completed once a week; for equipment in other work environments such as offices and shops this only needs to be carried out once a year.

The formal visual inspection consists of:

- operating the equipment to make sure it works properly and according to the manufacturer's instructions
- establishing that the equipment is suitable for its working environment
- opening the plug and checking the fuse and connections, if the equipment has a plug which isn't the sealed type
- speaking to the equipment owner or operator to find out if there have been any problems with the equipment
- ensuring that the correct switching is in place:
  - a functional switch on the equipment itself
  - an isolator to cut the supply to the equipment for maintenance and repair work
  - an emergency stop, if applicable, for example if the equipment is an item of machinery in a factory.

Combined inspection and testing

Combined inspection and testing of equipment consists of a formal visual inspection and a series of instrument tests. A Portable Appliance Tester usually incorporates all the necessary test facilities in one device. Appliances and equipment can be plugged into the machine.

The tests are shown in Table 2.6.

Safe working

Before carrying out a functional test, make sure it is safe to switch the equipment on and that you understand how to operate it safely. If necessary, ask the person who normally operates the equipment to switch it on and run it during the check.

Test	Description
Earth continuity	If the equipment has an earth connection (Class I equipment), the path from the equipment's exposed metalwork to its contact point with the installation earth must be continuous. The contact point may be the earth pin of a 13 A, for example. Acceptable reading: 0.1 Ω + the resistance of the protective conductor in the appliance lead (if it has one).
Insulation resistance	Normally carried out at 500 V d.c., this test confirms that the insulation of the equipment is sound and that there are no earth faults. The test is taken between the line and neutral (which are connected together) and the equipment metalwork. It is important to check that the equipment can withstand the 500 V test voltage. All switches should be on and covers in place. Acceptable reading: 1 MΩ or 0.3 MΩ for appliances with heating elements.
Touch current	If the 500 V d.c. insulation resistance test cannot be carried out, a touch current test can be used to test equipment insulation. Care has to be taken as this test is done while the equipment is switched on and operating. The test is carried out between the internal live parts of the equipment and its metalwork and insulation. Acceptable reading: 3.5 mA or 0.75 mA for appliances with heating elements.
Functional	Once all the tests are complete, the equipment is operated to check that it is functioning correctly.

Table 2.6: Tests to be carried out on electrical equipment

Faulty or damaged cables

There will be a lot of electrical cables on a construction site – equipment and tool leads, extension leads, and wiring for temporary lighting and supplies to site huts and offices.

All leads, whether connected directly into electrical equipment or used as extension leads, must be regularly inspected for damage. The main hazards posed by cables are as follows.

- Electric shock – the cable used must be of a standard and type designed to withstand the **mechanical damage** likely on a construction site. If the sheath and insulation is damaged the conductors will be exposed, posing a shock hazard. Also, cables need to be connected securely so that live conductors do not slip out of the terminal and become exposed.
- Fire – caused by short circuit or overheating. The correct-sized cable must be used because, if the conductor cross-sectional area is too small for the current it carries, it will heat up and pose a fire risk. Likewise, a loose connection will arc and become hot.

Tripping hazards and trailing cables

Extension leads and equipment leads tend to be strewn about the site and can be a tripping hazard. If a lead has to be run across a walkway or road, it should be covered by a cable protector of some sort. If the cable is to be in place for some time then it is a good idea to secure it with cable ties.

Proximity of cables

When working, you must always be aware of any nearby electrical cables. It is easy to cut or damage a cable.

Buried/hidden cables

When digging trenches or drilling or chopping into walls, you must be aware that there might be cables hidden under the surface. All underground cables should be marked on a layout drawing so that they are not damaged when excavation takes place.

For cables buried in walls, check before drilling that there isn't a switch or socket or other item of electrical equipment directly below. If there is, be aware that it is probably near a cable passing through the area you are about to work on. It is not always possible to be sure where a cable might run, so it may be an idea to use a cable detector to locate any hidden cables.

Sources of electrical supply for tools and equipment

One way of reducing the risk of electric shock from power tools is to supply them at a reduced voltage. The three classes of voltage are:

- extra low voltage – 0 V to 50 V (not considered harmful)
- low voltage – 51 V to 999 V (this includes our 230 V mains supply which is considered harmful and potentially fatal)
- high voltage – 1000 V and higher (extremely dangerous and harmful).

By using an extra low voltage, power tools are rendered safe because even if there is a fault, the voltage will be too low to cause injury. The two main methods of reducing power tool voltage are described on the next page.

Key term

Mechanical damage – physical damage to cables or electrical equipment. Sheaths, enclosures and containment, cable armouring and toughened casing are all called mechanical protection.

 Safe working

When using extension leads:

- always use them fully extended because they will become hot if left wound onto the reel
- do not plug one extension lead into another. Get a longer extension lead instead.
- always use a three-core extension lead.
- fit an RCD plug if the lead is longer than 3 m.

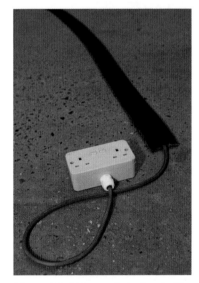

Figure 2.9: Cable cover for leads and cables laid across a road or walkway

 Safe working

Only good quality cable detectors should be used for finding hidden cables. Ask an electrical wholesaler for advice. Look for a BS, EN or ISO number on the detector.

Figure 2.10: A battery-powered drill

Safe working

- Always use the correct charger for power tool batteries.
- Always use the correct battery with each tool.

Safe working

Never try to recharge non-rechargeable batteries as they could heat up and catch fire.

Figure 2.11: A centre-tapped transformer

Battery-powered supplies

The simplest method is to supply the tools with a battery. These are usually rechargeable. Most battery-powered power tools will be supplied with two batteries and a charging unit. Two batteries means that one can be kept on charge while the other is in use.

Apart from safety, the other advantage of battery-powered tools is that there is no power lead which can be damaged or constantly plugged in and unplugged as you move around the work area, and which can become a tripping hazard.

110 V supplies

An alternative method of supplying power tools at a reduced voltage is to transform the supply down to 55 V (see Chapter 1 for more detailed information on transformers and how they work). This is achieved by plugging the power tool into a 230 V/110 V, centre-tapped transformer. This means that:

- the supply is 230 V
- the output to the power tool is 55 V.

The 110 V transformer produces an output of 55 V (half of 110 V) because it is centre-tapped. Figure 2.11 shows a centre-tapped transformer. Their sockets and associated plugs are coloured yellow to distinguish them from other voltages.

The advantage of this type of voltage reduction method is that it is constant and does not run out like a battery. The main disadvantages are:

- although there are portable versions available, the transformers are heavy
- transformers need power leads
- a 230 V supply is present, both in the transformer and in the transformer's supply lead.

Generators

Chapter 1 describes generators and how they work in detail, but basically a generator is a mechanical machine which produces electricity. Smaller, portable generators are available for supplying power to construction sites. You may find yourself in an old building or remote site which has no electricity supply. In this case you would need a portable generator.

These machines are usually supplied by petrol or diesel and can supply electricity at various voltages. Because fuel is needed there is a risk of fire, so the generator fuel must be stored and handled carefully. Remember that, although it is a petrol-driven machine, it is producing electricity and electricity is always dangerous! Another hazard is fumes from the generator exhaust, so make sure it is placed in a well-ventilated area.

Safe isolation procedure

When switching off supplies prior to working on a circuit, the safe isolation procedure must be followed.

1 Obtain an approved voltage indicator that conforms with Guidance Note 38.

2 Test voltage indicator on a known supply or a proving unit – if the indicator is working correctly then continue with the next step.

3 Identify circuit to be switched off and isolate, normally by removing or operating the protective device.

4 If the protective device is a fuse, remove and place in a secure place. If the protective device is a circuit breaker, fit a lock to prevent it from being switched back on before the work is complete. Erect warning notices and barriers.

5 Test isolated circuit.

6 Re-test voltage indicator on known supply or proving unit.

7 If the indicator is working correctly and has shown the circuit to be dead, work can begin.

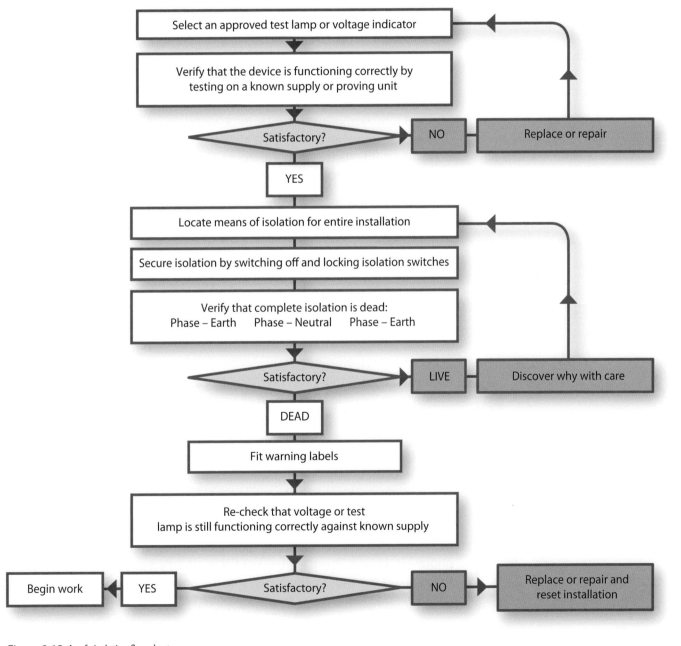

Figure 2.12: A safe isolation flow chart

Back-up supply hazards

Most electrical equipment and components are safe when isolated from the supply. However, some components are dangerous even when the power is switched off.

Many organisations are equipped with back-up power so that vital systems continue to run if there is a mains power failure. Typical back-up systems are:

- generators
- uninterrupted power supply (UPS).

Generators

Generators are often installed to start up automatically and provide back-up power in the event of a power failure. When shutting down and working on electrical circuits in generator-supported areas, care must be taken to isolate the generator from the installation. If not, the generator might start and cause circuits to become live.

Uninterrupted power supply (UPS)

Even if there is a generator installed as a back-up for mains power, there is often a delay of several seconds before the machine starts and emergency power is fully restored. The UPS is designed to supply emergency power during the intervening time. It consists of a bank of batteries which are under constant charge while the mains power is available. When mains power fails, the batteries are switched in.

A UPS is not intended to supply power for long periods of time but to 'fill in' during the generator start-up. Because it is essentially battery power, a UPS supply has to be converted to a.c. using invertors.

As with generators, care must be taken when working on UPS-protected circuits because the UPS will automatically switch in once mains power is lost.

Batteries

Batteries cannot be switched off and, as a result, can be dangerous. While charged a battery can deliver an electric shock at any time.

Banks of batteries must be well ventilated because some types give off toxic fumes and contain harmful chemicals. Occasionally, batteries will explode. Appropriate PPE must be worn at all times when working with batteries. The PPE consists of:

- goggles
- gloves
- a protective overall or apron.

Back feed hazards

Photovoltaic arrays

More and more houses and other premises are being supplied using photovoltaic cells. These are banks of cells, usually mounted in the roof, that convert sunlight to electricity. Like the power supplied from a UPS, this electricity is created as d.c. and then converted to a.c.

Photovoltaic arrays are always producing electricity as long as there is light, so without the correct equipment installed, or if handling the panels directly, there is a risk of electric shock.

Figure 2.13: Batteries can be dangerous as they cannot be switched off

Capacitors

Capacitors are designed to charge while the power is on. Once the power is switched off they will discharge and sometimes this discharge can be high voltage. Banks of capacitors are used for power factor correction in large industrial installations and as part of electronic equipment. They are also used for smoothing voltage when it is converted from a.c. to d.c. by use of rectifiers.

Transformers

Transformers (see Chapter 1) consist of conductors wound around an iron core and this can cause **back emf**.

1 When the supply is switched on the iron core becomes an electromagnet.

2 When the supply is switched off the current in the windings drops away almost instantly.

3 The magnetic field in the core, however, takes longer to fade.

4 As the core magnetic field collapses it induces a current back into the windings, which could cause a shock to anyone in contact with them.

> **Key term**
>
> *Back emf (electromotive force)* – a voltage produced within electromagnetic windings (such as those in motors and transformers). As well as producing the voltage required, the magnetic field in the iron core produces this second, opposing voltage. The current from back emf flows against the supply current. This opposition is called reactance, which reduces the amount of power produced. The effect of this inefficiency is to cause a low power factor.

Working practice 2.1

An electrician was called to a cottage to wire some extra lights in the kitchen. The main light was a standard pendant type and the new spotlights were to be fed from this light. The apprentice was given this job. The apprentice had not been working in the industry for long and was pleased to be given the responsibility of doing this work unsupervised. The electrician himself was only partly concerned to help improve the apprentice's confidence. He was very busy and had underestimated the amount of time this job would take, so he thought he could save time by letting the apprentice do some of the work by himself.

The apprentice switched off the light switch and climbed a pair of metal steps to the light fitting. He disconnected the neutral and switch-wire and then the two brown conductors connected to the middle terminal of the ceiling rose. He didn't understand what they were for, but the fitting needed to come down so the cable needed to be disconnected. He slid the plastic base of the ceiling rose over the cables, then grabbed at them to pull them out of the way while he made the hole in the ceiling large enough for the new cable.

The two mystery conductors were still live. The apprentice was suddenly in agony, unable to shout for help and unable to let go. The householder happened to come in, panicked and ran to fetch the electrician who immediately switched off the supply at the main switch. The apprentice needed to go to hospital, suffering from burns to his hand, and was badly shaken up.

1 Were the two brown conductors still live even though the light was switched off?

2 What should the electrician have explained to the apprentice before he allowed him to start work?

3 What should the apprentice have done when he saw the cables and didn't know what they were for?

4 What procedure should the apprentice have been taught to follow?

Electric shock procedure

If a colleague receives an electric shock, this procedure should be followed.

1 Do not touch the victim – or you will receive an electric shock as well.

2 Switch off the power – when someone is hurt, shutting down an area or operation no longer matters so, if necessary, the main isolator can be operated.

3 If safe to do so, remove the victim from the supply. You would have to use a non-insulating object to do this such as a wooden broom handle or piece of wood.

4 If able and safe to do so, place the victim in the recovery position (see Figure 2.14).

5 Call for help – nearly everyone carries a mobile phone so dial 999 for an ambulance as quickly as possible. Also send someone, or go yourself, for a first-aider.

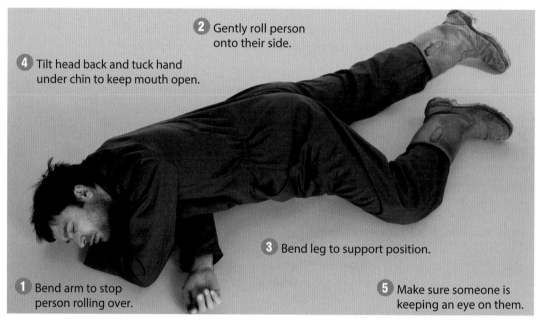

2 Gently roll person onto their side.

4 Tilt head back and tuck hand under chin to keep mouth open.

3 Bend leg to support position.

1 Bend arm to stop person rolling over.

5 Make sure someone is keeping an eye on them.

Figure 2.14: Steps to the recovery position

Cardiopulmonary resuscitation (CPR)

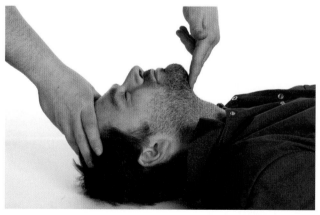

Figure 2.15: Opening the airway

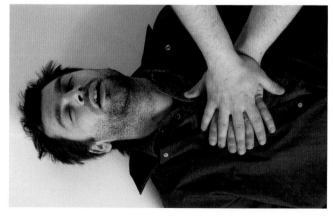

Figure 2.16: Chest compression

One of the results of a severe electric shock is stoppage or interference with the heartbeat. Because of this, an electric shock victim may be unconscious with no pulse and not breathing. If this is the case, CPR can be applied while you are waiting for an ambulance to arrive. CPR consists of chest compressions and rescue breaths which keep blood and oxygen circulating through the body.

CPR (based on NHS guidelines)

1   Place the heel of your hand on the breastbone at the centre of the person's chest. Place your other hand on top of your first hand and interlock your fingers.

2   Using your body weight (not just your arms), press straight down by 5–6 cm on their chest at a rate of 100 chest compressions a minute.

3   After every 30 chest compressions, give two breaths. Tilt the casualty's head gently and lift the chin up with two fingers. Pinch the person's nose. Seal your mouth over their mouth and blow steadily and firmly into their mouth. Check that their chest rises. Give two rescue breaths, each over one second.

4   Repeat this until an ambulance arrives.

When you call for an ambulance, telephone systems now exist that can give basic life-saving instructions, including advice on CPR. These are now common and are easily accessible with mobile phones.

Treatment of minor burns

Contact with a live part can cause burns to the skin. This can be from either prolonged contact or as a result of an electrical flash or explosion. Serious burns must be treated in hospital but the following first aid can be applied to minor burns.

1   Immediately get the person away from the heat source to stop the burning.

2   Cool the burn with cool or lukewarm water for 10 to 30 minutes. Do not use ice, iced water or any creams or greasy substances, such as butter.

3   Remove any clothing or jewellery that is near the burnt area of skin, but do not move anything that is stuck to the skin.

4   Make sure the person keeps warm – for example by using a blanket – but take care not to rub it against the burnt area.

5   Cover the burn by placing a layer of cling film over it.

6   Use painkillers, such as paracetamol or ibuprofen, to treat any pain.

Reporting to supervisors

Electric shock must be reported to your supervisor using the RIDDOR procedure discussed on page 69.

Progress check 2.6

1   What are the two main hazards associated with electricity?

2   What do you check when you carry out a visual inspection of an electric drill?

3   What should you do and not do when using an extension lead?

Chapter
2

WORKING WITH GASES AND HEAT-PRODUCING EQUIPMENT

One of the most immediate sources of fire is gas used for operations such as welding and brazing. This type of equipment must only be used by trained, skilled and competent people and always used correctly and following safe procedures.

Types of gas used on site

Gases fall into two main categories.

- Flammable – gases which burn if subjected to heat or high pressure.
- Inert – gases that will not burn and can sometimes be used to put out fires, for example carbon dioxide. Inert gases are not necessarily safe. They can affect breathing or cause ice burns.

The main types of gas used on site are described in Table 2.7.

Gas	Cylinder colour	Uses	Hazards
Propane	Red tank	• Barbecues • Portable stoves • Oxy-gas welding • Fuel for engines • Residential central heating	• Explosion and fire
Butane	Blue tank	• Propellant gas used to drive other gases at pressure • Cigarette lighter fuel • Bottled fuel used for gas barbecues and cooking stoves • Propellant in aerosol sprays such as deodorants • Refrigerants – the chemical that converts heat to cold • Cordless hair irons are usually powered by butane cartridges	• Explosion and fire • Temporary memory loss • Frostbite when the skin is exposed to butane at high pressure • Asphyxiation – another name for suffocation, caused by breathing in butane • Ventricular fibrillation – irregular heartbeat that can stop the heart and cause death
Oxygen	Black with white shoulder	• Used in oxy-gas welding and cutting operations to increase the flame temperature	• Explosion and fire
Acetylene	White with maroon shoulder	• Used in oxy-gas welding and cutting as a fuel gas and can burn at about 3500°C	• Explosion and fire
Nitrogen	White with black shoulder	• Nitrogen is an inert gas used as a propellant, fertiliser in its solid state and as a coolant	• Asphyxiation • Freeze burns when in liquid state

Table 2.7: Gases used on construction sites. Note that the colours are changing at the time of going to print, so may differ.

Safely transporting, storing and using bottled gases and equipment

Because most gases used in the building services engineering environment are hazardous in some way, they must be handled carefully. A damaged gas cylinder, for example, can cause explosion, fire or release of poisonous fumes into the air.

Gas cylinders

Gas is normally contained in cylinders which incorporate a valve for filling and releasing the contents. This valve must be in good working order and checked regularly.

The PPE that must be worn when cylinders are filled or refilled comprises:

- eye protection
- protective overalls
- gloves
- ear protectors.

All gas cylinders and associated equipment such as valves and regulators and safety equipment must be given a visual inspection before use. The things to look for are:

- bulges in the cylinder
- scorch marks from fire damage
- deep scratches or other damage.

Only cylinders made by approved companies can be used. All new cylinders must also be inspected by an approved inspection body. The inspection body stamp must be visible on the container. Once in use, all cylinders and their valves and other equipment must be tested regularly.

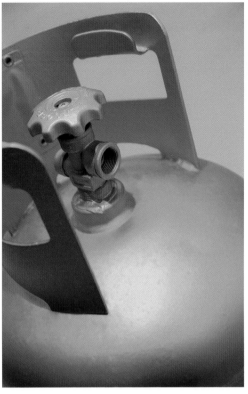

Figure 2.17: Valves on a gas bottle

Transporting

Gas cylinders have to be moved around and delivered to site, but this is a very dangerous operation. It is vital that the following precautions are followed.

- Clearly mark the cylinder to show what it contains.
- Only use suitable lifting equipment such as cradles and slings.
- Never lift a cylinder by holding its valve.
- Secure cylinders to stop them falling or moving; this will usually be upright although some cylinders can be stacked horizontally.
- All regulators and hoses must be disconnected and protective valve caps fitted.
- The cylinders must not stick out from the sides of the vehicle that is carrying them.
- The vehicle must be marked to show what sort of gas it is carrying.
- The driver of the vehicle must be trained.
- There must be documentation in the vehicle stating the types of gas being carried.

Storage

Great care must also be taken when storing gas cylinders. If they are kept in damp or hot conditions, some gases will explode or change from a relatively safe substance to a more dangerous one.

- Store the cylinders in an area designed for the purpose. It should be flat and well ventilated with no chance of damage.
- If necessary, secure the cylinders in place so that they do not fall over.

- Make sure the cylinders are clearly marked to show what is inside.
- Do not store gas cylinders for too long. This means that you should only buy the amount of gas that you need for each job.
- When organising the storage area, keep the oldest cylinders at the front and use these first.
- Keep the cylinders away from extreme heat or naked flames.
- Do not store the cylinders where they will be standing in water.
- Keep the valves closed.

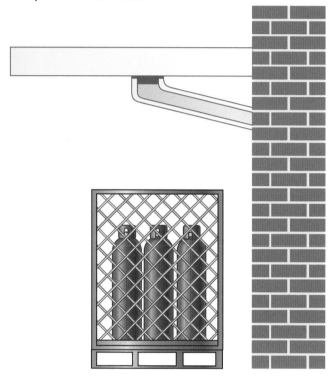

Figure 2.18: Correctly stored gas cylinders

How combustion takes place

Fire and combustion need three elements. If one of these elements is not present then the fire will not burn.

The dangers of working with heat-producing equipment

Heat-using equipment

The main types of heat-using equipment seen in the building engineering environment are:

- welding sets
- cutting equipment
- gas torches
- soldering irons
- electric paint removers.

The main safety points when using this type of equipment are as follows.

- Do not use the equipment near flammable materials.
- Make sure the work area is well ventilated.

Figure 2.19: A fire triangle

- Fire-fighting equipment such as extinguishers should be available in case of an accident.
- Wear the correct PPE which should include gloves, protective overalls and eye protection.
- You should be trained to use the equipment. For some equipment it may simply mean being shown and supervised for a short while; for others, such as welding and cutting equipment, you would have to be fully qualified.

Welding and cutting equipment

There are complete sets of regulations and guidelines for using and handling welding and cutting equipment. The three types of welding and cutting equipment are:

- oxy-gas – a combination of pure oxygen mixed with a flammable gas such as acetylene
- electric arc – the heat is created using a high electric current between the welding handle and the workpiece
- combination electric and gas – examples of this are MiG and TiG.

The main safety points for heat-using equipment must be observed, plus the following points.

- You must be fully trained and qualified to carry out these operations.
- PPE must include a metal face protector, protective overalls and gloves.
- Other people must be protected from the heat, and in the case of arc welding, the bright light from the welding process.
- All equipment must be given a visual inspection before use; this includes inspection of the valves, cables and hoses.
- All equipment must be tested for leaks using an approved leak detection chemical.
- Starting the welding process and closing it down afterwards must be carried out in the correct order.
- Only approved hoses and other equipment must be used. Oxygen and fuel gas hoses must never be swapped around.

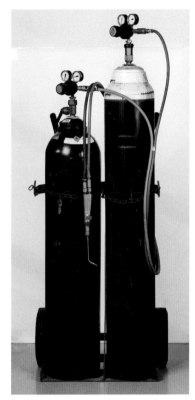

Figure 2.20: An oxy gas welding set

Procedures on discovery of fires on site

Many companies carry out fire alarm exercises so that the workforce can practise evacuating a building in the event of a fire. It is important to know what to do if the alarm does sound and to familiarise yourself with fire exits and escape routes. Below are the main steps in evacuating a building during a fire.

1 If you discover a fire, sound the alarm. You can do this by smashing the nearest 'break glass' fire alarm button. You should also shout a warning to anyone in the area and, if you have a mobile phone, dial 999 and call for the fire brigade as soon as you are able.

2 If you feel confident enough, and it is safe to do so, attempt to fight the fire. It is important that you use the correct extinguisher. Using the wrong one can result in injury or even making the fire worse. (See Table 2.8 overleaf.)

3 On hearing the alarm, leave whatever you are doing immediately.

4 If possible and safe to do so, shut all the windows. This will help limit the amount of oxygen available for the fire.

5 Leave the area as quickly as you can. Do not try to gather up your belongings. Things can be replaced, human lives cannot.

6 Head for your assigned assembly point. Your fire warden (see page 74) will meet you there and take a register of all those from your particular area. This is to make sure that everyone is accounted for.

7 Do not go back into the building under any circumstances.

Activity 2.3

Walk around the department at your college and locate all the fire alarm pushes and exits. Write down the locations: obtain a copy of the layout drawings for the building or mark them on a sketch. Is there a pattern? What reason can you think of for the positions chosen for the exits and alarm buttons?

Classes of fire and types of fire extinguisher

The class given to a fire depends on its fuel. It is important to recognise the class of a fire so that the correct extinguisher is used and the fire dealt with efficiently and as safely as possible.

Class of fire	Description	Fire extinguisher type
A	Wood, paper, textiles, etc.	Water – red
B	Flammable liquid, petrol, oil, etc.	Foam – cream label Dry powder – blue label Carbon dioxide – black label
C	Flammable gases	Dry powder (multi-purpose) – blue label
D	Burning metals, for example magnesium	Dry powder (special purpose) – blue label
E	Fire caused by electrical faults and where there is still an electrical supply present	Carbon dioxide – black label
F	Cooking oils and fats	Wet chemical – yellow label

Table 2.8: Classes of fire

Figure 2.21: The extinguishers shown may be used to fight fires of A, B, C, D and E classes. The class used to fight class F fires has a yellow label

---

Progress check 2.7

1 What are the three elements needed for combustion?
2 What class is a petrol fire?
3 Which extinguisher would you use to tackle a paper or wood fire?
4 What must be considered when transporting a gas cylinder?
5 What should be done with a gas welding set before use?

USING ACCESS EQUIPMENT

LO5&7

A considerable amount of building services engineering work is at height. This means that access equipment is needed. This can be anything from a step-up to a full set of scaffolding. Because of the risk of collapse or falling, the equipment must be right for the job, in good condition and erected correctly by people who have been trained to do so. The Work at Height Regulations 2005 is the statutory document covering this type of work.

Types of access equipment and their safe use

There are several types of access equipment, each designed for a different type of job. The main types are outlined on the next few pages.

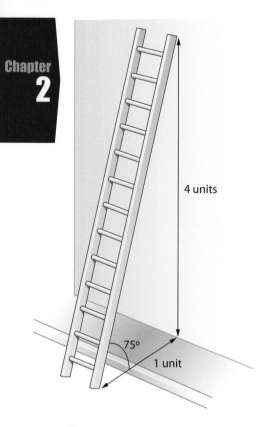

Figure 2.22: A correctly erected ladder

Ladders

The best known type of access equipment, ladders enable you to climb to the work area but limit you to the direct area of the job. The HSE recommend that you only work at the top of a ladder in one position for a maximum of 30 minutes. If you have to reach across from the top of a ladder to continue working, move the ladder, because stretching sideways can cause the ladder to fall. Apply the belt buckle rule. If your belt buckle hangs over either side of the ladder when you are trying to carry out a job, move the ladder. The HSE also recommends 10 kg as the maximum weight to be carried up a ladder without a manual handling assessment.

Ladder safety is covered by the HSE guide, *Safe use of ladders and stepladders*. Below are the main points to remember when erecting a ladder.

- Ladders must be leant against solid walls or structures. The ideal angle is 75°.
- Ladders must be placed on an even, solid surface.
- If possible, a ladder should be secured at the top. This cannot be done, for example, if the ladder is leaning against a wall. If this is the case, you might be able to tie the ladder at some other point, for example a window halfway between the top and bottom of the ladder.
- Do not use a metal ladder near electrical equipment.
- Only one person should be on a ladder at any time.
- Only carry equipment or materials up a ladder if you can carry it in one hand. Use your free hand to help you to climb.
- Do not climb a ladder if your footwear is slippery, for example covered with mud.
- Only use a ladder if the weather is suitable, for example do not use a ladder outside if there is a strong wind.

Activity 2.4

Working in pairs, transport and erect a ladder. If possible, select a structure which allows the ladder to be tied at the top. Make sure the ladder is at the correct angle. Each member of the pair should climb the ladder to tie it off while the other foots the ladder. When both team members have completed the exercise, take the ladder down and take it back to the starting point for the next pair.

Extension ladders

Extension, or telescopic, ladders are designed to be fully adjustable. The ladder must be of a high quality because there is a lot of strain on the ladder when extended. Here are a few points to remember when using an extension ladder.

- The two parts of the ladder must not move against each other when it is extended.
- The two sections must be secured while in the extended position so that the top part does not slide back down.
- The point at which the two ladders meet is called the overlap. Extension ladders should have a minimum overlap label on the ladder. The general rule, however, is to allow at least 1 m overlap.

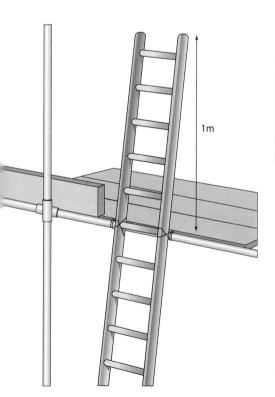

Figure 2.23: A ladder secured to scaffolding

Step ladders

A stepladder is a set of short ladders that can stand on their own. There is usually a small platform at the top. This platform is not designed for you to stand on but forms part of the structure of the step ladder and provides a place to rest tools and materials. Here are the main points to remember when using a stepladder.

- Do not use the top two steps of a stepladder. If it isn't high enough to carry out the work then use a taller one or consider a different type of access equipment.
- Do not place a stepladder on a moveable base such as a pallet or mobile elevating platform.

Roof access equipment

Working on a roof has its own special hazards. A roof may look solid but many are constructed from materials that cannot take your weight and will give way if you try to stand or crawl on them. The first thing to consider is: can the work be done without getting onto the roof? If not, then the correct equipment and techniques must be used.

Figure 2.24: A stepladder

Roof ladder

The top end of a roof ladder is formed into a hook which should be laid over the ridge and against the opposite side of the roof. Once securely in place, the ladder provides a set of steps which you can use to climb up and down the roof.

Crawling boards

Crawling boards can be laid across the roof, each end located over a joist. The board will then spread your weight. Crawling boards must be secured to stop them sliding down the slope of the roof.

Edge protection

If you are working on a roof and are within 2 m from the edge, you must erect edge protection. This prevents both you and your equipment and materials falling and injuring someone below.

Edge protection can consist of boards erected around the perimeter of the roof or on the edge of scaffolding. Edge protection boards should be supported from ground level and not fixed to the roof itself.

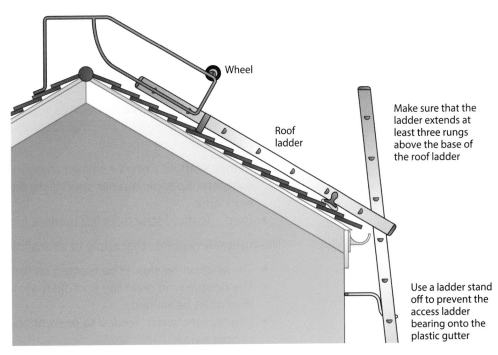

Wheel

Roof ladder

Make sure that the ladder extends at least three rungs above the base of the roof ladder

Use a ladder stand off to prevent the access ladder bearing onto the plastic gutter

Figure 2.25: A roof ladder

Safety nets

Safety nets are recommended for roof work, both to break a fall and to catch any falling materials or equipment. Safety nets must be:

- as close as possible under the roof surface
- securely attached and strong enough to withstand the impact of a human body
- installed by a competent person.

Case study

An electrician and apprentice were working in an industrial unit. The building was brick with a high apex roof. The roof was supported by metal beams. New lighting was to be installed, fed by metal conduit which was fixed to the beam by means of conduit saddles and girder clips. The only access equipment the pair had was a large extension ladder. The ladder was erected to almost its full length and the top was rested on the girder at the point where it reached a peak. There was no surface in front of the ladder for most of its length.

Because he was the lighter of the two, the apprentice was sent up the ladder to screw in the next length of conduit using a running thread. He struggled up the ladder, conduit in hand. The climb was frightening as there was nothing but empty space around him and the ladder bowed alarmingly with each step.

Once at the top he had to hang on very tightly while he manhandled the conduit into position and struggled with the thread. All the time he could feel the ladder moving. He had to overreach to secure the conduit to the saddle and was becoming increasingly frightened. At last he had finished and, relieved, clambered down the ladder.

As he reached the floor and walked away, a fork lift truck swung round the corner and crashed into the ladder, knocking it over.

If the apprentice had been on the ladder he would have fallen from a great height onto a concrete floor and been seriously injured or even killed.

There were no barriers, the ladder was not secured at the top and was certainly not the correct access equipment for the job. The near accident – a dangerous occurrence – brought the electrician to his senses and a cherry picker (see pages 99–100) was arranged to finish the work.

Scaffolding

Scaffolding should be used if you need to work at height for a long period of time or the job is a major one that requires access to a large area above ground level. The two types are:

- mobile scaffold towers – smaller towers that can be erected by competent people but not specialists, and can be moved around the work area
- fixed – scaffold structures left in place for the duration of the work.

The main safety points that apply to all scaffolding are as follows.

- The scaffolding should be erected on firm level ground.
- The people who erect the scaffold may not need to be specialists but they must be trained.
- Edge protection is needed to prevent objects from falling onto people below.
- Safety rails must be put in place to prevent people falling from the scaffolding.

Mobile scaffold towers

At some point many building services engineering jobs require the use of a mobile scaffold tower. These are supplied as a kit which includes:

- sections
- poles and cross bars
- a platform
- outriggers
- wheels.

Never try to erect a mobile scaffold tower on your own. This is at least a two person job. There should be a set of instructions. The tower will consist of a set of interlocking sections. The board can be placed onto the sections to enable you to fit the next level. The platform is then lifted onto the next section and so on. When carrying out this part of the job, do not place the board on top of the upper section. Guard rails must be fitted first to prevent falls.

Most platforms will have a trapdoor fitted. Climb onto the platform through the trapdoor, not by scrambling around the outside of the tower.

The outriggers must be fitted once the first section is completed. These stop the tower from toppling over as you carry on building it to its full height. The wheels must also be locked.

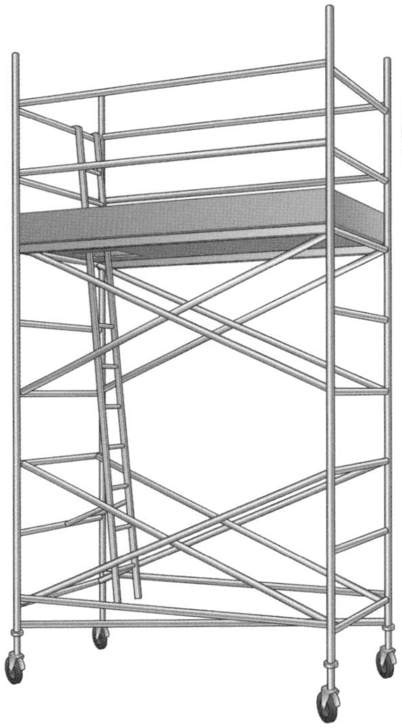

Figure 2.26: A mobile scaffold tower

Fixed scaffold

The same general rules about guard rails and edge protection apply to fixed scaffolding. Other points to remember are as follows.

- Scaffolding must be erected by people trained to do so.
- All scaffolding over 38 m high must be designed by a professional engineer.
- Platforms must be a minimum of 5.4 m wide and fully planked.
- The vertical struts (standards) must be fitted with side plates. The plates should then rest on a spreader board, which distributes the weight of the scaffolding.
- Any ladders fitted to the scaffolding must extend above the resting point by at least 1 m.

Mobile elevated work platforms

Mobile elevated work platforms (MEWPs) are mechanical machines which allow you to work at height, known as scissor lifts and cherry pickers. They are usually towed by, or mounted on, the back of a vehicle and consist of a platform and lifting mechanism. You will often see them being used for work on street lamps and overhead wiring.

Here are the main safety points.

- Make sure the ground is stable and solid enough to take the weight of the MEWP without it tipping over or sinking.
- The MEWP should be fitted with outriggers for stability.

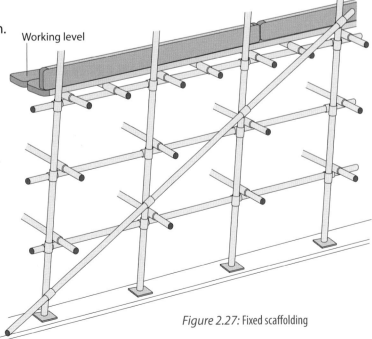

Working level

Figure 2.27: Fixed scaffolding

Chapter 2

Figure 2.28: A cherry picker

- The platform must be fitted with guard rails and edge protection.
- Place the MEWP close enough to the work so that you don't have to lean out.
- Cordon off the area around the machine using barriers or cones.
- Never use a MEWP on your own. There must be another person on the ground to go for help if the elevator mechanism fails or there is some other problem.

Working practice 2.2

An electrician and apprentice were rewiring a house. Wiring the lighting circuits involved the electrician working upstairs, with the floorboards up, installing cables in the floor space. A hole needed to be cut in a structural brick wall so that a cable could be run from one room to the next. The electrician set to work and cut out large pieces of masonry. He laid them on the plasterboard ceiling below. The weight of the masonry was too much and it fell through the plasterboard, onto the shoulder of the apprentice who was working directly below. Fortunately the apprentice received bruises but no serious injury, which could have occurred if it had hit his head.

- How could the accident have been avoided?

Progress check 2.8

1 How could you secure a ladder if you cannot tie it off at the top?

2 What is edge protection?

3 Who needs to be involved if scaffolding needs to be more than 38 m high?

LO6

WORKING IN EXCAVATIONS AND CONFINED SPACES

While building services engineering does not normally require you to carry out major excavation work, there are times when you may have to work, for example, in trenches, cable jointing or installing water or gas services. An excavation can be a hazardous environment with danger from collapse or falling. Remember, one cubic metre of soil can weigh as much as one tonne.

On the other hand, as a building services engineer, you will certainly be required to work in a confined space. This type of work is physically and mentally demanding, so preparation and robust health and safety procedures are needed before you start.

How excavations should be prepared for safe working

The main hazards when working in excavations are:

- collapse of the excavation, burial or injury
- equipment, people or material falling into the excavation
- the presence of electrical or gas services that could cause fire, explosion or shock.

If possible, it is always better to carry out work without using a trench. However, if trenches have to be dug, preparations should be made before any actual digging work takes place. These are described below.

Locate any underground services in the area

Obtain drawings showing any pipe and cable runs in the area; this could prevent a breakage or power cut caused by digging equipment. If drawings are not available, use locators to trace any services. Mark the ground accordingly. Look around for obvious signs of underground services.

These are:

- valve or manhole covers
- electrical and telecommunications boxes
- patching of the road surface.

Training

Make sure that all the people involved in the work are familiar with:

- safe digging practices and emergency procedures
- safe access into the excavation.

Trench support

Decide what temporary support you will need for the trench. Make sure this equipment is ready and on site before work starts. Examples of support equipment include props, trench sheets and edge protection.

Props

The sides of the trench are covered with boards or metal sheets. These are held in place by adjustable steel poles called props. This stops the sides of the trench collapsing.

Trench sheets

The boarding or metal lining a trench is called a trench sheet. These sheets can be lowered into place in the trench without the need for anyone to enter the unprotected excavation first.

Edge protection

Edge protection is needed not only to stop equipment, materials and even people falling into the trench but also to prevent the edges of the trench collapsing onto the people below. Edge protection is provided by:

- guard rails set up along the side of the trench
- extending the trench sheet above the lip of the trench.

Warning notices and barriers

Before excavation begins, the area must be sealed off so that no one can walk or drive close to the trench. This prevents both falling and collapse accidents. Barriers and warning notices must be placed around the work area and access given only to trained or competent people.

Check for undermining

At all times the excavation should be checked to make sure it does not undermine scaffolding, footings or the foundations of nearby buildings. If this is a possibility, then extra support for the structure is needed before excavation begins. A **structural engineer** must be called in to conduct a survey and give advice on what must be done.

Safe excavation work

Once the work begins, safety must be the most important aspect of the job. Here are some of the good safety practices that must be observed.

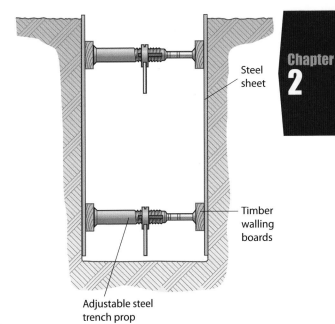

Steel sheet

Timber walling boards

Adjustable steel trench prop

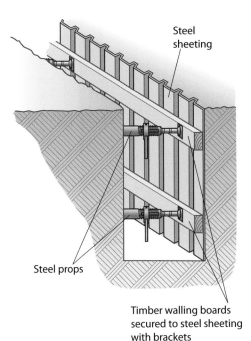

Steel sheeting

Steel props

Timber walling boards secured to steel sheeting with brackets

Figure 2.29: Trench safety equipment

> **Key term**
>
> ***Structural engineer*** – a structural engineer analyses, and contributes to the design of, any structure that will support a load. For example, a structural engineer would be heavily involved in the design of a new bridge, as the main purpose of the construction is to take the weight of the traffic that passes over it.

Battering

Battering refers to the angle of the trench wall. In other words, instead of digging the trench straight down and creating vertical walls, the walls are sloped away from the bottom. If the soil is loose, the battering angle of the slope should be less. If the soil is wet, a considerably flatter slope should be created.

Inspection

All excavations must be inspected by a competent person at the start of each working day or shift. In this case, a competent person means someone who understands the dangers and safety precautions you have put in place.

If any incidents occur that might weaken the excavation or safety equipment, then it must be inspected. Any fault found must be put right before work can continue. All inspections must be recorded.

Working in a confined space

The HSE defines a confined space as 'a place which is substantially enclosed (though not always entirely), and where serious injury can occur from hazardous substances or conditions within the space or nearby (e.g. lack of oxygen).' The Confined Spaces Regulations 1997 is the relevant legislation for this type of work. Examples of confined spaces are:

- drainage systems
- plant rooms
- main service duct rooms
- tanks
- cylinders
- boilers
- cisterns
- under suspended timber floors
- roof spaces.

Figure 2.30: Typical situation where work has to be carried out in a confined space – in this case, an underground cable duct

Safety considerations when working in confined spaces

Although working in confined spaces is to be avoided as far as possible, there are times when it has to take place. A risk assessment must be drawn up and a permit to work (see page 78) issued before any work can be carried out in a confined space.

Confined space hazards

There are a number of hazards connected to working in a confined space, as shown in Table 2.9.

Hazard	Effect
Fire and explosion	• Presence of flammable substances • Excess oxygen, e.g. from oxy-gas welding
Toxic gas, fumes or vapour	• Fumes from previous contents of the area • Fumes from outside, e.g. vehicle exhaust
Lack of oxygen	• Insufficient air supply
Ingress liquid and other material	• A leak can cause water or other materials to enter and fill the space
Excessive heat	• A confined space can become very hot, especially if there is nowhere for the heat to escape • Can cause heat stroke or unconsciousness

Table 2.9: Potential hazards of working in a confined space

The main points to consider when working in a confined space are:

- freedom of movement such that the work can be completed safely
- ventilation so that there is a constant air supply and any gases resulting from the work can escape
- escape routes for the people working in the space.

Ventilation

A good oxygen supply is needed. This can be achieved by making sure that the space is kept open. If necessary it can be pumped in and out of the space. Breathing apparatus might be needed in some cases.

Lighting

These areas can be very dark, so good lighting must be provided. This serves two purposes – it:

- illuminates the work area itself so work can be safely executed
- illuminates escape routes so that you can get out quickly and safely if something goes wrong.

PPE

As with all work in the building services engineering sector, everyone involved with the excavation work must be issued with, and wear, the correct PPE. This includes:

- a hard hat
- a hi-vis jacket or waistcoat
- protective footwear
- gloves
- eye protection
- ear protection when using power tools such as pneumatic drills to break hard surfaces.

Depending on the work, ear defenders, dust masks or even respirators may be needed. The PPE to be worn will be specified in the risk assessment (see page 77).

Evacuation procedures

There must be a set procedure for evacuating the space. Before starting work you must know what this is. There must be clear and unobstructed escape routes. Someone must be posted outside the space to assist the person to get out in the event of an emergency.

Medical conditions

Working in a confined space can be stressful. It is difficult to move and can be hot, frustrating and claustrophobic. Anyone with a medical condition such as heart disease or a respiratory problem should not work in a confined space.

Lone working

You should not work in a confined space alone. There are, of course, spaces where only one person can enter and carry out the job, but there must be someone else present and also the permit to work will notify your supervisor:

- that you are working in a confined space
- of the start time
- of the finish time
- of the work you are carrying out.

Progress check 2.9

1 What are the signs that there may be underground services?

2 What is a prop?

3 What are the basic precautions needed when working in a confined space?

Chapter 2

APPLYING SAFE WORKING PRACTICE TO MANUAL HANDLING

It is not just lifting heavy items that can cause injury. Light but awkward objects can strain our backs and arms. When it comes to heavy items, we must remember that our bodies are not lifting machines and they can be damaged by attempting to lift heavy weights.

There is no shame in asking for help! Even better, why not use lifting equipment, such as a simple sack barrow, or a trolley or wheelbarrow? If you do have to lift an object of any sort, remember to use your legs to lift and not your back. Page 105 shows the correct method for lifting. The statutory document dealing with safe manual handling is Manual Handling Operations Regulations 1992.

The main points for manual handling

If at all possible, avoid hazardous manual handling operations. This can be done by:

- rethinking the job or by using mechanical lifting equipment
- carrying out a risk assessment. This does not have to be a written assessment – just look at the job and try to think about the risks to yourself and others, and how they can be avoided.

Team handling

Some loads need to be carried by a team of people. The team should work together. There are a number of points to remember when lifting and carrying as a pair or team.

- Make sure there is enough space for the team to work.
- All team members should be able to take hold of the load. Lifting and carrying equipment should be used if the load is small or difficult to hold.
- One person should take charge of the operation. This person must make sure that the group work as a coordinated team.
- Good communication between team members is needed.
- Ideally the team members should be of similar build and strength.
- If the weight of the load is distributed unevenly, it is better for the heavier part to be lifted by the strongest members of the team.

Activity 2.5

Working in pairs, carry a full length of conduit from one end of a workshop to the other. Obstacles should be set up to make the route as difficult as possible. The pair must work together to safely transport the pipe.

Progress check 2.10

1 What sort of injuries can be suffered from manual lifting?
2 How can you reduce the risk of injury when lifting a heavy object?
3 How do you ensure that a lifting team works in a coordinated way?

1 Always check the nature of the object to be lifted before going straight in to lifting it; objects can be heavier than they appear. If you think that lifting aids and/or assistance might be required, do not attempt to lift unaided.

2 Clear your path to where the object is to be moved to and make sure that you have adequate space to set it down. Adopt a secure stance with the legs shoulder-width apart either side of the object.

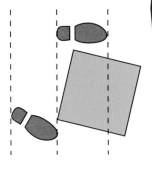

3 Keeping your back straight, bend your legs and get a solid grip on the object with arms straight out.

4 Begin to lift the object by extending the legs straight and keeping the back straight.

5 Move smoothly to your destination, keeping the object close to the body at waist height. Avoid jerky movements and never run.

6 When putting down, do so steadily, bending the legs rather than flexing the back (as with lifting). Remember that putting down can be as hazardous as lifting!

Figure 2.31: Step-by-step guide to manual lifting

Knowledge check

1 The Health and Safety at Work Act places responsibility for health and safety on:

a   employers and shareholders
b   employers and employees
c   employers and the government
d   employers and the local council

2 The Electricity at Work Regulations is intended to ensure:

a   electricians are the only people allowed to use an electrical system
b   electrical installations are in the control of a technician
c   tools and equipment are inspected and tested regularly
d   the electrical system in a workplace is safe for everyone who works there

3 Guidance Note 3 and the *IET On-Site Guide* are:

a   non-statutory codes of practice
b   statutory legislation
c   training documents
d   technical design documents

4 Construction site power tools should be powered using:

a   a UPS system
b   a three-core lead
c   a 110 V transformer
d   a capacitor

5 When transporting gas cylinders, the vehicle should:

a   be registered with the Gas Licensing Board
b   marked to show what sort of gas it is carrying
c   driven by a gas engineer
d   have a sealed trailer for the cylinders

6 Stored gas cylinders should:

a   be laid down and stacked as tightly as possible
b   have their valves removed
c   be kept for as long as possible
d   be stored with the oldest at the front

7 The type of ladder used for accessing a roof is:

a   a roof ladder
b   an extension ladder
c   a step ladder
d   a ladder rack

8 The lining placed into a trench is called:

a   trench plating
b   battering
c   trench sheeting
d   baulk

9 When should an excavation be inspected?

a   Every six weeks during the excavation work
b   Once a month during the excavation work
c   At the start of each day of excavation work
d   Only when an incident or accident occurs

10 What hazard can fumes present in a confined space?

a   Delayed working
b   Asphyxiation or explosion
c   Heat exhaustion or heat stroke
d   Stress and restricted movement

Electrical installations technology

This chapter covers:

- implications of electrical industry regulations

- technical information

- electricity supply

- wiring systems and circuits

- earthing systems

- micro-renewable energy.

Introduction

The modern world works almost entirely on electricity. Whether a bedroom light or the processes and equipment in a massive industrial complex, nothing will function without an electrical supply. Flick a switch or press a button and light appears, machines start up, your computer is ready for use. In fact, it is so simple we hardly notice it at all.

Actually getting an electrical supply to that lamp, computer or giant factory machine requires planning and often complicated engineering, and not only within a building or premises. Electrical power has to be generated in industrial quantities then transmitted over vast areas using equipment that is exposed to extreme weather conditions and given hard use.

And what about the installation itself? Will it be a construction site or a house, a harsh environment or an environment where the installation will be used carefully and the main wiring left undisturbed? Which wiring system should be used? Can we run unprotected cables through roof spaces and under floors, or should they be encased in metal trunking? Will we need armoured cables, or a fireproof version designed to keep emergency systems operating while the building burns?

This chapter explores the process of installing an electrical system, from generation and transmission to entry into premises and the selection of the right wiring systems for the job.

Note that practical exercises related to this chapter are described in *Chapter 4: Installation of wiring systems and equipment* and also in the accompanying Training Resource Pack. The chapter follows this process in a logical order and covers all the unit requirements set out in the City and Guilds 2365 Level 2 syllabus.

IMPLICATIONS OF ELECTRICAL INDUSTRY REGULATIONS

LO1

Working with electricity is extremely dangerous, not least because it is a completely invisible form of energy. It cannot be seen or smelled and it is impossible to tell if a conductor or terminal is live by simply looking at it. It is vital, therefore, that you treat electrical work and systems with utmost respect and caution. The Electricity at Work Regulations 1989 recommend that you do not work on live circuits or equipment unless it is absolutely impossible not to. Checking that a circuit is dead before commencing work should be achieved by carrying out the safe isolation procedure. Only test instruments that conform to HSE Guidance Note 38 should be used for testing and taking readings.

As well as specific electrical hazards, there are other hazardous tasks and situations that an electrician may have to perform or encounter. A lot of electrical work is carried out on construction sites with the associated risks that arise from working at height, in confined spaces, from fire and chemical hazards. Most contractors require that everyone due to work on their sites should receive health and safety training before being allowed through the gate.

PPE must always be worn (see Figure 2.3 on page 70). Hard hats, hi-vis jackets and protective footwear are mandatory on most construction sites. Eye and ear defenders and gloves should be available when needed. Safety signs must be heeded. They are not erected around work areas as decoration but as real warnings of real hazards. Many require you to wear some form of PPE or behave in a certain way in that area and the instructions they give must be followed.

Health and safety at work, and working life in general, is governed by sets of statutory and non-statutory regulations.

Statutory regulations are law and you can be prosecuted if you do not conform to their requirements. The main regulations that relate to you as an electrician, and as someone engaged in the workplace in general, are described in *Chapter 2: Health and safety in building services engineering*.

Non-statutory regulations are not law but an interpretation of the law, often written in a more specific and practical style. They are often guidelines and codes of practice. These include BS7671:2008, which are the requirements for electrical installation and testing. A further set of guidance notes has been issued by the Institution of Engineering and Technology, each one dealing with a specific aspect of electrical work such as earthing, protective devices and inspection and testing. Chapter 2 describes these in more detail.

TECHNICAL INFORMATION

LO2

To get an electrical installation job started, the electrician will need a certain amount of technical information. It is no use turning up at an empty building with a van filled with materials but no idea as to what is required or where anything goes.

The most basic information source is the drawing. There are a number of basic types available.

- Layout drawing – a scaled, bird's-eye view of the building that shows the positions of the accessories and equipment you are going to wire.
- Wiring diagram – shows how the installation is actually going to be wired.
- Circuit diagram – does not always show the wiring and connections exactly as you will see them but shows how an electrical system will work.
- Block diagrams – usually used to describe a process.
- Exploded views – drawn as if an item of equipment is in the process of being blown apart. It will show how the equipment is put together and which part goes where.
- Schematic drawing – a single-line representation of how a system or circuit works. The map of the London Underground can be said to be a schematic drawing because it shows no detail and is not to scale, but gives a readable, easy-to-follow picture of how the Tube system is structured and how to travel from one station to another.

Standard symbols have been developed for drawings so that:

- the drawing does not have to be made unreadable by written descriptions of the equipment and accessories it shows
- there is a standard reference which everyone understands and can work to.

Technical information is described in detail in *Chapter 5: Understand how to communicate with others within building services engineering*.

LO 5&6

ELECTRICITY SUPPLY

Until recently, virtually all electricity used in Britain was created in power stations. These are complexes built around huge generators (see *Chapter 1: Principles of electrical science*). In most power stations the generators are driven by steam turbines. Steam is produced using a number of methods, and power station types, such as coal-fired and nuclear, are named after these methods.

Fossil fuels

At present, the majority of power stations are powered by fossil fuels. These are:

- coal
- oil
- gas.

Using the fossil fuel, purified water is heated in a boiler until it becomes **super-heated steam**. The steam is pumped at high pressure through the blades of a turbine, which in turn operates the generator and produces electricity.

One of the most famous fossil fuel power stations in Britain is Ratcliffe-on-Soar just outside Nottingham. This coal-fired power station produces 2000 mW of electrical power.

Key term

Super-heated steam – when water is heated beyond its boiling point of 100°C, it is considered to be 'dry' because all the moisture has been converted to steam.

Figure 3.1: Ratcliffe-on-Soar power station is coal-fired

Nuclear power

Nuclear power stations work on the same basic principle as fossil fuel versions. The difference is in the way the water is heated. In a nuclear power station this is achieved using fission. **Radioactive materials** such as uranium are placed in a giant vessel called a reactor. The uranium atoms are split and this split releases the energy which turns to heat. The two main types of nuclear power station are:

- pressurised water reactor – water is heated directly by the heat source and converted to steam to drive the turbines
- advanced gas cooler reactor – heat is transported from the heat source in the reactor to the water system where it heats the water into steam.

Renewable energy

The contribution to our electricity supply made by renewable generation is slowly increasing. The main examples of this are:

- wind turbines
- photovoltaic arrays
- hydroelectricity
- geothermal.

Wind turbines

Wind farms have become a familiar feature of our landscape, often on high ground where the wind is unobstructed and stronger. Wind turbines are also placed out at sea where they can catch the strong air movements caused by the differences in sea and land temperatures.

A wind turbine is a large propeller set on a tower. The blades are designed to catch moving air and rotate a turbine, which, in turn, drives a generator. They are usually mounted at 30 m or more above ground level so that they can take advantage of faster, less turbulent wind. The rotor speed produced by the propeller blades is approximately 10–20 rpm. This is too slow to generate sufficient electricity, so a gearbox is fitted which converts the speed to approximately 1 500 rpm.

Photovoltaic arrays

Photovoltaic arrays are collections of cell modules used to convert light into electricity. A photovoltaic cell consists of a semiconductor wafer which has been treated to form an electric field. One side of each cell is positive, the other side negative. Light energy knocks electrons from their orbit about the atoms in the semiconductor material. Photovoltaic cells produce d.c. electricity which has to be converted to a.c. via an inverter. Excess electricity generated by photovoltaic panels can be sold back to the grid or stored in batteries.

Hydroelectricity

A slow-moving stream can be used to produce usable amounts of electricity because it is 800 times denser than air. Sources of hydroelectric energy are described in Table 3.1 on the next page.

> **Key term**
>
> **Radioactive materials** – radioactive materials contain unstable elements. Particles are constantly flying from their surfaces. These particles can cause serious damage to living cells and result in illness and even death.

Chapter 3

Figure 3.2: Wind turbines may be placed at sea or on high ground where wind is at its strongest

Hydroelectric source	How it works
Hydroelectric dams	A dam is built across a large fast-flowing body of water – the water is allowed to pass through the dam to rotate the turbines.
Waves and tide	Produced by harnessing the action of the ocean tides, which change direction twice a day. These changes result in the movement of large amounts of water. The world's largest tidal power station is in the Rance estuary in northern France.
Pumped storage	Water is pumped to a storage area during times of low demand. When the demand peaks the water is released to operate the generators. An example of this type of storage hydroelectricity can be found at Dinorwig in Wales.

Table 3.1: Sources of hydroelectric power

Geothermal

Heat energy is stored in the earth itself and this energy can be tapped to produce electricity. This energy is heat conducted towards the surface from the earth's core. The heat energy is transferred to the generators using water which is pumped into the ground where it is heated into steam which then flows back up to the generator turbines.

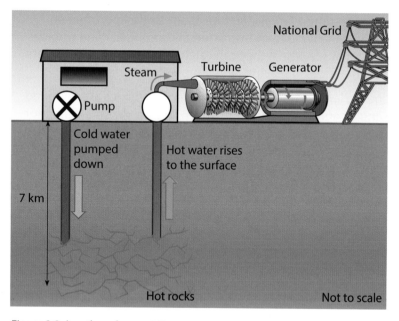

Figure 3.3: A geothermal power station

Micro-generation and renewable energy

As well as being used for electricity production on an industrial scale, renewable generation and energy sources are now available for use by ordinary householders and small- and medium-sized businesses. Called micro-generation and renewable energy technology, these smaller, domestic-type systems are described in more detail in *Chapter 10: Fundamental principles and requirements of environmental technology systems* but as an introduction we will briefly describe their principles and uses. Two examples of this type of renewable energy source are biomass burners and heat pumps.

Biomass

Biomass is a carbon-neutral heating system. Biomass boilers and stoves consume wood as fuel and can be used for both space and water heating. The wood fuel is available as logs, pellets or chippings. The carbon dioxide exhaust from biomass usage is converted to oxygen by trees, as part of their respiration cycle. These trees are then used as fuel.

Heat pumps

A heat pump works like a reverse refrigerator. The heat sources for a heat pump can be:

- the earth
- water
- air.

An element containing a refrigerant chemical is circulated through the heat source and brought to boiling point at a relatively low temperature. The fluid is then pumped into the heat pump unit and used to heat a second refrigerant. The second refrigerant is compressed, which raises its temperature even further, and is then used to heat water for central heating or underfloor heating, or air for blow heating. Heat pumps are considered to be low-carbon, rather than carbon neutral, because they use electricity as part of their cycle.

Transmission and distribution

Once generated, the electricity supply needs to be transmitted around the country. This is achieved using a network of cables and sub-stations. This network is divided into three main parts, called:

- transmission
- distribution
- local distribution.

Transmission

The national electrical transmission system, known as the grid, is fed directly by the main power stations and transports electricity around the country.

Most power stations generate at 25,000 V (25 kV). Using transformers (see *Chapter 1: Principles of electrical science*), the voltage is stepped up to 400 kV. This extremely high voltage is necessary because the transmission lines carry electricity over vast distances. The longer the conductor, the more voltage is dropped (for more information on volt drop, see page 149). High voltage helps reduce volt drop and also counters the wastage of energy as heat.

Distribution

Supply companies are supplied from the transmission grid. They then distribute it to their customers through a series of supply networks and step-down transformers. Eventually the supply will be reduced to the 400 V three-phase supply used by domestic (single-phase 230 V), commercial, public and small-to-medium industrial customers.

Progress check 3.1

1 Name three methods of producing electricity at a power station.

2 What is renewable energy?

3 Describe the main principles of hydroelectric generation.

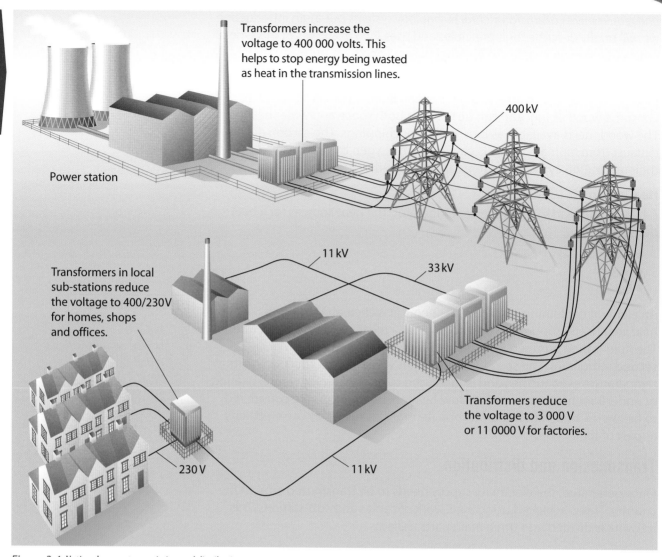

Transformers increase the
voltage to 400 000 volts. This
helps to stop energy being wasted
as heat in the transmission lines.

400 kV

Power station

11 kV

33 kV

Transformers in local
sub-stations reduce
the voltage to 400/230 V
for homes, shops
and offices.

Transformers reduce
the voltage to 3 000 V
or 11 0000 V for factories.

230 V

11 kV

Figure 3.4: National power transmission and distribution

Local distribution

Local distribution begins at a sub-station. There will a sub-station somewhere
near to where you live. They are often hidden behind a wooden fence, and a
yellow warning notice, 'Warning: Danger to Life', is usually displayed on the
outside of the fence. This complex consists of a transformer which is fed at
33 KV. The output is 400 V three-phase. *Chapter 1: Principles of electrical science*
explains how three-phase is generated and how it works.

- U_L is the voltage between any two of the lines in a three-phase system.
  For local distribution, this is 400 V a.c.
- U_{PH} is the voltage between any of the lines and neutral. This is
  calculated using the formula below.

$$U_L = U_{PH} \times \sqrt{3} \qquad 400\,V = 230\,V \times 1.7$$

Many commercial, small industrial and larger installations such as colleges
are fed directly via a three-phase supply. We will look at this in more detail
on pages 119–120.

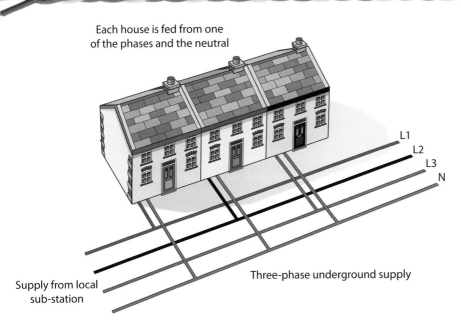

Each house is fed from one of the phases and the neutral

L1
L2
L3
N

Three-phase underground supply

Supply from local sub-station

Figure 3.5: Three-phase to single-phase distribution to domestic properties

Domestic installations are fed with a single-phase supply. This is obtained from the three-phase mains run from the sub-station. Each house or row of houses is fed from one of the phases and the main neutral.

Activity 3.1

Try to locate your local sub-station. It will be protected by a wooden fence or wire-mesh fence. What signs are displayed on its fence and who does it belong to?

Progress check 3.2

1 What are the transmission and distribution networks?

2 At what voltage is the distribution network fed from the transmission grid?

3 A local sub-station supplies electricity at what voltage?

Components of the network

The transmission and distribution networks are vast and bring electrical supplies to every part of the country. These systems are made up of a number of components, the main ones being:

- power stations
- transformers
- pylons
- sub-stations.

We have already looked at power stations and sub-stations in the previous sections. In this section we will describe the purpose and construction of transformers and pylons.

Transformers

Chapter 1: Principles of electrical science describes the working principles of transformers in detail. This section will look at their role in transmission and distribution. The main purpose of a transformer is to change voltage by either stepping up a lower voltage to a higher one or stepping down a higher voltage to a lower one.

Transmission and distribution transformers are large units designed to cope with high voltages. Power station transformers, for example, which transfer

Equipotential
bonding wire

Supporting
arms

Steel
tower

Figure 3.6: Features of a pylon

electricity from the main generators to the transmission grid, have a 25 kV input and a massive 400 kV output.

The main problem with the network transformers is heat. To counter this problem the transformers are filled with inert oil. As the iron core heats up, convection currents form inside the unit. These currents force the hot oil out through a series of pipes or fins mounted on the outside of the transformer, where it cools sufficiently to keep the core temperature constant.

Supply transformers are poly-phase types. The windings on the transmission and higher voltage end of the distribution system are all wound in delta configuration. At the sub-station, the secondary side is wound as star. The star configuration provides a central connection for one end of each line. This is called the star point. Phase voltage, U_{PH}, is measured between any line conductor and the star point. The sub-station transformer star point is connected to earth and is also the connection point for the installation neutral (see Figure 3.38 on pages 154–155).

Pylons

Electricity is carried around the transmission – and much of the distribution – network using overhead wiring systems mounted on pylons. These are the familiar metal towers that dot the countryside and are seldom out of view.

Most of the grid system was built in the 1930s and the pylons used today are virtually the same design as those used on the original grid. The main components of a pylon are shown in Table 3.2.

Constructed from steel, pylons can be anything from 15 to 55 metres. The world's tallest pylons are the 370 m towers that carry power to Zhoushan Island.

Component	Make-up	Function
Tower	• The central tower has a set of arms protruding from each side. • Pylons are connected together by a bonding wire which is connected to the top of each tower. • Each leg of the pylon is fitted with a barbed wire skirt that prevents anyone climbing up to the arms and getting access to the cable.	• The arms carry the cable and are high up to prevent any contact with the cables. They also hold the cables well away from the steel structure of the pylon itself. • Bonding wire ensures that all the pylons are at the same electrical potential and also helps disperse the effects of lightning strikes.
Insulators	• They consist of a series of plates and are usually made from glass, porcelain or composite polymer materials. • Insulator plates are formed into a curved, umbrella shape. The higher the voltage carried, the more plates there will be in the insulator set. Each plate represents approximately 10 kV.	• The job of the insulators is to provide a barrier between the cable and the metal suspension arm of the pylon. • The umbrella shape prevents a build-up of dirt on their surfaces, which could lead to flash-over – a conductive path forming on the insulator between the cables and pylon arm.
Cables	• The cables used for transmission and high voltage (HV) overhead distribution are un-insulated. • The cables themselves are made from aluminium with a central steel core. This is called a catenary wire. • When being installed, the cable is run from giant drums and not allowed to touch the ground.	• The amount of insulation needed would make the cables extremely heavy and would also be extremely expensive. Because of this the cables are run high above the ground. • The purpose of the core is not to conduct current (although it does) but to provide support. • Cables cannot touch ground because they could damage the aluminium and cause electrical discharge, which in turn can further weaken the conductor.

Table 3.2: Components of a pylon

Distribution supplies

Distribution supplies are also carried overhead on smaller pylons or on wooden poles. In rural areas you will notice that some poles are also fitted with a transformer and act as sub-stations.

In urban areas the distribution supplies are run underground. This is more expensive but does not affect the landscape. Supplies are taken from the main distribution cables by means of a cable joint. Cable jointing is a specialist skill within the electrical industry.

> **Safe working**
>
> Never attempt to climb an electricity pylon; the conductors it carries are un-insulated and can be supplied at voltages as high as 400 kV.

Chapter 3

Progress check 3.3

1 Describe how HV transformers are cooled.
2 What is a pylon insulator?
3 Why are HV overhead transmission cables un-insulated?

WIRING SYSTEMS AND CIRCUITS

LO3

There are many types of installation, from basic domestic light and power to complex industrial processes and alarm systems. Each of these requires its own wiring systems and circuitry and, in some cases, specialist skills to install. In this section we will look at the different types of wiring system, their selection and components and how installations are protected both from mechanical damage and electrical faults.

There are a number of different types of installation. Table 3.3 lists and describes the main ones.

Installation type	Description
Domestic	Houses and flats – this is a relatively non-hazardous environment for electrical installations. Cables and equipment used are not designed to provide a high level of mechanical protection.
Commercial	Shops, cinemas, pubs – although the environment is not immediately harsh for electrical installation, the presence of the general public has to be taken into consideration, for example cables with low smoke emission insulation should be used.
Hazardous	Environments where there is a particular physical hazard to electrical installation – this might be a construction site, an explosive or corrosive atmosphere or where there is a presence of water, such as a swimming pool or marina. Part 7 of BS 7671:2008 deals with 'Special Locations', i.e. locations where there is a particular hazard to that electrical installation.
Industrial	Factories – these are harsh environments, and cables, electrical fittings and equipment must have a high level of mechanical protection.
Agricultural	Farms – a harsh environment, with particular problems of damp, exposure to weather, plus danger from rodents. Livestock is also at risk. Electrical installations must be able to withstand the damp atmosphere and both protect livestock from shock as well as protect the installation from rodents.

Table 3.3: Typical installations undertaken by an electrician

For an electrical supply to work it must be formed into a circuit. *Chapter 1: Principles of electrical science* describes the components, principles and properties of the circuit in detail but essentially a circuit is a closed loop. In a.c. systems used to supply UK domestic, commercial and industrial installations, the current is brought to the load on the line conductor, then the circuit is completed by the neutral which carries the current back to the sub-station distribution transformer (see pages 154–155).

Some circuits are simple loops; others, such as control and alarm circuits, are more complex. No matter how complicated the circuit becomes, in the end it will work on the basic circuit principle. The electrician is expected to work on a number of different circuit types, the main ones including:

- radial
- ring final
- lighting
- alarm and emergency
- heating
- control
- data and communication.

All circuits start at the supply end. This will be different for different kinds of installation. We will look at the basic domestic supply and then at a poly-phase industrial version.

Domestic supply

The supply end of a domestic circuit consists of the following components, as shown in Figure 3.7.

- Supply cable – the type of cable used to bring an electricity supply into an installation depend on the locations. Urban installations are usually fed underground by means of an armoured cable jointed to the main supply cable in the street outside. A rural installation may well be fed by an overhead supply.

- Main fuse – this fuse belongs to the supply company and protects the whole installation. This would be rated at 100 A for domestic installations and is a BS-88 cartridge-type fuse.
- Meter – the electricity meter measures how much energy is used by the installation and this is used to calculate the cost of energy to the customer. The meter reads how much power is used each hour and displays it as kW/h. Most modern meters are electronic, though many mechanical meters still remain in use.
- Main earth terminal – all the main earth conductors in an installation are terminated at a connector block called the Main earth terminal or MET.
- Consumer unit – the consumer unit acts as the source of all the circuits in the installation.

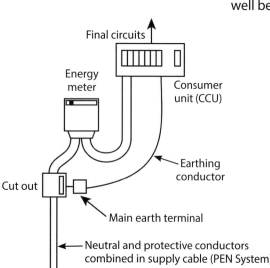

Figure 3.7: A domestic supply point

It consists of:

- a **double-pole switch** – enables power for all the circuits supplied by the unit to be switched on and off
- a neutral bar – connection point for all the neutrals in the circuits supplied by the unit
- an earth bar – connection point for all the earth conductors in the circuits supplied by the unit
- protective devices – connection point for the line conductors that supply the circuits, there is one **overcurrent** device, usually a circuit breaker, for each circuit. There may also be a Residual Current Device (RCD) which gives protection against shock (see page 148). Combined circuit breaker and RCD devices are available and are called RCCBs and RCBOs.

Key terms

Double-pole switch – a double-pole switch is a switch that makes and breaks both the line and the neutral conductors.

Overcurrent – when too much current flows in the event of a fault such as a short circuit.

Chapter 3

Safe working

NEVER switch the earth conductor.

Activity 3.2

Take a look at the supply intake position for your home.

- Sketch a diagram of the layout of the meter, main fuse, etc.
- Examine the consumer unit.
- What type of protective devices does it contain, and what are their ratings? How many circuits are there and what do they feed?

Industrial supply

The set-up for an industrial supply is very similar to that of a domestic one. The main difference is that an industrial supply will be a poly-phase rather than a single-phase system.

Poly-phase consists of three line conductors: L1 (brown), L2 (black) and L3 (grey). There will also be a neutral and an earth conductor. *Chapter 1: Principles of electrical science* describes the generation and principles of poly-phase systems in detail.

Because it is a poly-phase system, the main fuse arrangement will consist of three fuses, one for each line. The meter will also be designed to record kW/h from all three lines.

Main switch

All main isolating switches in a poly-phase supply will either be triple-pole or triple-pole-and-neutral. This means that they will make and break all three line conductors, as well as the neutral conductor if there is one on that particular circuit.

Bus-bar

The main switch will feed into a bus-bar. This is a set of copper or aluminium bars designed to distribute power through the supply equipment. Bars, rather than cables, are used because they are bigger and able to carry more current. **Switchgear** is connected to the bus-bars by means of either short cables (tails) or solid connection links.

Fused-switch

This is a switch which physically connects and disconnects a set of fuses when operated. Fused-switches are used for larger installations as part of the switchgear at the supply end.

Safe working

While all live parts are considered hazardous, 400 V three-phase is particularly dangerous. Remember that there are 400 V between any two of the phases in a three-phase system.

Key term

Switchgear – the collection of equipment such as fused-switches, bus-bars and distribution boards used to supply an industrial or large installation. In newer installations, the switchgear is usually collected into a panel.

Poly-phase distribution board

As with the domestic consumer unit, the distribution board is the source for a set of circuits within the installation. In this type of installation some of the circuits will be single-phase and others poly-phase. The single-phase circuits must be distributed evenly across the three phases.

- Single-phase circuit breakers can be fitted into the distribution board.
- Poly-phase circuit breakers are designed to fit across all three phases and will be operated by a fault on any of the lines or a short circuit between any of the line conductors.

Poly-phase fuseboard

If the circuits are protected by fuses rather than circuit breakers there will be a fuse for each line. If you are isolating a three-phase circuit fed by this type of board, care must be taken to remove all three fuses, otherwise two of the lines will remain live.

Progress check 3.4

1 What are the two ways a supply cable can be run to an installation?
2 What is the line conductor of a circuit connected to in a consumer unit?
3 What is a:
   - fused-switch
   - meter
   - MET?

Radial circuit

A simple radial circuit feeds a single load, as shown in Figure 3.8.

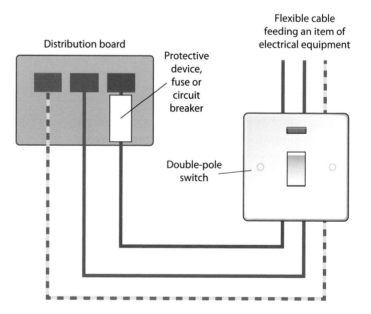

Figure 3.8: A simple radial circuit

Examples of a simple radial circuit include:

- a water heater
- an electric shower
- a cooker
- 16 A or 32 A socket outlet.

Ring final circuit

13 A socket outlets are fed using a ring final circuit. There are a number of requirements for ring final circuits.

- The protective device must be rated at 32 A.
- The cable's live conductors must have a minimum csa of 2.5 mm^2.
- A ring final circuit can feed an unlimited number of 13 A sockets within a floor area of 100 m^2.
- Branch supplies, or spurs, can be taken off the ring final circuit. Remember, it is better to add any extra 13 A sockets to the ring itself rather than wire it as a spur.
- A spur can feed a maximum of two outlets (a double 13 A socket counts as two outlets).
- Any other item fed by a spur from a ring final circuit must be protected by a fused-switch or outlet.

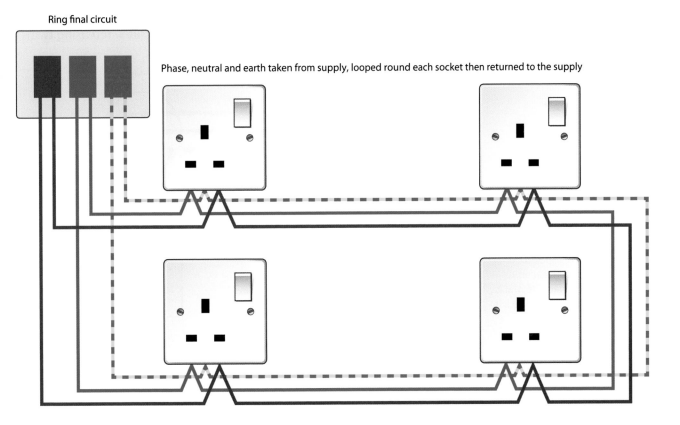

Ring final circuit

Phase, neutral and earth taken from supply, looped round each socket then returned to the supply

Figure 3.9: A ring final circuit

Working practice 3.1

An office in a telecommunications research centre was to be converted into an open plan office. Every desk was equipped with a desktop computer. All the computers were connected to a local area network (LAN) run through dado trunking around the walls. The electricians were brought in to install two ring final circuits – one for general use, the other dedicated to the computers.

The computer ring was equipped with a special 13 A socket which had a T-shaped earth slot. This was to prevent any other equipment from being plugged into these sockets. The circuit protective conductor for this circuit was colour-coded cream rather than green and yellow.

The reason for this was that the earth conductors in a computer circuit carry a small amount of current during normal use. This means that other equipment must not be plugged into these circuits. This type of earth is called functional earth. Because it carries current, a functional earth cannot be protected by a residual current device (RCD). If there is a fault, an alarm will be sounded.

1   How can both the ring final circuit and LAN cables be run through the same dado trunking?

2   How are 13 A computer sockets distinguished from general purpose ones?

3   What is a functional earth?

Lighting circuits

One-way lighting

A one-way lighting circuit consists of a single switch that operates one or more lights. Figure 3.10 shows a one-way lighting circuit wired using twin-and-earth cable.

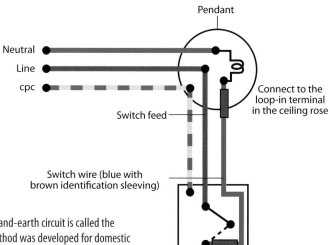

Safe working

Only the line cables should be switched in a lighting circuit, never the neutral. If the neutral is switched, the light will still be live, even when the switch is off.

Figure 3.10: The twin-and-earth circuit is called the 'loop-in' method. This method was developed for domestic wiring and is a very quick, efficient and easy way to wire lighting circuits

Two-way lighting

In a two-way lighting circuit, two switches are used to operate one or more lights. These two switches are connected in such a way that each one will change the state of the light. In other words, if one of them switches the light on (for example, at the bottom of a stairway) the other switch, the one at the top of the stairs, will switch it off. As well as stairs, two-way lighting can be in a room where there are two entrances and exits.

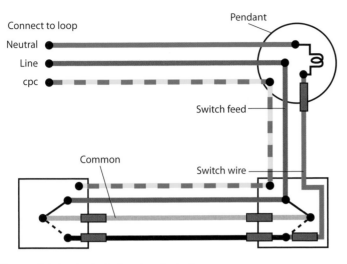

Figure 3.11: Two-way lighting circuit wired as a loop-in circuit

Intermediate switching

Intermediate switches are connected into the strappers between two switches of a two-way system. There is no limit to the amount of intermediate switches that can be installed in a single system. They are used to operate one or more lights when there are more than two entrances or exits, or more than one flight of stairs.

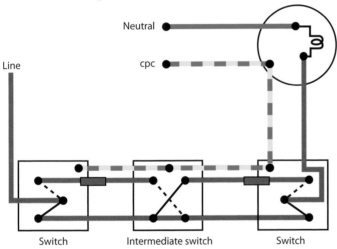

Figure 3.12: Intermediate switching circuit, singles method

Alarm and emergency circuits

The most familiar alarm circuit for most people is the fire alarm system. Alarm circuits are designed to operate some sort of audible and/or visual alarm in the event of an emergency, for example if there is a fire, gas leak or an intruder on the premises.

An alarm circuit consists of:

- sensors or detectors
- bells or other sounders
- an emergency operating button
- a control panel.

Progress check 3.5
1 What are strappers?
2 What type of circuit is wired using the loop-in method?
3 What is the maximum number of 13 A sockets allowed in a ring final circuit covering 120 m²?

Fire alarms

In the case of a fire alarm system the components would also include:

- emergency lighting
- emergency door opening or closure
- sprinkler or inert gas system designed to put out the fire.

Alarm systems are often divided into areas or zones. The control panel then acts as the central processing and monitoring point for all the zones. Figure 3.13 below shows a typical fire alarm system.

Table 3.4 shows the main components of a fire alarm system.

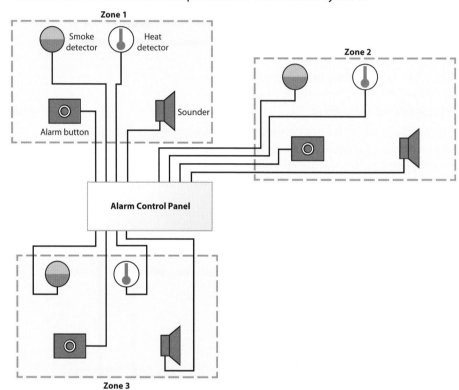

Figure 3.13: A typical fire alarm system

Component	Description
Emergency lighting	Fire can cause the failure of the main power supply and also produce smoke which makes it difficult for people to find their way to the exit. Emergency lighting provides illumination when the main supply has failed. The two types of emergency lighting system are: • maintained – a maintained fitting is either permanently on, or switched via a normal light switch • non-maintained – a non-maintained fitting only comes on when there is a power failure.
Heat detectors	Heat detectors send an alarm signal to the control panel when the surrounding temperature reaches a certain level. The two main types are: • fixed temperature – this will operate when the temperature reaches a certain level • rise of heat – this operates when the temperature begins to increase rapidly.
Smoke detectors	Smoke detectors send an alarm signal to the control panel when there is smoke in the atmosphere. The two main types are: • photoelectric – contains an infrared beam that sets off the alarm when it is broken by smoke • ionisation – detects smoke particles in the air.
Fire alarm buttons	The most familiar type of fire alarm buttons are the 'break glass' type. The glass front holds the button in. When the glass is broken the button pops out and operates the alarm.

Table 3.4: The main components of a fire alarm system

Intruder alarms

The main component of an intruder alarm is an infrared beam. When the beam is broken, it operates the alarm system. A simple example is an outside light containing a passive infrared detector (PIR). The light will come on if the beam is broken and act as a deterrent to anyone trying to gain entrance.

Detection can also be by vibration detectors and by pads on windows and doors that complete an extra low voltage circuit. When they are opened the circuit is broken and the alarm operated.

Heating circuits

Electrical supplies to heating circuits are either direct supplies to heating elements, for example an electric fire, water heater or the hot plate on a hob, or part of the control circuit. An example of this would be the control circuit for a gas or oil-fired central heating system, or the ignition circuit for a gas cooker or gas fire. Types of heating include:

- electric fires
- fan heaters
- hot water elements
- kettle elements
- underfloor heating
- oil-filled radiators
- cooker hotplates, and oven and grill elements.

Heating elements

All electric heating devices use an element to provide heat. This is usually a conductor, often wound into a spiral and made of a material that can heat up with no damage to itself. The temperature can be set, either by a simple high, medium or low control, or more finely to actual temperatures, e.g. 150°C, 200°C.

Thermostats

Temperature is normally controlled using a thermostat. This is a device that will allow the element to reach the required temperature, then switch off the electrical supply. Once the element has cooled below the required temperature, the thermostat switches the power back on.

The main components of a basic room thermostat are a bimetallic strip and heating coil. Because a bimetallic strip is made up of two different metals which expand and contract at varying rates, heating and cooling will cause it to bend. This bending action can be used to make and break a switch.

Rod-type thermostats expand and contract lineally and are used for water heater control.

Oven thermostats use the expansion and contraction of liquid to operate. The liquid is contained in a tube inserted into the oven heating area.

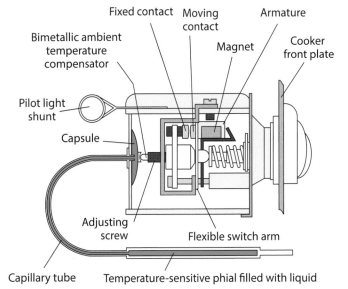

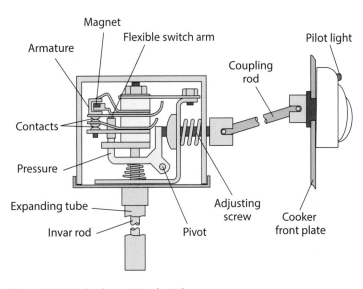

Figure 3.14: Cooker thermostat and switch

Three-setting switch

A three-position switch is often used for heat control in cooker grill elements. The elements themselves are arranged into banks and the switch connects the elements in three different configurations. These are:

- series – low heat
- series and parallel – medium
- parallel – high heat.

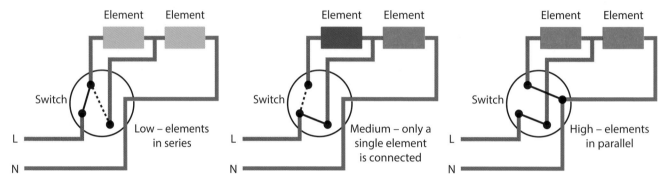

Figure 3.15: A three-position switch circuit

Electronic heat control

Increasingly electronic heat control is being introduced. An example of this is the thermocouple. This is a sensor probe made up of two different materials which will produce voltage at the point where they touch. Differences in the surrounding temperature will affect the amount of voltage they produce. This variation in voltage is then used to open and close or control the output through the electronic circuitry in the heat controller.

Control circuits

Complex processes require complex electrical switching and these types of circuits are called control circuits. Examples of control circuits are:

- electric motors
- conveyer belts
- complex machines
- heating systems
- air conditioning systems.

Programmable controller (PLC)

A programmable controller (PLC) allows you to set up equipment or a system to work in a certain way. For example, central heating systems include a programmable controller. From here you can set the temperature, and the times and even the days, when the heating comes on and goes off.

Another use is for a production line which needs to stop and start in a set sequence.

The PLC is operated by information sent to the main processing unit in the form of signals which it then translates into actions for the system it controls. There are three main components.

- Central processing unit (CPU) – the brain of the controller. The main programme is held here. The CPU receives information, makes decisions based on its programme, then issues instructions to the system or process it controls.
- Input module – relays external information to the CPU. This might be signals from temperature or load sensors.
- Output module – this is the CPU's link back to the system or process it controls, through which it will issue its instructions, for example to stop, start, slow down or speed up.

Relays

Relays are automatic switches that can be used as part of the control circuit. The heart of the relay is the electromagnetic coil. While *Chapter 1: Principles of electrical science* describes how these coils work in more detail, we will look at applications. Once the coil is energised, it produces a magnetic field which can be used to operate a switch, or set of switches.

There are three kinds of switches operated by a relay. These are:

- normally open – will close when energised
- normally closed – will open when energised
- normally closed/normally open – a changeover switch that is designed to re-route current round a circuit when operated, a bit like the points in a railway system.

This switching can be used in control circuits for routing current around the circuit in different ways, bringing one component online while shutting down another. Also, the coil usually requires a low current to operate. This means that a high-current load can be switched using low-current sensors and controllers. These low-current devices are connected to, and therefore switch, the coil, which in turn operates the switching for the high-current components of the circuit.

Contactors

The contactor is similar to the relay. In this case the switching mechanism works on the solenoid principle. A plunger is drawn into the centre of an electromagnetic coil. The plunger is connected to a set of switches which can be normally on, off or changeover type. (Contactors are sometimes used to operate large banks of lighting, for example in a sports hall or large workshop.) Because the total current taken by the lights is too high for a normal light switch, the light switch is used to operate a low-current coil. The coil then operates a set of high-current switches that energise or de-energise the lights.

Figure 3.16: A conveyor belt typical of the type of system controlled by a PLC

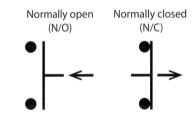

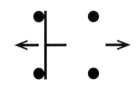

Figure 3.17: Symbols for N/O, N/C and N/C–N/O switches

Industrial electric motors are started using contactors. This is to:

- provide a facility for low-current controls such as sensors and timers
- provide a facility for remote control, such as the hand-held controller for gantry cranes
- reduce start-up current, which tends to be high when motors start.

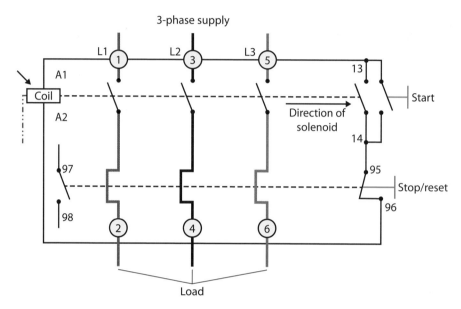

Figure 3.18: A direct-on-line starter

Motor current control

There are two main types of contactor used to limit electric motor start-up current – star delta and auto-transformer. They are both used for three-phase motors.

Star delta

1 On start up, the starter connects the motor connections in a star configuration. Because there is a star point, the voltage in each winding is 230 V.

2 Once the motor is running and does not need so much current, the starter switches the windings back into delta. In delta there is no star point so the voltage in each winding is 400 V.

3 Using Ohm's law it can be seen that the voltage divided by the resistance of the winding will give a lower current reading when in star than in delta.

Activity 3.3

To prove this point, calculate the following:

$$\frac{230\,V}{350\,\Omega} = ?\,A \qquad \frac{400\,V}{350\,\Omega} = ?\,A$$

Auto-transformer

The starter contains an auto-transformer (see pages 48–49). At start up the switching in the starter connects the motor windings to a lower voltage output in the transformer windings. At run the starter reconnects to the full 400 V position.

Data and communication circuits

Electricians are increasingly required to install data as well as electrical circuits. Also, as electrical circuits grow more complex and electronics are playing a greater role in electrical equipment and systems, there is sometimes a mixture of the two types of circuit within an installation.

Data and communication circuits carry information from a sender to a receiver. This information can be in the form of electrical or light pulses and is packaged into small units called bytes.

A digital system, which is what most data systems are, works on a binary code. In other words, there can only be zero or one. When the input message or operating instruction is zero, then no current or light pulses flow. When the input is one, the system or component is energised. This effect is used to create a code which tells a component or a circuit when to operate and when not to.

Local area network (LAN)

A common type of data circuit is the local area network (LAN). This is used in commercial and public premises to transport data to and from its computers. It can also be used to operate the telephone system. (WAN is a wireless area network.)

The cables are interlinked in a patch panel, which is a large junction box into which the cables are all terminated. Routers feed the information into the system and are located at a point where various networks connect. The information from the provider is received by the router which, in turn, sends it to your computer in a language the computer can understand.

Data and communication circuits must be run separately from main electrical circuits because the electromagnetic fields around main cables can cause interference in the data cables.

Category 5 and 5e cables

Category 5 and 5e are **twisted pair** cables designed to carry signals for video and telephone and are used as the cabling for computer networks such as ethernet. The maximum length of run for a category 5 cable is 100 meters. However, longer runs are possible if equipment, such as a repeater, is used to boost the signal.

Telephone cables

Telephone cables are also twisted pair cables and connect a telephone to the network. They are pvc-sheathed with copper conductors and are identified as two-pair, three-pair, etc.

Progress check 3.6

1. What is a coil used for in a relay and a contactor?
2. What is meant by a zone in a fire alarm system?
3. What is a thermostat?

Chapter 3

Key term

Twisted pair – twisted pair is a type of wiring in which two conductors (the forward and return conductors of a single circuit) are twisted together for the purposes of cancelling out electromagnetic interference.

Key term

Dielectric – a dielectric is an insulating layer that is affected by the presence of an electric current. The current will cause positive and negative charges to shift into the two sides of the material. Dielectrics are used in capacitors (see page 57).

Safe working

Never look into the cores of a fibre-optic cable. The cores are sharp and the light pulses they carry are bright enough to cause eye damage.

Progress check 3.7

1 What are the hazards to be considered when carrying out an agricultural installation?

2 What is a PLC?

3 What is meant by twisted pair?

Coaxial

These cables have a single copper conductor which carries the signal. This conductor is contained in a layer of insulation called a **dielectric** which is designed to allow an electrical charge to pass between the conductor and an outer layer of woven conductor (the screen). The purpose of the screen is to stop the signal leaking from the cable. Television aerial cables are coaxial cables.

Fibre-optic

Fibre-optic cables carry data in the form of light pulses that are reflected back and forth along the inside of the conductor wall. Care must be taken when installing fibre-optic cables because they are delicate and cannot be bent sharply.

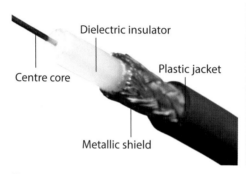

Dielectric insulator

Centre core

Plastic jacket

Metallic shield

Figure 3.19: Coaxial cable

Figure 3.20: Fibre-optic cable

LO3

WIRING SYSTEMS

The type of wiring system used in an installation depends on the environment. For example, a twin-and-earth wiring system would not be suitable for a harsh factory environment. On the other hand, a heavy duty wiring system, such as armoured cable, would be both ugly and unnecessary in a house.

Cables

The basis of any wiring system is the cable itself. We will look first at the main parts of a cable.

Remember, these are cables, not wires!

Conductor, usually copper, and the part that actually carries the current. Note: some conductors are solid, some are stranded

Circuit protective conductor (cpc), the conductor that carries the earth current if there is a fault. This is not normally live and does not need to be insulated

PVC insulation, colour-coded and preventing short circuit and electric shock

Outer sheaf contains and protects the components of the cable

Figure 3.21: The main components of a cable

Solid and stranded conductors

There are two types of conductor: solid and stranded.

Solid conductors

These are either copper or aluminium and consist of a single conductor. Solid conductors are generally used for cables of 2.5 mm^2 and smaller. However, the conductors in larger cables are sometimes solid. For example, high voltage cables often have a single, solid core. These conductors are often made of aluminium because it is both lighter and cheaper than copper.

Stranded conductors

A stranded conductor is a cable conductor made up of a collection of smaller conductors. Stranded conductors make a cable more flexible. Generally, cables of 4 mm^2 and over are stranded type.

Cross-sectional area

We have already mentioned cable sizes such as 4 mm^2. This measurement is the cross-sectional area (csa) of the conductor. Cross-sectional area is the face area of a conductor. The larger the cross-sectional area of a conductor, the more current it will carry.

Current-carrying capacity

As we saw earlier, the amount of current a cable can carry mainly depends on its size. However, there are other factors involved. These include:

- ambient temperature – temperature surrounding the cable
- bunching – is the cable installed with other cables?
- whether it is enclosed or on the surface – is the cable run through conduit or trunking, or is it clipped to the surface of a wall?
- type of protective device – some protective devices will reduce the amount of current that can be safely carried by a cable
- length of cable run – because voltage is the pressure that pushes current around a circuit, it is reduced by distance. This means that if the cable run is a long one, it will suffer from a large volt drop. The voltage measured at the load end will be less than the voltage at the supply end. Appendix 4 of BS 7671 includes a set of tables from which the volt drop and current-carrying capacity of a cable can be calculated. This calculation helps the electrician decide which size cable to use.

Fixing cables

Cables must be securely fixed or installed in a place where they are hidden from view and not likely to be damaged.

Surface fixing

The usual method of fixing cables to a surface is to clip or cleat them. Clips are used for twin-and-earth and flexible cables.

Safe working

Never cut strands from a stranded cable to make it fit into a terminal. This will reduce the cross-sectional area of the cable and, in turn, reduce the amount of current it will carry.

Figure 3.22: Clipping a cable to a surface

Cleats are used for heavier duty cables such as armoured or fireproof cables. The cleats are usually screwed into the wall or secured to a cable tray (see page 144) using nuts and bolts.

Hidden cables

Cables are not visible in most houses. This is because they have been run in the roof space or under the floor.

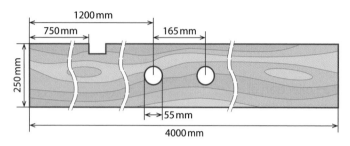

Figure 3.23: Joist drilling rules

- When cables are run in the attic, the joists should ideally be drilled or notched so that the cables are not laid on the top and subject to damage.
- Under the floor, the joists will need to be drilled so that the cables can be pulled through. Figure 3.23 shows the rules regarding holes and notches in joists. The holes need to be far enough down so that the nails or screws that secure the floorboards do not damage the cable. It is also best to drill the holes in the middle of the gap as floorboards tend to be fixed at the edges. Care must be taken when lifting floorboards as many of these will be tongued-and-grooved. This means that each one has a groove along one edge and a projecting 'tongue' on the other so that the floorboards slot together. Before lifting these boards the tongue will need to be cut between the boards. This can be done with:
  - a pad saw
  - a retractable knife
  - a circular or jig saw, although the blade must be set at a very shallow setting so that the saw doesn't cut into the joist as well.

Floorboards must always be replaced securely.

- Chipboard floors present more of a problem because it is difficult to lift the whole board. In this case, a 'trap' needs to be cut into the boards. The trap has to be cut carefully and neatly and when it is replaced, wooden 'noggins' secured in place as a fixing point for the trap.
- False floors – many office environments are fitted with false floors. These are tiles mounted on legs which provide a cavity through which cables can be run. 'Power' or 'service' tiles are then fitted next to each desk position. These tiles have a removable lid under which are power and data supply points.

Cables are also concealed in walls. If the wall is made of brick or block then the cable will be 'chased'. A cable chase is a groove cut into the wall deep enough so that all cables are below the finished surface. The cable is secured in the groove by running it through a plastic conduit or under a plastic capping. Although chases can be cut using a hammer and bolster, most modern electrical companies now use chasing machines which cut down on the time and effort taken.

The accessories fed by these concealed cables, such as 13 A sockets and light switches, are also installed 'flush' with the finished wall surface. A metal back box is cut into the wall just deep enough so that it is flush with the finished surface.

Cables can also be dropped through plasterboards or stud walls. This is usually carried out in a new installation when the wooden frame or stud work is fitted but before the plasterboard is nailed in place.

Working practice 3.2

A large stately home called Glemham Hall was due for a complete rewire. The existing lighting was wired in very old, rubber-insulated single cables, installed in 16 mm steel conduit. An electrician and apprentice were sent to begin work. They walked around the building to get their bearings and mark the positions of switches and lights on a set of layout drawings. They then set to work trying to locate the conduit boxes. However, when they began pulling up floorboards they discovered that there was another floor underneath. The original floorboards, now decayed, ran at right-angles to the top layer. This job was going to take longer and prove more awkward than they had anticipated.

1 Why did they mark all the switch positions and lights on a drawing?

2 Why did they need to locate all the conduit boxes?

3 How did they gain access to the conduit boxes under the original flooring?

Overhead

Cables can also be run overhead. Overhead cables are a specialist type, usually single-insulated with stranded aluminium conductors. Running through the core of the cable will be a catenary wire. This is made of steel and designed to take the weight of the cable.

Cable identification

Cable cores are usually identified by the colour of their insulation. There are standard colours for cables. Table 3.5 shows the main colours.

Type of circuit and conductor	Current cable colour	Pre-2000 colours – you will encounter these colours when working in older installations
Single-phase		
Line	Brown	Red
Neutral	Blue	Black
Earth	Green and yellow	Green and yellow, or plain green
Three-phase		
Line 1	Brown	Red
Line 2	Black	Yellow or white
Line 3	Grey	Blue
Extra low voltage		
Circuits 50 V and below	Violet	

Table 3.5: Cable colours

There are other colour schemes, for example for electrical control panels and high voltage (1000 V and above) but the colours shown in Table 3.5 are the ones most electricians will deal with.

When wiring control circuits or lighting circuits, there will often be collections of cables which are all the same colour. It is important that the ends of each cable are matched up. This can be done by using numbered or lettered tags pushed onto either end of each cable.

Activity 3.4

Make up a loom of cables. They must all be the same colour. Using a continuity tester, identify each end of each cable and mark it with a numbered tag or tape.

Types of cable

Table 3.6 shows the main types of cable, along with their main uses and a description.

Cable type	Description	Uses
Single	A single conductor protected by one layer of insulation.	Must only be installed in conduit or trunking.
PVC/PVC multicore (twin-and-earth)	Three or more insulated conductors enclosed in a PVC outer sheath.	Used for most domestic and light commercial installations. Resilient enough to be run without mechanical protection, both on the surface or concealed in floor, roof and wall spaces.
Flexible cable	Two or more insulated, stranded cores enclosed in an outer sheath. Many variations available including multicore and heat-resistant versions.	The purpose of a flexible cable is to link an item of equipment to the main supply. The basic method is via a 13 A plug and socket. Flex should not be used as the fixed wiring method for an installation.
Armoured cable	Insulated and sheathed cable. Extra mechanical protection given by a layer of steel wire or tape. A low-smoke version has a sheath that will not emit excessive smoke or poisonous fumes in the event of a fire.	For harsh environments, a tough resilient cable is needed. Armoured cable gives the type of mechanical protection needed in these situations. Can also be used as underground cable. Armoured flex is also available. Low-smoke versions are used in public premises.
FP Gold	Fire-resistant, low smoke-emitting outer sheath extruded over a layer of aluminium tape. The tape acts as both a moisture barrier and screening. The conductors are insulated with a fire-resistant material called Insudite.	FP Gold is the main type of fire-resistant cable currently in use and is very simple to install and connect.
Mineral insulated	Copper sheath filled with tightly packed magnesium oxide powder, which acts as the insulation around the cable's bare conductors. The copper sheath can also be used as the cable's circuit protective conductor (CPC).	The first type of fire-proof cable in general use.
Tri-rated cable	Single core, high voltage and high temperature, flame-retardant electrical cable designed for use in panel building.	Often referred to more generally as Panel Wire or BS 6231 Cable. Manufactured in a wide variety of insulation colours, including brown, orange, yellow, pink and dark blue.

Table 3.6: The main types of cable and their uses

Working practice 3.3

A major upgrade to the sewage system in East Anglia was to be centred around a new processing plant. The design called for the whole electrical installation to be wired in mineral-insulated cable. This was not only for the fire alarm circuits but also for power, lighting and pump motor circuits. The installation work was started by an electrician and an apprentice, who had only just started her training. The electrician knew that the apprentice would not be able to help in any way unless she was able to fit a termination gland to mineral-insulated cable. This is not a straightforward job. Mineral-insulated cable is made up of a number of layers. These are:

- bare copper conductors
- a magnesium dioxide powder insulation
- a copper outer sheath.

The first task was to learn how to strip the outer sheath. This is accomplished using a specialist stripping tool which resembles a crank handle fitted with a blade, adjustable to fit various cable sizes. Once the outer sheath is stripped back to the required length, the pot is screwed onto the outer sheath. A moisture-proof compound is squeezed into the pot, and a seal is pushed over the conductors and fitted into the mouth of the pot. Once seated, the seal is crimped into place.

After stripping her first cable, the apprentice was about to blow away excess insulation dust when the electrician shouted at her to stop! The powder insulation is highly moisture-absorbent, and will lose its integrity even if simply exposed to the air.

After some practice she mastered the technique and went on to be very adept at this particular task. She was also able to make a very useful contribution to the job.

1. Why must you never blow away the excess dust when fitting a mineral-insulated cable?
2. What is the outer sheath of the cable made of?
3. Are the conductors insulated?

Progress check 3.8

1. What is a CPC?
2. What is FP Gold cable used for?
3. What is meant by a cable's cross-sectional area?

Conduit and trunking

There are many types and variations of conduit and trunking, from basic steel versions to the small plastic mini-trunking used in domestic installations. The type chosen depends on the environment and the purpose it serves. The three main reasons for using electrical enclosures and support systems are:

- mechanical protection
- cable routing
- aesthetics.

Mechanical protection

Some environments can be extremely harsh for an electrical installation. In a factory, for example, there are often high temperatures, toxic and corrosive fumes and chemicals, and the risk of mechanical damage from machinery and general rough treatment. In this case cable must be contained in protective enclosure.

The cable installed in trunking or conduit is normally single-insulated, non-sheathed because this is flexible, easy to handle and smaller than sheathed, multi-core types.

Cable routing

There are times when it is difficult or unacceptable to run cables on the surface of a wall or ceiling. This might be because:

- the walls have been decorated and the customer does not want the decor disturbed by chasing
- the walls are finished brick or block and therefore cannot be chased
- the walls are concrete or some other hard surface which is difficult to fix clips to
- there is a gap or open space, such as the roof area of a factory or warehouse, through which the cables have to be run.

Aesthetics

Aesthetics means the look of something. For example, a cable clipped down a wall will spoil the look of a room. Trunking and conduit is not always pleasing to the eye, but it can be painted over and tucked away in the corner of a room or run along the top of skirting, and is not as noticeable as bare cable would be.

Conduit

Conduit is a pipework system designed to carry wiring and protect it from mechanical damage. Conduit is typically used for electrical installations in harsh working environments such as factories, workshops and farms. Single cables, which are cables with only one layer of insulation, are the type usually drawn into conduit.

The two main types of conduit are:

- PVC – available as heavy or light gauge
- steel – available as:
  - o black enamel
  - o galvanised
  - o stainless.

The first two types of steel are the most common types in use. Stainless steel is more expensive and is used mainly where appearance is important and where cleanliness is vital – such as a hospital or milking dairy.

The *IET On-Site Guide* provides capacity and information about a number of conduit sizes – from 16 mm up to 75 mm. However, the two sizes normally in use at present are:

- 20 mm
- 25 mm.

The measurement is the outside diameter of the pipe.

Because it is not carrying gas or liquid, conduit does not have to be watertight. Nonetheless, all conduit joints must be secure so that there is no movement which would damage the cables inside.

Metal conduit

Metal conduit is exposed metalwork and must be at the same electrical earth potential along its entire length.

It is supplied in 3 m lengths with either a galvanised or black enamel finish. Lengths and fittings are threaded and metal conduit systems are assembled by screwing the parts together. Metal conduit can be formed into angles using a conduit bending tool. *Chapter 4: Installation of wiring systems and enclosures* describes how to bend and form conduit and also contains a number of exercises.

A variety of fittings are available for conduit systems. Conduit fittings are divided roughly into:

- conduit and inspection boxes
- general fittings.

Conduit boxes

Conduit boxes have a number of purposes.

Figure 3.24: Conduit boxes

- They provide a means of joining different parts of a conduit system together, for example using intersections and Ts.
- Conduit boxes act as connection points for ceiling roses and other fittings.

Types available include:

- through
- 90°
- T
- cross or intersection
- terminal (a single spout).

Through boxes act as wiring points in both long, straight runs of conduit and runs which include bends and angles.

Through, T and 90° boxes are available as:

- plain – the threaded spouts are in line with the centre of the box
- tangent – the threaded spouts are set along the edge of the box.

Another type of conduit box is the inspection box. No longer widely used, these are a similar diameter to the conduit itself and are used when there is no room for a standard conduit box. Through, T and right-angle inspection boxes are among the inspection conduit boxes available.

General accessories

As well as inspection conduit boxes, a selection of other general fittings are needed to construct a conduit system.

- Coupler – a short, threaded tube used to join lengths of conduit together.
- Male bush – a brass bush that screws into couplers and the spouts of conduit boxes. Its main use is to join conduits to the back boxes of accessories such as 13 A sockets and light switches. Male bushes are available with either short or long threads.
- Female bush – a brass fitting designed to screw onto the end of a conduit to cover the sharp edge at the end of its thread.
- Lock nut and ring – screwed onto a bush and conduit thread to secure them in place.

Figure 3.25: A selection of conduit fittings

- Nipple – male threaded joining piece, designed to be screwed into a coupler or conduit box spout. Used as the main component of a running thread (see page 185).
- Saddles – used for fixing conductor. The two main types are:
  o spacer bar – the most common version – the spacer bar saddle holds the pipe close to the wall
  o deep spacer bar – this has a thicker base and holds the conduit off the wall. A bigger version of this is the hospital saddle, so called because it can be used in areas where cleanliness is essential and enables the space behind a conduit to be kept clean.

 Safe working

Conduit is supplied in bundles of ten lengths. These bundles are heavy so correct lifting procedures must be used. A single length of pipe is 3 m long so awkward and careless handling could cause injuries to others. Ideally, carrying bundles and single lengths of conduit should be a two-person job.

Specialist tools for metal conduit

Conduit work requires a number of specialist tools.

- Combination vice and bender – a combination tool with a vice with three sets of gripping teeth for holding the conduit secure while cutting and threading. The conduit bending part of the tool consists of a bending arm, a block against which the pipe is secured while being bent and a former around which the pipe is bent. The former comes either as 20 mm or 25 mm.
- Stocks and dies – lengths of metal conduit are joined together by screwing one to the other. Full lengths are sold with a machined thread on either end of each pipe. Once sawn to length, however, a new thread must be cut onto the pipe. Conduit threads are cut using stocks and dies.
  o The stocks are the handles used to turn the tool and the frame into which the dies are secured. There is also a guide fitted to the back which slides over the conduit and holds the tool straight as the thread is cut.
  o The dies consist of a steel ring inset with cutting teeth which cut the thread onto the outside surface of the pipe.
  o Cutting compound should be used to protect the teeth of the dies and to ease the cutting operation.
    - Reamer – once the conduit is cut and threaded, the sharp inner edges of the pipe left by the operation need to be smoothed away before any cables are installed. These sharp edges are called burrs and will damage cable insulation. The reamer is a tool designed to be pushed into the end of the pipe and twisted to remove any burrs. A round file can also be used for this job.

Plastic conduit

Plastic conduit is supplied as heavy and light gauge. The gauge is given by the thickness of the conduit wall. The common sizes for plastic conduit are:

- 20 mm
- 25 mm.

Figure 3.26: A combination vice and bender

Plastic cable ducting is also available. This is large piping that can be installed underground as a route for supply and data cables.

Plastic conduit is used in situations where a measure of extra mechanical protection is needed. An example of this would be an agricultural installation where it can be damp and where cables can suffer from rodent damage.

It is also used in commercial installations, schools and colleges as a means of running cables in areas where they cannot be clipped directly to the surface or hidden in cavity walls.

Fittings

The fittings used with plastic conduit are similar to those used for metal conduit. Plastic conduit fittings are not normally threaded but glued together. The exception is the adaptor.

An adaptor is a coupler that has one glued end and one threaded end. This is used to secure plastic conduit to accessories such as switches and sockets. A plastic bush is used instead of a brass one. Some adaptors have a male thread which means that a plastic lock ring is used to secure the conduit to the accessory.

Another conduit fitting unique to plastic conduit is the expansion coupler. The purpose of this coupler is to allow for the expansion of the pipe in hot conditions.

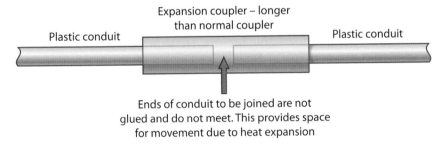

Ends of conduit to be joined are not glued and do not meet. This provides space for movement due to heat expansion

Figure 3.27: A plastic conduit expansion coupler

Safe working

When gluing plastic conduit, make sure the area is well ventilated. The fumes can be strong and as a chemical it is subject to COSHH regulations.

Specialist tools

Plastic conduit is not formed into bends using a vice and bender but with a spring. There are versions for both light and heavy gauge 20 mm and 25 mm plastic conduit.

A rope or cord should be attached to the back end of the spring so it can be pulled out of the pipe once the bend is complete.

There are a number of specialist cutting tools for plastic conduit which give a clean straight cut. A junior hacksaw can also be used.

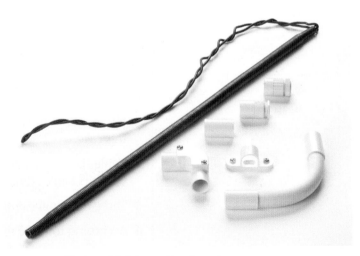

Figure 3.28: Plastic conduit fittings and bending spring

Fixing plastic conduit

Plastic conduit saddles are installed in much the same way as metal conduit. As well as conventional saddles there are clip-fit type fixings similar to the type used for copper water pipes.

Flexible conduit

Flexible conduit is tubing used to connect electrical equipment to the supply. It is used where it is inappropriate or impossible to use flexible cable, for example where the control system for the equipment is complex and requires a large number of cables. It is also used in situations where there is a risk of mechanical damage.

Flexible conduit is supplied in lightweight plastic as well as metal reinforced versions. The pipe is connected to the isolator and equipment using glands which provide a secure joint – important in order to avoid damage to the cable inside.

Conduit capacity calculations

According to the *IET On-Site Guide* to BS 7671:2008, there is a limit to the amount of cables that can be installed in conduit. The numbers of cables depend on:

- the size of the conduit
- the number of bends in a conduit run
- the length of the conduit run between inspection boxes.

Appendix E of the *IET On-Site Guide* contains a set of tables from which conduit capacity can be calculated.

- Tables E1 and E3 show cable **factors**. These are factors relating to various sizes and types of cable.
- Tables E2 and E4 show conduit factors. These factors relate to various lengths and numbers of bends in a run of conduit.

Key term

Factors – numbers used in engineering and scientific calculations. They represent a property or effect on a cable.

Worked example

Conduit factor calculation
Two ring final circuits are to be drawn into a conduit run.
The conduit run is 9 m long and includes three bends.
What size conduit is needed for this job?

Step 1: Calculate total cable factor for the circuit

1. Refer to Table E3 which shows cable factors for circuits drawn into conduit runs which include bends.
   Note: it does not matter whether the cables are stranded or solid in this case.

2. Cable size for ring final circuits is 2.5 mm². Each circuit will be made up of six cables (two line, two neutral and two CPCs).

3. Three ring final circuits will consist of $2 \times 6 = 12$ cables.

4. According to Table E3 the factor for 2.5 mm cable is 30.

5. Find the total factor value for our example: $12 \times 30 = 360$

Step 2: Refer to Table E4

1 The top line shows:

- conduit diameters
- number of bends – two bends in this case.

2 The left side shows lengths of run – 9 m in this case.

3 Cross reference between the 9 m length of run and the two bends columns until you find the nearest number to 360.

4 The nearest number is 500.

5 Refer back up this column and you will see that its conduit size header is 32 mm.

6 32 mm conduit should be used for this job.

Activity 3.5

Select the conduit size for the following cable looms.
1 Two radial 4 mm² circuits in 5 m of conduit with one 90° bend.
2 Two one-way lighting circuits in 4.5 m of conduit with three bends.
3 Three radial circuits: 2.5 mm², 4 mm² and 1.5 mm² in 9 m of conduit with two bends.

Progress check 3.9

1 What are stocks and dies used for?
2 What are the main purposes of a conduit box?
3 When would you use flexible conduit?

Trunking

Trunking is used as containment for large amounts of cable. It provides varying levels of mechanical protection, depending on which type is installed. Trunking is also used where it is not possible to clip cables directly to the surface or hide them by chasing them into walls or running them through ceiling and floor voids.

Metal trunking

Usually made from sheet steel, metal trunking is available in 2 m lengths which are riveted or bolted together. The joints between lengths of trunking run need to be mechanically secure, both to protect the cables inside from damage and also to provide a good earth connection throughout the run. As with metal conduit, trunking is classed as exposed metalwork and, as such, needs to be at the same electrical potential as earth throughout the entire installation. Because of this, small copper earth bars must be fitted across every joint.

Figure 3.29: Metal trunking

Fittings

An assortment of end caps, angles and Ts are available, as well as joining brackets and copper earth bars. It is also possible to purchase sections of lid pressed out so that accessory face plates such as 13 A sockets and spur outlets can be fitted into the trunking.

Wiring capacity

As with conduit capacities, there is a set of tables in Appendix E of the *IET On-Site Guide* from which trunking capacity can be calculated.

Key terms

BS6004 – BS6004 cables are single cables with standard insulation.

BS7211 – BS7211 thermosetting insulation cables are single cables with low smoke-emission insulation. This means that in the event of a fire they will not give off smoke or poisonous fumes.

The tables for trunking capacity are split into:

- E5 cable factors for various cable sizes and types. Solid and stranded conductors are separated and each has its own sets of factors. There are also separate factors for different types of cable insulation, PVC **BS6004** and thermosetting **BS7211**.
- E6 trunking factors.

To use the trunking capacity tables:

- establish the types and sizes of the cables to be run in the trunking
- count up the factors for the cables
- compare the total with the same or nearest (larger) trunking factor. This gives the minimum-sized trunking that can be used for that run.

Worked example

Four ring final circuits, and four 4 mm^2 and six 6 mm^2 radial circuits are to be run in trunking. The cables are all stranded, with thermosetting insulation.

Step 1: Calculate factors from cable factor Table E5

Ring final circuit
The ring final circuit will be wired in 2.5 mm.
Four ring circuits = 24 2.5 mm cables
The factor for 2.5 mm cables is 13.9, so the total ring final circuit factor is $24 \times 13.9 = 333.6$

4 mm radials
The factor for 4 mm^2 cable is 18.1
There are three cables in a radial circuit, therefore the total factor will be $4 \times 3 = 12$, $12 \times 18.1 = 217.2$

6 mm radials
The factor for 6 mm^2 cable is 22.9
There are three cables in a radial circuit, therefore the total factor will be $6 \times 3 = 18$, $18 \times 22.9 = 412.9$

Step 2: Calculate total factor

$333.6 + 217.2 + 412.9 = 963.7$

Step 3: Refer to trunking factor table

The minimum-sized trunking able to contain these circuits is 100 mm $\times$ 25 mm which has a factor of 993.

Activity 3.6

Calculate the minimum acceptable trunking sizes for the following (the cables all have BS6004 insulation).

1. Six ring final circuits (solid conductors), five 2.5 mm^2 radial circuits (solid conductors) and six 4 mm^2 radial circuits (stranded conductors).
2. Eight 6 mm^2 radial circuits and four 10 mm^2 radial circuits.
3. Seven 16 mm^2 sub-main supplies.

Plastic trunking

Like plastic conduit, plastic trunking is used in lighter, commercial installation where a measure of mechanical protection is needed, or where cables cannot be clipped to the surface or chased into the walls.

It is installed in much the same way as metal trunking. The joints between lengths can be glued together. The lid is usually a snap-on type.

Multi-compartment trunking

This is typically used in offices and classrooms where it is necessary to run both mains electrical cable and data cabling. Multi-compartment trunking consists of two or more separate compartments which enable different types of wiring to be run in the same enclosure.

The separation between compartments is enough to protect the data cables from electromagnetic interference by the mains cabling. Multi-compartment trunking can be either plastic or metal.

Figure 3.30: Multi-compartment trunking

Dado trunking

This is a version of multi-compartment trunking designed to be run along the walls of an office or a classroom. Accessories such as 13 A sockets and data outlets can be fitted flush into the front of the trunking.

Lighting trunking

Lighting trunking is similar to basic metal trunking. The difference is that:

- it has a wide lip
- the lid is usually clip-on plastic.

Lighting trunking is used as a method for supplying and installing large numbers of light fittings. Typically this would be a factory or large workshop.

- The trunking is normally suspended from the roof metalwork by means of threaded studding and brackets.
- The light fittings are fixed to the trunking by means of a bush-type fitting which incorporates two metal feet designed to slide under the trunking lip. The bush is secured, using some form of a washer-plate and lock ring.
- The bushes are spaced to coincide with the fixing holes in the back of the fitting, which is then secured to the bushes using another set of lock rings.
- The bushes are hollow so that the lighting supply cables can be run through them and into the light fitting.

Bus-bar trunking

Bus-bar trunking is a method for distributing power supplies around an area and providing flexible take-off points. This means that machines, equipment or work stations can be positioned exactly where they are needed and fed straight from the bus-bar trunking via some form of take-off connection.

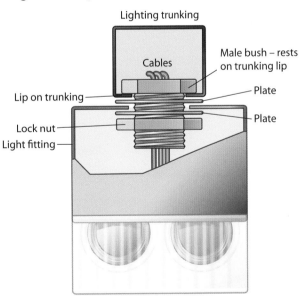

Figure 3.31: A cross-sectional diagram of lighting trunking

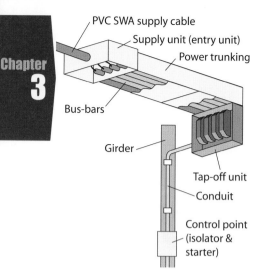

PVC SWA supply cable
Supply unit (entry unit)
Power trunking
Bus-bars
Girder
Tap-off unit
Conduit
Control point
(isolator &
starter)

Figure 3.32: Bus-bar trunking

The trunking itself usually consists of a metal or plastic trunking body pre-fitted with bus-bars. As well as joining the trunking securely, the bus-bars also need to be connected to ensure continuity of supply.

In a factory-type installation, bus-bar trunking is often suspended above the machinery. In an office area it can be installed under a raised floor. The floors consist of carpeted tiles above a concrete floor. Services are run under the floor and can be accessed easily.

When power and data is required by a workstation or desk, a power tile is fitted. This incorporates a lidded box containing 13 A sockets and telephone outlets. The power tile is then fed, in turn, from the bus-bar trunking under the raised floor.

Bus-bar trunking can also be used as a riser-type power supply in a multi-storey building such as an office block. In this case power is routed via bus-bar trunking from the main intake position in a service area at the bottom of the building. The supply for each floor is then taken from the bus-bar trunking and fed to a local distribution board.

Mini-trunking

Mini-trunking sizes range from 16 mm × 10 mm to 40 mm × 40 mm. It provides minimal mechanical protection and is not acceptable as a wiring enclosure for single insulated cables.

Typically, mini-trunking is run in domestic installations in situations where the cable cannot be clipped direct to the surface, or chased into the wall, or run in the floor and ceiling voids.

It can be installed using standard fixings such as screws, cavity fixings and rawlplugs. Versions are available that have double-sided tape on the back for quick and easy fitting.

A selection of angles and other fittings are available.

Working practice 3.4

An electrician needed to run a set of 25 mm² single cables (three-phases, neutral and earth) into an already crowded trunking. The cables would provide a three-phase feed to a refurbished workshop. The job was arduous and the trunking was at high level, with little spare capacity. It took a whole afternoon and ended with the cables coiled up into the service cupboards next to the distribution board. A week later, the electrician returned to the service cupboard to make the final connection. However, try as he might, he could not make the specification and drawing match up with the actual switchgear in the service cupboard. He returned to the main layout drawings and realised, to his horror, that he had pulled the cables to the wrong service cupboard. There was no choice but to pull them out and re-route them to the correct place.

How could this mistake have been avoided?

Cable tray

Cable tray is used to support cables in industrial, public and commercial premises. It is a very workman-like wiring method and usually kept out of sight. It consists of lengths of steel tray with its edges formed into lips to prevent cables falling off. The tray is stamped with holes and slots which can be used for cable fixings such as p-clips, cleats and cable ties.

Figure 3.33: Cable tray

Types of cable tray

- Standard – this is supplied in a range of widths between 50 mm to 900 mm.
- Basket – rather than being a flat steel tray, basket tray is formed into a V-shape into which cable can be laid. The sides are generally made from lengths of wire supported from two parallel rails. Basket tray is not intended to provide fixing facilities for cables and is generally used for data wiring.
- Ladder – intended for large cables, it is supplied in sections that resemble a ladder.

Fixing cable tray

Standard fixings can be used to install cable tray. However, a gap must be left behind the tray to allow cable fastenings to be fitted. For example, if the cable is being secured using cable ties, the tie must be passed through the back of the tray, then out again on the other side of the cable. Spacers should be used to raise the tray off the surface to which it is being fixed.

Bending cable tray

Cable tray can be formed into angles in much the same way as conduit. Tray bending machines are available for linear (flat) bends. These consist of a bending arm and two formers into which the lips of the cable tray are fitted. The formers can be adjusted to account for the width of the tray.

For lateral (sideways) bends the tray has to be cut and formed in much the same way as trunking. Bolts can be inserted into the holes of the tray to secure the bends.

Figure 3.34: A tray bender

Progress check 3.10

1 What is dado trunking?
2 What type of trunking is used for running both mains voltage and data cables?
3 Where would you use mini-trunking?

Activity 3.7

What wiring systems would you use for the following types of installation?

1 Extra 13 A sockets in a house that has been decorated – the owner does not want the décor disturbed as far as possible.
2 A feed out across a garden to feed a three-way distribution board in the garage.
3 A large grid of fluorescent fittings to light a large motor vehicle workshop.

Protective devices

As we saw earlier, an electrical installation is protected by a main fuse. Once the supply enters the premises, it is divided into separate circuits, each one protected by either a fuse or a circuit breaker.

These are protective devices and their job is to prevent damage to the cables, equipment and the building as a whole, from fire caused by electrical faults.

Definitions

BS 7671:2008 has a number of definitions connected to electrical faults.

Term	Meaning
Fault	When current flows along a path that it is not intended to flow along, for example if a line conductor snaps off and touches the frame of a metal cooker
Fault current	Current that flows when there is a fault – this is usually an extremely high current
Overcurrent	When too much current flows for the cable and equipment – a fault current can be an overcurrent
Overload current	When too much current flows down otherwise normal and functioning cable and equipment
Short circuit current	The result of live or neutral conductors coming into direct contact with each other or with earth
Shock current	Electrical current flowing through a living body, such as a human being
Fault protection	A method of protecting people from electric shock if there is a fault

Table 3.7: Terms relating to electrical faults

Prospective fault current

When there is a fault, either a short circuit or an earth fault, it will result in a high current. This is because there is no longer a load to limit the amount of current that can flow. The source of the current is the main sub-station transformer that feeds the installation. The fault current can be many thousands of amps and can do considerable damage if allowed to flow at that level for any length of time.

The two types of prospective fault current are:

- prospective short circuit current – the fault current that flows when live conductors come into direct contact with each other
- prospective earth fault current – the fault current that flows when live conductors come into direct contact with earth.

The prospective fault current of an installation is considered to be the highest value of the two.

Protective devices such as fuses and circuit breakers must be able to tolerate a high prospective fault current for the few moments it takes for them to operate. Any protective device not designed to tolerate such a current should be backed up with one that is.

Discrimination

BS 7671:2008 requires that only the protective device nearest to the fault should operate if a fault occurs. As seen on page 118, a house is protected by a 100 A fuse. It would be a nuisance if this fuse blew every time there was an electrical fault in the house. Therefore, a chain of protective devices is installed at various points between the supply and the load. Each protective device in this chain is rated lower than the previous one.

The fuse

A fuse consists of a thin wire that has a low resistance and a low melting point. When the high current caused by a fault flows through this wire, or element, it will melt and open the circuit.

Fuses are rated at certain currents. The rating of a fuse is the amount of current (plus a little more) it can carry before it overheats and ruptures. Because of this, fuses can only be used once. When the wire, or element, has been ruptured by a fault it cannot be repaired. Fuse elements are made of:

- zinc
- copper
- silver
- aluminium
- **alloys** which provide stable and predictable characteristics.

The fuse should carry its rated current for many years, and rupture as soon as the current begins to rise. Fuses are designed not to corrode or change their structure, no matter how long they are in place. Occasional current surges must not damage the element.

Fuse elements are sometimes formed into shapes which increase heating effect. In large fuses, elements consist of a number of thin metal strips rather than one single conductor. Elements may be supported by steel or nichrome wires, so that no strain is placed on the element. Sometimes a spring is included to increase the speed at which the element breaks once hot.

> **Key term**
>
> *Alloy* – a substance made from a mixture of metals.

>  **Safe working**
>
> Never replace a fuse with one of a higher rating. If a fuse continually blows then there is a fault that must be attended to.

Type of fuse	Description
BS88	This is a cartridge-type fuse which has a high rupture capacity, or HRC. This means that it will be able to tolerate and contain a high fault current. The fuse element may be surrounded by air, or by materials intended to quench the explosive arc caused by a high current rupture. Silica sand or non-conducting liquids are sometimes used. They can also withstand a high current for a short while, for example the BS88-2 Type gM, motor fuses, which can tolerate the high current generated by an electric motor when it starts.
BS3036	No longer installed, the BS3036 is the rewirable fuse. You will still find these in older installations. Fuse wire is connected to two terminals at either end of the fuse holder and replaced when it ruptures. This type of fuse is relatively slow and can cause damage when it ruptures.
BS1362	Plug top fuses and the fuses found in spur outlet units are a small cartridge type called the BS1362. Standard ratings for plug top fuses are 3 A and 13 A.

Table 3.8: Main types of fuses

Circuit breakers

Like fuses, circuit breakers have specific current ratings and will automatically switch off the supply if that rating is exceeded. Unlike fuses, circuit breakers are automatic switches and do not need to be replaced when they operate.

The two main types of circuit breaker operation are:

- heater: the main working part in this type of circuit breaker is a bimetallic strip. If excessive current flows, the strip will heat up and bend and operate the tripping mechanism. It takes anywhere from a few minutes to several tens of minutes to trip in this mode. This makes it suitable for loads that draw a momentary high current when they start (e.g. electric motors).
- magnetic: in this type of circuit, current flows through a magnetic coil. This creates a magnetic field (see *Chapter 1: Principles of electrical science*) which operates the trip arm, tripping the breaker. The operation of this type of circuit breaker is almost instantaneous.

The main type of circuit breaker in general use is the BSEN60898 moulded case circuit breaker. The three types are:

- type B – domestic installations, heaters, showers, cookers and socket outlets
- type C – electric motors, power supplies and general lighting circuits
- type D – transformers, motors, **discharge lighting** circuits and computers.

Residual current device (RCD BSEN61008)

The residual current device, or RCD, provides shock protection and is not intended to operate if there is an overcurrent. An RCD will operate and shut off the supply to a circuit when a fault, or leakage, current flows between a live conductor and earth.

An RCD should operate if a fault current flows for 40 ms (milliseconds) and before the fault current reaches 30 mA.

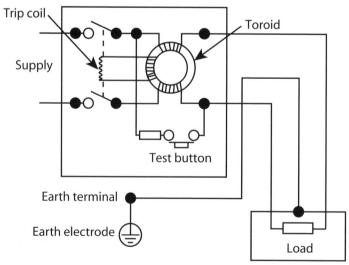

Figure 3.35: How an RCD works

Key term

Discharge lighting – this includes fluorescent fittings and is a method of creating light by discharging a current through a gas.

Types of RCD include:

- high sensitivity (HS): 6 – 10 – 30 mA (for direct contact shock protection)
- medium sensitivity (MS): 100 – 300 – 500 – 1000 mA (fire protection)
- low sensitivity (LS): 3 – 10 – 30 A (e.g. for protecting machines).

Residual current circuit breaker with overload protection (RCBO BSEN61009)

The RCBO is a combined RCD and overcurrent protection. RCBO types are B, C and D. These types apply to the overcurrent part of the device (see page 148).

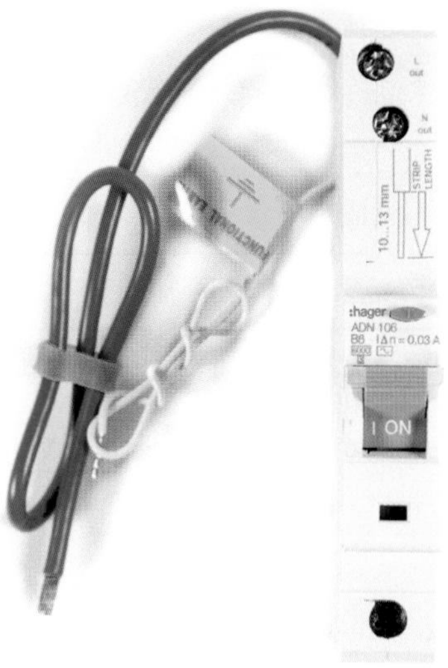

Progress check 3.11

1 What does HRC stand for?
2 Where would you expect to find a BS1362 fuse?
3 What type of circuit would be protected by a type C circuit breaker?

Figure 3.36: An RCBO

Current-carrying capacity – cable selection

Cables are selected to carry certain amounts of current. The larger the cross-sectional area of a conductor, the more current it can take. There are, however, other factors which affect the amount of current a cable can carry. These include:

- the type of environment the cable is run in
- whether the cable is run through thermal insulation
- the amount of cables bunched with the cable
- the wiring method, e.g. conduit, clipped to the surface, etc.

Because they all cause an increase in cable temperature, these factors can reduce the amount of current a cable can carry. As mentioned in *Chapter 1: Principles of electrical science*, the warmer a conductor the higher its resistance and therefore the less current it can carry.

A further factor in the current-carrying capacity of a cable is the type of protective device.

Calculation tables

Appendix 4 of BS 7671:2008 is made up of tables from which the cable size for a particular load can be calculated. The tables show the following information.

Millivolt drop/amp/metre

The cable selection tables in Appendix 4 of BS 7671:2008 show the voltage drop (mV/A/m) for various cable sizes installed in particular ways.

Volt drop is the amount of reduction in voltage measured at the load end of a cable compared with the voltage at the supply end. Think of it in terms of a hosepipe: the longer the pipe, the lower the water pressure. The same applies to cables: the longer the cable run, the higher the volt drop.

The maximum volt drops allowed are shown in Table 3.9.

Type of circuit	Maximum voltage drop allowed as a percentage	Maximum voltage drop allowed as a voltage 230 V single-phase	Maximum voltage drop allowed as a voltage 400 V three-phase
Lighting	3%	6.9 V	N/A
Other	5%	11.5 V	20 V

Table 3.9: Maximum volt drops allowed

This means that the volt drop for a cable selected for a particular load must not exceed these numbers.

Remember, the table shows millivolts (mV) not volts, so any calculation needs to be divided by 1 000.

Some volt drop tables will have three values for each cable.

- r – resistive load such as a heater.
- x – reactive load such as an electric motor or transformer.
- z – high-impedance load where the reactance is greater than the resistance.

Current-carrying capacity

This is the table from which an initial cable selection is made. This cable choice is then tested by a set of calculations which will show the resulting volt drop. If the answer is lower than the BS 7671:2008 maximum values, the chosen cable can be used safely for that circuit. If it is higher, a larger cable should be selected and tested using the volt drop calculation.

Reference method

The reference method describes how the cable is installed. A set of tables and diagrams in the first half of Appendix 4 of BS 7671:2008 show the various reference methods. Examples are:

- clipped direct to the surface
- enclosed in conduit or trunking
- enclosed in conduit in a wall.

Type of cable

Cable type has a bearing on current-carrying capacity. Conductor material will also have a significant influence – aluminium conductors, for example, carry less current than copper conductors.

Activity 3.8

Find the following, using the tables in Appendix 4 of BS 7671:2008.

1 What is the current-carrying capacity for 16 mm^2 copper single-core, single-phase, non-armoured cable with thermoplastic insulation? The cable is installed in conduit in a thermally insulated wall.

2 What is the current-carrying capacity of 4 mm^2 multi-core armoured copper cable with 70°C thermoplastic insulation? The cable feeds a three-phase circuit and is installed on a cable tray.

The volt drop formula

The total amount of voltage dropped by a cable is calculated using the formula below.

$$\text{Total volt drop} = \frac{\text{Length of run} \times \text{current} \times \text{mVd/A/m}}{1\,000}$$

- Total volt drop – the amount of voltage dropped by the cable you select for your circuit.
- Length of run – length of cable feeding the circuit (in metres).
- Current – current taken by the load.
- mV/A/m – the amount of voltage dropped for each amp of current over each metre of its run.
- 1 000 – the answer has to be divided by 1 000 because the volt drop for the cable is stated in millivolts.

Worked example

A 3 kW, single-phase heater is fed using armoured multi-core cable which is clipped directly to the surface. The conductors are copper, with 70°C thermoplastic insulation. The length of run of the cable is 8 m. What is the minimum-sized cable that can be used for this circuit?

Step 1: Calculate load current

I = P
I = 3000 W = 13 A
V = 230 V

Step 2: Find a suitable cable size

Table 4D4A gives current-carrying capacities for multi-core armoured cables.

Remember the following points.

- The cable is single-phase, so it will have two cores.
- It is clipped directly to a surface.

Using this information, you will find that the first cable to carry 13 A is 1.5 mm^2.

Step 3: Find the mV/A/m for the chosen cable

Table 4D4B gives mV/A/m for the multi-core armoured cables.

Two-core 1.5 mm^2 armoured cable has a mV/A/m of 29 mV (0.029 V)

Step 4: Carry out the voltage drop calculation

$$\text{Total volt drop} = \frac{\text{Length of run} \times \text{current} \times \text{mVd/A/m}}{1\,000}$$

$$\text{Total volt drop} = \frac{8 \times 13 \times 29}{1\,000} = 3.016 \text{ V}$$

This is within the acceptable limit of 6.9 V which means that 1.5 mm^2 can be used for this circuit.

Activity 3.9

1 What is the millivolt drop/amp/metre for a 16 mm² single-core, non-armoured copper cable feeding a single-phase circuit? The cable is clipped to a tray, touching other cables, as per Reference Method C. The insulation and sheath is 70°C thermoplastic.

2 What is the millivolt drop/amp/metre for a 2.5 mm² multi-core, non-armoured copper cable feeding a three-phase circuit? The insulation and sheath is 70°C thermoplastic.

Activity 3.10

Find the minimum-sized cables for the following circuits. The insulation in each case (except Question 4) is 70°C thermoplastic and the conductors are all copper.

1 A 5 kW three-phase electric motor is fed using non-armoured, non-sheathed single cables. The cables are run in a conduit, which is fixed to the surface of a wall. The length of run is 12 m.

2 A single-phase, 7 kW oven is fed using a multi-core, non-armoured cable. The cable is run through conduit which is buried in a thermally insulated wall. The length of run is 7.5 m.

3 A three-phase 275 kW high impedance supply is run using single-core armoured cables which are clipped to a horizontal cable tray according to Reference Method F. The cables are all touching. The length of run is 13 m.

4 A three-phase transformer is fed by a multi-core armoured cable with aluminium. The total distribution board load is 46 kW. The armoured cable has 90°C thermosetting sheath and 70°C insulation. The load is reactive. What is the minimum-sized cable that can be used if the cable run is 11 m and it is enclosed in duct which is buried in a wall according to Reference Method A?

Cable correction factors

When selecting a cable there are sometimes several other factors to be taken into account as well as length of run, current and volt drop. These are called correction factors. These figures can be found in BS 7671:2008 Appendix 4 (Tables 4B1 to 4C6). They take into account other influences that could reduce the current-carrying capacity of a cable. The correction factors are shown in Table 3.10.

Key terms

Ambient temperature – the temperature of the surrounding air.

Resistivity – all matter has a certain amount of resistance. Resistivity is the amount of resistance per metre of a material. Its symbol is ρ (the Greek letter r (rho)) and it is measured in Ω/m. Insulators have a high resistivity while conductors have a low resistivity.

Rating factor code	Description
Ca	**Ambient temperature**
Cc	Buried circuit
Cd	Factor applied for the depth the circuit is buried
Cf	Circuits protected by rewirable fuses; the factor is always 0.725
Cg	Grouping factor, applied when a number of cables are grouped together
Ci	Applied to circuits run through thermal insulation
Cs	Thermal **resistivity** of soil

Table 3.10: Correction factors

Correction factors are applied to the design current, *In*, the protective device current for a circuit.

A 3 kW single-phase load would have a design current (*Ib*) of 3000 W which is 13 A.

This means that the protective device rating (*In*) would be 16 A, as in a 16 A circuit breaker.

When using correction factors in a cable selection calculation, you will need this equation:

$$It \geq \frac{In}{Ca \times Cc \times Cd \times Cf \times Cg \times Ci \times Cs}$$

It will be the current value you look up in the cable current tables in BS7671:2008 Appendix 4. As you can see, it needs to be equal to, or greater than, the protective device rating when it has been divided by the appropriate correction factors. (Note that *Iz* is used to represent the current-carrying capacity of the cable after other factors have been taken into account, such as grouping, ambient temperature, thermal insulation and protective device.)

You may not need all the correction factors. It depends on the circuit. For example, if it is protected by a rewirable fuse and run through a roof space where there is thermal insulation, then you will only need Cf and Ci for your calculation. In this case the equation will look like this.

$$It \geq \frac{In}{Cf \times Ci}$$

EARTHING SYSTEMS

The mass of the earth beneath our feet acts as a negative terminal for the electrical current used in most electrical installations. The amount of current that flows to earth, via the neutral conductor, is limited by the resistance of the loads in the installation. However, if that resistance is removed, a high current will flow. This is called the prospective earth fault current.

If a live conductor comes into contact with the metal casing of electrical equipment, the casing will become live and present a shock hazard. To prevent this kind of shock and injury, a system of cables, connections and components is included in the installation. The purpose of this system, the earthing system, is to draw the current away from the point of fault and automatically disconnect the supply.

Progress check 3.12

1   What is mV/A/m a measurement of?
2   What is a reference method?
3   What is *It*?

LO4

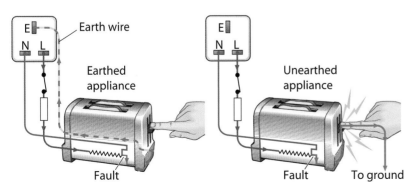

Figure 3.37: Faulty toaster – earth fault

There are three earthing systems designed to provide what is called an earth fault loop impedance path. These are:

- terra-terra (TT) – the connection between the installation and the earth is provided by an electrode driven into the ground
- terra neutral separated (TN-S) – connection made using a separate conductor
- terra neutral combined-separate (TN-C-S) – connection made by terminating the installation earth conductor to the neutral on the supply company side of the supply.

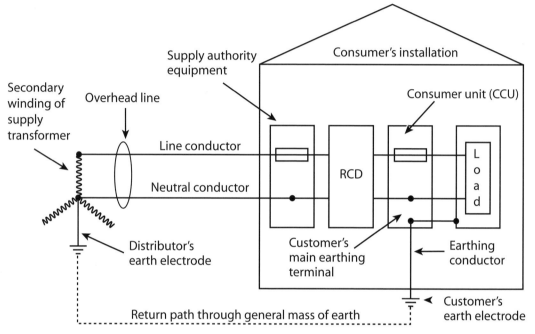

TT system

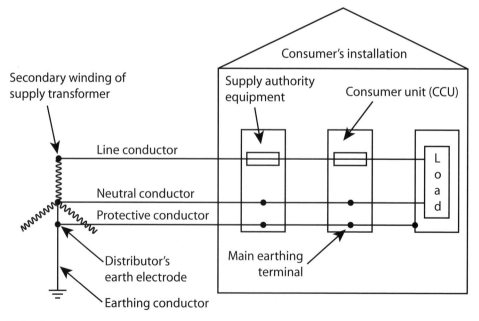

TN-S system

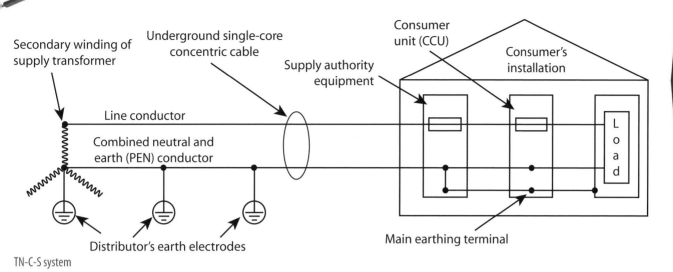

Consumer unit (CCU)

Consumer's installation

Secondary winding of supply transformer

Underground single-core concentric cable

Supply authority equipment

Line conductor

Combined neutral and earth (PEN) conductor

Distributor's earth electrodes

Main earthing terminal

TN-C-S system

Figure 3.38: The three earthing systems

Activity 3.11

Examine the supply intake position for your home and establish what type of earthing system is being used.

TT if earth conductor is run outside to an earth stake.

TN-S if earth conductor is connected to the armoured sheath of supply cable.

TN-C-S if earth conductor is connected to supply neutral.

Earth fault loop impedance path

The earth fault loop **impedance** path is the path taken by a fault current from the point of fault, through the installation earthing conductors to the star point earth connection at the main sub-station transformer, then back to the point of fault along the line conductor. The resulting current increase and imbalance will operate a protective device. This can be either:

- a circuit breaker – which operates when a current passes through it which is higher than its rated current
- a RCD – which operates when the line and neutral currents in its windings are sent out of balance (see pages 148–149).

Zs and Ze

The two parts to the earth fault loop impedance path are:

- Zs – the impedance of the full fault path
- Ze – the impedance of the fault path on the supplier's side, from the main earth terminal to the sub-station and back to the customer intake point.

The earth fault loop impedance path should be as low as possible to enable a high fault current to flow. The higher the current, the quicker the protective device will operate. The value of the earth fault loop impedance path is measured when an electrical installation is tested.

> **Key term**
>
> *Impedance* – the total, current-limiting effect in an a.c. circuit. Impedance is made up of resistance and reactance (see *Chapter 4: Installation of wiring systems and enclosures*). Its symbol is Z and it is measured in ohms (Ω).

Chapter 3

R_1 and R_2

Zs can be measured, or calculated, using R_1 and R_2.

- R_1 – resistance of line conductor.
- R_2 – resistance of circuit protective conductor (CPC).

R_1 and R_2 are taken during the continuity of protective conductor test. This will be looked at in more detail in *Chapter 8: Electrical installations: inspection, testing and commissioning*.

To calculate Zs from these readings the formula shown below must be used. Note that the Ze value can either be measured (see Chapter 8) or the value obtained from the supply company.

$$Zs = Ze + (R_1 + R_2)$$

Components of automatic disconnection of supply (ADS)

The earthing system for an installation is made up of a number of components. These are described below.

CPC

The earth conductors within a circuit are called circuit protective conductors (CPCs). The CPC in a multi-core cable such as twin-and-earth is usually one size smaller than the live conductors. So, for example, a 2.5 mm^2 twin-and-earth will have a 1.5 mm^2 CPC.

When wiring a circuit in single cables, the CPC will be the same size for smaller circuits, or calculated using the adiabatic equation for circuits with larger line conductors.

The first instrument test carried out during inspection and test procedures is the continuity of protective conductor. Using a low reading ohmmeter, the resistance of the CPC for each circuit is recorded. This continuity test is only carried out on the CPC because it does not carry current under normal conditions, so if there was a break in the cable it would remain undetected until there was a fault. If the CPC is broken, the earth protection system would not work properly and someone could receive a potentially fatal electric shock.

Main earth terminal (MET)

This is the connection point for all the earth conductors in an installation. The supply earth is also connected to the MET.

Earth conductor

The earth conductor is the cable that connects the MET to the supply earthing system.

Main protective bonding conductor

The main protective bonding conductor connects the metallic pipework for the main services such as water and gas to the installation earth, usually at the MET. The connection to these services must be made as near as possible to the point at which the pipe enters the building. This is usually a 10 mm^2 or 16 mm^2 single cable. Because it does not carry any current under normal conditions, it does not need to be enclosed in conduit or trunking.

Supplementary protective bonding conductor

Any premises will contain metalwork which, potentially, could become live in the event of an electrical fault. There are, of course, metal switch plates, conduit or trunking. But pipes and general metalwork are also potential threats, so these must be bonded to earth as well. This can either be achieved by connecting them directly to the installation earth system or by connecting one part of the metalwork system to the installation earth, then connecting the remaining metalwork together so they are all at the same electrical potential. The conductor used for this is the supplementary protective bonding conductor.

The minimum cross-sectional areas for supplementary protective bonding conductors are:

- 2.5 mm if the cable is sheathed or installed in conduit, trunking or similar
- 4 mm and larger if installed with no mechanical protection.

Protective devices

Protective devices are dealt with on pages 145–149.

Exposed conductive parts

An exposed conductive part is the metalwork directly associated with an electrical installation. Examples of this are:

- metal switch and socket plates
- metal conduit
- metal trunking
- metal light fittings
- the frame of a cooker or other household appliance.

This metalwork must be connected to the installation earthing system. The connection point is usually a terminal. In the case of tray or trunking, copper joining bars are usually fitted to joints between lengths and one end of the system is generally connected to the installation earth using a main protective bonding conductor.

Extraneous conductive parts

Extraneous conductive parts are the metalwork in an installation that is not directly part of the electrical installation but could potentially become live in the event of a fault. This includes:

- water, gas and oil pipes
- steel duct work
- structural steel.

As with exposed conductive parts, this metalwork must be bonded to ensure all parts are at the same potential, then connected to the installation earth using a supplementary protective bonding conductor and main protective bonding conductor.

Progress check 3.13

1 What is connected to the MET?
2 Where would a TT earthing system normally be used?
3 What is the Zs of a supply if $R_1 = 3.7\ \Omega$, $R_2 = 5.18\ \Omega$ and Ze = 4.2 Ω?

LO6

MICRO-RENEWABLE ENERGY

Micro-renewable energy sources and systems are described more fully in *Chapter 6: Electrical principles*. However, we will introduce them here and give a brief description of their purpose and working principles.

Micro-renewable systems are used for heating and for producing electricity in small domestic premises. Although expensive and often complex to install, these systems can be seen as a long-term investment, paying for themselves and eventually saving the owner money by providing low-cost or even free resources. At the same time they cause no damage to the environment and help to preserve resources.

Each system has its own particular requirements. For example, a water source heat pump will need a body of water as its heat source, and roof-mounted collector panels require the roof to be in good condition.

Building and regulatory requirements

For the most part, renewable and micro-generation equipment can be installed without the need for planning permission. There are, however, cases when it may need to be sought, for example if the sizes or locations for the equipment, such as solar panels, need to be larger than the permitted size, or for an air-source heat pump system which can produce noise pollution. Listed buildings will also require permission before being fitted with this type of system.

Because they produce smoke, biomass boilers and stove installations will require local authority permission due to clean air regulations, which the authority has a responsibility to enforce. The building regulations will need to be applied to structural, fire and electrical issues.

Types of micro-renewable energy

Type	Description	How it works
Solar thermal	Heating derived from sunlight	Panels mounted on a roof or free standing. Uses a transfer medium with a low boiling point to transfer the heat from the collector panels to the heat load, e.g. water.
Heat pump	Heating derived from external heat source. This can be: • ground • air • water	Heat pump unit works on the refrigeration cycle. This is a method of obtaining usable heat from a low-heat source, e.g. the ground or a body of water, by passing a low-boiling point fluid through the heat source. This chemical is then used to heat a refrigerant (also low boiling point) which is turned to gas, pressurised to raise its temperature further, before transferring its heat to the load (water, central heating, underfloor heating).

Biomass	Boiler or stove that produces heat by burning wood or other organic material	Used to heat water, central heating or as a space heater, biomass burners are fuelled by wood in the form of either logs, pellets or wood chips.
Photovoltaic	Electricity derived from sunlight	Panels mounted on a roof or freestanding convert light to electricity. Excess power can be fed back into the grid and earn the owner a discount or repayment.
Micro-wind	Electricity derived from small wind turbines	Micro-wind turbines work exactly the same way as larger turbines. Wind turns the blades, which, in turn, drive a turbine and generator. The feed-in tariff from excess electricity applies to micro-wind generation. An alternative way of using excess electricity is battery storage.
Micro-hydro	Electricity derived from water-driven turbines	As with micro-wind, micro-hydro generation is a smaller version of the hydroelectric dam. Moving water drives a turbine and generator. There are water extraction issues as well as environmental ones to be considered. An environmental audit has to take place before micro-hydro systems are installed.
Micro-combined heat and power	Electricity derived from domestic heating	Micro-combined heat and electricity uses an existing heating system to drive a generator.
Rainwater harvesting	Capture and reuse of rainwater	Rainwater is diverted to a water system and used for: • car washing • clothes washing • flushing toilets • watering the garden. Rainwater is considered to be a Class 5 risk (the highest risk level) and must be kept separate from the domestic wholesome water system.
Grey water reuse	Reuse of waste water from washing	Limited to: • car washing • flushing toilets • watering the garden – but not any produce that is eaten raw (e.g. fruit and salad).

Table 3.11: Types of micro-renewable energy

Progress check 3.14

1 What is micro-renewable energy?
2 What are the three heat sources for a heat pump?
3 What is grey water and what can it be used for?

Knowledge check

1 At what voltage is electricity usually generated at a power station?

a 250 kV

b 250 V

c 25 kV

d 25 mV

2 Which part of the electricity supply system belongs to the supply companies?

a Distribution grid

b Super grid

c Transmission grid

d Hydro grid

3 At what point does the supply enter an installation?

a The meter

b The MET

c The consumer unit

d The main fuse

4 Which of the following is a domestic installation?

a Factory

b Caravan park

c Flat

d Hospital

5 A fire alarm system is divided into:

a zones

b segments

c quadrants

d sections

6 The cable used to connect an item of electrical equipment to its supply is:

a twin-and-earth

b flex

c tri-rated

d singles

7 The cable used as a television aerial is:

a coaxial

b category 5

c fibre optic

d twisted pair

8 Metal containment systems such as conduit and trunking must be:

a airtight

b watertight

c earth-tight

d volt-tight

9 Shock protection is provided by a:

a circuit breaker

b residual circuit breaker

c HRC fuse

d isolator

10 In which earthing system is the earth conductor connected to the neutral on the supply company side of the intake position?

a TT

b TN-S

c TN-C-S

d TN-C

Installation of wiring systems and enclosures

This chapter covers:

- choosing tools and preparing to install wiring systems safely
- installing wiring systems
- bonding mains services to main earthing terminals
- inspecting and testing a dead electrical installation.

Introduction

Installing and testing electrical installations requires a number of specialist and general practical skills. You need to know which tools to use and how to use them properly. Also, you must look after them. Broken or damaged tools are not only difficult to use, they produce bad workmanship and can be dangerous.

While *Chapter 3: Electrical installations technology* looked at the basic wiring systems and installation methods in detail, this chapter describes how to actually install them and work with the materials. Once an installation is complete it has to be inspected and tested. This chapter will show you the basic dead tests you need to carry out and also the functionality checks that are the final part of the testing regime. The complete testing procedure is covered later in *Chapter 8: Electrical installations: fault diagnosis and rectification.*

This chapter does not contain specific practical exercises, but shows the basic skills involved such as cutting metal trunking, threading and forming a bend in conduit, and how to clip and strip twin-and-earth cables.

LO1&2

CHOOSING TOOLS AND PREPARING TO INSTALL WIRING SYSTEMS SAFELY

As with all practical tasks, tools are needed. Some are basic, others will require training and practice before their correct use can be mastered. The important rules for tools are outlined below.

- Use the right tool for the job.
- Do not use damaged or faulty tools.
- Make sure you know how to use the tools.

This section describes some of the tools used for electrical work, the hazards to look out for and the **PPE** you should use when working with each tool.

Key term

PPE – personal protective equipment.

Tools, hazards and PPE

Screwdrivers

One of the tools most widely used by electricians is the screwdriver. The main purpose of a screwdriver is to tighten or loosen a screw. This can be a terminal screw for a light switch or 13 A socket, or a fixing screw used to secure a distribution board or conduit saddle to a wall.

There are three main types of screwdriver tip. Table 4.1 lists them and describes their uses.

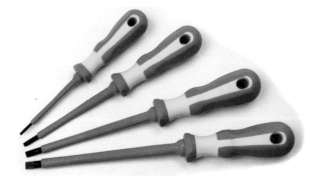

Figure 4.1: Screwdrivers

Type of screwdriver	Description and uses
Flat-blade	Older screws tend to have a slotted head. Flat-blade screwdrivers are designed to fit into this type of screw head. Available in all sizes, from the small terminal driver type to large heavy duty versions.
Phillips	Most fixing screws now have Phillips-type heads. Phillips screwdrivers tend to be available in two sizes: smaller for terminal screws and larger for fixing screws.
Pozidriv	Similar to Phillips, Pozidriv screws are often used as terminal screws. The Pozidriv's star shape makes for a firmer grip between the screwdriver and the screw itself.

Table 4.1: Screwdriver tips

Ratchet screwdrivers

Ratchet screwdrivers are fitted with mechanical gearing, which means that you can turn the screwdriver bit by either:

- pumping the handle
- turning a handle which will only engage when rotated in one direction.

Magnetic screwdrivers

Magnetic screwdrivers and gripping screwdrivers will hold a screw steady if it is being inserted into an awkward position where you cannot grip the screw with your fingers. Magnetic screwdrivers can also be used for picking up screws that have dropped into small or difficult spaces.

Electric screwdrivers

Electric screwdrivers are being used more and more. Although they can be purchased as a specialist power tool, most modern electric drills will accept screwdriver bits and are geared so that they can not only tighten fixing screws, but also loosen them by being run in reverse.

Safe working

Never point a ratchet screwdriver towards your own, or anyone else's, face if the shaft is retracted into the main body. The shaft will spring out with a lot of force and could cause a serious eye injury.

Tool	What to check for	Optimum performance	PPE to be worn when using these tools
General purpose hand-held screwdriver	• Damage to handles • Shaft loose in the handle • Damage to insulation on handles • Damage to insulation on screwdriver shaft	• Flat-headed – blade is sharp and undamaged • Phillips and Pozidriv – end is intact and not worn	There is no PPE specific to using screwdrivers. The main method of controlling risk is careful use.
Ratchet screwdriver	• Weakened catch that will not hold retracted shaft	• Keep retractable shaft lubricated	
Electric screwdriver	• Insulation is intact • Shaft and tip is undamaged • No sign of overheating • On/off switch or trigger is working • Lead and plug are undamaged (if plug-in type)	• Battery is fully charged if battery type • Tips are sharp with no wear on edges	

Table 4.2: Using screwdrivers safely

Cable cutters, strippers and pliers

Cables need to be cut cleanly and quickly, then stripped of their insulation without damage to the conductor inside. There are specialist tools for these jobs.

Tool	Description and uses
Side cutters	Side cutters are specifically designed for cutting cable. Make sure they are sharp and not damaged. Nicks in the blades will cause a ragged cable end. Side cutters are not suitable for large cables. These should be cut using a hacksaw or a specialist tool.
Cable strippers	It is important to strip insulation from a cable core without damaging the conductor. If the stripping tool cuts into the conductor it will be weakened. This could cause a heat spot when the conductor is live, or can cause it to snap off completely. Stripping tools are usually adjustable. Make sure that they are set correctly – too loose and the insulation will not be removed cleanly, too tight and the conductor will be damaged.
Combination pliers	Combination pliers consist of a pair of gripping jaws and cutting blades. The gripping jaws can be used for bending conductors, twisting them together or for tightening small nuts and bolts. Experienced electricians can often use the cutting blades to both cut and strip cables and this means that they only need to carry a set of combination pliers instead of two separate tools – side cutters and cable strippers.

Table 4.3: Cable cutters, strippers and pliers

Tool	What to check for	Optimum performance	PPE to be worn when using these tools
Side cutter	• Damage to handles • Damage to insulation on handles • Nicks and other damage to blades	• Keep lubricated • Only use side cutter with sharp, functioning blades	There is no PPE specific to using side cutters, cable strippers and combination pliers. The main method of controlling risk is careful use.
Cable stripper	• Damage to handles • Damage to insulation on handles • Nicks and other damage to blades	• Keep lubricated • Only use cable stripper with sharp, functioning blades	
Combination pliers	• Damage to handles • Damage to insulation on handles • Broken or damaged jaws • Nicks and other damage to cutting blades	• Pivot point lubricated • Only use undamaged combination pliers	

Table 4.4: Using cutters, strippers and pliers safely

Progress check 4.1

1 What are the three main types of screwdriver?

2 Why is it important to correctly adjust cable strippers before use?

3 What single tool can be used to replace cable cutters and cable strippers?

Knives

There are times when it is easier or more effective to remove a cable sheath or insulation with a knife. The main knife used by electricians is the retractable type. The blades in these knives are pushed out of, and drawn back into, the knife handle using a slider. The advantage of the retractable bladed knife is that there is no exposed cutting edge when the tool is not in use.

Figure 4.2: A retractable bladed knife

Tool	What to check for	Optimum performance	PPE and other safety advice
Knife – retractable blade type	• Nicked and blunted blades • Handle parts not clamping together securely • Blade not retracting fully	• Always use a sharp blade • Keep blades dry or wrapped in an oil or grease wrapper – some blades are supplied in this type of wrapper	• Gloves • Always move the blade away from yourself when using a knife to cut or strip something

Table 4.5: Using knives safely

Safe working

Never play around with knives – it may appear as if a retractable blade is hidden fully in the knife body but a small tip can be exposed. Also, the blades used for retractable knives are extremely sharp.

Case study

A practical workshop lesson was about to start at a further education college. The learners were on a Level 2 electrical installation course and the lesson was to practise terminating steel wire armoured cable glands. The outer and inner sheaths of the cable had to be stripped using a knife.

The learners gathered round the centre bench for the demonstration and were then issued with retractable knives, each one signed for. The learners were instructed in their safe use and warned of the dangers.

One learner, however, began to play around and slashed at another student with one of the knives. Thinking the blade was retracted and safe, he stabbed at his mate's throat. Unfortunately a small amount of blade was exposed and the learner was horrified when a deep gash appeared on his mate's neck. No malice was intended and it was a terrible accident. However, it shows very clearly why you must NEVER play around with any type of knife.

After being rushed to the Accident and Emergency department of the local hospital, the learner was given stitches and, thankfully, recovered fully. However, the jugular vein passes down the side of the neck, and if this had been severed, the learner could have died.

Saws

Electricians use saws for many jobs, for example cutting trunking, conduit and large cables. In addition, an electrician has to carry out other jobs that support the main electrical work: joists and beams might have to be cut to allow for cable access, or wooden noggins produced when replacing floorboards or chipboard sheeting.

Type of saw	Description and uses
Wood saw	The wood saw is the most recognisable type of saw. It is made up of a wooden or plastic handle fitted with a large tapered blade. A wood saw should only be used for cutting wood as any other material will blunt or damage the teeth.
Hacksaw	Hacksaws consist of an adjustable metal frame and handle. The blade can be fitted and changed according to the material to be cut. Hacksaws are normally used for cutting metal but can also be used on other materials, such as plastic. The blades should be inserted with the teeth facing forward. The junior hacksaw is a smaller version of the hacksaw. When using a hacksaw, hold it with two hands, one on the handle the other on the other end of the frame. Use the whole blade and adopt an easy fluid motion. Let the blade do the cutting.
Pad saw	As with the hacksaw, pad saw blades can be removed and swapped for types suitable for the material to be cut. Pad saws consist of a handle and a projecting blade. The pad saw can be used for cutting holes in plasterboard walls and cutting notches out of wooden joists and studwork in awkward places.
Jigsaw	The jigsaw is a hand-held power saw. It has a short, narrow blade which is driven in a pumping motion. Different blades can be fitted to suit the material to be cut. Jigsaws can be used to cut wood, plastic or metal. One of the main uses electricians have for the jigsaw is cutting slots, for example in the edge of trunking where it meets the entrance to a distribution board. Figure 4.3 describes how to cut a slot in metal trunking using a jigsaw.

Table 4.6: Different types of saw

Check list

PPE	Tools and equipment	Source information
• Eye protection • Gloves • Ear defenders	• Jigsaw with metal cutting blade – either 110 V or battery operated • Power drill – either 110 V (with transformer) or battery operated • Hole saw • Tape measure • Straight edge • Set square • Scribe • File	• Worksheet, plan or specification

1 Mark out slot. Use a set square and straight edge to make sure the slot is neat.

2 Using one of the holes place the jigsaw blade against the slot line.

3 Drill either end of the slot with a hole saw. If the slot is large then drill a hole in each corner.

4 Cut slowly and carefully. Hold onto the jigsaw firmly but there is no need to apply excessive pressure.

Figure 4.3: Using a jigsaw to cut a slot in trunking

Circular saw

Circular saws are available either as a fixed saw bench or as a portable power tool. The bench type has a flat, smooth worktop. The circular blade projects from a slot in a surface. The workpiece to be cut is moved over the flat surface and across the rotating blade. This is an extremely dangerous piece of equipment and should only be used by those trained to do so.

 Safe working

Always wear eye protection when using a jigsaw. The cutting action throws up dust and debris, and also, if the blades break, they can sometimes fly upwards towards your face.

Chapter 4

The Health and Safety Executive (HSE) has issued an information sheet called 'Circular saw benches – Safe working practices'. The main points are covered below.

- Saw benches should be fitted with a brake that stops blade rotation in less than 10 seconds.
- Do not use worn or faulty saw blades because they produce poor quality work and make feeding harder which increases the risk of accidents.
- Never try to clean a blade while it is running.
- The diameter of the smallest saw blade that can be used safely must be marked on the machine.
- A push-stick should always be used to move the workpiece across the blade if the cut is less than 300 mm long. It should also be used for feeding in the last 300 mm of a longer cut. Push-sticks should be 450 mm long with a small slot, or 'bird's mouth', at one end. This gives a better grip on the workpiece.
- If the finished cut piece is less than 150 mm, the push-stick should always be used to remove it from the saw blade.

The blade on the portable version is attached to a handle which is pulled down and through the workpiece. A variation of this is the grinding wheel which has a rotating carborundum cutting wheel instead of a saw blade. This type is often used for cutting lengths of metal such as strut channelling. Another type of portable circular saw is held in both hands and pushed through the workpiece.

Remember the following safety points when using a portable circular saw.

- Wear eye protection, protective gloves and a disposable dust mask.
- Replace dull or burned saw blades.
- Lift the saw from the cut only after the blade stops.
- Use the correct blade for the material you are cutting.
- Disconnect the saw from its power source before cleaning or changing blades, or making adjustments.
- Place materials on a firm surface for cutting (never on your hands, arms, across the knees or feet).
- Cut the materials beyond the end of a support so that the waste falls clear.

Tool	What to check for	Optimum performance	PPE and other safety advice
Wood saw	• Damaged handle • Blade loose in handle	• Always use a sharp blade • Keep blade dry and rust free • Do not store with teeth resting on a hard surface, as this can damage or blunt them	• Gloves • Keep workpiece securely fixed while sawing • Eye protection
Hacksaw	• Worn or weakened blades • Loose blade clamps • Damaged frame • Damaged insulation on the handles	• Make sure teeth are pointing in the right direction – forwards • Do not use worn blades	
Pad saw	• Worn or weakened blades • Loose clamping screws which hold the blade in the handle	• Do not use worn blades • If using hacksaw blades in a pad saw, keep exposed blade as short as possible	

Tool	What to check for	Optimum performance	PPE and other safety advice
Jigsaw	• Damaged lead and plug • Damaged insulation • General damage • Faulty on/off switch/trigger • Loose blade clamping screws	• Do not use worn blades • Use the right blade for the material • If possible, use a guide to keep the saw on the cutting line	• Eye protection • Gloves • Ear protectors • Secure the workpiece • Always cut away from yourself
Circular saw	• Damaged lead and plug • Damaged insulation • General damage • Faulty on/off switch/trigger • Check guard is in place and working properly	• Do not use worn blades • Use the right blade for the material • Use a guide to keep the saw on the cutting line	• Eye protection • Gloves • Ear protectors • Secure the workpiece • Always cut away from yourself if using a hand-held version

Table 4.7: Using saws safely

Hammers

The hammer is probably the most basic item in the electrician's toolbox and is used for a wide variety of jobs, such as knocking in nails and driving a chisel through masonry or wood. Hammers are generally graded according to the weight of their head.

Type of hammer	Description and uses
Claw hammer	The claw hammer is two tools in one. The hammer face is used for general work and the claw is used for removing unwanted nails and other levering tasks. To remove a nail, the head is gripped between the two blades of the claw and the top of the hammer used as the fulcrum of a lever action.
Ball pein hammer	Like the claw hammer, the ball pein is a combination tool with a standard striking face and a rounded second face. The ball pein was originally designed for driving in rivets.
Lump hammer	A lump hammer is a block-headed tool usually used in combination with a cold chisel. This is a heavy tool so make sure you select a weight that is comfortable. A large two-handed version of the lump hammer is the sledge hammer, used for breaking up concrete and for general demolition type work.
Scutching hammer	Used by electricians for cutting chases, the scutching hammer has replaceable blades or teeth that fit into each end of its head.
Soft face mallet	The mallet can be either wood or a nylon-type head. Mallets are used mainly in combination with wood chisels. The softer versions are also used for dressing cables, particularly mineral insulated, although they must be used with care when carrying out this type of job because striking the cable too hard could damage or misshape its outer sheath.

Table 4.8: Different types of hammer

Progress check 4.2

1 Why is a retractable bladed knife a safe option for the electrician?

2 In which direction should the teeth of a hacksaw blade face?

3 What type of saw is used for cutting holes in plasterboard walls?

 Safe working

Always remember the following safety guidelines.

• Never use a hammer if the head is loose.

• Do not use if the handle is cracked or greasy.

• Check the impact face of a hammer – if the edges are cracked or curled back, do not use.

• Wear gloves and eye protection when using a hammer – nails can break off or fly out if mishit.

Tool	What to check for	Optimum performance	PPE and other safety advice
Hammer	• Damaged handle • Loose head • Damaged head	• Use the right hammer for the job • Hold the hammer near the bottom of the handle – let the weight of the hammer head do the work • If you are buying a hammer, check the different weights and choose one that is comfortable	• Gloves • Eye protection • Secure the workpiece • Foot protection

Table 4.9: Using hammers safely

Chisels

Hammers are often used in combination with chisels. There are three basic types: wood chisels, cold chisels and bolsters.

Type of chisel	Description and uses
Wood chisel	Wood chisels are very sharp and used for woodworking jobs. They should be looked after so that their blades are protected from damage and blunting.
Cold chisel	Cold chisels are used for masonry work such as knocking holes into brick or block work. The cold chisel types are: • flat chisel • cross-cut • half-round • diamond point.
Bolster chisel	Bolster chisels have a flared blade that can be used for cutting straight edges into masonry, for example, the edges of a wall chase. Bolsters are also used for levering up floorboards.

Table 4.10: Different types of chisel

Tool	What to check for	Optimum performance	PPE and other safety advice
Cold chisel and bolster chisel	• Broken, split, ragged impact point at the top of the chisel • Damaged hand-guard	• Use the right chisel for the job • Keep blades sharp	• Gloves • Eye protection • Foot protection
Wood chisel	• Damaged handle • Blade is loose in the handle	• Use the right chisel for the job • Keep blades sharp • Store in a chisel roll or other protective covering; this prevents blunting of blades and corrosion	• Gloves • Eye protection • Foot protection

Table 4.11: Using chisels safely

Progress check 4.3

1 Give two safety reasons not to use a hammer.
2 What are two uses for a claw hammer?
3 What is a bolster chisel used for?

Files

Available in different sizes and levels of coarseness, files are generally used by an electrician for smoothing metal after cutting or drilling. Table 4.12 shows the main types of file.

Type of file	Description and uses
Flat file	Used for flat surfaces, tapers in thickness and width at the front. A thinner version is known as a warding file and is used for filing narrow slots.
Hand file	Has a safe edge with no teeth that can be used up to a shoulder without marking it.
Round file	Sometimes called a rat-tail file for opening up holes or filing internal radii.
Square file	Used for opening up square holes.
Three square file	Used for filing square corners, sometimes called a triangular file.
Knife file	Used for tapered narrow slots and for very precise work.
Half round file	Used to file the inside of curved surfaces.

Table 4.12: Different types of file

Using a file

When using a file to smooth metal, remember the following points.

1 The workpiece should be secure and at a comfortable height. If it is a small workpiece, secure it in a vice.

2 Grip the file with two hands – one on the handle, one on the tip of the file.

3 Apply pressure on the forward stroke of the file. The file teeth face forward and only cut when the file is moved in that direction.

4 Do not apply pressure on the return stroke. This is because it will have no effect on the workpiece and can also damage the file.

5 If the piece to be filed requires a coarse file to correct, finish off with a fine version.

Figure 4.4: Correct file use

Tool	What to check for	Optimum performance	PPE and other safety advice
File	• Damaged handle • Loose handle	• Use the right file for the job • Do not use worn files • Adopt correct filing method	• Gloves • Eye protection

Table 4.13: Using files safely

Reamers

Cutting metal conduit leaves a jagged edge which can damage cables when they are drawn into the conduit. A reamer is a cone-shaped tool with integral cutting blades. It is inserted into the end of the conduit and twisted so that the blades cut off any jagged edges and smooth the inner lip of the conduit.

Gripping tools

There are times when parts or fixings need to be held tight, for example for securing or tightening or loosening. Electricians might need a gripping tool for tightening lengths of conduit or securing conduit fittings to the ends of pipe.

Type of gripping tool	Description and uses
Slip grip	Slip grips have a number of different names such as pipe grips or pterodacs (because they look a little like the head of a pterodactyl!). The jaws can be quickly adjusted to fit the workpiece. This is done by sliding the lower jaw along a slot in the base of the upper jaw.
Mole grips	Mole grips are adjusted using a screw in one of the handles. The grips can be pre-adjusted to fit the workpiece or adjusted once they are fitted around the workpiece. Once adjusted to size, the handles can be squeezed together so that the gripping jaws lock onto the workpiece.
Stilson	Stilsons are a heavy-duty adjustable gripping tool. Like mole grips, they can be pre-adjusted. They have only one handle and are not intended to grip parts. They can be used for tightening by placing onto the workpiece, pulling it round, releasing the tool then sliding the jaws round for a re-grip.

Table 4.14: Different types of gripping tool

Spanners

Spanners are used for securing and loosening nuts and bolts. They are supplied in set sizes, although adjustable spanners are also available. It is important to use the correct-sized spanner. If it is too small it will not fit the bolt or nut. If it is too big it will slip, damaging the corners of the bolt head or nut and possibly injuring your hands.

Figure 4.5: An adjustable spanner

Socket sets

A socket is a type of spanner in the form of a short cylinder. The open end is placed over the head of a bolt. The socket is operated by a ratchet handle which acts as a lever for loosening and tightening bolts. Sockets usually come in sets of various sizes.

Tool	What to check for	Optimum performance	PPE and other safety advice
Mole and slip grip, and stilson	• Damaged handle • Damaged gripping mechanisms • Damaged jaws	• Use the correct grips for the job • Set mole grips and stilsons so they grip the workpiece securely	• There is no PPE specific to using gripping tools and spanners. The main method of controlling risk is careful use
Spanner and socket set	• Worn jaws or sockets • Damage to handles	• Do not use spanners or sockets with worn jaws as they will slip. This will round the corners of the nut or bolt head and make it difficult to remove • Loosen rusted or over-tight bolts and nuts using a lubricating spray	

Table 4.15: Using gripping tools and spanners safely

Allen keys

The screws used for assembling equipment are often Allen type. The socket in the screw head is hexagonal and can only be secured using the correct-sized Allen key. These are usually supplied with the equipment and should be kept safe and labelled so that if any repair work has to be carried out, the equipment can be dismantled and reassembled easily.

Progress check 4.4

1 What is the difference between mole grips and stilsons?
2 When filing, pressure should be put on which stroke?
3 What is an Allen key used for?

Gauges

A gauge is a measuring tool and tends to be used when an extremely fine and accurate measurement is required. Examples are:

- bore gauge – measures the depth of holes
- callipers – when measuring distance between two sides of an object
- feeler gauge – measures gaps (e.g. a spark plug gap)
- vernier – measures height
- wire gauge – measures thickness of wire.

Figure 4.6: A gauge

Chapter 4

Figure 4.7: A tape measure

Safe working

Be careful when retracting a tape measure into its cassette. The tape is made of thin metal and will slide home very quickly. The edges of the tape can make deep cuts on your fingers or hand.

Safe working

Never look directly into a laser beam. The light is extremely intense and can seriously damage your eyes.

Tape measure

The tape measure is a vital part of an electrician's toolkit. The positions of accessories must be correct – in some cases an incorrectly placed switch or socket could end up hidden behind a cupboard and therefore rendered unusable. Also, conduit, trunking and cables all need to be measured to the correct length before cutting. Most tape measures are retractable types marked in centimetres and millimetres.

Levelling tools

Even though many of the accessories or items of equipment fitted by an electrician are small, they still have to be level. A crooked switch or consumer unit looks unsightly and gives the impression that it was installed by a poor quality electrician. When fitted close to straight edges or on the front of wall tiles, the fact that a switch or socket has not been levelled will be obvious to everyone who sees it. Conduit and trunking runs, and cable runs, must also be installed horizontally and vertically.

Type of levelling tool	Description and uses
Spirit level	Spirit levels are available in a number of lengths. They consist of a long wood or metal straight edge fitted with a bubble capsule. The capsule contains oil, but is not completely filled. A small space is left and forms a bubble. When the level is perfectly horizontal or vertical, the bubble will sit in the centre of the capsule.
Laser level	A modern way for marking a straight line is to use a laser level. A laser is a tightly focused beam of light. This light will always be exactly horizontal or vertical in relation to the laser projector. Once the projector is set up and adjusted to the required measurement, the light will create a perfect line on the wall or ceiling.
Chalk line	When installing long lengths of trunking, conduit or cables, a straight fixing line can be drawn on the wall using a chalk line. The chalk line body contains a coil of string which can be pulled out or rewound with a handle, like a fishing reel. The container is filled with powdered chalk, usually a highly visible colour like blue. When the string is pulled out it will be covered with the chalk dust.

Table 4.16: Different types of levelling tool

To make a chalk line for fixing, follow the directions below.

1 Mark a measurement at each end of the fixing run.

2 If there are two of you, one person will hold the end of the chalk line string on one of the marks. If you are using the chalk line on your own, insert a screw into the mark and attach the end of the string to the screw.

3 Walk back with the chalk line, unwinding the string as you go.

4 Hold the string on the other mark and pull it tight.

5 Pull the string out and 'ping' it against the wall.

6 The chalk-encrusted string will make a thin line on the wall between the two marks.

The plumb line is another method for marking a vertical line. It consists of a string weighted at one end. The string will always hang vertically. Plumb lines are seldom used now.

Progress check 4.5

1 How does a spirit level work?

2 What tool can be used for marking a straight line along a complete wall?

3 Which tool uses light to provide a straight fixing line?

Electric drill

One of the tools in constant use by electricians is the electric drill. Although mainly used for drilling holes in masonry and metal, many drills can also be fitted with a variety of attachments for sanding and cutting, etc.

Extra low voltage power tools are required on most construction sites. This is because hand-held tools get hard usage in this environment and damaged power tools can lead to electric shock. Drills, therefore, will be either 110 V or battery operated.

Hammer drill – drilling masonry

Most modern electric drills are fitted with a hammer function. This is percussive action which drives the drill bit back and forth as it cuts its way at a fast speed into a wall or concrete floor. This gives the drill bit extra force and also helps clear debris from the drill hole. A masonry bit has a hardened tip which is shaped like an arrow head. When drilling a hole in brick or similar material, jiggle the drill bit backwards and forwards to clear the hole of dust.

The hammer function must only be used when drilling masonry, never for metal or wood.

Figure 4.8: Screwdriver drill bits

Figure 4.9: A masonry bit

Pillar drill

Pillar drills are fixed drilling machines used mainly in workshops and factories. The workpiece is secured in place using a vice which can be adjusted so that the workpiece is in the exact drilling position. The drill is then lowered onto the workpiece using either a handle or wheel. Modern pillar drills are automated or even programmable so that they can carry out repeated operations with the same degree of accuracy.

A clear, see-through guard must be fitted to the drill to prevent eye injuries from flying **swarf** and other debris thrown out from the drill area.

> **Key term**
>
> **Swarf** – fine chips or filings produced by machining.

> **Activity 4.1**
>
> Carry out a visual safety inspection on a selection of tools. Demonstrate this to your tutor. Explain what you are looking for, e.g. damaged handle, blunt edges, damaged leads and plugs on power tools.

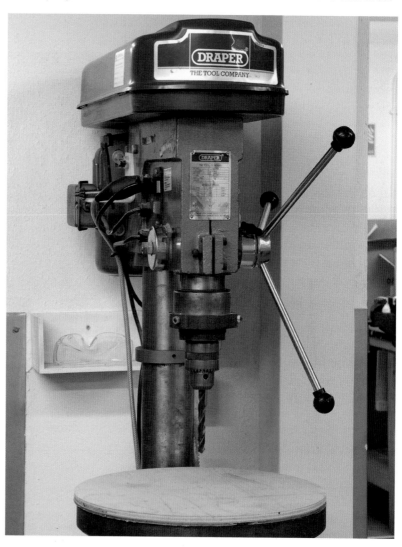

Figure 4.10: A pillar drill

Drilling metal

Metal should be drilled at a slow speed using a twist bit. Make sure the workpiece is secure before attempting to drill. A centre punch should be used first to avoid skidding and to create a more accurate drill hole. The pointed end of the centre punch is pressed against the centre of the drill site. Some centre punches are then struck with a hammer to make a small dent in the metal surface. Spring-loaded centre punches are also available. The drill bit is placed into the dent made by the punch.

Figure 4.11: A centre punch

A hole saw should be used for cutting large holes in metal. Hole saws, or tank cutters as they are sometimes called, are metal cylinders with teeth cut into the open end. The saw is then screwed onto a mandrill. The mandrill is also fitted with a twist bit which drills a pilot hole into the workpiece and which anchors the saw while cutting takes place.

Figure 4.12: Metal bit and hole saw

Drilling wood

Wood can be drilled using a metal twist bit, but there are a range of specialist wood bits available. These are spade-shaped with a sharp pilot point. Care must be taken when using these types of drill bit because they can become jammed and cause the drill itself to twist round.

If no power is available, a brace-and-bit is used for drilling wood. This is shaped like a crank shaft and fitted with auger bits (much like twist drills). A brace-and-bit is usually fitted with a ratchet handle so that it can be used in a confined space where the crank cannot be rotated fully.

Figure 4.13: Wood bit

Tool	What to check for	Optimum performance	PPE and other safety advice
Electric drill	• Damaged lead and plug • Damaged insulation and casing on the drill • Free movement of the chuck • Chuck grips the drill bit securely	• For battery drills, keep batteries fully charged and always keep a spare charged battery • Only use sharp drill bits	• Eye protection • Gloves • Ear protection • Secure the workpiece

Table 4.17: Using electric drills safely

Progress check 4.6

1 What type of drill bit is used for drilling a small hole in metal?

2 What is a pilot drill?

3 What tool can be used for drilling wood if no power is available?

INSTALLING WIRING SYSTEMS

LO3

Select materials from drawings

The main diagram the electrician will work to is the scaled layout drawing, often called a plan. From this the electrician can count up the amount of accessories and items of equipment needed for the job. Cable lengths can also be calculated using the scale and likely cable routes shown on the drawing.

Marking out

Before starting actual installation work, the job has to be marked out. The drawing is used to show where all accessories and items of electrical equipment are to be fitted. It is important to install all equipment in the correct position. If a socket or switch is not placed where it should be it could end up hidden behind a cupboard or radiator. When marking out, the position of accessories, equipment and cable drops are drawn on to the wall with a pencil or chalk.

Cables

Cables form the nervous system of an electrical installation. There are many different types of cable, each one designed for a certain environment and the job they are expected to do. There are also installation methods specific to each cable type. Some cables are equipped with their own mechanical protection while others need to be run in enclosures of some sort. Some are delicate and require careful handling while others can withstand harsh treatment and conditions. This section looks at some of the main types of cable.

Twin-and-earth

If it is necessary to run a twin-and-earth cable on the surface, then it should be clipped. When clipping a twin-and-earth it is important to run the cable in a straight line, smooth it out and make the run as neat as possible.

Check list

PPE	Tools and equipment	Source information
• Eye protection • Protective footwear • Hi-vis jacket	• Tape measure • Chalk or laser line • Spirit line • Hammer	• BS7671: 2008 • IET On-Site guide • Layout drawing

1 Mark a straight line to clip to.

2 Straighten the cable.

3 Clip at either end of run (in this image one clip is shown in place and the other is being hammered in).

4 Fill in with clips at regular intervals.

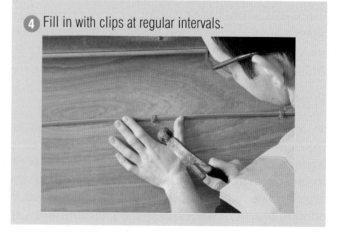

Figure 4.14: Clipping twin-and-earth cable

Stripping twin-and-earth

The earth, or circuit protective conductor (CPC), in a twin-and-earth is not insulated. The CPC can, therefore, be used for stripping the cable.

Check list

PPE	Tools and equipment	Source information
• Eye protection • Protective footwear • Hi-vis jacket	• Cable side cutters • Pliers • Cable strippers • Terminal screwdriver	• BS7671: 2008 • IET On-site Guide • Wiring or circuit diagram

① Use side cutters to bite into the end of the cable.

② Peel open the outer sheath so that the CPC can be gripped using pliers and pulled back to cut through the outer sheath. Your tutor may suggest using a retractable knife to cut the sheath instead.

③ Once the required length is reached, peel away the sheath and cut neatly.

④ Strip the insulation from the conductors using a cable stripping tool. Don't strip away too much insulation – 10 mm is about right for most terminations.

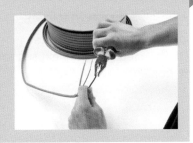

⑤ Cut the exposed cores to 1.5 times the width of the box. Too short and it will be difficult to connect, too long and it will be difficult to fix the face plate to the back box without damaging cables.

Activity 4.2

Practise stripping twin-and-earth cable. Use the method recommended and described here. No insulation should be damaged and the conductors must be inspected for pressure marks from the cable stripping tool used to remove the insulation.

Figure 4.15: Stripping twin-and-earth cable

Single cables

Single, or non-sheathed, cables have only one layer of insulation and no outer sheath. BS 7671:2008 Regulation 521.10.1 makes it clear that they must be run inside conduit or trunking. Singles are often used in industrial environments because metal conduit and trunking provides extra mechanical protection for the electrical installation.

Figure 4.16: Jointing cables ready to be drawn into a conduit

Figure 4.17: Single cables being drawn into conduit

Installing single cables

Conduit runs can be long and often include a number of angles and bends, so installing single cable into conduit is at least a two-person job. One person will usually have to pull the cables at one end of the conduit run while the other person feeds them in at the other.

The standard method for wiring conduit is to push a draw wire or draw tape into one end of the run. This tape is usually made of steel or nylon. While flexible enough to work its way round the various bends and angles, it is strong enough not to buckle as it is pushed into the pipe. Once the draw wire or tape reaches the far end and is pulled out, the cables need to be connected or 'made-off'.

A secure method to prepare a loom of cables ready to be drawn into conduit is as follows.

1 Strip one cable back about 200 mm – call it cable A.

2 Strip the other cables back to 10 mm.

3 Wrap the bared conductor of the first of these other cables neatly around the bared conductor of cable A.

4 Wrap the bared conductor of the next conductor tightly around the bared conductor of cable A. Start this at the point where the previous conductor winding finished.

5 Repeat for the remaining three cables.

6 The remaining length of bared conductor from cable A should be attached to the draw wire and wound tightly back on itself.

The advantages of this method are:

- the joint is very secure and will not give way while being pulled into a conduit
- the cables are layered rather than bunched into a cumbersome clump on the end of the draw wire.

It is important to pull all the required cables through a conduit run at the same time. Trying to install more cables into a pipe that already contains other circuits is both difficult and can also damage the existing cables.

Conduit runs will include a number of inspection boxes. It is at these points that the cables should be pulled out, then fed into the next section. The cables should be laid carefully into the box and not twisted around each other.

Magnetic effect

When current flows, a magnetic field is set up and spirals along the outside of the conductor, clockwise with the current. Because of this, a conduit or trunking run should not be filled with cables that all carry current in the same direction. The combined magnetic fields of such a set of cables will cause heat and vibration. This effect must be countered by including cables with opposing current flows. This also applies to entries into metal distribution boards and other enclosures.

Armoured cable

Fixing armoured cable

Steel wire armour can be run on cable tray using either cable ties or cleats, depending on the size of the cable. If fixing directly to a wall, use a chalk line or laser level to make sure your cable is straight and neat.

Terminating armoured cable

Armoured cables require specialist termination kits. This ensures that the cable, which is sometimes large and heavy, is securely fixed to whatever equipment it supplies or is fed from. The other purpose of the termination gland is to connect the armouring to the installation earth. The armouring is often used as the CPC, so it has to have a secure connection.

> **Progress check 4.7**
>
> 1 Which conductor can be used for stripping twin-and-earth?
> 2 What is used to pull single cables into conduit?
> 3 Why are armoured cables fitted with glands?

Check list

PPE	Tools and equipment	Source information
• Protective footwear • Hi-vis jacket	• Tape measure • Knife with retractable blade • Junior hacksaw • Adjustable spanner and grips	• BS7671: 2008 • Wiring diagram

1 Score the cable and strip the outer sheath.

2 Saw halfway through the armouring wires.

3 Bend the wires back and forth until they snap off.

4 Strip outer sheath another few mm, then push the shroud and female gland onto the cable (this can be added before stripping the cable).

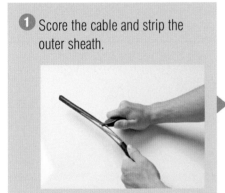

5 Push the male gland onto the cable. The dome section is seated under the armour wires.

6 Secure the female gland onto the male thread, clamping the armour wire securely.

Figure 4.18: Termination of armoured gland

Terminating conductors

There are different types of cable termination. The main ones are described in Table 4.18.

Terminal type	Termination method
Standard clamped terminal	Usually a brass cylinder in which the conductor is secured using a grub screw. If a single conductor is to be terminated, strip the insulation away to twice the required length, then bend the conductor double. This creates a larger conductor for the retaining screw to grip on to.
Post type	A threaded post onto which the conductor is secured using a set of nuts and washers. Wrap the conductor tightly round the post in the same direction as the locknut is turned for tightening (usually clockwise).
Crimped termination	A fitting that can be pushed onto the conductor, then crimped tight with a specialised crimping tool. The smaller versions are usually colour-coded – the larger ones are bare metal. These have a variety of terminal types such as a ring, spade and fork. The ring type is a good method for securing cables to post terminations.
Soldered	Often used for very small conductors in electronic circuits. The conductor is secured to the terminal using solder which is melted onto the terminal using the tip of a soldering iron. Precautions must be taken because solder can give off toxic fumes.
Push type	Some terminals are designed to allow the conductor to be pushed home. A spring-loaded clamp secures the conductor. Check that the termination is tight because a loose termination can create a hot spot.

Table 4.18: Different types of cable termination

Connecting a 13 A socket

Follow these simple guidelines when connecting a 13 A socket.

1 There are three terminals. These are:

- L terminal – for line conductor
- N terminal – for neutral conductor
- E terminal – for CPC.

2 Strip the cable's outer sheath.

3 The three cores should be cut to a length that is approximately one-and-a-half the width of the 13 A socket back box.

4 Strip the insulation on the line and neutral cores.

5 Cover the CPC with green and yellow earth sleeving.

6 Make sure:

- the outer sheath is taken into the socket box
- no copper conductor is showing
- the terminal screw is not nipping the insulation instead of the conductor.

Note: if a single conductor is connected into a terminal, bend the conductor double and squeeze tight with a pair of pliers. This makes the conductor bigger and easier to secure into the terminal.

Connecting a light fitting

The light in Figure 4.19 is being wired using the loop-in method.

- L terminal is for the blue switch wire that connects the lamp to the switch. Brown sleeving is used to identify that this blue core is NOT being used as a neutral.
- N terminal – neutral.
- E terminal – CPC.
- Loop – for the supply to the circuit. The feed to the switch is also connected into this terminal.
- Cord grip – the flex cores must be passed underneath these anchor points. The flex grip will take the weight of the lamp and the shade.
- The cable sheaths must be taken into the ceiling rose.
- Lay the conductors neatly in the ceiling rose.

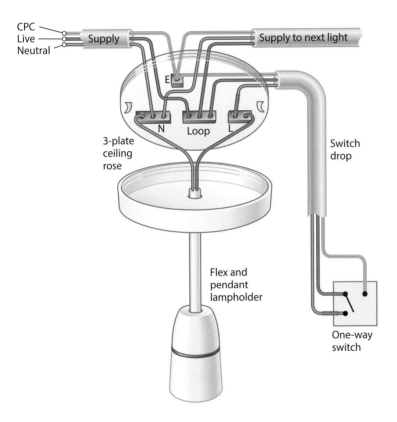

Figure 4.19: A correctly connected light fitting

Connecting a junction box

- The junction box has no specific line, neutral or earth terminals.
- The cable sheath must be taken into the junction box.
- No conductor should show outside the terminal.
- Lay the conductors neatly in the box.

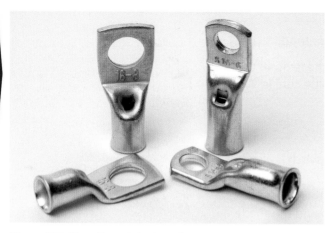

Figure 4.20: Crimped lug and crimping tool

Crimped termination

Crimped terminations can be fitted onto the end of cable conductors. They can be used to make a more secure connection into a terminal. Some terminals consist of a threaded post and a set of washers and nuts. A crimped lug like the one shown is the easiest way to connect a conductor to this type of terminal.

The correct-sized lug must be fitted to a conductor and be compressed as tightly as possible. Small crimped termination sizes are colour-coded (see Table 4.19). Larger versions are not insulated and require heavy duty compression tools.

Colour code	Conductor size
Red	1.5 mm²
Blue	2.5 mm²
Yellow	6 mm²

Table 4.19: Small crimped termination sizes and their colour codes

Progress check 4.8

1 What is a loop-in terminal in a ceiling rose?
2 Which conductor is connected to the N terminal in a 13 A socket?
3 What is good practice when terminating a single conductor into a standard terminal?

Conduit

When cutting and threading conduit, the pipe must always be secured in the conduit vice.

Cutting metal conduit

1 Measure conduit to required length.

2 Mark the pipe clearly, running the mark all the way round the pipe.

3 Cut with a hacksaw.

4 Ream the end of the pipe to remove burrs and jagged internal edges.

Threading conduit

1 Approximately 300 mm of pipe should protrude out of the vice.

2 Place the stocks and dies over the end of the pipe, then turn them clockwise using the handles.

3 Push hard with each turn until the teeth bite – small, sharp turns work best.

4 Once the dies are cutting, make about three to four full turns, then unthread the stocks and dies and remove from the pipe.

5 Smear cutting compound on the end of the pipe.

6 Continue threading until approximately three to four new threads protrude from the end of the dies (about 20 mm).

7 When threading, occasionally reverse the direction to allow swarf to drop away – this will make sure you do not get **cross-threading**.

8 Unscrew the dies, clean up the thread and ream out any burrs.

Key term

Cross-threading – this occurs when swarf interferes with the cutting process and stops components being attached by thread appearing not to be compatible.

Running thread

There are times when a new piece of conduit has to be fitted to an existing installation. If the new piece of conduit has been formed into a bend of any sort, it may not be possible to simply turn the new pipe and screw it into place. If this is the case, a running thread should be manufactured, using the following steps.

1 Cut a thread onto the new conduit – this thread should be approximately 35 mm long.

2 Screw a lock ring all the way up the thread.

3 Screw a coupler onto the thread until it is past the end of the pipe.

4 Hold the new pipe against the end of the existing conduit and bring the coupler down until it secures the two pipes.

5 Lock the coupler in place using the lock ring.

6 Some thread will be exposed – paint this thread to prevent corrosion.

Bending conduit

Conduit should always be formed using the bending equipment on the combination conduit vice and bender. The former stops the pipe from flattening and gives the bend a smooth shape. The two measurements in relation to conduit bends are:

- the back of a bend
- the centre of a bend.

90° set in metal conduit

Carry out the following to bend conduit into a 90° set.

1 Mark the required measurement on the conduit.

2 Push the conduit into the bending machine.

3 Lift the bending arm and the block.

4 Push the pipe along the slot in the top edge of the former and through the bending arm.

5 Push the pipe under the block.

6 For a centre-of-bend measurement the mark should be 3 × pipe diameter from the front edge of the former.

7 Pull the bending arm down to form the bend.

8 If the bend is a right-angle, check the bend using a straight edge and set square, or by placing the workpiece in a door frame.

If there is more than one bend, or a double set or bridge set to be made in a length of conduit, make sure that the bends are in line with each other.

Once the first set has been formed into the pipe, check that it is sitting at the correct angle before making the second bend.

Chapter 4

⚠ **Safe working**

Care should be taken when threading and bending conduit as it can be physically strenuous.

- Face the workpiece square on.
- Do not twist your back while you work.
- Do not run your fingertip around the inside of the pipe, as the burrs can be sharp and cause cuts.

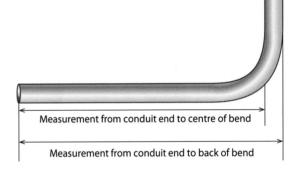

Measurement from conduit end to centre of bend

Measurement from conduit end to back of bend

Figure 4.21: The difference between back and centre bends

Forming a double set

Figure 4.22 shows a double set – to clear a 45 mm obstacle.

Check list		
PPE	**Tools and equipment**	**Source information**
• Hard hat • Protective footwear • Hi-vis jacket	• Tape measure • Scribe • Straight edge • Combination vice and conduit bender	• Layout drawing showing conduit runs • Diagram of bend with measurements

1 Bend the conduit to the required angle (conduit being bent).

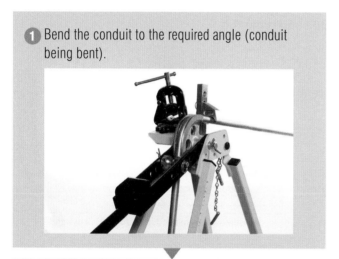

2 Lay conduit against a straight edge and measure from the straight edge. The bending point of the second angle will be the point at which the bottom of the conduit is 50 mm from the straight edge.

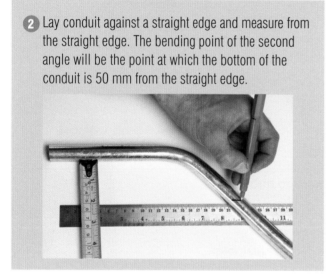

3 Insert conduit back into the bending machine and check that the first angle is in line with the bending arm.

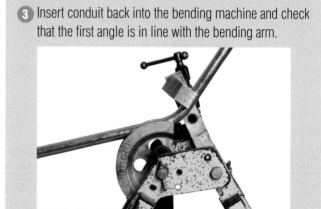

4 Bend the conduit to the same angle as the first bend.

Figure 4.22: A double set – to clear a 45 mm obstacle

Forming a bridge set

This is basically two double sets back-to-back.

1 Great care must be taken to keep the angles the same for each half of the set. One method would be to form the first double set, then lay it on a board and draw round it to create a template.

2 Check the second double set against the template.

Forming a hump set

A variation of the bridge set is the hump set, in which the top of the set is formed out of a single angle and not two as in the bridge set.

To form conduit into a hump set, follow the sequence below.

1 Bend the top angle first. This will be a 90° set.

2 Lay the conduit on the floor.

3 Measure from the apex to the point at which each 'leg' needs to be bent outwards.

4 Lay a straight edge across the two legs. Each point at which it crosses the legs should be the same distance from the peak of the apex.

5 Mark the two bending points.

6 Make the bends, being very careful to keep the conduit in line each time.

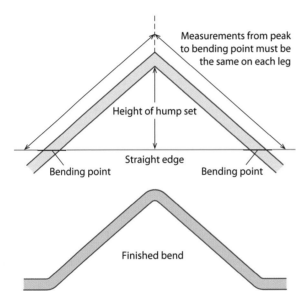

Figure 4.23: Method for marking the bending points on the hump set legs

Fixing metal conduit

Saddles are normally used to fix conduit to a surface. BS 7671:2008 states that conduit fixings must be no more than 1 m apart. A chalk or laser line should be used to mark a straight line for a long conduit run.

If it is necessary to pass conduit or trunking through an existing wall, the hole must be refilled around the containment to:

- prevent the spread of fire
- maintain the structural strength (integrity) of the wall.

Bending plastic conduit

Plastic conduit is formed into bends and sets using a bending spring. There are versions for both light and heavy gauge 20 mm and 25 mm plastic conduit. A rope or cord should be attached to the back end of the spring so it can be pulled out of the pipe once the bend is complete.

If the bend has to be formed in the conduit some way from the end, follow this sequence.

1 Mark the centre of bend on the pipe.

2 The middle-point of the spring should be held against the measurement.

3 Knot or tape the spring's draw cord level with the end of the pipe.

4 When the spring is pushed into the pipe, the tape or knot will indicate when the spring is in the correct position inside the conduit.

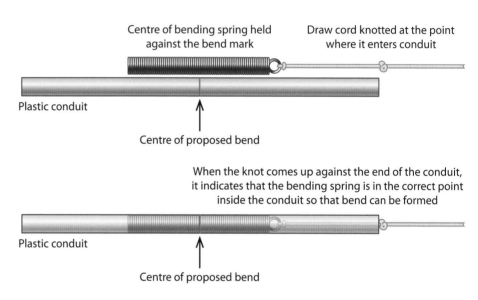

Centre of bending spring held against the bend mark

Draw cord knotted at the point where it enters conduit

Plastic conduit

Centre of proposed bend

When the knot comes up against the end of the conduit, it indicates that the bending spring is in the correct point inside the conduit so that bend can be formed

Plastic conduit

Centre of proposed bend

Figure 4.24: Measuring a spring against a bend in plastic conduit

Figure 4.25: Conduit being formed over the knee

Bending plastic conduit

Push the correct-sized bending spring into the pipe to the required position. The pipe can be formed around the knee. The bend must not be too tight, otherwise it will be difficult to pull cables round the set and could put extra strain on the conductors. There is also a chance that the pipe will wrinkle or collapse if the bend is too tight.

Bending plastic conduit is easier if the pipe is warmed before bending. This can be achieved by rubbing it with a cloth or even running it under hot water.

As with metal conduit, multiple bends should be in line so that the pipe is not twisted. Because the conduit is being formed around the knee, it is more difficult to achieve this than with metal conduit which is secured in the bending machine and can be easily sighted before continuing with the next bend. Therefore care must be taken to ensure that all the bends are in line.

Bend past the required angle because plastic conduit will try to re-straighten itself once the bending spring is removed.

Fixing plastic conduit

Plastic conduit saddles are installed in much the same way as metal conduit. As well as conventional saddles there are clip-fit type fixings similar to the type used to install copper water pipes.

Trunking

There are no specialist tools used for installing trunking, although you will need:

- hacksaw
- tape measure and set square
- drill
- twist bits and hole saws
- file
- spirit level
- chalk line or laser level.

Because it has three sides which have to be cut, metal trunking can be difficult to cut straight. This section gives helpful tips on cutting and working with metal trunking.

Cutting metal trunking

When you cut a length of metal trunking, follow the guidelines below.

1 Always secure the trunking before cutting. If possible, use a vice.

2 Insert wooden blocks into the trunking to prevent damage when tightening in the vice.

3 Measure and mark the trunking all the way round. Use a scribe as a marker. Avoid felt tip pens or permanent markers. These produce a thick line and make it difficult to achieve an accurate cut.

4 Hold the hacksaw with both hands, one on the handle and the other on the end of the frame. This will help keep the saw straight while cutting. Adopt a steady, fluid motion. Use all the blade and let the blade do the work.

5 Cut on the waste side of the line – on the side that is not going to be used for this job.

6 Because the trunking is three-sided, the cut will need to be constantly checked to make sure it is not wandering off the line.

7 File the cut edge to remove burrs and jagged metal.

Forming trunking

Although Ts and angle pieces are available, there may be times when trunking has to be formed by the electrician, for example when:

- no fittings are available
- an awkward, non-standard angle has to be formed.

Trunking can be formed by cutting, bending and riveting or bolting the bend together.

Progress check 4.9

1 What is the name of an extra-long thread cut onto the end of a conduit?

2 Which tool is used to cut threads onto conduit?

3 Which tool is used to form bends in plastic conduit?

Check list

PPE	Tools and equipment	Consumables	Source information
• Protective footwear • Hi-vis jacket • Gloves • Hard hat • Goggles	• Tape measure • Set square and scribe • Hacksaw • Electric drill • File • Twist drill bit for metalwork • Rivet tool	• 50 mm × 50 mm galvanised steel trunking • Rivets	• Layout drawing showing trunking runs • Diagram of bend with measurements

1 Cut a 300 mm length of 50 mm × 50 mm trunking. Divide the trunking into three equal sections and mark the sections using a scribe and set square.

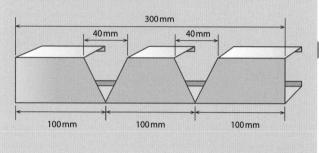

2 Measure 20 mm on either side of 100 mm marks and draw a line using a scribe. Mark angles on the back of the trunking. Ensure that you allow for the thickness of the metal.

3 Cut the triangle-shaped sections out of the trunking. Remember the following.

- Cut on the waste side of the line, that is, the side to be discarded.
- Because you are cutting two sides at the same time, you must constantly check your saw blade as you work to make sure that the saw is not wandering from the line.
- Once the section has been cut out, file the sawn edges because they will be sharp and slightly ragged. This is to both protect yourself from cutting your hand and also to protect the cables that will be installed in the trunking.

4 Cut the small lip in the open side of the trunking with a junior hacksaw, then bend the trunking and secure with a plate. Fix plate with rivets or M5 nuts and bolts.

Figure 4.26: How to cut and form metal trunking into a 90° angle

Fixing trunking

Trunking can be fixed to walls and other surfaces using standard fixings such as woodscrews, Rawlplugs and cavity fixings. For long runs use a chalk line or laser level to mark out a level run. Trunking can also be fitted to girders using girder clamps. It can also be suspended using threaded studding. In this case the trunking is secured to the studding using nuts and washers. Suspension brackets are also available.

Trunking is not designed to be installed outdoors or in damp conditions.

Activity 4.3

Practise cutting metal trunking using scrap and off-cuts. Practise:

- a straight cut on trunking lid
- a straight cut on a piece of metal trunking
- a 45° cut on lid
- 45° cuts on trunking.

Use a combination set square to test the accuracy of the cut.

Progress check 4.10

1 Give two occasions when bends have to be manufactured in trunking by an electrician.

2 On which side of a line should the blade rest when cutting trunking?

3 What is the final task to be carried out after cutting metal trunking?

BONDING MAINS SERVICES TO MAIN EARTHING TERMINALS

LO4

Chapter 3: Electrical installations technology describes the mains intake position for an electrical installation. One of the components of a typical mains position is the main earthing terminal (MET). The MET is a simple connection block into which the earth conductors from the installation are connected to the supply earth.

It is vital that the connection is sound because the earthing system provides protection against electric shock. A loose connection can result in the protection system not working and the result could be serious, even fatal.

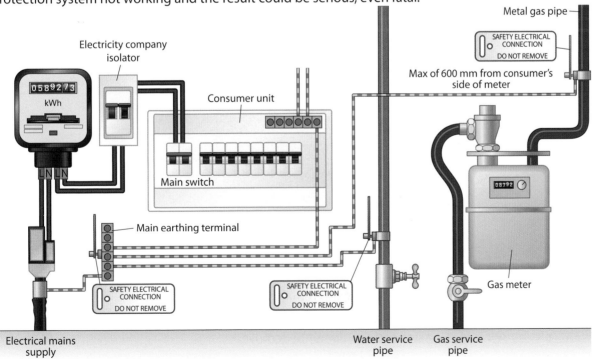

Figure 4.27: Earth connections at the mains position

Earthing conductor

The earthing conductor connects the distribution board or consumer unit earth to the supply earth.

Main bonding conductor

The main bonding conductor connects the main services to the installation and supply earth. Examples of these services are:

- water
- gas.

The main bonding conductor must be connected to the pipe as close to its point of supply as possible. In accordance with Regulation 544.1.2, 'as close as possible' means within 600 mm.

If the supply is **protective multiple earth** (**PME**) then the main bonding conductor must be selected in relation to the size of the neutral conductor. Table 4.20 shows the minimum sizes for a main bonding conductor in an installation with a PME supply.

> **Key term**
>
> ***Protective multiple earth (PME)***
> – a supply system in which the earth from the installation is connected to the supply neutral. It is also called TN-C-S (see pages 154–155).

Supply neutral conductor	Main bonding conductor
Up to 35 mm^2	10 mm^2
Over 35 mm^2 to 50 mm^2	16 mm^2
Over 50 mm^2 to 95 mm^2	25 mm^2
Over 95 mm^2 to 150 mm^2	35 mm^2
Over 150 mm^2	50 mm^2

Table 4.20: The minimum sizes for a main bonding conductor in an installation with a PME supply

Supplementary bonding conductors

Supplementary bonding conductors connect extraneous metalwork to the installation earth. Examples of the type of metalwork might be:

- metal ducting
- structural metalwork.

Supplementary conductors are also connected between pipes to ensure that they are all at the same potential. Table 4.21 shows how to calculate the minimum sizes for supplementary bonding conductors.

Connection	Size
Between two exposed conductive parts	Not less than the size of the smallest CPC connected to the exposed conductive part
Between an exposed conductive part and an extraneous conductive part	Not less than half the size of the CPC connected to the exposed conductive part
Between two extraneous conductive parts	Not less than 2.5 mm^2

Table 4.21: How to calculate the minimum sizes for supplementary bonding conductors

Bonding clamps

Supplementary conductors and main bonding conductors are connected to pipework using bonding clamps. An example is shown in Figure 4.28. These must be pulled tight around the pipe. If there is paint on the pipe then it should be scraped off to expose the bare copper underneath. The clamp should be fitted to the bare metal.

The bonding clamp includes a label warning that it is part of the earthing system and should not be removed.

Figure 4.28: A typical bonding clamp

Activity 4.4

Obtain a copy of BS 7671:2008, turn to Part 5 and look up all the references to earth bonding. Find out the following.

1 On which side of a gas or water meter must a bonding connection be made?
2 What are the minimum sizes for buried earth conductors?
3 What material should the conductors of main earthing conductors be?
4 What do the requirements say about the connection of a main bonding conductor to pipework?
5 What must **not** be used as an earth electrode?

Progress check 4.11

1 At which point are all the installation earth conductors connected to the supply earth?
2 Which conductor connects gas and water supply pipes to the installation earth?
3 Which conductor connects extraneous metalwork to the installation earth?

INSPECTING AND TESTING A DEAD ELECTRICAL INSTALLATION

LO 5&6

Once an electrical installation has been completed it should be inspected and tested. This is to make sure that the installation works correctly and is safe. *Chapter 8: Electrical installations – inspection, testing and commissioning* describes the tests required for electrical installations in detail. However, Table 4.22 shows the basic procedure and the instruments required.

Test	Purpose	Test instrument	Acceptable reading
Visual inspection	Check for damaged cables and fittings and incorrect wiring methods.	None	n/a
Continuity of protective conductor	Because the protective conductor does not normally carry current, it has to be tested to ensure there are no breaks either from cable damage or loose connections.	Low reading ohmmeter	Low reading checked against tables B1 and B2 in IET Guidance Note 3
Insulation resistance	Proves the integrity of the cable insulation.	Insulation resistance tester	$>1\ M\Omega$
Ring final circuit tests	A set of tests to ensure: • the continuity of the ring final circuit • polarity • secure connection of conductors.	Low reading ohmmeter	Low readings
Polarity	Test to make sure that conductors are connected into the correct terminals.	Low reading ohmmeter	Low readings
Functionality	Final check to make sure that the circuit is working correctly. This will include light switches, start and stop buttons and tests to confirm that an RCD operates correctly.	RCD tester	RCD operates within the time specified for the device and for the installation

Table 4.22: Basic procedure and instruments required for inspecting and testing a dead electrical installation

Progress check 4.12

1 Why should an electrical installation be inspected and tested?

2 What instrument is used to test for continuity of protective conductors?

3 What is the minimum acceptable reading for an insulation resistance test?

Activity 4.5

1 Working in teams, construct a ring final circuit consisting of three 13 A sockets. These do not have to be fixed to a board but wired together using singles cables (perhaps even scrap cable).

2 Once complete, carry out the full set of ring final circuit dead tests: end-to-end, r1 + rn and r1 + r2, and insulation resistance.

3 Introduce a fault into your circuit.

4 Retest to see how the fault has shown up on your test instrument.

5 Swap your circuit with another team and try to find the fault by carrying out the full set of ring final circuit tests on their circuit.

Knowledge check

1 A slot can be cut into metal trunking using a:

  a circular saw

  b jigsaw

  c wood saw

  d junior hacksaw

2 The cutting wheel on a grinder is made from:

  a carborundum

  b aluminium

  c steel

  d iron

3 Why must extra low voltage tools be used on construction sites?

  a To reduce risk of fire

  b To reduce risk of electric shock

  c To reduce risk of short circuit

  d To reduce risk of mechanical damage

4 Which hammer is used with a cold chisel for cutting masonry?

  a Ball pein

  b Mallet

  c Scutching

  d Lump

5 Which type of diagram is used for counting up amounts of materials needed?

  a Assembly drawing

  b Layout drawing

  c Blueprint

  d Block diagram

6 Drawing the positions of equipment and cable runs on walls before starting the work is called:

  a a site survey

  b marking out

  c a quantity survey

  d laying out

7 The length of cable cores left at a socket or switch box for connection should be:

  a 1.5 times the width of the box

  b 2 times the width of the box

  c 2.5 times the width of the box

  d 0.5 times the width of the box

8 Which of these cables is normally run in conduit?

  a Twin-and-earth

  b Mineral insulated

  c Main bonding conductor

  d Singles

9 What colour is a crimp lug for a 2.5 mm conductor?

  a Yellow

  b Red

  c Blue

  d Green

10 The main bonding conductor should be connected:

  a as far from the point of supply as possible

  b to E terminal in a 13 A socket outlet

  c as close to the point of supply as possible

  d to the loop-in terminal in a ceiling rose

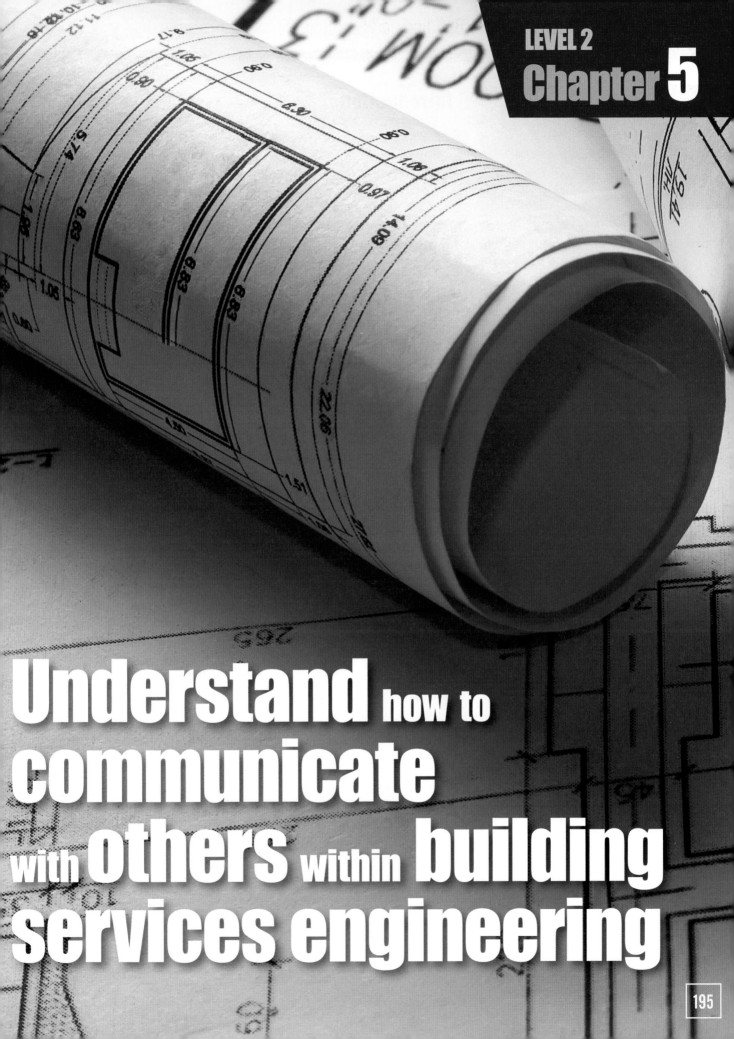

Understand how to communicate with others within building services engineering

Chapter 5

This chapter covers:

- the construction team members and their role within the building services industry
- applying information sources in the building services industry
- communicating with others in the building services industry.

Introduction

For a building services engineer to work in a consistent, efficient way with others on site, it is essential to be able to communicate well. This may mean understanding exactly who is on site but also what specific role each person has. It is easy to make assumptions about a person's role but this could be misleading and end up in an expensive misunderstanding, or even danger to those on site.

In this chapter you will look at the various methods of communication that an engineer in the building services industry can expect to use.

THE CONSTRUCTION TEAM MEMBERS AND THEIR ROLE WITHIN THE BUILDING SERVICES INDUSTRY

LO1

You will come into contact with many people when you are working on site; it is very important to understand exactly what they do and what they do not do.

Key roles of the site management team

A construction project is a very complex and involved process that requires managing all the way through from initial planning to final handover to the customer. The site management team has overall responsibility on site for the people working on the project. This is because, on every construction site, there is a team of people with different roles and responsibilities. Each part of this site management team manages an aspect of the installation and may have other teams below them with further managers looking after specific tasks.

Figure 5.1: A construction team

However, the client has overall responsibility as they hire the architect and sign the contract with the main **contractor**.

It is the responsibility of the main contractor to employ sub-contractors and specialist companies to build what the client wants.

Key term

Contractor – an individual or a company that has a specific job function within a project.

Chapter
5

Architect

Buildings need to be designed so they can be used for specific functions. The architect's responsibility is to carry out the design based on a client's request but also make sure that it is within the rules and regulations set out for that type of building. An architect may also call upon the services of more specialist design engineers or consultant engineers if the project is very large or complex. He or she will be employed directly by the client and may also oversee the whole construction project on their behalf.

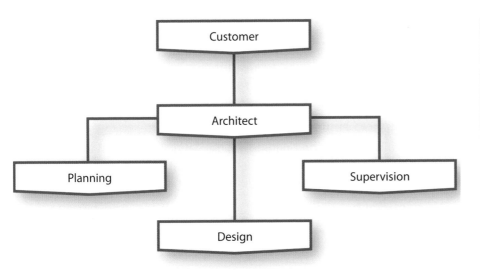

Figure 5.2: An architect may need to work with several teams

Figure 5.3: An architect must liaise with many people

Project manager or clerk of works

These two roles can be quite similar but generally a project manager will have a specific project management qualification such as PRINCE 2 or APM (Association of Project Managers). These formal qualifications may be required to progress to senior roles such as senior project manager or project director, or to work on specific contracts within large corporations. Some government contracts will insist on the project manager holding this level of formal qualification. A highly trained project manager will be able to run most types of project without being a specialist in the subject they are delivering – solid project management principles are all they require.

A clerk of works will have day-to-day responsibility for quality assurance. This may involve monitoring the quality of workmanship and materials. The architect may also give them responsibility to issue instructions for changes. For very large construction projects, there may be a clerk of works/project manager for different aspects such as electrical work or air conditioning/ventilation. There may also be a need for an overall project director who manages a team of project engineers/managers.

Structural engineer

Not only must a new building look good but it must be designed so that it is safe and able to withstand the weather and continual use that it is meant for. It is the responsibility of the structural engineer to make sure this is the case. Structural engineers:

- provide technical advice
- select the most appropriate materials for the job
- inspect the property to check the condition
- inspect foundations
- inspect damage.

Structural engineers are highly trained, the highest level being a chartered structural engineer.

Surveyor

A surveyor will work closely with an architect and structural engineer to assess the condition of a building. The surveyor will have knowledge of, and be able to advise on, the construction and materials used in a building. He or she will also be aware of the up-to-date regulations and commercial aspects of the building such as market value for insurance or rebuild requirements.

Building services engineer

Imagine a fully functioning building. All the services that exist within that building have been designed, installed and maintained by a building services engineer. The range of services is shown in Figure 5.4.

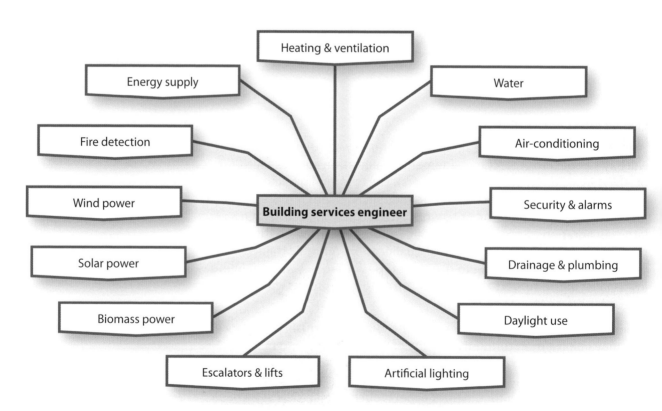

Figure 5.4: A building services engineer is responsible for many services in a building

More information can be found on the website of the Chartered Institution of Building Services Engineers: www.cibse.org.

Quantity surveyor

The costs of a new installation can vary but it is the job of the quantity surveyor to monitor, control and take action if costs change. The quantity surveyor will be involved at the beginning of a project by creating an initial bill of quantities from the design plans. This will then be used to set the budget for construction and will be monitored during the project roll-out.

Buyer

Professional buyers are often part of a large contract process. The buyer's responsibility is to source the most cost-effective products for the job. Increasingly this may also involve other issues to meet customer demands; for example a customer who is very environmentally aware might demand only ecologically sustainable products or locally sourced products to reduce their carbon footprint. Customers might also specify that only carbon neutral products can be used.

Estimator

Before a contract can be agreed between a customer and the contractor doing the work, initial costs need to be decided. The estimator will be responsible for producing the costs that will go into the customer tender response.

Case study

During the 2012 Olympics, cost-efficient LED technology was used to make patterns and text around the 80,000-seat Olympic stadium. In total 70,000 LED modules were fitted next to seats. Each LED panel contained nine LEDs.

1 What other types of lights were used in the Olympic park?

2 Use the Internet to find the final construction costs for the main Olympic stadium.

Figure 5.5: Seating at the Olympic park used LED technology

Contracts manager

Large contracts require a contracts manager. A contracts manager will be involved at the beginning of a contract when it is being set up and at the end when it is handed over as completed to the customer. The contracts manager is also responsible for the day-to-day management of the contract engineers. This can involve managing the ongoing contract engineer costs of the project, making sure they are closely monitored against any budgets.

Construction manager

Sometimes also referred to as a construction project manager (CPM), the construction manager will have overall responsibility for planning, coordinating and controlling a major building project. This responsibility includes the start of the build project and only ends once the building has been handed over, completed, to the client. This role is different to that of a project manager (see page 197), who only has responsibility for the construction project.

Principal contractor

On construction sites, work that takes over 30 days or 500 man hours is considered 'notifiable work' under the rules of CDM (Construction Design and Management) law. With construction work of this size, a principal contractor must be involved and as soon as they are appointed, details need to go to the Health and Safety Executive (HSE).

The principal contractor has legal duties for certain aspects of project management, contractor engagement and workforce engagement. An example of this says: 'Contractors must provide the workforce with the required training and information to carry out work safely. The principal contractor must check that this happens and every worker is provided with an induction on health and safety and training.'

Progress check 5.1

1 On a large construction project, who is the customer's main point of contact?
2 What are the main responsibilities of the contracts manager?
3 What does a quantity surveyor do?

Key roles of individuals who report to the site management team

Sub-contractor

For specialist work a contractor may need another company to come in and complete specific tasks for them. These sub-contractors may be specified by the client due to their personal preferences or they may have won a tender process issued by the main contractor. Clients often have specific specialist companies in mind due to their reputation, specific skills or if they have worked successfully with them before. Electrical installation companies will generally be a sub-contractor.

Site supervisor

On all major building projects there will be a site supervisor to oversee all day-to-day activities. Their role will be to carry out regular inspections of security and the environment and take action where required. The site supervisor may have several other roles that are unique to each site, but main responsibilities may include:

- ensuring the site is run in compliance with all company and statutory health and safety regulations
- induction of all temporary workforce and contractors on site, including visitors
- running team meetings and liaising with different groups and management on site.

Trade supervisor

Within the specific trade areas a day-to-day supervisor is required to oversee that particular area. For electrical, the trade supervisor would be a fully qualified electrician or senior electrician grade. The trade supervisor will be aware of what stock and materials are on site and available as the job progresses. An amount of liaison will form part of their role to make sure any special instructions from the contracts engineer or architect are implemented once the relevant paperwork (such as variation orders) has been processed.

Trades

Groundworker – at the beginning of a construction contract, particularly on a new building, there is a requirement to prepare the ground. Groundworkers are responsible for preparing the site foundations and ground before other trades come in to complete their tasks. This could involve digging foundations, trenches, footings or simply clearing an area in preparation for construction.

Figure 5.6: Groundworkers must prepare the ground at the beginning of a construction contract

Figure 5.7: A joiner at work

Bricklayer – following on from groundwork, the bricklayer is responsible for building the structure before other trades start their work.

Joiner – on most construction sites, wooden structures need to be built, including door frames, doors, stud work and window frames. This is the responsibility of the joiner, a specialist carpenter.

Plasterer – ideally, when all conduit, cable chasing and first fixing of electrical back boxes has been completed, a plasterer will cover over this containment. The plasterer may need to put up plasterboard first and then it is their responsibility to finish off the wall to a good level so that decorators can sand, paint or put up wall covering. Careful liaison between trades is required at this stage to make sure the socket back boxes are clearly marked, or they may disappear forever!

Tiler – tiles are another type of finish for a wall. It is the responsibility of the tiler to fit the tiles near the end of the construction stage. This stage will often coincide with the decoration stage when all the walls have been finished.

Electrician – there are different job functions within the general term 'electrician', as shown in Table 5.1.

Heating and ventilation (H&V) fitter – heating and ventilation fitters are specialists that install systems to control the **ambient temperature** in a building. As technology changes and becomes more available, more buildings are moving towards climate control. The developing technology has seen trades getting closer and closer together, with many of the most sought-after engineers in building services having multiple qualifications and skills.

Gas fitter – in most industrial, commercial or domestic installations there will be gas as a source of energy. Gas may be installed for heating or hot water purposes. Registered engineers will need to be involved in the installation and maintenance of all gas fittings and appliances. Historically,

Key term

Ambient temperature – the temperature of the air where equipment or cable is installed.

Electrician grade	Technician	Approved electrician	Electrician	Electrical improver	Labourer
Qualifications	L3 electrical certificates + NVQ L3 + 5 years' experience	Been through apprenticeship or structured training + NVQ L3 or minimum L2 technical certificate and AM2	Been through apprenticeship or equivalent training program. NVQ L3 or L2 technical certificate and AM2	A registered apprentice; completed FE electrical course	No specific qualifications apart from the site requirements for health and safety
Duties and responsibilities	This is a highly skilled operative capable of giving installation advice and solving electrical issues and problems	A skilled operative that can work without immediate supervision; the operative will also have a good working knowledge of the regulations and codes of practice	Be able to carry out all electrical work proficiently and with a good working knowledge of the regulations and codes of practice	Apprentice or operative that has completed training; they are able to carry out supervised work up to first fix but not final connection or test and inspection	Assist in all electrical first-fix work as directed by qualified electrician, such as installing cables, chasing and construction of containment

Table 5.1: Electrician grade specific responsibilities

gas fitters would have been registered with the Council for Registered Gas Installers (CORGI). This registration proves the engineer has a level of competence and experience relevant to the work they are contracted for. This Accredited Certification Scheme has now been taken over by the Gas Safety Register and is run by the Health and Safety Executive.

Figure 5.8: A decorator will be involved in the later stages of an installation

Decorator – responsible for painting, filling in cracks in plaster, sanding window frames, doors and any wood finishes. The decorator will mainly be involved in the last stage of an installation and will generally have to work around all the other trades, negotiating when he or she can come in and finish off an area.

Chapter
5

Figure 5.9: Visitors to a construction site could be doing various tasks

Key roles of site visitors

Visitors to site are commonplace and may come from different local authority or government departments, interested in various aspects of the construction work. Even though these visitors may be wearing a suit and seem important, they should always be challenged and asked to go through the same site induction program and signing-in process as other contractors or visiting trades. All visitors should expect this as the site will become unsafe if there are two sets of rules. If an incident occurs on site you will need to know who is there and who can be accounted for by the emergency services.

Building control inspector – building control inspectors provide advice on request to help building projects comply with the building regulations. Inspectors normally work directly for the local authority. An inspector will come to the site at the beginning of the build project to make sure all the regulations have been considered and planned for. Inspectors can visit a building site as many times as required to satisfy any questions they may have but will generally come again at the completion stage to sign off the installation.

Water inspector – very occasionally you might see a water inspector from the Drinking Water Inspectorate (DWI). These inspectors are in place to maintain the quality of 'wholesome' water or drinking water provided by the water companies. They have the power to go to water companies and test at source but they also have the power to go on site and measure at the tap. They will be testing for microorganisms, chemicals and metals, as well as what the water looks and tastes like.

HSE inspector – health and safety law is enforced by inspectors from the Health and Safety Executive (HSE) or inspectors from the local authority. Inspectors have the right to go into any workplace without notice to look at the management of health and safety and compliance with the law. Inspectors will come to a site if an incident has been reported or they can just simply turn up unannounced. Inspectors have the right to close a site down for very serious health and safety breaches or issue a notice for improvement.

More about the process of HSE inspections and the consequences is covered in *Chapter 2: Health and safety in building services engineering*.

Case study

During the Olympics over 40,000 people worked on site, with a peak of 13,000 at any one particular moment. The Olympic Delivery Authority set about with the aim of being the safest construction project ever. At one point, the project went over 3 million hours without a single reportable injury.

Electrical services inspector – approved companies registered with, for example, the NICEIC, NAPIT or ECA will have visitors from time to time to inspect the level of work. Domestic installers that self-certify their own work under self-certification schemes will require periodic checks. These inspectors are there to check on the quality and adherence to the wiring regulations. Electrical services inspectors are qualified in the electrical trade with the experience to make judgements on installation standards and workmanship.

For a company to comply with a scheme they must be inspected on a regular basis depending on the terms of the scheme. Local authority inspectors may also be involved and come to a site at various stages of a build to make sure the regulations are followed, or for final sign-off just before completion and handover.

Activity 5.1

1 You are a self-employed electrician and you have been injured. The injury has meant you are unable to work for eight days. Go to the HSE website and find out what you must do.

2 You are the manager of a construction company and you are going to build five houses on the open space next to a college. Go to the HSE website and download and print a 'notification of construction work' form. Complete it for the construction of five new houses.

Progress check 5.2

1 What are the main responsibilities of the site management team?

2 When can a building control inspector visit a site?

3 What are the responsibilities of a building control inspector?

4 What powers do HSE have when visiting a construction site?

Chapter
5

APPLYING INFORMATION SOURCES IN THE BUILDING SERVICES INDUSTRY

Types of statutory legislation and guidance information that apply to the industry

Legislation means literally 'to make law'. Laws exist to protect the rights of individuals and businesses. Each country has different legislation and processes but the UK is heavily influenced by European law. An area that is particularly affected is employment law. In the UK the laws that govern what you must and must not do are influenced by three main sets of legislation. These are:

- common or contract law
- UK legislation
- European legislation and judgements from the European Court of Justice.

Some of the relevant legislation in this country that sets out the minimum rights and responsibilities of an employer and employee are given below.

The Race Relations Act 1976 and Amendment Act 2000

When originally passed, the Race Relations Act 1976 made it unlawful to discriminate on racial grounds in relation to employment, training and education, the provision of goods, facilities and services, and certain other specified activities. The 1976 Act applied to race discrimination by public authorities in these areas, but not all functions of public authorities were covered.

The 1976 Act also made employers vicariously (explicitly) liable for acts of race discrimination committed by their employees in the course of their employment, subject to a defence that the employer took all reasonable steps to prevent the employee discriminating.

The Commission for Racial Equality (CRE) proposed that the Act should be extended to all public services and that vicarious liability should be extended to the police. The main purposes of the 2000 Act were to:

- extend further the 1976 Act in relation to public authorities, thus outlawing race discrimination in functions not previously covered
- place a duty on specified public authorities to work towards the elimination of unlawful discrimination and promote equality of opportunity and good relations between persons of different racial groups
- make Chief Officers of police vicariously liable for acts of race discrimination by police officers
- amend the exemption under the 1976 Act for acts done for the purposes of safeguarding national security.

The Sex Discrimination Act 1975

The Sex Discrimination Act 1975 makes discrimination unlawful on the grounds of sex and marital status and, to a certain degree, gender reassignment. The Act originated out of the Equal Treatment Directive, which made provisions for equality between men and women in terms of access to employment, vocational training, promotion and other terms and conditions of work.

The Equal Opportunities Commission (EOC) has since published a Code of Practice. While this is not a legally binding document, it does give guidance on best practice in the promotion of equality of opportunity in employment, and failure to follow it may be taken into account by the courts.

The Sex Discrimination Act was amended to ensure compliance with the Equal Treatment Directive, with all changes being effective from April 2008. The definition of harassment is extended so that if, for example, a male supervisor makes disparaging comments about women, it is no longer a defence to show that he makes similar comments about men. In addition if someone witnesses sexual harassment of a colleague, they can bring a claim of harassment themselves if they felt it made their work environment intimidating.

Employer liability has also been extended to make organisations liable if they haven't taken reasonable steps to prevent harassment by a third party such as a visitor or customer.

Employment Relations Act 1999 & 2004

The 1999 Act is based on the measures proposed in the White Paper: Fairness at Work (1998), which was part of the Government's programme to replace the notion of conflict between employers and employees with the promotion of partnership.

As such it comprises changes to the law on trade union membership, to prevent discrimination by omission and the blacklisting of people on grounds of trade union membership or activities; new rights and changes in family-related employment rights, aimed at making it easier for workers to balance the demands of work and the family, and a new right for workers to be accompanied in certain disciplinary and grievance hearings.

The Employment Relations Act 2004 is mainly concerned with collective labour law and trade union rights. It implements the findings of the review of the Employment Relations Act 1999, announced by the Secretary of State in July 2002, with measures to tackle the intimidation of workers during recognition and de-recognition ballots and provisions to increase the protections against the dismissal of employees taking official, lawfully organised industrial action.

The Human Rights Act 1998

The Human Rights Act 1998 covers many different types of discrimination – including some not covered by other discrimination laws. However, it can be used only when one of the other 'articles' (the specific principles) of the Act applies, such as the right to 'respect for private and family life'.

Rights under this Act can only be used against a public authority (such as the police or a local council) and not a private company. However, court decisions on discrimination will generally have to take into account what the Human Rights Act says.

The main articles within this Act are: right to life, prohibition of torture, prohibition of slavery and forced labour, right to liberty and security, right to a fair trial, no punishment without law, right to respect for private and family life, freedom of thought, conscience and religion, freedom of expression, freedom of assembly and association, right to marry, prohibition of discrimination, restrictions on political activity of aliens, prohibition of abuse of rights, and limitation on use of restrictions on rights.

The Employment Rights Act 1996, Employment Acts 2002 & 2008

Subject to certain qualifications, employees have a number of statutory minimum rights (such as the right to a minimum wage). The main vehicle for employment legislation is the Employment Rights Act 1996 – Chapter 18. If you did not agree certain matters at the time of commencing employment, your legal rights will apply automatically. The Employment Rights Act 1996 deals with many matters such as:

- right to statement of employment
- right to pay statement
- minimum pay
- minimum holidays
- maximum working hours
- right to maternity/ paternity leave.

The Employment Act 2002 amended the 1996 Act to make provision for statutory rights to paternity and adoption leave and pay. The Employment Act 2008 makes provision for the resolution of employment disputes including compensation for financial loss, enforcement of minimum wage and of offences under the Employment Agencies Act 1973 and the right of trade unions to expel members due to membership of political parties.

The Race Relations Act 1976 (Amendment) Regulations 2003

The Race Relations (Amendment) Regulations 2003 modify the Race Relations Act 1976.

- Indirect discrimination on grounds of race, ethnic origin or national origin is extended to cover informal as well as formal practices.

- The concept of a 'Genuine Occupational Requirement' is introduced for situations where having a particular ethnic or national origin is a genuine requirement for the employment in question.

- The definition of discriminatory practices is extended to cover those who put particular groups at a disadvantage, rather than only those where there is proof that a disadvantage has been experienced.

- The Act is extended to give protection even after a relationship (such as employment in an organisation, or tenancy under a landlord) has finished.

- The burden of proof is shifted, meaning an alleged discriminator (such as an employer or landlord) has to prove that he or she did not commit unlawful discrimination once an initial case is made.

Case study

Avneet works in a small electrical company that specialises in electrical work for the healthcare sector. Over the last year, whenever Avneet has gone to the depot to collect stores she has suffered verbal racial abuse from her manager. Because this is a small company and she loved her work so much, she said and did nothing. After a year the comments became more frequent and Avneet eventually put in a grievance against her manager. The company upheld the grievance and admitted that racial abuse had occurred. The manager was reprimanded but no further action was taken by the company. Avneet later found herself on a poor performance target list and was put forward for redundancy even though her work had always been highlighted as 'good practice' in her yearly reviews.

1 What laws may have been broken here?

2 What course or action could have been taken

Racial and Religious Hatred Act 2006

The Racial and Religious Hatred Act 2006 makes inciting hatred against a person on the grounds of their religion an offence in England and Wales. The House of Lords passed amendments to the Bill that effectively limit the legislation to 'a person who uses threatening words or behaviour, or displays any written material which is threatening ... if they intend thereby to stir up religious hatred'. This removes the abusive and insulting concept, and requires the intention – rather than just the possibility – of stirring up religious hatred.

Employment Equality (Religion or Belief) Regulations 2003

These regulations make it unlawful to discriminate against, harass or victimise workers because of religion, or religious or similar philosophical belief. They are applicable to vocational training and all aspects of employment, recruitment and training.

Equality Act 2006

This amends the Sex Discrimination Act and places a statutory general duty on employers when carrying out their functions to have due regard to the need to eliminate unlawful discrimination and harassment, and also to promote equality of opportunity between men and women.

Equality Act 2010

From 1 October 2010, the Equality Act replaced most of the Disability Discrimination Act (DDA). However, the Disability Equality Duty in the DDA continues to apply. The Equality Act 2010 aims to protect disabled people and prevent disability discrimination. It provides legal rights for disabled people in the areas of:

- employment
- education
- access to goods, services and facilities including larger private clubs and land-based transport services
- buying and renting land or property
- functions of public bodies, such as the issuing of licences.

The Equality Act also provides rights for people not to be directly discriminated against or harassed because they have an association with a disabled person. This can apply to a carer or parent of a disabled person. Also, people must not be directly discriminated against or harassed because they are wrongly perceived to be disabled.

Protection from Harassment Act (PHA) 1997

Harassment is defined as any form of unwanted and unwelcome behaviour (ranging from mildly unpleasant remarks to physical violence) that causes alarm or distress by a course of conduct on more than one occasion (note that it doesn't need to be the same course of conduct).

The PHA is the main criminal legislation dealing with harassment, including stalking, racial or religious intimidation and certain types of anti-social behaviour such as playing loud music. Significantly, the PHA gives emphasis to the target's perception of the harassment rather than the perpetrator's alleged intent.

Employment Equality (Age) Regulations 2006

The Employment Equality (Age) Regulations 2006 is a piece of legislation that prohibits employers from unreasonably discriminating against employees on grounds of age.

Data Protection Act 1998

Information about people can also be subject to abuse, so the Information Commissioner enforces and oversees the Data Protection Act 1998 and the Freedom of Information Act 2000. The Commissioner is a UK independent supervisory authority reporting directly to the UK Parliament. It has an international role as well as a national one.

The principles put in place by the Data Protection Act 1998 aim to ensure that information is handled properly. Data must be:

- fairly and lawfully processed
- processed for limited purposes
- accurate, adequate, relevant and not excessive
- not kept for longer than is necessary
- processed in line with your rights
- secure and not transferred to other countries without adequate protection.

'Data controllers' have to keep to these principles by law.

Regulations

Regulations are the practical part of a law that tell the participants exactly how they are meant to act. The regulations give control or rights and also allocate responsibilities. A regulation can put legal restrictions on a company, contractor or individual that will be monitored by a government body. Regulations can also be self-regulated by industry, such as through a trade organisation.

Examples of regulations that exist to control and allocate responsibilities are the Electricity at Work Regulations (EAWR). The 33 regulations are specifically designed to reduce the risk of death or injury from electricity in the workplace. The EAWR covers voltages over 1000 V a.c. and 1500 V d.c. More detail about the most important Electricity at Work Regulations is covered in *Chapter 2: Health and Safety in building services engineering.*

BS 7671: Requirements for Electrical Installations (The IET Wiring Regulations)

The 'regs', as they are known by technicians in the electrical industry, are not actually regulations and they differ slightly in their approach when compared to the Electricity at Work Regulations. Whereas EAWR cover all electrical work, BS 7671 only covers specific areas of electrical installation and maintenance. BS 7671 is periodically reviewed and new editions are brought out every few years depending on the scale of the changes and amendments required. The early editions over 100 years ago consisted of only a few pages with details of what could and could not be done. The latest 17th edition is very different and relies heavily on the technician to have the skill, knowledge and understanding of the document. A fully qualified electrician will be able to read and interpret the wiring requirements. BS 7671 has been developed this way to allow flexibility with all the developing technologies and products available but still stay within the guidelines dictated by EAWR.

Detailed work with BS 7671 will be covered later, but it is worth becoming familiar with the layout and how to use the wiring requirements.

BS 7671 – layout

The wiring requirements are laid out in a specific and logical way to enable a code to be given to a specific piece of electrical information. This code can then be referred to and used. The document is broken down into seven parts containing all the main guidance (see Table 5.2). Any additional information or reference material is then found in one of 16 appendices. For work in areas not covered by BS 7671, there are other standards that the technician needs to be qualified and trained in.

Activity 5.2

Look up the Kitemark website and write down the main advantages a customer has when employing a Kitemark electrician.

Figure 5.10: The British Standards Kitemark

Part	Title	Summary
1	Scope, objectives and fundamental principles	Amongst other things, this sets out what BS 7671 covers and, more importantly, what it does not cover. As a technician, if you are asked to complete a piece of installation work, look here to see if you are trained to be there.
2	Definitions	This describes a majority of the terms and electrical phrases used within the industry – a very good starting place if you are not sure of a term being used. 'Definitions' is also regulation 2 of EAWR.

Part	Title	Summary
3	Assessment of general characteristics	As an electrician, you know if you are meant to be on site and you also understand all the terms – so now you need to understand the location you are about to work in. Understanding the location means understanding the purpose of the installation and understanding the supply and how it is delivered, and also how compatible the system is likely to be with the equipment likely to be used. Other considerations are the maintainability of the system, the need for safety services and considering the continuity of service.
4	Protection for safety	Protection for safety is a natural progression from the previous part. This part covers the essential requirements to protect against electric shock, including basic and fault protection of persons and livestock. Other protection is also required and covered within this part, including protection against burns, protection against fault or over-currents and protection against voltage and electromagnetic disturbances, to name a few.
5	Selection and erection of equipment	You are now ready to start the design stage of your installation. You need to select the correct and most appropriate equipment and conductors before you start the installation work. This part details exactly how to make the correct choices based on conditions and design criteria.
6	Inspection and testing	Once the installation has been designed, selected and installed the next natural progression stage is to inspect and test. All the details required to make sure your installation is safe can be found in this part. Detailed specific guidance on test and inspection is also found in Guidance note 3 – a separate guide just on this subject. Through all of these parts reference is made to various appendices. This allows the main parts to remain, as far as possible, uncluttered with tables, forms and details.
7	Special installations or locations	For all normal domestic work the first six parts are adequate but there will be occasions when the installation falls slightly outside of this. Special installations or locations will cover these instances. They may include bathrooms, wet rooms, solar, pools, construction sites, marinas, medical locations and many more.
A1–14	Appendices	The 14 appendices contain all the detailed material required and referred to throughout the seven parts of the wiring requirements. Design information or simply forms can all be found in this part.

Table 5.2: Summary of BS 7671, The IET Wiring Regulations

British Standards

Technical committees and specialist boards set up by the British Standards Institute (BSI) approve British Standards – of which there are currently 27,000. Having a British Standard means that certain specifications are met while encouraging manufacturers to standardise the way they produce products. The Kitemark can be used to indicate something is certified by the BSI but it is mainly used for safety and quality. A competent electrician can also apply for a Kitemark from BSI.

Harmonisation with Europe has meant that some BS numbers have been superseded by the EN number (**BS EN**), showing that a British Standard has been replaced by the European Norm.

Key term

BS EN – British Standards European Norm.

Examples of the standards related to electrical installation work can be found in Appendix 1 of BS 7671, Requirements for Electrical Installations. Some of the relevant standards are shown in Table 5.3.

BS or EN number	Title	Reference in the 'regs'
BS 88	Low-voltage fuses for use by mainly authorized persons (mainly industrial applications)	432.4 Tables 41.2, 4, 6
BS 3036	Specification for semi-enclosed electric fuses (ratings up to 100 amperes and 240 volts to earth)	Table 41.2,4
BS EN 60698	Specification for circuit-breakers for over-current protection for household and similar installations	Table 41.3, 6

Table 5.3: Relevant standards in Appendix 1 of BS 7671

Codes of practice

The main purpose of a code of practice is to provide extra help and guidance at work when applying current legislation. Codes of practice are specifically aimed at areas that require more help and a little more detail. Examples of this are earthing, lightning, explosive atmospheres (such as petrol stations and factories) and lighting levels in offices. The approved codes of practice are government-approved advice that is sanctioned by both the HSE and the Secretary of State. Although failure to comply does not mean you are officially breaking the law, if it came to a criminal court case and you were found not to be following them, you would have to prove you had an equally good alternative. In some cases statutory regulations may be accompanied by codes of practice approved under Section 16 of the HASAWA 1974. These codes do have a legal status and this is defined fully in Section 17 of HASAWA.

Although codes of practice do not have to be British Standards, most of them are. Some of the codes of practice are listed in Table 5.4.

Standard	Description of code
BS 5266 part 1–10, also BS EN 50172	Code of practice for emergency lighting
BS 5839 parts 1–11, also PD6531:2010	Code of practice for system design, installation, commissioning and maintenance of fire detection and alarm systems for buildings
BS 8519	Code of practice for the selection and installation of fire-resistant cables and systems for life safety and firefighting applications
BS EN 62305, 4 parts	Code of practice for protection of structures against lightning
BS 7375:2010	Code of practice for distribution of electricity on construction and building sites
BS 7430:2011	Code of practice for earthing
BS 7671	Requirements for electrical installations
BS EN 60079	Electrical apparatus for explosive gas atmospheres
BS EN 50281 and BS EN 61241	Electrical equipment for use in the presence of combustible dust
BS 7909	Code of practice for temporary electrical systems for entertainment and related purposes
BS EN 60335–2–96	Electric surface heating systems
IEC 60479 Parts 1–4, also PD6519	Guide to effects of current on human beings and livestock
BS EN 60529	Specification for degrees of protection provided by enclosures (IP codes)

Table 5.4: A sample of codes of practice

Key fact

BS 6651:1999 has been replaced by BS EN 62305 but you may still find it referred to.

Figure 5.11: You may need to protect your installation against lightning strikes

There are also codes of practice and guidance on electromagnetic compatibility and electrical appliances. You will specifically need to know about three codes that deal with special electrical considerations when dealing with:

- lightning protection
- explosive atmospheres
- earthing.

Code of practice for lightning protection (BS EN 62305)

Lightning protection is a very specialist area of electrical work. For this reason it has its own code of practice that gives extensive design and installation advice. BS 7671 only goes so far with guidance and then the code of practice adds the full detail.

Lightning can occur as positive polarity and negative polarity. This means there can be a build-up of electrons at the cloud or the ground level with a positive charge at the other. When the charge difference becomes large enough, an electrical conductive path can be created and the strike happens between the cloud and the ground. Strikes also occur in the atmosphere between clouds and within clouds but generally this is not an issue for the building services engineer. The current that can be generated on a lightning strike can rise to 120 kA and 350 coulombs of charge but on average strikes are 30 kA and transfer 15 coulombs of charge and 500 MJ of energy. This is a considerable amount of electricity that your installation will need to be protected against.

Code of practice for equipment in explosive gas atmospheres (BS EN 60079)

There are many electrical installations that can be considered dangerous for various reasons and which require special consideration. Locations that are exposed to flammable atmospheres require specific guidance and their own code of practice. Such locations might include petrol stations, flour

Key fact

Health and Safety Guide 41 is a specific guide produced to help with petrol station installations.

Key terms

Zone – this can be considered '0' where the explosive atmosphere is continuously present down to a '2' if the explosive atmosphere, if it occurs, will only be present for a short time.

Intrinsically safe – this means there can be no risk of sparks from installation equipment. This grade of equipment is assessed and given a special mark to show it is suitable for the grade of area where it is installed.

Figure 5.12: Petrol stations can present an electrical hazard

mills or other dusty atmospheres, or anywhere where flammable products are present or processed in some way, e.g. paint. Petrol is the most obvious flammable substance that an electrician might be required to work near, as it explodes easily. Any electrical equipment used in specific **zones** in a petrol station must therefore be considered **intrinsically safe**. The zone will be determined by the expected level of explosive gases present.

Battery rooms still exist in large commercial/industrial locations to add an extra level of back-up power for operation-critical applications such as emergency lighting or telecommunications. A by-product of battery acid is hydrogen which is highly flammable and requires special consideration if electrical or any other installation work is to take place nearby.

This area of installation is very specialist due to the risks involved. Because of this, a new regulation was created in 2002 called Dangerous Substances and Explosive Atmospheres Regulations (DSEAR).

Case study

On 11 December 2005 a number of large explosions occurred at Buncefield oil storage depot in Hemel Hempstead. One of the explosions registered 2.4 on the Richter scale and to this day counts as Britain's most costly disaster. Following the prosecution of five companies, the main reasons for the disaster were shared in a report by the HSE so that other companies could learn from the experience. Poor maintenance and management were amongst the main factors that led to the explosion and large fire.

Figure 5.13: Explosions at the Buncefield oil storage depot

Code of practice for earthing (BS 7430:2011)

This approved code of practice gives basic principles and guidance on earthing of electrical supply systems. The purpose is to make sure electrical installations and any connected equipment can operate safely to protect human life and livestock. Because this area is very complex, other standards are also referred to within this code of practice.

Other guidance has been produced to help the electrician such as GS38 and Guidance note 3 for test and inspection.

Guidance from manufacturers

Manufacturers' documentation and guidance notes are a very valuable source of information to a building services engineer. This information must always be fully read and understood before installation of specific equipment. Manufacturers' guidance will be supplied with all the equipment a building services engineer has to work with. A manufacturer has a duty of care to provide all the installation fitting details that might be required. This may include any specific details on the positioning of equipment, load and weight restrictions, fittings, heating or cooling requirements or simply what the product contains and is made of. Manufacturers' guidance is essential reading for the engineer and cannot be skipped over, even though it might take several hours to read and digest fully. Reading all the manufacturers' information may save hours, days and a considerable amount of money so you should get used to taking the time to do this.

Two key sets of information are generally produced by manufacturers in the form of technical information or data sheets and operational/functional instructions. The data sheet will contain as much information as the manufacturer believes you will need. Operational/functional instructions are used in the installation process but will also be useful for the customer after the installation has been completed. This type of information will form part of the handover pack to the client.

With technology developments and the boundaries becoming blurred between the different specialisms and trades, installation work has become very complex. Often, further advice will be required and engineers will no longer be able to rely on basic information and guidance. Engineers should expect to have regular contact with manufacturers. There will be cases when the installation specification cannot be completely matched and advice about tolerances on the installation method will be required. Any extra information you receive as the installing engineer is potentially valuable to the customer and must be kept and presented to the client as part of the handover pack. Remember you may not be the next engineer that works on site. This extra information will be very important if your company is to look good and maintain a good relationship with the client.

Working practice 5.1

As part of a large installation, a building services engineer has been contracted to run in the electrical supply for a telecommunications room in a major department store as part of a refurbishment. The building services engineer has been given basic instructions, a set of drawings and some manufacturers' equipment guides. The electrician has fitted this type of equipment before and decides to save time and start straight away with the installation work in the next available project management slot while the store is closed. The installation work is completed and the job is handed over to the test and inspection team. It is found that the equipment specification has changed from the previous installation and there is now a requirement for a clean uninterrupted separate earth all the way back to the main board. The issue: there are no more installation slots available in the project plan because the shop needs to be closed so the cable route can be opened up again. The project go live date has to be postponed a week and the customer is not happy as customer service is affected and the store director receives complaints. The building services company has to provide compensation under the terms of their contract.

1   What makes a contract binding?
2   What else can happen if a contract is broken?

Key term

Controlled document – a changing document that is reviewed and then reissued by the project manager so that all the people involved in a project have exactly the same information and are using the most current version.

The purpose of information used in the workplace

The amount of documentation that a building services engineer can expect to see on site is vast. A lot of companies use standard format documents and forms such as risk assessments, but many companies create their own to meet their specific needs or ways of doing things. For large projects documentation must be controlled so that all the engineers and trades involved on site have the same common information. For this reason a project manager or project director will issue **controlled documentation**.

Any changes that are required will then have to go through the project manager in a process called 'change control' (note that project changes also go through a change-control process). As an engineer you will receive your instructions via documentation generally. You will also have to produce or complete documentation throughout the job as it is very rare that you will be working in complete isolation. You will always have to pass information on to either the customer or to other groups of engineers/site operatives so they can continue and finish their stage of the installation or repair. For some types of information it is acceptable to pass the detail on verbally – for other information, which needs to be tracked, a written document is the preferred way of communication. Wherever possible it is a good idea to write information down – this is considered best practice.

Job specifications

When an installation is planned, certain details are required so that any engineer or operative knows exactly what it is they are expected to do. A job specification will give that detail. Job specifications will also be used in conjunction with many other types of documents.

A specification can be in many forms based on the purpose. You might have a general specification that details the materials, construction, weight, colour and power requirements of a particular product. A performance specification will be slightly different and may include detail on how well a product performs in different conditions.

Input voltage:	110 to 220 V AC 50 to 60 Hz
Output voltage:	13.5 V DC
Maximum load:	5 A
Regulation:	Better than 2.5% at full load
Weight:	1.5 kg
Dimensions:	155 × 85 × 65 mm
Connector:	Standard IEC
Compliance:	CE

Figure 5.14: Example of a job specification

Plans/drawings

As an engineer in the building services industry you will come into contact with many different types of drawings, plans and diagrams. Each one will give slightly different information to build up the whole picture of what needs to be achieved. Some of this information will only be required during the installation and some will be required on an ongoing basis for maintenance and installation growth. The need for engineers to be familiar with all types of plans and drawings is key to a successful installation or finding and repairing faults.

Assembly drawings

Assembly drawings are very common and are used for complex machines to show how something is put together. A control panel or a motor might have an assembly drawing. An assembly drawing is also sometimes known or referred to as an exploded diagram and is very good at showing all the components and the order in which they are put together. As you can imagine, if you had to take a boiler control panel to pieces to replace a control board, knowing the order and how the panel is put together and taken apart will save a considerable amount of time and effort. An assembly drawing is also a very handy way to identify why you have one bolt left after you have put something together!

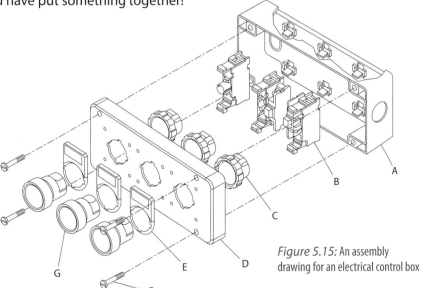

Figure 5.15: An assembly drawing for an electrical control box

Wiring diagrams

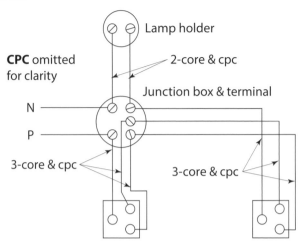

Figure 5.16: A wiring diagram

Although wiring diagrams do not specifically use circuit symbols, they do show the actual layout of a circuit with the physical connections. For this reason they are a very common diagram that you as a building services engineer will use.

Layout drawings

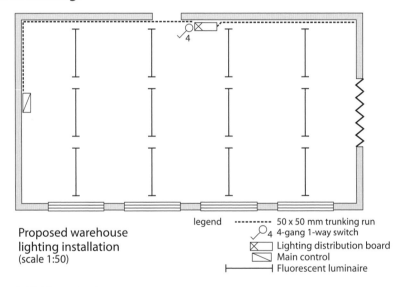

Figure 5.17: A layout drawing

A layout drawing shows exactly how everything in the installation is laid out. This is often used in combination with other forms of diagram to give more information that might help clarify the installation. A layout drawing will be based on the site drawings provided by the architect. A common practice is to use the standard BE EN 60617 graphical symbols for electrical power, telecommunications and electronic diagrams. However, if these diagrams are produced by a computer-aided design software package there can be some variation in the symbols used. Layout drawings are drawn to scale so that measurements can be taken off for installation positions of specific electrical equipment and points. The main symbols you will use are shown in Figure 5.18.

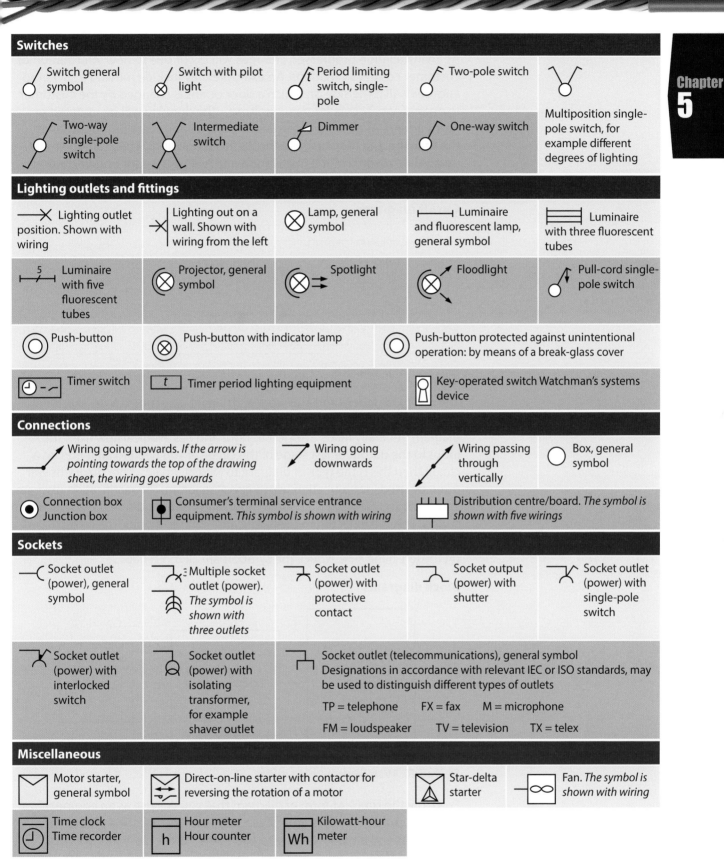

Figure 5.18: The standard symbols used in installation drawings

As fitted drawings

As an installation progresses, changes will occur due to new information or customer contract variations. An example may include moving a socket to another wall if the side on which a door opens is changed by the customer to enable easy access for new furniture – the drawing will be updated or notes added. At the end of the job all these changes are put together on the drawing and they become the 'As fitted drawings'. These are the most up-to-date records of the installation and will be used for ongoing work such as maintenance or upgrades.

Circuit diagrams

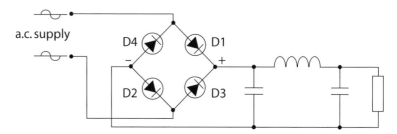

Figure 5.19: A circuit diagram

The ability to understand and work with circuit diagrams is a key skill for all engineers as they show how to connect circuit components. This type of diagram generally reads from left to right and has a logical flow from the input to the output or last stage of the circuit. If you need to fault find on a circuit then this is the type of diagram to look at. Again, other diagrams will often be used in conjunction with a circuit diagram. The rectification diagram shows the logical steps from left to right. The right-hand side shows the input supply. This is connected to the rectification bridge diodes. The next part of the circuit is the smoothing capacitors and then final connection to the load on the left-hand side. Notice the symbols are all BS EN 60617.

Block diagrams

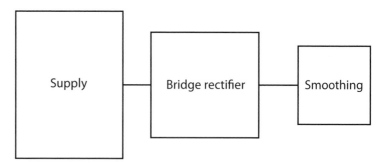

Figure 5.20: A block diagram of a bridge rectifier

Probably the simplest form of diagram that you will use as an engineer is the block diagram. This type of diagram shows stages in a process or operations and how they relate to each other. Block diagrams are very useful to get a general understanding of how something works. An example is the block diagram for the bridge rectifier. Figure 5.19 gives the actual components and how they are connected but it is difficult to understand exactly what is going on without some kind of basic overview.

The block diagram simplifies this so the different stages of rectification can be understood. This would be very helpful when trying to fault find in this circuit as a particular condition of the fault will point to a particular stage of the rectification. By identifying a particular stage of the rectification, the engineer can focus on one area of the circuit and this leads to quicker fault finding. A block diagram will not contain any detail of connections or wiring and is not a scale drawing.

Schematic diagrams

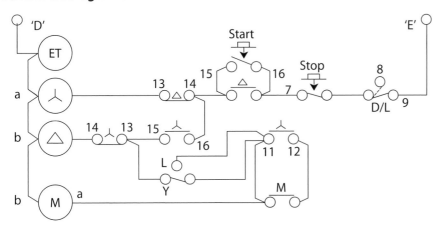

Figure 5.21: A schematic diagram

Often used in conjunction with a wiring diagram, a schematic diagram shows how a circuit is intended to work. Contactor controls are often best shown using schematic diagrams as they can seem quite complex. The schematic diagram in Figure 5.21 gives an indication of how the circuit is made up and also the logical connection of start/stop switches.

Work programmes

An important part of installation work is understanding what is required in stages. A work programme will give you all the work required in a construction project and show the interrelationships between the different construction activities. A works programme is also referred to as a programme of works. This information can feed into a project plan or be part of the main project planning process.

Job sheets

Job sheets can also be known as job cards or works instructions. A typical job card is shown in Figure 5.22.

A job card gives details about a specific task. The job card will be given out to an engineer to perform a task, typically giving the time allocated and most of the information required for successful completion of the job. Once the job is finished, the job card is completed and handed back to the relevant supervisor. Job cards are often automated computer applications held on laptops or handheld devices. A job is defined, issued to an engineer, completed and then handed back to the job controller. This type of process allows larger tasks to be broken down, allocated to a range of differently skilled operatives and then carefully managed.

Job Sheet	**Evan Dimmer** Electrical contractors
Customer	Dave Wilkins
Address	2 The Avenue Townsville Droopshire
Work to be carried out	Install 1 x additional 1200 mm fitting to rear of garage
Special conditions/instructions	Exact location to be specified by client

Figure 5.22: A typical job sheet

Delivery notes

Communicating with suppliers and wholesalers effectively makes the difference between success and failure. Arranging for stock and materials to turn up on site at a specific time slot can be fraught with issues. On a large construction site, deliveries are constantly coming in and if the received goods are not carefully checked, a particular part of the project might not go ahead on time and contract penalties will be issued to sub-contractors. A method of checking the delivered stock is the delivery note. This needs to be checked off against the order list or purchase order by the authorised electrician or supervisor. Any variations between the delivery note and the purchase order need to be confirmed with the delivery person and then the wholesaler, so that any missing materials can be arranged, reordered or reworked.

Delivery note		**A. POWERS** *Electrical wholesalers*
Order No.		Date
Delivery address 2 The Avenue Townsville Droopshire		Invoice address Evan Dimmer Electrical Contractors
Description Thorn PP 1200 mm fit fitting 1.5 mm T/E cable	Quantity 1 50 m	Catalogue No.
Comments		
Date and time of receiving goods		
Name of recipient		Signed

Figure 5.23: A delivery note for electrical equipment

Architect's Instruction

Issued by: John Smyth
Address: Church Lane
Employer: J. G. Associates
Address: South Easterly
Contractor: A. Spark
Address: Crowland
Works: New build dwelling
Site address: Southend

Job Reference: KL/Zx2319

Variation Order No: 0002
Issue date: 12/01/13

Under the terms of the above mentioned contract, I/we issue the following instructions

	Office use: approximate costs	
	£ omit	£ add
1. Reroute ring final circuit on ground floor to new architect plan	00.00	200.00
2. 12 additional sockets on ground floor to new specifications	00.00	300.00
3. Under floor kitchen heating	00.00	600.00
Approximate costs	00.00	1100.00
Signed: John Smyth		

Figure 5.24: A typical variation order

Variation orders

The customer is always right, but on the odd occasion when things do not completely go to plan, extra work may need to be ordered. This extra chargeable work is over and above the original contract agreement and must be paid for by the customer. A variation order is a good way to make sure there are no misunderstandings or conflicts when the final bills are due to be paid as all extra work is accounted for. Variation orders are controlled by a senior member of the site team that may include the lead project manager or architect. Any work over and above an agreed contract must be requested through a controlled process and a variation order issued for the work.

Working practice 5.2

A team of engineers start work on a school dormitory refurbishment at the beginning of the summer holidays. The programme is fairly tight with only a six-week time slot to remove all the old wiring and fuse boards. Within the six weeks, the team have to also install new sockets to all 20 bedrooms, new lights and wiring to all rooms, put up cable basket to go above a new false ceiling and fully test and commission the work before handing over to the other trades. The other trades include:

- carpenters, who are putting in new door frames and skirting
- plasterers, who are giving the whole building a skim
- tilers, who are retiling the four bathrooms
- gas fitters and plumbers, who are installing a new boiler and bathrooms
- ceiling contractors for the new ceilings throughout the three-storey building
- carpet fitters, who are levelling all floors before fitting special lino and carpet.

The customer site manager is in constant contact with all the trade supervisors and trades. When the site manager inspects one of the student dormitory rooms she decides there are not enough sockets for the amount of beds in a room and asks, while the dado trunking is being cut, for three extra sockets to be put in for each room. The site manager discusses this with the supervisor and promises to sort it out with the project manager. This never happens but the work is still done by the engineers. When the time comes for the electrical stage to be handed over to the ceiling specialist, the contract is two weeks later than agreed. The ceiling contractors had allowed two weeks to complete the task before they move on to a contract somewhere else. A new contractor is required at short notice. The project is late and more expense has been incurred. Because no variation order was issued and no contractual evidence was available, the electrical sub-contractor had to pay the penalty as they had breached their contract.

1 From the trades listed, in what order should the work be carried out?

2 Who should have overall responsibility to make sure the project is delivered on time and within budget?

Time sheets

Time Sheet				**Evan Dimmer** Electrical contractors		
Employee				Project/site		
Date	**Job No.**	**Start time**	**Finish time**	**Total time**	**Travel time**	**Expenses**
Mon						
Tue						
Wed						
Thu						
Fri						
Sat						
Sun						
Totals						
Employee's signature			Supervisor's signature			
Date						

Figure 5.25: A typical time sheet

As a member of staff you will often be required to complete a time sheet. Larger companies run electronic schemes but the principles are the same. As an engineer you will often be working away from your official company office. You will need to keep your employer informed of your work on site. A time sheet helps your employer keep track of your work on site, including hours, extra expenses incurred, any travel time and the actual job address you are working at. Time sheets are regularly completed and handed in to the employer so that contracts can be tracked and project costs amended to reflect what is actually happening on site. The time sheet is the main method by which an employee gets paid. Different types of installation work might also mean overtime work – work outside of the normally contracted hours. Extra work may mean special payments are made and again the time sheet is the method by which these payments are made. Some companies will not actually pay an employee until the time sheet has been completed, so treat them with respect and always complete them when told to!

Policy documentation

Policy documentation can come from many different sources. A government policy comes from a law being passed, and then instructions follow detailing how something must be done or not done. Alternatively, a company will have their own policy documentation to guide employees.

Most successful companies are unique because they have a unique proposition that makes them marketable. This leads to many companies having their own unique policies. These policies define how the company operates in different areas of their business or how they want their staff to work. A policy statement may state how a safe working environment is achieved or there could be a company policy stating that only certain ecologically friendly products can be used.

As a contractor or sub-contractor, your manager, supervisors or company directors will be aware of specific customer policies and will manage or negotiate how you work with these.

The purpose of information given to customers

Quotations

For a contract to be won, a quotation must have been given to a customer. This can be via a tender process or simply a request from a customer for a quote for a specific piece of work. Large contracts often go through a formal tender process that involves a customer writing out a requirement formally and then asking a list of companies to bid for the business. The response to tender will not only give the costs in the form of a quotation but highlight any extra reasons why that particular company should win the bid. The quotation contained within the tender response must be as close to the actual requirement as possible as this will generally form the basis of the resulting contract.

The tender process can be a very formal process with set response timescales and quote formats. Some companies keep the process very simple and have a set list of preferred suppliers that get invited to tender. Once the tender process has been completed, and the customer is happy with the quote, there could be a further process to refine the detail down to a more focused quote.

An example of a simple quotation would be an electrical company asking their wholesaler for a best price quotation on a list of required stock. This quotation may arrive in the form of a letter or an email directly from the wholesaler to the company director.

Estimates

A less formal process of getting costs for a job or stock involves asking for an estimate. A potential customer could ask you for an estimate. Although this is only an estimate, it is intended to give the customer a general idea of costs so they can set expectations for others in their company or possibly set a budget. Estimates do need to be as accurate as possible but they can only be as good as the information they are based on. A medium-sized electrical company would have a person or people with the necessary skills to go to a customer's site and estimate the costs of a job. The estimate might be the basis of the contract or used within a tender response. The customer might also be happy with the estimate and sign a contract for the work without further negotiation.

Invoices/statements

An invoice is used to help finalise payment. It is sent to the individual or company that is paying for the work to be completed.

Invoice No. 000457

A. Spark

Sparks Electrical Building Services Ltd

84 Church Lane
Southgate

INVOICE

Customer

Name	Mrs R Khan	Date	13/01/13
Address	2 Ealing Road	Order No.	0002
Town	Southall	Rep	
Phone		FOB	

Qty	Description	Unit Price		TOTAL	
10	Hours labour	£	20.00	£	200.00
1	Extractor fan	£	65.00	£	65.00
2	Electrical towel rails	£	40.00	£	80.00
20m	2.5mm twin & earth cable	£	15.00	£	15.00
25m	1.5mm twin & earth cable	£	15.00	£	15.00
7	Bonding clamps	£	1.50	£	10.50
		Sub total		£	385.50
		Shipping		£	00.00
		20.00% VAT		£	77.10
		TOTAL		£	462.60

Figure 5.26: A typical electrical invoice

Chapter 5

Key term

Contract – a legally binding agreement between two or more parties. For a contract to exist there has to be an offer, acceptance and consideration by both parties. Contracts do not need to be written.

A customer might require a fully itemised statement for their accounts department. This will show all of the costs of the project and be dated. Invoices are generally sent to a company for payment and the period for payment is agreed. Many companies have a stated payment period after being invoiced that is controlled by the company accountant. Failure to meet the payment date after invoicing can result in legal action or a review of the company credit rating.

Statutory cancellation rights

When you buy goods or services from a supplier or trader, you are entering into a contract with them. Contract law in the UK states that under this **contract** you now have a set of implied rights known as statutory rights. The word 'statutory' means laws are involved and these include the Sale and Supply of Goods to Consumers Regulations 2002 and the Unfair Contract Terms Act 1977. These laws state that the consumer has the right to goods that are deemed to be of satisfactory quality, fit for the purpose they are intended for, free from faults, safe and which have a satisfactory finish and appearance.

Contracts can be cancelled at any point up until the offer has been accepted but the contractor must notify the customer of their intention to withdraw the offer. Some contracts are often time limited. This means if the offer is not accepted within a certain time, the contract is not valid. Other reasons for a contract not becoming legally binding are rejection of the contract by the customer and death of a contractor before the offer is accepted. However, it must be proved that the customer was informed of the contractor's death for the contract to be made invalid.

When a contract has actually been made, it is difficult to get out of this legally binding deal unless a breach of contract can be proved. In certain circumstances, the customer is given the right to cancel a contract over a specific period of time. This time is called the 'cooling off period' but it depends on what was purchased and how. In the UK, buying goods or services online, by phone or by mail is subject to the Distance Selling Regulations which give the consumer a seven-day cooling off period in which they can cancel without financial penalty. There are also certain circumstances which mean the cooling off period can be extended by a further three months.

Handover information

With every completed installation, a certain amount of information will need to be given to the customer so they are able to operate any new equipment or carry out basic maintenance. The handover pack should be presented to the customer and each aspect run through individually with them until they understand all the various aspects of the installation. Handover information will typically include:

- as fitted diagrams
- drawings
- operation manuals
- flow diagrams/system block diagrams
- test certificates
- maintenance contract
- technical data sheets
- any health and safety specific information.

This list is not exhaustive and will be different for each installation and customer.

Activity 5.3

You have just completed a house rewire on a three-bedroom, detached house with all the latest up-to-date technology. Write a list of what you believe would be included in a handover pack.

The importance of company policies and procedures that affect working relationships

Company working policies and procedures

As mentioned before, all companies are unique and some companies will have written down versions of their company policies. These may cover all aspects of the company operation. If your company has a set of policies, it is your responsibility to make yourself aware of the detail. A company may have induction days that include detailed training on how the company policies are put into operation.

Typical company policies may include:

- behaviour expectations, on site and off
- timekeeping
- acceptable dress code
- contract of employment.

Some policies are legal responsibilities such as health and safety and equal opportunities, but others may be dictated by the nature of the business, for example an environmental energy provision company which will only use 100 per cent recyclable products.

If you have signed a contract with a company then you need to understand the company policies and work within them.

Working within your limits

If you are trained to be able to complete a specific job and you have all the relevant equipment (including safety kit) then it is not unreasonable for you to complete the job in hand. If, however, you do not have the training or the correct equipment then you are working outside your limitations. You have seen the training requirements for the various grades of electrical engineer on page 203). The general responsibilities at each level of grade and training are detailed in Table 5.1 (on the same page). If you are asked to complete work that you know you are not qualified to do, you should not continue with the job and clarify your position with your supervisor. If something goes wrong, ultimately it will be you who pays the price – this could be a very costly experience for the sake of not clarifying a point!

Working practice 5.3

An apprentice is put in a team of electricians who are contracted to change SON sodium street lights in a large car park. The only way to change the lights is to use a cherry picker-type lift from the back of a rented truck. The apprentice is told that he must complete all the changeovers before 4 p.m. that day while the rest of the team prepare stock for another project the following day.

The apprentice and one other technician get to work but the cherry picker gets stuck and the apprentice is trapped 40 ft in the air – the rest of the team who are fully trained cannot be found. As the day comes to an end the apprentice decides to climb down but he falls, breaking his ankle. The HSE are called and the company receives a prohibition notice, a large fine and are closely monitored for the next two years for repeat offences and breaches of the Health and Safety at Work Act.

1   What training courses are available for working at height?

2   What does the HSE recommend about working at height?

Chapter
5

Supervisor and management responsibilities

Specific roles are allocated to specific members of staff within a company. Generally they are in that role because they are the most qualified and suitable for the tasks that come with their responsibilities. A supervisor in an electrical team will have the relevant technical qualifications and also the experience to make decisions about day-to-day activities. A manager is appointed following either specific management training or years of experience in the business. Some managers are there because it is their business but, either way, clear responsibilities for each role within a company exist, based on the company job description. A job description will define the role and responsibilities and the staff member will have been interviewed with those in mind (see *Chapter 11: Career awareness in building services engineering*).

A manager will understand the nature of the company's business. He or she will also be in a position to make decisions that affect the company, company contracts and its employees. The supervisor will take their instructions from the manager. These instructions will then be passed down one stage further to the employee completing the work. The supervisor will then monitor the work and the manager will monitor the supervisor's performance and the contract. If any amendments to a contract are required as it runs, the employee may need to pass information up the chain via the supervisor to get something changed. This chain of command should always be followed unless there is an immediate health and safety risk.

Activity 5.4

Use the Internet to research lighting products for a new house. The customer has requested that you only use products that are kind to the environment. Summarise the products in a short letter to your customer giving your estimate and key benefits of the products you have found.

Progress check 5.3

1 Describe how a contract is made.
2 What might be included in a customer handover pack?
3 Give three examples of when a variation order could be used.

COMMUNICATING WITH OTHERS IN THE BUILDING SERVICES INDUSTRY

Methods of communication available

Communications skills are a highly prized skill in industry. A person's ability to communicate effectively can mean the difference between success and failure on a job. Some people are natural communicators and others take a long time, if ever, to pick up the skills.

Oral communication

You are sure to know someone who communicates well verbally – so what is their secret? It is the ability to know their audience, and to understand quickly the best words, phrases and type of language to use. Some people prefer very short instructions whereas others require a full discussion to go into finer detail. But remember not to get too complicated or you will lose your message.

Engineers can be the worst offenders at using shortened terms, phrases and acronyms for tools and commonly used products. This is a steep learning curve for all new staff so make sure you ask for an explanation when a term comes up you are not familiar with. It could be very costly if you do not understand.

Activity 5.5

Some experienced tradesmen have their own terms for tools or materials that have been passed down to them. Some tradesmen use their own terms that have come about due to not listening, mishearing or misinterpreting. This is usually as a result of poor verbal communication. You can hear some bizarre things.

- 'John, pass me the 32s and the jobblers' – 32s probably indicates the gland size required to make a cable joint and jobblers comes from mishearing 'croppers'.
- 'Pass me the croppers' – cable shears (or it could mean something else if you were rewiring a farm facility).
- 'Mims, MICC, Pyro' – all refer to the same basic thing – mineral-insulated cable.

Use the experience of the people around you to discuss acronyms and terms that you have been too afraid to ask about, or the names given to electrical products, and write a list with definitions.

Written communication

Written communication is good for putting down information that needs to be referred to at a later date. This can be in the form of a letter, report, text message, fax or an email. Because of the advances in readily available technology, increasing numbers of tradesmen have advanced phones or portable data devices. These devices can process the same information as personal computers. Some manufacturers are also developing software for phones, especially for the building services industry, which makes information a lot more accessible. The different forms of written communication were covered on pages 218–228.

Activity 5.6

Write a brief instruction sheet for a new recruit on how to change a hacksaw blade.

Communicating effectively with others

Building services and construction have a very multicultural environment and you can expect to work with a diverse mix of people. There are a wide range of barriers that need to be addressed when working and communicating with different people. Common sense must prevail to overcome any awkwardness and reservations.

Figure 5.29: A diverse mix of people work in building services and construction

Language differences

If you travel abroad for work, you may find yourself in an isolated position. You cannot understand your colleagues and they cannot understand you. Signs, symbols and pointing often work – don't be tempted just to speak louder! Speak clearly, concisely and slowly as a lot of European languages have the same common roots – some words will be understood both ways. If you find yourself working with a team, and English is not the first language, take the opportunity to pick up some new words and develop your language skills. If accent or dialect is the barrier, ask your colleague to slow down a little and repeat the word.

Physical disabilities

Communicate with people with physical disabilities in the same way as you communicate with anyone else. You will be led by the disabled person but listen to the feedback you are given. There are, however, some common sense points to think about.

- If you offer someone assistance, wait until your help is accepted before you carry out the action.
- If someone has a visual impairment, make sure you identify yourself and others and use names if talking in a group.
- Call people with disabilities by their first names only if you are on first name terms with the rest of the group.
- If someone has a speech impediment, wait for them to finish speaking before you start – don't interrupt or speak over them.
- Speak directly to someone rather than going through a work colleague or interpreter.
- If someone has a physical disability, still offer your hand to shake.

Above all, relax and do not be embarrassed if you accidentally say, for example, 'Have you seen this?' to someone who has a visual impairment.

Learning difficulties

Communicating with someone with learning difficulties may have its own unique challenges which depend on the complexities of the individual's needs. For example, communicating with a person who has profound and multiple learning disabilities (PMLD) may mean that using speech, symbols or signs will not be adequate. Specialist help and training may be required in this instance. The main considerations are for you to think about your tone, voice and body language and again take guidance from the individual you are trying to communicate with. Communication is a basic human right and allows an individual to interact with other people, show their feelings and make decisions that affect their lives. A considerable amount of guidance can also be found from bodies such as Mencap or disability-specific organisations.

Dealing with conflict in the workplace

Customer conflict

Conflict in the workplace is not good, although sometimes it can clear the air between people. Occasionally it cannot be avoided as there will always be people who fundamentally disagree with the way something is being done or they simply do not get on. However, if the conflict is between the customer and an operative, this can cause complications that are bad for the contract and business.

Contractual disputes are the biggest source of conflict between customers and contractors. If a contract has not been set out in enough detail, and the detail has not been discussed and agreed formally in writing, there may be conflict. Issues like this are always best settled by a meeting with all the parties involved and a chair person who is independent of both groups if possible.

If the conflict cannot be solved by simple 'round the table' discussion, a mediator might be required. The Ministry of Justice is responsible for developing Alternative Dispute Resolution (ADR) to avoid costly use of the court system. Mediation is where both parties come together and discuss their needs and concerns in the presence of an independent person (the mediator) to reach an agreement. This type of approach is good for small claims, civil disputes and contractual disputes. However, if the conflict is contractual and cannot be agreed in this way, lawyers can be involved.

The best way to avoid conflict with a customer is to keep them fully informed at all times, stay professional and bring any issues to their attention as quickly as possible. If there is a problem, go to the customer and tell them about it, but also offer solutions as the conflict could simply arise due to frustration. If you offer a solution at the same time you may be able to avoid conflict.

Conflict with co-workers

Conflict between co-workers can be due to individuals not getting on, personal competitiveness, jealousy, lack of information, lack of training or a whole range of other possibilities. Conflict between workers on site will not reflect well on the company, as the last thing a customer wants to see on site is arguments. This would be bad for the business and it will obviously affect productivity and potentially the quality of the work being done.

Conflict with supervisors and operatives

Unfortunately there will always be a manager in your working life that you do not get on with. This can be for a number of reasons but this situation can become very difficult and lead to a lot of unhappiness in the workplace. Your manager or supervisor may be asking you to do something you are not trained to do. You may not want to admit that you do not know how to do it as you want to protect your job. Your manager is frustrated because every time he or she asks you to do something, you take too much time and don't seem willing to ask questions. This can lead to bad feeling on both sides.

A situation like this can arise when people are not honest with each other and do not take the time to discuss matters. If this occurs, you need to call a meeting with your supervisor and discuss the points. You could set up some ground rules that both parties sign up to, focusing on better ways of working together. Some examples are described in Table 5.5.

Activity 5.7

Conflict is a part of life that everyone has to deal with at some time but it is your ability to manage conflict without escalation that can set you apart from others. Discuss with another learner an example of conflict that you have managed successfully.

Employee commitment	Manager/supervisor commitment
I will ask questions when I need clarification.	I will not get angry when I am asked questions.
I will ask for feedback after work.	I will make time at the end of the day to review work and give constructive comments.
I will attend all the training offered to me.	I will look at training requirements of individuals and offer advice and courses.
I will attend monthly reviews.	I will arrange monthly reviews.

Table 5.5: Ground rules for working together

If disputes cannot be resolved by talking together, a third party might be required. If a more senior manager or supervisor cannot resolve the issue then the use of an alternative dispute resolution service such as the Advisory, Conciliation and Arbitration Service (ACAS) might be required. As part of ACAS services, a third party is invited into the meeting. The level of dispute resolution will be determined by the level of control the two parties are willing to give up. Obviously if the dispute cannot be resolved in this way, and the courts are involved, the final decision will be bound by the law and neither party will have a say.

Method	Descripton
Conciliation	This is where both parties retain all the power. There is a facilitated discussion where all points, needs and concerns are discussed and all reach an agreement that is 'honour bound'.
Mediation	Some of the powers are given up by both parties and this is agreed beforehand and signed up to. The rest of the process is the same as conciliation.
Arbitration	Both parties give up their power to the independent arbitrator beforehand and sign up to this. Although arbitration is voluntary, once signed up to both parties have to stick to the decision

Table 5.6: Alternative methods of dispute resolution (ACAS)

A lot of arguments start from a simple misunderstanding between workers. To avoid this situation, simply ask your colleague for clarification, stay calm and, if necessary, check with another colleague or manager.

Effects of poor communication on an organisation

Effects of poor communication between operatives

Poor communication can cause a number of issues. In a team working to tight deadlines, good communication becomes even more essential. Sometimes there is only limited time to give and get instructions on site. If these instructions are passed down the line of management to a team, and then the operative carries out the activity, he or she has to be confident that it is correct. If you are not completely sure about an instruction that you have been given, do not start work – check again until you feel confident. If you don't, it could cost the company a great deal of money or damage the company's reputation.

Effects of poor communication between operatives and management

With large contracts, instructions can be passed through a management chain. These must be written down to ensure the message does not get changed or misinterpreted. Any official changes will be controlled by a project manager, principle contractor or architect. These instructions will be written and again, if there is any doubt, no one will mind if you stop to check.

Case study

John was given his first installation job as part of a very small team in a redevelopment of a warehouse that was being converted into offices. He was given the task of running the cables along the corridors in high level basket and into each of 15 rooms ready for connection. The senior electrician and another apprentice were in another part of the warehouse. They were tracing old cables, removing them after safely isolating them, and preparing the consumer unit. John had a meeting with the senior electrician at the beginning of the first day and was given verbal instructions – he did not write them down or ask any questions as he did not want to appear stupid. The instructions were to run 4 mm² twin and earth cable in each room to each socket outlet from the sub main at the end of the corridor on the first floor. John was not totally sure but thought he would figure it out – after all he had seen this in college.

The senior electrician got caught up in a difficult safe isolation on an unknown live cable and didn't manage to speak to John at length until the end of the second day. By this time John had run a separate cable to each socket all the way back to the consumer unit – in total he had run and cut 30 cables to length. He was only supposed to run two. This cost the company 600 m of cable, two man days plus a third day to pull back, wind up and store the cable that could be saved. John wasn't dismissed but it took a few months before he was allowed to take on this kind of responsibility again.

The case study above is real so how could this have been avoided?

1   Describe a set of actions that could have been taken to avoid this situation.

2   Write a set of 'open' questions that could have been asked to avoid this kind of poor communication between a manager and apprentice.

Consequences of poor communication with a customer

Poorly managed installations with no clear lines of communication are certain to end up in difficulties. A customer pays the contractor and has the final say in what is to be done on site. Regular structured conversations and communication are required between a company and customer to make sure the installation work stays on track, within budget and finishes on time. If the customer is not kept informed of changes, difficulties, setbacks, stock issues or workforce issues, the contract could finish late. If the customer is unable to start business on time because of this, it could lead to a legal case being brought against the company.

Formal project management is one very good method of controlling communication in a structured way. Meetings are held on a regular basis and any changes to instructions are written down in minutes and a change control process is put in place.

Activity 5.8

You are in the process of trying to win a tender for a large installation job in a private faith school. The potential customer has asked for more information before they are willing to award the contract to you. If your company wins you will be the sub-contractor responsible for installing a complete rewire of the electrical installation. Write a short report (no more than 500 words) explaining to the customer your company's health and safety policy and how you will carry out the work maintaining a safe environment. You must take into account any considerations and requirements for a faith school (you can choose the particular faith of the school) and be prepared to debate your assumptions.

Knowledge check

This chapter is assessed via online multiple choice questions. Please attempt the following knowledge check questions. If you do not get all the questions correct first time, read the relevant parts of the chapter again and re-attempt the questions until you are confident of the answers.

1 What does APM stand for?

   a Associate Post Manager
   b Alternative Project Management
   c Association of Project Managers
   d Alternative Practical Measure

2 On construction sites, 'notifiable work' has to be registered when what conditions are reached?

   a A project is 30 days or 500 man hours long
   b A project is 15 days and 50 man hours long
   c A project is 10 days and 10 man hours long
   d A project is 5 days and 5 man hours long

3 What does ECA stand for?

   a Electrical Contractors Academy
   b Edwards Contractors and Artisans
   c Electrical Contractors Association
   d Elevator Contractor Association

4 How many regulations are in the EAWR?

   a 10
   b 12
   c 25
   d 33

5 Part 5 of BS 7671 covers what specific subject area?

   a Definitions
   b Protection for safety
   c Scope
   d Selection and erection of equipment

6 What is the British Standard for the semi-enclosed rewireable fuse?

   a BS 88
   b BS EN 60698
   c BS 399
   d BS 3036

7 You are handed the standard BS EN 62305 – what are you about to work on?

   a Emergency lighting
   b Fire protection systems
   c Earthing
   d Protection of structures against lightning

8 What diagram uses BS EN 60617 symbols?

   a Assembly drawing
   b Schematic diagram
   c Circuit diagram
   d Wiring diagram

9 What is the best diagram to show you simply how a system works?

   a Block diagram
   b Schematic diagram
   c Wiring diagram
   d Assembly drawing

10 What does ACAS stand for?

   a Alternative Conciliation and Service
   b Advisory, Conciliation and Arbitration Service
   c Alternative Combat Advisory Service
   d Advisory Contact Arbitration Service

Principles of electrical science

Chapter 6

This chapter covers:

- the principles of a.c. theory
- illumination, lighting and lighting design
- three-phase systems
- electrical machines
- electrical devices
- electrical heating systems and control
- electronic components and applications.

LO1

Key term

Phasor diagram – a line diagram that has magnitude and direction, i.e. 4 cm long at an angle of 30° to some named reference point. This is also called a vector. Note: scalar means magnitude only.

Introduction

In Chapter 1 you were introduced to the basic science concepts of electricity and mechanics. To be a competent engineer in the building services industry a deeper understanding is required. As an engineer you may be required to go back to science principles to complete a complex design. Identifying and correcting complicated faults often require a higher level of knowledge. Besides practice, the other key element is a sound understanding of the science principles involved. This chapter aims to give you those skills.

THE PRINCIPLES OF A.C. THEORY

Effects of components in a.c. circuits

Electrical components act differently in an a.c. circuit when compared to a d.c. circuit. An example is the coil or inductor. In an a.c. circuit there is a certain amount of current opposition that occurs, whereas in a d.c. circuit the coil simply acts like a long piece of wire once the current has settled to a steady level. In this section you will look in detail at how different components act under different a.c. conditions. You will be able to predict the reaction of a circuit to current and voltage, and prove the results mathematically and graphically using **phasor**/vector **diagrams**.

One important concept needs pointing out again at this stage.

- In a series circuit, the current is constant and the voltage varies.
- In a parallel circuit, the voltage is constant and the current varies.

This is exactly the same principle for an a.c. circuit as it is for a d.c. circuit and you will need to refer to this when considering more complex a.c. circuits.

Resistance

A resistor is what it says – it resists current flow. A voltage pressure caused by a potential difference between the two ends of the resistor causes electrons to flow through it. The UK supply frequency is 50 Hz. This means current flow direction will change 50 times a second as the electrons (and current) change direction. A resistor will not affect the relationship between the current and the voltage. The current will rise at the same time as the voltage rises. If you look at an oscilloscope trace of voltage and current, the two sine waves would track each other exactly. This is only the case if the resistor is a pure resistor and had no other capacitive or inductive properties.

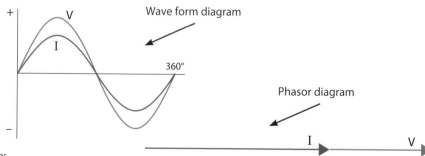

Circuit diagram

a.c.   Lamp

Wave form diagram

360°

Phasor diagram

I   V

Figure 6.1: Wave form, circuit and phasor diagrams

You have already covered Ohm's law in Chapter 1 and the methods to work out overall resistance values for resistors connected in series and parallel. Ohm's law will not be covered again here. (Please see page 17 and try the knowledge checks again if you are in any doubt!)

Worked example

If a 3 Ω resistor is connected in series with a 4 Ω and a 12 Ω resistor, what is the total resistance and the voltage dropped across each if the supply is 24 V a.c.?

$R_t = R_1 + R_2 + R_3 = 3 + 4 + 12 = 19\ \Omega$

$I = \dfrac{V}{R}$

$I = \dfrac{24}{19}$

$I = 1.26\ A$

Now we have the total current through the series resistors, you will be able to work out the individual voltage dropped across each resistor.

$V_1 = R_1 \times I$

$V_1 = 3 \times 1.26$

$V_1 = 3.78\ V$

$V_2 = 4 \times 1.26$

$V_2 = 5.04\ V$

$V_3 = 12 \times 1.26$

$V_3 = 15.12\ V$

Inductive reactance

From the introduction to a.c. theory you will remember that a coil or any component that contains a winding of wire will behave in a different way when connected to an a.c. supply instead of d.c. This inductive effect causes a 'back emf' and induces a current that opposes the main current generating it.

There are several factors that can influence the amount of opposition to current flow. The reaction of a coil to an a.c. current is called 'inductive reactance'. If the frequency of the supply is increased from 50 Hz to 100 Hz this will double the opposition to current flow. Inductive reactance is also directly affected by the inductive properties of the coil, measured in henrys. The total inductive reactance of a coil can be found using the formula:

$X_L = 2\pi fL$

where X_L is the inductive reactance measured in ohms, f is the frequency measured in Hz and L is the inductance of the coil measured in henrys (H).

The direct effect of frequency on the opposition to current flow can be seen in Figure 6.2.

A point to note is the unit of inductive reactance is ohms, the same as for a pure resistance.

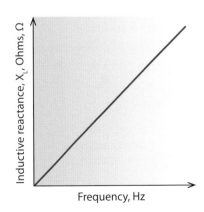

Figure 6.2: The graph shows that inductive reactance increases with frequency

Chapter
6

Worked example 1

If a coil is connected to an a.c. supply of 50 Hz and has an inductance of 24 mH, what is the inductive reactance?

$X_L = 2\pi fL$

$X_L = 2\pi \times 50 \times 24 \times 10^{-3}$

$X_L = 7.54\ \Omega$

Worked example 2

If the frequency in Worked example 1 is now doubled, what is the new inductive reactance?

$X_L = 2\pi fL$

$X_L = 2\pi \times 100 \times 24 \times 10^{-3}$

$X_L = 15.08\ \Omega$

As you can see, by doubling the frequency the inductive reactance opposition to current flow has doubled.

Activity 6.1

1   A coil of 65 μH is connected to a variable frequency a.c. source. The frequency is switched initially to 40 Hz and then to 60 Hz. What is the range of inductive reactance?

2   If the coil in Question 1 is changed to a 2.4 H coil, what is the new range of inductive reactance?

3   If a coil with an inductance of 26 mH has an inductive reactance 3.4 Ω, what is the frequency of the supply?

4   What frequency is required to cause a 35 mH coil to have a reactance of 300 Ω?

5   If a 50 mH coil is connected to a variable frequency supply and the reactance ranges from 31.42 Ω to 47.12 Ω, what is the approximate frequency range?

The sine wave used to represent this inductive circuit would look like this.

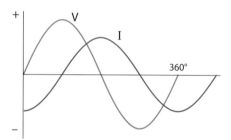

If we represent this as a phasor diagram, we end up with this.

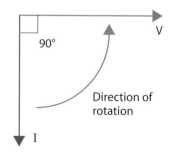

Figure 6.3: Sine wave and phasor diagrams in an inductive circuit

Voltage and current in a series inductive, resistive circuit

You have already looked at what happens to current in relation to voltage in a circuit with just a coil. The current will lag behind the voltage by up to 90°. It will not be any more than 90°. In reality, current will lag less than 90° as a coil will have some resistance due to it being a long piece of conductor. So, a coil will have resistance and inductive reactance in an a.c. circuit and the phase angle between current and voltage will be something equal to or less than 90°. If the coil has very little inductive reactance, the current will be closer in phase to the voltage. If the current was able to get in phase with the voltage, there would be no phase shift or phase angle. The phase relationship between current and voltage can be represented by a wave form diagram or a phasor diagram, as shown in Figure 6.3.

Capacitive reactance

As with inductors in an a.c. circuit, capacitors also have an effect on current flow. Capacitors were briefly described in Chapter 1. Capacitors have the opposite effect to inductors and this can actually be very useful, as you will see later on in this chapter.

Capacitors in an a.c. circuit will have an inverse relationship with frequency – as frequency goes up, capacitive reactance (the opposition to current flow in a capacitive a.c. circuit) will go down. Capacitive reactance will also go down as the capacitance of the circuit increases. The formula to calculate capacitive reactance is:

$$X_c = \frac{1}{2\pi f C}$$

where X_c is capacitive reactance measured in ohms, Ω, C is capacitance measured in farads and f is frequency measured in Hz.

Worked example 1

A 64 μF capacitor is connected to a 50 Hz supply – what is the capacitive reactance?

$$X_c = \frac{1}{2\pi f C}$$

$$X_c = \frac{1}{2\pi \times 50 \times 64 \times 10^{-6}}$$

$$X_c = 49.74 \ \Omega$$

Worked example 2

If the frequency in Worked example 1 was increased to 100 Hz, what is the new capacitive reactance?

$$X_c = \frac{1}{2\pi f C}$$

$$X_c = \frac{1}{2\pi \times 100 \times 64 \times 10^{-6}}$$

$$X_c = 24.87 \ \Omega$$

In this example, you will notice that as the frequency doubles, the capacitive reactance halves. This means that as frequency increases, there will be less opposition to current flow.

Activity 6.2

1 A 24 μF capacitor is connected to a 50 Hz supply. What is the capacitive reactance?

2 A capacitive circuit measures a reactance of 50 Ω when the supply frequency is turned to 150 Hz. What is the value of the capacitor?

3 What frequency is the supply if a 64 μF capacitor gives a capacitive reactance of 12 Ω?

4 A 2.4 μF capacitor is added to a circuit with a 50 Hz supply. What capacitive reactance can be expected?

5 A capacitor of 10 μF and 20 μF are placed in parallel to a 30 Hz supply – what is the total reactance of the circuit?

Note that to find total capacitance in a parallel circuit you simply add them up – you will recall that this is the opposite to when you are adding resistors in parallel.

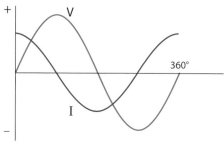

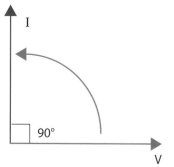

Figure 6.4: Sine wave and phasor diagrams in a capacitive circuit

Voltage and current in a capacitive, resistive circuit

From the introduction to a.c. theory you will remember current leads voltage by up to 90°. Just like an inductive circuit, this will not be any more than 90° and, in reality, it will be something less than this. The phase relationship between current and voltage can be represented by a wave form diagram or a phasor diagram, as shown in Figure 6.4.

Impedance

An a.c. circuit has more than one of these reactive components present at any one time. There is also a strong possibility that all three components will exist: capacitive, inductive and resistive. If this is the case there must be an overall effect that is a combination of reactance. This overall effect is called 'impedance' because current flow is impeded.

If a pure capacitor will cause current to lead voltage by up to 90° and a pure inductor will cause current to lag the supply voltage by up to 90°, then it can also be seen that they are exactly opposite in what they bring to an a.c. circuit. This can be seen in Figure 6.5.

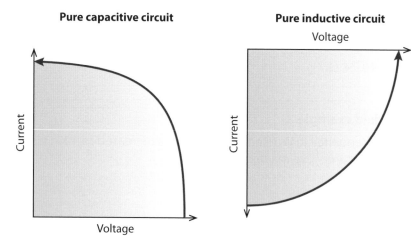

Figure 6.5: Current leading and current lagging voltage in pure capacitive and inductive circuits

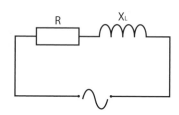

Figure 6.6: Impedance triangle for resistance and reactance in series

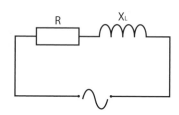

Figure 6.7: Series resistive/inductive reactance circuit

For a circuit with resistance and reactance (capacitive or inductive) the overall impedance can be found by drawing an impedance triangle. The horizontal axis represents the resistive part of the circuit and the vertical line represents the reactance. By drawing these two components to scale using a right angle, the hypotenuse represents the actual impedance, as can be seen in Figure 6.6.

Triangles are used extensively in electrical science. For a resistor and inductive series circuit (shown in Figure 6.7) Pythagoras' theorem can be used to find the overall effect or impedance of the circuit.

Pythagoras states: 'the sum of the sides squared is equal to the square of the hypotenuse'.

This gives the formula:

$$Z = \sqrt{R^2 + X_L^2}$$

The formula can be rearranged to make the subject either the adjacent side (R) or the opposite (X_L):

$$R = \sqrt{Z^2 - X_L^2}$$

or

$$X_L = \sqrt{Z^2 - R^2}$$

Worked example 1

A resistor of 6 Ω is connected in series with a coil of reactance 2.5 Ω. What is the circuit impedance?

$$Z = \sqrt{R^2 + X_L^2}$$

$$Z = \sqrt{6^2 + 2.5^2} = 6.5 \ \Omega$$

The solution to this can also be found by drawing the values to scale, as follows.

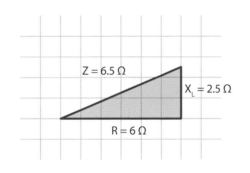

Z = 6.5 Ω

X_L = 2.5 Ω

R = 6 Ω

Worked example 2

A resistor of 2 kΩ is connected in series with a coil of reactance 2.9 kΩ. What is the circuit impedance?

$$Z = \sqrt{R^2 + X_L^2}$$

$$Z = \sqrt{6^2 + 2.9^2} = 3.52 \ k\Omega$$

Note: there is no need to include the 000s as they can be added in at the end.

Progress check 6.1

1 In an inductive circuit, does the current lead or lag the voltage?

2 In a capacitive circuit, does the current lead or lag the voltage?

3 Explain the term 'impedance'.

Activity 6.3

1 A 27 mH inductor is connected in series with a resistor of 6 Ω. The circuit is then connected to a supply frequency of 50 Hz. Find the circuit impedance by calculation.

2 For the circuit above, prove your calculation by scaled drawing.

3 A 350 mH inductor is connected to a series 100 Ω resistor. What is the circuit impedance if the supply is 50 Hz?

Ohm's law in an a.c. circuit

Up until now you have only used Ohm's law with d.c. circuits.

$$I = \frac{V}{R}$$

An a.c. circuit can also use Ohm's law but, instead of using just R, you need to replace this with Z.

$$I = \frac{V}{Z}$$

(It is worth noting that if an a.c. circuit only has resistance then Z = R.)

Worked example

A resistor of 6 Ω is placed in series with an inductor of 2.5 Ω. If the supply is 230 V, what is the circuit current and the voltage dropped across each component?

With this type of question it is a good idea if you have time to draw a quick sketch of the circuit, as shown below.

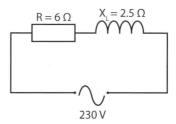

$$Z = \sqrt{R^2 + X_L{}^2}$$

$$Z = \sqrt{6^2 + 2.5^2}$$

$$Z = 6.5$$

$$I = \frac{V}{Z}$$

$$I = \frac{230}{6.5}$$

$$I = 35.38 \text{ A}$$

Now work out the voltage drop across each load by applying Ohm's law.

$$V_1 = I \times R$$
$$V_1 = 35.38 \times 6 = 212.28$$

$$V_2 = I \times X_L$$
$$V_1 = 35.38 \times 2.5 = 88.45 \text{ V}$$
$$V_T = 88.45 + 212.28 = 300.73? \rightarrow \text{NO, it must equal 230 V!}$$

Notice how the two voltages do not simply add up, as would be the case in d.c. circuits! The two voltages must be added up taking into account the phase angle that has been created between them. This can be done by phasor addition. You will come back to this example later on in this section when you have the little bit of extra knowledge required to solve this.

Activity 6.4

This worked example is similar to a fuse, with the coil being the fuse element and the resistance simply the resistance of the wire.

1 Explain what would happen if the coil was stretched out so it was no longer a coil but became a long conductor.

2 What would be the new current?

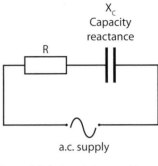

Figure 6.8: Series resistive capacitive circuit diagram

Resistors and capacitors in series

In a d.c. circuit, a capacitor is a block to current as there is no conductive circuit between the capacitor plates. However, this is not the case with a.c. as the plates are continuously charged and discharged 50 times a second (if the supply is 50 Hz) causing the current to lead the supply voltage by up to 90°. The circuit diagram can be seen in Figure 6.8.

The same triangle principle is used to calculate the total impedance of a resistive capacitive circuit. This gives the formula:

$$Z = \sqrt{R^2 + X_c^2}$$

The formula can be rearranged to make the adjacent or opposite the subject, as follows:

$$R = \sqrt{Z^2 - X_c^2}$$

or

$$X_c = \sqrt{Z^2 - R^2}$$

Worked example 1

A resistor of 4.5 Ω is connected in series with a capacitor of reactance 3.5 Ω. What is the circuit impedance?

$$Z = \sqrt{R^2 + X_c^2}$$

$$Z = \sqrt{4.5^2 + 3.5^2}$$

$$Z = 5.7\ \Omega$$

The solution to this can again be found by drawing to scale, as follows.

This is called an impedance triangle.

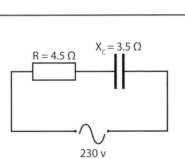

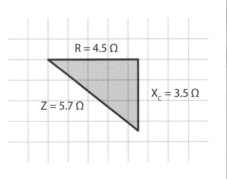

Activity 6.5

1 A 24 µf capacitor is connected in series with a resistor of 300 Ω. The circuit is then connected to a supply frequency of 50 Hz. Find the circuit impedance by calculation.

2 Now prove your calculations by completing a scale drawing of an impedance triangle. Remember, at this level you should now be able to work with scales.

Resistor, capacitor, inductor series circuits

A pure capacitor will cause current to lead by up to 90° and a pure inductor will cause current to lag by up to 90°. They have an opposite effect and so to solve a circuit that has all three components you simply need to take the smallest reactance away from the largest and then complete the calculations as follows.

$$Z = \sqrt{R^2 + (X_L - X_c)^2}$$

It does not matter which way round the reactance numbers are, as the square of a negative is a positive number anyway.

Worked example

A resistor of 4.5 Ω is connected in series with a capacitor of reactance 2.5 Ω and an inductor of reactance 6.5 Ω. What is the circuit impedance?

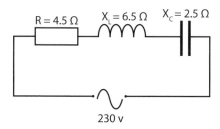

$$Z = \sqrt{R^2 + (X_L - X_c)^2}$$

$$Z = \sqrt{4.5^2 + (6.5 - 2.5)^2}$$

$$Z = \sqrt{4.5^2 + (4)^2}$$

$$Z = 6.02 \ \Omega$$

The solution to this can again be found by drawing to scale, as follows.

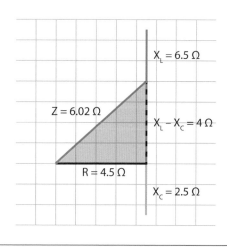

Phasor diagram for voltage in an a.c. series circuit

It is also possible to find the total supply voltage by phasor diagram. As mentioned previously, current is constant in a series a.c. circuit. It is the voltage that can vary. The current will be common in all three components but the voltage drop across the inductor will lead the current (this is a different way of saying that the current lags the voltage when the current is taken as the reference). The voltage will lag the current across the capacitor and the voltage and current will be in phase across the resistor.

Using the impedance previously found, the circuit current, I_s, can be found using:

$I_s = \dfrac{V_s}{Z}$

$I_s = \dfrac{230}{6.02}$

$I_s = 38.21\ A$

The voltage drop across the resistor is in phase with the circuit supply current and becomes the reference when drawing a phasor, as can be seen in Figure 6.9.

Ohm's law can now be applied to each component to find the voltage drop on the resistor, capacitor and the inductor.

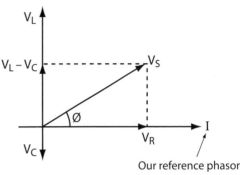

Figure 6.9: Phasor diagram

Resistor voltage drop:

$V_R = R \times I_S$

$V_R = 4.5 \times 38.21$

$V_R = 171.95\ V$

Inductor voltage drop:

$V_L = X_L \times I_S$

$V_L = 6.5 \times 38.21$

$V_L = 248.37\ V$

Notice how X_L replaces R but it is still basic Ohm's law.

Now consider the voltage drop across the capacitor using the same method:

$V_C = X_C \times I_S$

$V_C = 2.5 \times 38.21$

$V_C = 95.53\ V$

These are now ready to be drawn on a phasor diagram. The capacitive and inductive voltages are simply taken away from each other to give a resultant reactive voltage that can then be plotted against the resistor voltage. The resultant voltage of the supply, V_s, is as expected approximately 230 V. The circuit power factor angle can also be taken by direct measurement with a protractor from the phasor diagram.

> **Progress check 6.2**
>
> 1 What is inductive reactance?
> 2 What is capacitive reactance?
> 3 What is impedance?

Activity 6.6

1 A 640 μF capacitor is connected in series with a resistor of 3 Ω and a 22 mH inductor. The circuit is then connected to a supply frequency of 50 Hz. Find the circuit impedance by calculation and then prove it by completing a scale drawing of an impedance triangle.

2 A 240 μF capacitor is connected in series with a 50 Ω resistor and a 20 mH inductor. What is the overall impedance of the circuit if the frequency is 50 Hz?

3 If the frequency is changed to 100 Hz in Question 2 and the resistor is replaced with a 10 Ω resistor, what is the new circuit impedance?

4 An inductor of 12 Ω is connected in a series circuit with a capacitor of 159 μF and a 5.6 Ω resistor and a 3.9 Ω resistor. Assume the frequency is 50 Hz and calculate the circuit impedance.

5 What is the supply frequency if a resistive/capacitive series circuit has an overall impedance of 55.9 Ω? The resistor is 50 Ω and the capacitor is 25 μF.

6 A series a.c. circuit contains a 10 Ω resistor, 12 Ω capacitive reactance and an inductor with reactance of 6.9 Ω. Calculate:

- the total impedance

- the current in the circuit if the voltage supply is 400 V.

Parallel circuits – resistor/inductor/capacitor

In a series circuit, current is common to all components and voltage varies as it drops across each component. In a parallel circuit, current will vary as it will split and travel down the various paths open to it. When working with parallel a.c. circuits the current will still react to inductors and capacitors in the same way as in a series circuit. For inductors in parallel to a resistor the current will lag the voltage, for instance. When solving parallel a.c. circuit problems using a phasor diagram, the voltage is the reference as this is common to all parallel components. This parallel arrangement can be seen in Figure 6.10.

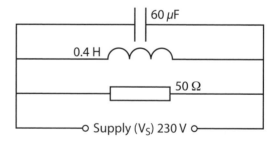

Figure 6.10: Resistor, capacitor and inductor in parallel connected to an a.c. supply

As the parallel branches in this example are pure inductors, resistors and capacitors, the currents can be added using a triangle with phasors representing the currents in each branch. The currents in the capacitor and inductor branches are completely opposite each other (capacitive current leading by 90° and the inductive current lagging by 90°). To resolve these two currents, simply take the smallest away from the largest. The result of this will also tell you if the overall effect is a leading or lagging current. The current through the resistive branch is completely in phase with the reference voltage.

The actual values of the currents can be found by applying Ohm's law and the formulae for reactance, as follows.

Current in the resistor branch:

$$I_R = \frac{V_S}{R}$$

$$I_R = \frac{230}{50}$$

$$I_R = 4.6 \text{ A}$$

Current in the inductive branch:

Don't forget you have to work out the inductive and capacitive reactance first. Assume, unless otherwise told, the frequency is 50 Hz.

$$I_L = \frac{V_S}{X_L}$$

$$I_L = \frac{230}{126}$$

$$I_L = 1.83 \text{ A}$$

Current in the capacitive arm:

$$I_C = \frac{V_S}{X_C}$$

$$I_C = \frac{230}{53}$$

$$I_C = 4.3 \text{ A}$$

The final value of supply current drawn can be seen in Figure 6.11.

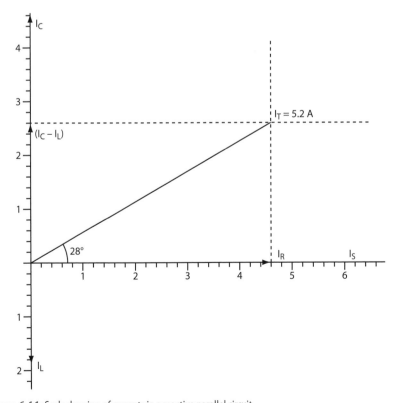

Figure 6.11: Scale drawing of currents in a reactive parallel circuit

Power quantities

Power is defined as the amount of energy used in time. Power is found in a d.c. circuit by a straightforward relationship between voltage and current, voltage and resistance, or resistance and current. In an a.c. circuit the reactive components have a large effect on the power used. As you saw earlier, when the inductive effect was removed by straightening out a coil in a fuse, the current rises. This is because the reactive coil causes a back emf and consequently when the coil is removed the outcome is a larger current. With reactive components in an a.c. circuit, the power requirement will change. If there are large reactive components, the power requirement will be larger. If there are less reactive components in the circuit, the power requirement will be smaller.

You have looked at the impedance triangle previously. There is also a power triangle to consider along with this. The impedance triangle is made up of a resistive part, reactive part and an overall impedance part that is used to calculate the current drawn in the circuit. The power triangle is also made up of the active power due to the resistive part of the circuit, the reactive power due to the reactive part and the overall apparent power due to the total impedance of the circuit. The idea in an a.c. circuit is to try and reduce the reactive effects to zero and therefore reduce the power requirements as far as possible. The effects of reducing the reactive components of an a.c. circuit can be seen from the impedance triangle in Figure 6.12. The smaller the reactive element becomes, the shorter the overall impedance becomes. This is the same for power, as can be seen in the power triangle in Figure 6.13.

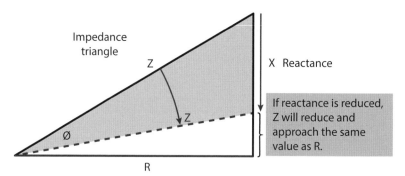

Figure 6.12: Impedance triangle

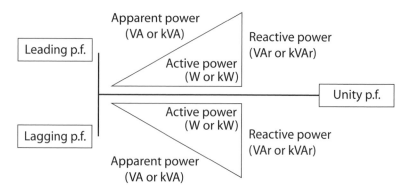

Figure 6.13: Power triangle

The beer or cappuccino analogy

If you are having difficulty understanding the different terms used with a.c. power, try considering this analogy. A pint of beer or a cappuccino is made up of several parts. The main body of beer/coffee can be thought of as the useful bit, whereas the head or froth on the top can be considered as the useless part, as invariably it is left at the end of the drink. The whole drink is what you actually pay for and you cannot get the main drink without at least a little bit of froth. Even though the froth is often wasted at the end, you still have to pay for its creation. This is the correct way to think about power!

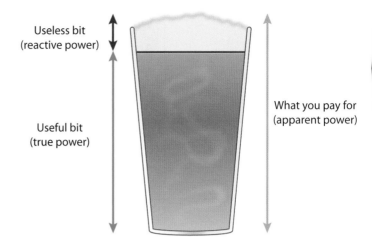

Figure 6.14: Beer 'power' analogy

True power or active power (W or kW)

True power is also known as active power. The pure resistive part of the circuit accounts for the true power requirement. In an ideal world, a load that requires 2 kW will just need 2 kW. This would be because the current has only resistance to contend with.

From your work with d.c. circuits you will remember that the power is the product of voltage and current. In a.c. circuits this is also the case. Consider the wave form diagram in Figure 6.15 for a purely resistive circuit. The instantaneous power can be found by looking at the instantaneous voltage and instantaneous current. Because in this example there are no reactive components, the voltage and current wave forms are completely in phase. When the individual readings are taken at any point in time and multiplied together, the power or heating effect is shown by the third dotted wave form. You will notice that when the current and voltage go negative, they are multiplied to result in a positive power curve. The average or root mean squared (rms) value of this will show the power taken in a purely resistive circuit.

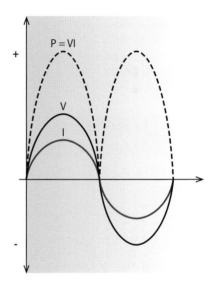

Figure 6.15: The average power of a purely resistive a.c. circuit

In a purely resistive circuit there is only active power. This is not the complete story as a purely resistive circuit is almost impossible to achieve in reality – there is always what could be termed contamination from coils and capacitance. These will throw the current and voltage out of phase and cause the power calculation to become a little more complicated.

True power can be found practically in a reactive circuit by using a wattmeter or by taking individual voltage and current readings and adjusting by the power factor using:

True power, $P = V \times I \times \cos \emptyset$

Note: the 'cos Ø' represents the power factor and can be likened to power 'percentage efficiency' or the amount the voltage has been thrown out of phase with the current by reactive components. Remember that in a real a.c. circuit generally speaking the power used will be greater than just the true power.

Worked example

Active power

An a.c. circuit draws 30 A and the voltage across the load measures 500 V. The reactive components in the circuit cause a phase angle between the current and voltage of 30°. What is the true power for this circuit?

15 kW is the power used taking into account the reactive components ($P = VI = 30 \times 500$). The power factor angle is 30°.

Power factor $= \cos \varnothing$

Power factor $= \cos 30°$

Power factor $= 0.866$

True power $= VI \times \cos \varnothing$

True power $= 500 \times 30 \times 0.866$

True power $= 12\,990$ W or 12.99 kW

This shows that although 15 kW is the apparent power required by the load, the true power at the load is only 12.99 kW. The difference in power is due to having to overcome the reactive components in the circuit.

Progress check 6.3

1 What type of power is there in a purely resistive circuit?

2 What is true power measured in?

3 What does the term power mean?

Figure 6.16: Capacitive reactive power

Reactive power (VAr or KVAr)

The reactive components, specifically coils, will make a considerable difference to the power requirements and this is something that an electrician can work with to reduce. This can be shown by looking at the instantaneous voltage and current in an example. Consider a capacitive circuit where the current is leading the voltage by up to 90°. The power is still the product of the voltage and current. In this case the instantaneous values are multiplied together and you can see that the power curve is very different (see Figure 6.16) – there is zero power. There is no power in a purely capacitive circuit.

If you also consider a purely inductive circuit, the same average power would happen. This means that for a purely capacitive or inductive circuit, there will be current and voltage but no heating effect. The reactive power is sometimes referred to as the 'useless' power or 'wattless' power. Reactive power can be found from the triangle rules using:

Reactive power $= \sin \varnothing \times$ apparent power

Apparent power (VA or KVA)

This is the actual power you end up paying for. Apparent power is measured in volt amps or, if it is a large value, kilo volt amps. In the power triangle this value is represented by the hypotenuse (longest side of the triangle) and is made up of the combined effect of active (or true) and reactive power. To measure the apparent power a voltmeter and ammeter is required (a wattmeter is used to measure the true power).

Apparent power $= V \times I$

Calculate the power factor

If there are reactive components in an a.c. circuit then all three types of power will also exist. The angle that exists on a power triangle between the apparent power and the active power is the power factor angle. If the reactive power is reduced by cancelling out some of the reactive components, the power factor angle will also reduce. Using the rules of triangles:

$$\cos \emptyset = \frac{\text{Adjacent}}{\text{Hypotenuse}}$$

$$\cos \emptyset = \frac{\text{True power}}{\text{Apparent power}}$$

So power factor (pf) can be given by:

$$\cos \emptyset = \frac{\text{True power (P)}}{\text{Apparent power (S)}}$$

Power factor, load and bulk correction

Throughout this section on a.c. theory you will have noticed the effect a capacitor has when placed in parallel with an inductor. They are opposite. By adapting an a.c. circuit so that the current is in phase with the voltage the circuit will become much more efficient. The purpose of power factor correction is to reduce the phase angle between voltage and current so power factor is said to be **unity**.

Various methods of correcting power factor have already been covered in *Chapter 1: Principles of electrical science*, page 59, but the main method is to add a parallel capacitance to the circuit. This correction can be via **load correction** or **bulk correction**. One other method used for large industrial sites with heavy inductive equipment is a synchronous motor. The synchronous motor introduces a leading power factor that corrects the effects of the inductors.

For a very inductive circuit, the exact value of balancing capacitive reactance can be worked out by calculation or by phasor diagram.

> **Key terms**
>
> *Unity power factor* – occurs when the phase angle between the voltage and current is '0'. Cos0 = 1 and hence unity.
>
> *Load correction* – the addition of a single parallel capacitor.
>
> *Bulk correction* – the addition of banks or parallel capacitors that are switched in to the circuit via contactor control.

Worked example

A coil of 0.15 H is connected in series with a 50 Ω resistor across a 100 V 50 Hz supply.

1 Calculate the power factor of the circuit.

2 Calculate the resultant current.

3 What is the power factor if a capacitor of 26 μF is connected in parallel?

4 What is the new resultant current?

Answer 1

Power factor $= \dfrac{R}{Z}$

But first you need to find the circuit impedance.

$$Z = \sqrt{R^2 + X_L{}^2}$$

continued

To find the circuit impedance you also need to find the inductive reactance.

$$X_L = 2 \times \pi \times 50 \times 0.15 = 47.13 \, \Omega$$

Now the inductive reactance is known, this can be used to find impedance.

$$Z = \sqrt{50^2 + 47.12} = 68.71 \, \Omega$$

Putting the resistance and the impedance into the formula for power factor gives:

$$\text{Power factor} = \frac{50}{68.71} = 0.73$$

$$\text{Power factor} = \cos \emptyset = 0.73$$

$$\emptyset = \cos^{-1} 0.73 = 43°$$

This means the current is lagging behind the voltage by a phase angle of 43° in this circuit.

Answer 2
The resultant current can be found using Ohm's law.

$$I_S = \frac{V}{Z}$$

$$I_S = \frac{100}{68.71} = 1.46 \, A$$

Answers 3 and 4
One way to find the new resultant power factor and current is to draw a phasor/parallelogram diagram. The only missing element is the current in the capacitive branch which can be found by the following.

$$X_C = \frac{1}{2 \times \pi \times f \times C} = \frac{1}{2 \times \pi \times 50 \times (26 \times 10^{-6})}$$

$$X_C = 122.43 \, \Omega$$

$$I_C = \frac{V}{X_C}$$

$$I_C = \frac{100}{122.43} = 0.82 \, A$$

Using an appropriate scale and protractor, the resultant current can be found after power factor correction has been applied.

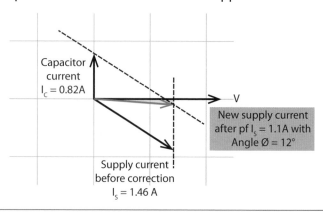

Activity 6.7

1 What is considered to be a perfect power factor?

2 What is the power factor of a series inductive circuit containing a 16 Ω resistor and an impedance of 22 Ω?

3 If this circuit is connected to a 230 V, 50 Hz supply, what capacitor could be added in parallel to the inductive circuit to improve PF to 0.98? (Hint – draw out the circuit and the phasor following the previous worked example.)

ILLUMINATION, LIGHTING AND LIGHTING DESIGN

LO2

Lighting systems have developed at a rapid rate due to advances in science and new technology. The first electric arc light more than a century ago has been updated numerous times over the years, with the very latest being LED technology. Along with the light source changes, the control systems have developed, with advanced electronic control systems achieving a much wider range of effects.

The quality and level of light provided determine what tasks can be carried out in a workspace or home. Therefore lighting is an important aspect of building and open space planning. Lighting design has become a very specialist area of the building services industry, with its own specialist bodies, rules and guidance.

In this section, you will be looking at the various technologies currently used in industrial, commercial and domestic installations. You will also look at the science required for designing lighting, the operation and the applications of different lighting systems.

Illumination quantities

To understand lighting design it is important to understand what light is. You will also need to understand the terms used in the industry.

The human eye can only see a very small **spectrum** of light in the range 380 to 740 nm (nanometres). This is the light range that is visible to you and can easily be reproduced by widely available technologies. The colour of light varies according to the **wavelength**. Visible light spectrum ranges from ultraviolet at the bottom of the scale to infrared at the upper limits of the human eye.

Key terms

Spectrum – a range of wavelength values. Light is electromagnetic radiation.

Wavelength – one full electromagnetic wave cycle is measured in metres. Visible light, as produced by the sun, is measured in nanometres, $\times 10^{-9}$ metres!

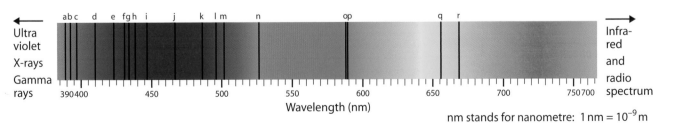

nm stands for nanometre: $1\,\text{nm} = 10^{-9}\,\text{m}$

Figure 6.17: Spectrums and wavelengths of visible light

Key terms

Colour rendition – the ability of a light source to recreate accurately the colour of an object when compared to how the colour of an object appears in natural light.

Luminous intensity (I) – the light power emitted by a light source in a particular direction. It is measured in candelas. A candle emits roughly one candela of light.

Luminous flux (F) – the useful light power emitted by a light source. It is measured in lumens (lm).

Luminous efficacy – the ratio of useful light power leaving a light source (luminous flux) to the input power supplied to the light. Luminous efficacy is measured in lumens per watt (lm/w).

Luminaire – a name the industry uses for the whole light unit including the fittings, fixings, reflector, diffuser, control gear and housing of the actual light.

Illuminance – the amount of light falling on a specified area. It is measured in lux (lx) or lumens per metre squared (lm/m²).

The colour of a light is very important to both building and open air lighting design. Examples of two completely different light requirements are a street light and an operating theatre. A street light requires objects to be seen for safety. You need to see where you are driving and parking, where pedestrians are, road signs and objects you need to avoid or head towards. There is no real need to see a full range of colour. Consider driving at night and passing several cars – can you tell the colour of every car that passes? You probably can't. That is because of a quality of light called **colour rendition**. Certain wavelengths of light, such as the orange light given off by a street light, limit how the colours of an object will appear to the human eye. If the light source is orange the likelihood is you will not make out many object colours at night. However, if you consider an operating theatre, colour rendition is a matter of life or death. You must see the full range of colours to make medical decisions, so colour rendition must be the best money can buy.

When making a decision about lighting there are other design aspects to consider. The light power at the source is called the **luminous intensity**. If you consider a car has horse power, a light source has candle power.

Because the eye is not totally efficient and cannot see all of the light produced by a source, the useful light power that is emitted by a light is called the **luminous flux**. The luminous flux level or lumens is often seen on light packaging to help customers make light level decisions.

Light shining from a source through the atmosphere to an object will lose some of its power on the way. In earlier work you discovered that no system is 100 per cent efficient. For a light source to be truly efficient, all the power supplied to the light must be converted into visible light. The ratio of luminous flux emitted by the light to the input power is called **luminous efficacy**.

The light source or **luminaire** efficacy is a design decision that must be considered when choosing the correct light solution for the application.

The light that leaves the light source is important but the most important design consideration is the light that actually arrives. The amount of light that reaches a surface that needs it is called the **illuminance** and is measured in lux.

Examples of typical light levels for different applications, work areas and situations can be seen in Table 6.1 below.

Application/work area/situation	Lux level
Eye inspection	5–10 000
Shop window	1 500–3 000
Typing/drafting	1 500–2 000
Library	750–1 500
Reading/studying	500–1 000
Wash room/toilet	150–300
Warehouse	100–150
Corridor/stairs	75–150

Table 6.1: Light levels for various applications, work areas and situations

Progress check 6.4

1 What does illuminance mean?

2 Define luminous efficacy.

3 What is meant by luminous intensity?

Other design considerations

Utilisation factor (UF)

When deciding exactly how a room should be lit there are some other points to consider. The actual physical design and the room purpose will make a difference. The colour of the walls, ceiling and floor will react with the light in the room. For instance, a room with very dark decoration will seem dark compared with an identical room decorated in white, even if the light levels are technically the same. An allowance for the decoration must be factored into the design. The room shape, size and height of the ceiling will also impact on the light levels. These factors are called **utilisation factors**. Lists of these factors are published by the Illumination Engineering Society. Utilisation factors will be considered further later in this chapter when you cover lighting calculations.

Maintenance factor (MF)

Lights are installed and stay in the same position for a number of years. The environment will affect their performance. If a room is very dirty or dusty this will settle on the lamp and the walls. The light performance will also gradually deteriorate with age – the light will burn less brightly. The conditions in the room might also shorten the life of the lamp and cause it to blow earlier, such as a machine room.

The **maintenance factors** are listed in Table 6.2 below.

Maintenance factor	Definition
LSF	Luminaire survival factor – percentage of lamp failures. In some harsh environments, lamps can suffer. A very hot smelting room will affect the life of any luminaire. A machine room with lots of vibration would also potentially shorten the life of any lighting system present and would need to be factored into the design.
LMF	Luminaire maintenance factor – due to age-related dirt and grime from the room settling on the lamp.
LLMF	Lamp lumen maintenance factor – after a long while burning, a luminaire starts to literally burn out and the light produced starts to reduce. The lamp will become less efficient with age.
RSMF	Room surface maintenance factor – the colour of the room will change over time and this will affect the way the light appears – paint, wallpaper and ceilings get dirty with age.

Table 6.2: Maintenance factors

> **Key terms**
>
> *Utilisation factor* – determines the illumination efficiency of a room and is affected by the colours, height of the ceiling and the type of luminaires. An average value is given as 0.6.
>
> *Maintenance factor* – found by multiplying a range of individual numbers that are allocated to lamp failure %.

> **Chapter 6**

> **Progress check 6.5**
>
> 1 Name three factors that can affect light levels in a room.
> 2 Explain the difference between maintenance factors and utilisation factors.
> 3 What factor code is specifically about dirt and grime reducing a light's brightness?

> **Activity 6.8**
>
> 1 Look around you and make a judgement about the maintenance factors for the lighting in your room.
> 2 Research light maintenance factors using the Internet and summarise them in a short report (no more than 500 words).

Laws of illumination

You have seen that the light leaving a luminaire is not the same as the light reaching the work surface being lit up. Losses occur due to many factors – the same as any machine or process. So how will you know how much light reaches the surface you are designing for? The recommended light levels for each type of activity, from stairs in an emergency staircase to the surface of a surgery table, are known. You need to work out what light source will create the light you require and which is fit for purpose.

Inverse square law

One method of calculating the light level required at a source is the inverse square law. Imagine looking at a candle 10 m away from you. Now imagine moving the candle 20 m away from you. You might think that because the distance has doubled, the light would have halved. This is not the case. The light is actually one quarter the illumination. If the light source is moved further away to 30 m, the amount of illumination that will reach you is one ninth. This concept is clearly shown in Figure 6.18.

The inverse square law is particularly useful for light sources that are directly above the surface being lit.

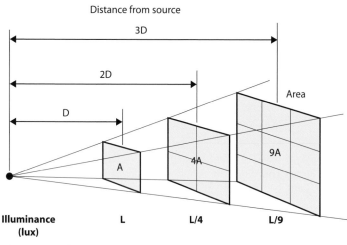

Figure 6.18: The inverse square law

The formula required to work out the illuminance at the work surface is:

$$E = \frac{I}{d^2}$$

where E is the illuminance measured in lux, I is the luminous intensity of the luminaire measured in candelas (cd) and d is the distance between the light source and the surface.

Worked example 1

A fluorescent light produces a luminous intensity of 500 cd and is fixed at a height of 4 m above a workbench where you are making off steel wire armour cable glands.

1 How much light is actually reaching your workbench?

2 Do you think there is enough light to do the job or should you move?

$$E = \frac{I}{d^2}$$

$$E = \frac{500}{4^2}$$

$$E = 31.25 \text{ lux}$$

This is not enough light for you. Ideally you need 150–300 lux. You should move to an area with more light.

Worked example 2

You are working in a cellar wiring a consumer unit and there is a temporary light 3 m directly above your working area producing 350 cd.

1 How much light is actually reaching the consumer unit you are working on?

2 Do you require better temporary lighting?

3 What luminous intensity would be required to give a working surface illuminance of 250 lux?

$$E = \frac{I}{d^2}$$

$$E = \frac{350}{3^2}$$

$$E = 38.89 \text{ lux}$$

This is not enough light for you. Ideally you need 150–300 lux. You should get a better temporary light.

To calculate a working light level of 250 lux, the formula needs to be rearranged.

$$E = \frac{I}{d^2}$$

$$I = E \times d^2$$

$$I = 2\,250 \text{ cd}$$

Activity 6.9

1   A work light is 4 m directly above a table and has a luminous intensity of 1 500 cd. What is the illumination at the table surface?

2   A bulkhead light is 3.5 m directly above a cable duct and is giving off a luminous intensity of 375 cd. What is the illumination at the cable duct?

Key term

Perpendicular – imagine a plumb line being dropped from a light so it hangs vertically.

Lambert's cosine law

A light is very rarely directly above where you are working. In fact a light directly above you would cast a shadow so it is not ideal. For lights that are positioned off centre you will need to consider a different calculation using Lambert's cosine law.

Imagine standing directly under a light at position A in Figure 6.19. The illumination on you is at its greatest as the light is **perpendicular** to you.

As you move away from this centre point to point B, the amount of light falling on you will decrease. The greater the angle between you and the perpendicular axis of the light, the less light falls on you.

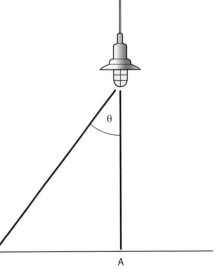

Figure 6.19: Lambert's cosine law

The light at a point away from the centre can be found using:

$$E = \frac{I \times \cos \varnothing}{d^2}$$

where Ø is the angle between the perpendicular (directly under the light) and the position you are standing in away from the light. The d is distance between the light source and the point of illumination being measured. In trigonometry this is called the hypotenuse (see page 13 for trigonometry).

Worked example 1

What is the illumination at points A and B in Figure 6.19 if the luminous intensity is 1 500 cd? The distance between A and B is 10 m (let's call that distance 'b'). The height of the light is 4 m from the floor (let's call that 'a').

The illumination at the point directly below the light, A, can be found by using the inverse square law, as follows.

$$E_A = \frac{I}{d^2}$$

The d in this instance is the height of the light above the floor level, 4 m.

$$E_A = \frac{1\,500}{4^2}$$

E_A is the illumination directly below the light.

$$E_A = 93.75_{lux}$$

Now moving away to a new point B you need to use Lambert's cosine law. There are some missing parts that need further calculation before Lambert's formula can be used. The distance, d, is no longer the drop distance from the light to the floor. This distance is now the hypotenuse of a triangle.

To calculate the distance, d, you need to use Pythagoras' theorem.

$$d^2 = a^2 + b^2$$

$$d^2 = 4^2 + 10^2$$

$$d = \sqrt{4^2 + 10^2}$$

$$d = 10.77 \text{ m}$$

Now you are in a position to complete the calculation.

$$E = \frac{I \times \cos \varnothing}{d^2}$$

Remember from Chapter 1:

$$\cos \varnothing = \frac{\text{Adjacent}}{\text{Hypotenuse}}$$

Adjacent to the angle = the distance 'a' = 4 m

Hypotenuse = d = 10.77 m:

$$\cos \varnothing = \frac{4}{10.77} = 0.37$$

$$E = \frac{1\,500 \times 0.37}{115.99}$$

$$E = 4.78 \text{ lux}$$

Worked example 2

What is the illumination at point B in Figure 6.19 if the luminous intensity of the light source is 2 500 cd? The distance between A and B is 450 cm (let's call that distance 'b'). The height of the light is 3.1 m from the floor (let's call that distance 'a').

To calculate the distance, d, you need to use Pythagoras' theorem again.

$d^2 = a^2 + b^2$

$d^2 = 3.1^2 + 4.5^2$

$d = \sqrt{(3.1^2 + 4.5^2)}$

$d = 5.46$ m

Now you are in a position to complete the calculation.

$E = \dfrac{I \times \cos Ø}{d^2}$

Remember:

$\cos Ø = \dfrac{\text{Adjacent}}{\text{Hypotenuse}}$

Adjacent to the angle = the distance 'a' = 3.1 m

Hypotenuse = d = 5.46 m

$\cos Ø = \dfrac{3.1}{5.46} = 0.57$

$E = \dfrac{2\ 500 \times 0.57}{29.81}$

$E = 47.8$ *lux*

Progress check 6.6

1 What is a reasonable light level for a general maintenance workbench?

2 Describe how you would work out the light level directly under a luminaire.

3 Describe how you would work out the light level a distance to the side of a luminaire.

Activity 6.10

1 What is the illuminance if you are standing at point B in Figure 6.19? The light source has a luminous intensity of 2 300 cd. The angle between the 4 680 mm hypotenuse and the adjacent is 30°.

2 What is the illumination directly under the light in Question 1?

Lighting design using the lumen method

For straightforward square or rectangle rooms it is possible to design the light positions and quantities. All you need to know are the dimensions and the light level required for the room. You have seen some examples in Table 6.1 on page 256 of typical light levels for different applications. This is a good starting point.

You have previously worked on individual lights and expected lighting levels directly under a luminaire and to one side. What you now have to take into consideration is the actual room and the effect that will have on the light. If you now use the light factors previously covered you will be able to complete a very simple design.

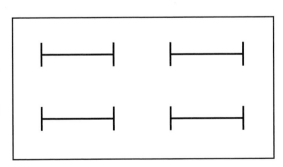

Figure 6.20: Plan of a uniform array of luminaires in a room

Key term

Lumen method – a simplified design technique used by most lighting design engineers. Sometimes this method is used to get an approximation before more detailed design work.

The technique that is used for square and rectangular rooms is called the **lumen method** and uses the following formula.

$$E = \frac{F \times n \times N \times MF \times UF}{A}$$

For most design instances you will want to know the actual number of luminaire fittings required. Remember, a luminaire might contain more than one actual lamp, such as a four-tube fluorescent fitting. To find the number of luminaires requires a bit of transposition to give:

$$N = \frac{E \times A}{F \times n \times MF \times UF}$$

where:

E = average illuminance

F = initial lamp lumens

N = the number of luminaires

n = number of lamps in each luminaire fitting

MF = the maintenance factor

UF = the utilisation factor

A = the area of the room being lit.

To fully understand, look at the worked example below.

Worked example

You have been asked to design the lighting for a classroom that is 8 m by 6 m. The typical average light level (illuminance) required at the learners' desks is 400 lux. How many luminaire fittings will be required when each luminaire contains two fluorescent tubes of 2 300 lumens per fluorescent tube? You can assume the maintenance factor is good at 0.8 and the utilisation factor is also good due to the room colour and ceiling tiles being just off-white, 0.85.

$$N = \frac{E \times A}{F \times n \times MF \times UF}$$

$$N = \frac{400 \times 8 \times 6}{2\ 300 \times 2 \times 0.8 \times 0.85}$$

N = 6.14 luminaires (rounded down to 6)

Progress check 6.7

1 What does maintenance factor mean?

2 What does utilisation factor mean?

3 What is the formula for calculating the number of luminaires in a square room?

Activity 6.11

A 20 m by 10 m workshop requires lighting. The typical average light level (illuminance) required at the workbench is 300 lux. How many luminaire fittings will be required when each luminaire contains four fluorescent tubes? You can assume the maintenance factor is average at 0.6 and the utilisation factor is also average due to the room colour and ceiling tiles being whitewashed but dirty, 0.65.

Space to height ratio (SHR)

Different manufacturers design luminaires to different specifications. For each particular luminaire, a manufacturer will therefore have different recommendations for how many are required to give specific light coverage in a room. The space to height ratio is the ratio of centre to centre distance between adjacent luminaires to the height above the work surface being lit. The manufacturer's recommendations for a particular luminaire should not be exceeded. To find the space to height ratio for a rectangular room, the following formula can be used.

$$\text{Space to height ratio} = \frac{1}{H_m} \times \sqrt{\frac{A}{N}}$$

where:

H_m = the height from the light to the work surface

A = the surface area of the floor

N = the number of luminaires.

Operation and application of luminaires

Luminaire design and technology is developing extremely fast due to consumer demand. The need for cheaper, more efficient lighting systems has been accelerated by the ecological demands of the customer. LED technology is currently a growth market with manufacturers working very hard to overcome issues such as cost, control and the actual light quality. However, other technologies will be around for a long time to come and are also discussed in this section.

Leaving LED lights until later, the other two main types of lighting that are covered in this chapter are incandescent and discharge lighting.

Incandescent lighting

By passing current through a conductor, a couple of things happen.

- The wire will get warm if there is resistance.
- If the wire gets very warm, it will glow.

This is the basic principle of an incandescent light. There are many types of incandescent lamps with variations in technology that give them slightly different properties suitable for different applications.

General lighting service (GLS) lamps

The GLS lamp is by modern standards an inefficient method of lighting. The reason GLS lamps have been around for a long time is largely due to the wide range and design of the glass case, colour and shape. When you change a blown GLS bulb at home, the first thing you do is check the fitting. There are different ways of connecting the lamp to the supply which means different types of **lamp cap** and socket. The two main types of cap are the bayonet and the Edison screw cap.

The Edison screw lamp has also been around for a number of years and comes in different cap diameters. The code given to each type of Edison screw lamp defines the size of the cap. Edison screw cap lamps are used in both domestic and industrial applications. The standard household Edison

Key term

Lamp cap – a light lamp cap is the part of the lamp that connects to the supply.

lamps are E14 and E27 (14 mm and 27 mm diameter fitting caps). Larger caps are available for industrial use. The giant Edison screw cap lamp is a discharge mercury lamp with the code E40 (40 mm diameter cap) – this will be covered later in the chapter on page 268.

The GLS incandescent lamp works by visible light being produced by current passing through fine gauge tungsten filament coils. The coils are wound into a coil and then the coils are further wound to make them more efficient. As the tungsten heats up a reaction happens and some of the tungsten evaporates, moving to the inside wall of the glass envelope, turning the inside darker. Over a period of time evaporation causes the filament to wear thin. Eventually the filament will fail just like a fuse.

Figure 6.21: Edison screw cap

Figure 6.22: Bayonet cap

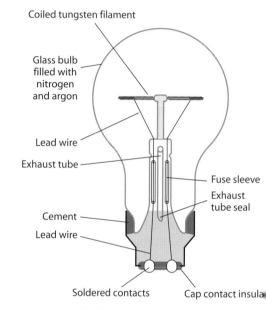

Figure 6.23: GLS lamp

The glass bulb is filled with various gases that have different functions. Argon helps reduce the evaporation and nitrogen reduces the chances of arcing. When a GLS 'light bulb' eventually does fail, a current surge can occur just as with any machine that is at the point of failing. A 40 W bulb in normal conditions will only require 0.17 A but this could peak for an instant close to the breaker rating of typically 6 A if other lights are on in that particular circuit. This surge can cause tripping on sensitive circuit breakers and RCDs. The glass envelope can get quite hot as the filament typically operates in the range of 2 500°C to 2 900°C. The colour rendering of a GLS lamp is fairly good but the efficacy is not as good as other light technologies with a rating of 14 lumens per watt – for every watt of power supplied, 14 lumens of luminous flux are emitted.

Tungsten halogen lamps

Another type of incandescent lamp is the **tungsten halogen lamp**. This lamp works on the same principle as the GLS but it operates at a much higher temperature. As with the GLS lamp, tungsten will evaporate from the filament as it gets hot. The tungsten molecules drift to the inside wall of the glass envelope. As the molecules get closer, a chemical reaction occurs. The

Key term

Tungsten halogen lamp – sometimes also referred to as a quartz iodine lamp as iodine is a halogen element that is added to the gas inside the lamp.

added halogen element, iodine or bromine, combines with the tungsten and drifts back to the filament. This **halogen cycle** occurs because of the higher operating temperature and gives the tungsten halogen lamp several advantages over the standard GLS lamp. Unlike the GLS lamp where after a period of use the inside starts to turn black and the light appears to dim, the tungsten lamp stays cleaner for a lot longer. The halogen cycle stops the black deposit on the inside of the quartz envelope by re-depositing the tungsten on the filament. The lamp will stay clearer longer and the efficacy is improved.

By operating at a much higher temperature, a brighter light is achieved. However, the high operating temperature does come with some problems. The quartz glass envelope will reach in excess of 250°C. This will obviously burn an unprotected hand. The grease from handling will also cause the lamp to fail. As the quartz heats up to operating temperature, any grease spots on the surface will cause the quartz to heat up unevenly. If the difference is great enough on the surface the quartz will crack and the bulb will blow.

The tungsten halogen lamp is much smaller than the GLS for the same amount of light. This means it has a wider range of applications, can operate at higher temperatures and gives a whiter light with good colour rendering. The efficacy of tungsten halogen lamps is 20 lumens per watt.

Key term

Halogen cycle – the reversible chemical reaction that allows the evaporated tungsten to deposit back on the filament.

Figure 6.24: Halogen capsule lamp

Safe working

Always wear gloves when changing or installing tungsten halogen bulbs to prevent them prematurely blowing (and to prevent burns). If you do handle them with a naked hand always make sure they are cleaned thoroughly with methylated spirit before being installed.

Figure 6.25: Linear tungsten halogen capsule lamp

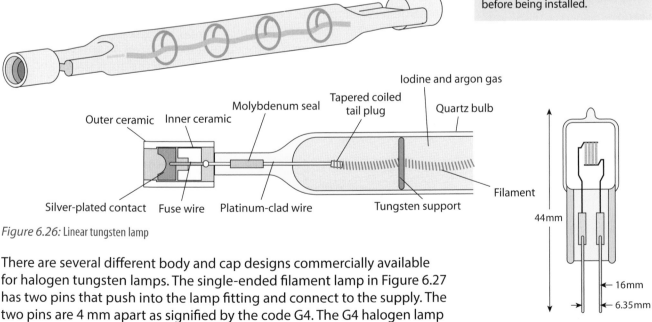

Iodine and argon gas
Tapered coiled tail plug
Quartz bulb
Molybdenum seal
Outer ceramic   Inner ceramic

Silver-plated contact   Fuse wire   Platinum-clad wire
Tungsten support
Filament

Figure 6.26: Linear tungsten lamp

There are several different body and cap designs commercially available for halogen tungsten lamps. The single-ended filament lamp in Figure 6.27 has two pins that push into the lamp fitting and connect to the supply. The two pins are 4 mm apart as signified by the code G4. The G4 halogen lamp is commonly used in low voltage applications such as under-unit kitchen lighting or living room lamps.

44mm
16mm
6.35mm

Figure 6.27: Single-ended filament lamp

Figure 6.28: Halogen spotlight

The single-ended filament lamp is also very common in domestic, industrial and commercial situations. Often used in floodlights, the linear tungsten lamp also has to be handled carefully as it suffers the same grease problem as the smaller capsule lamp. This type of lamp is available in a range of power ratings at mains voltage, with some security lights reaching 1 000 W or more.

Another variation of the tungsten halogen lamp is the halogen spotlight. The version most common in domestic and commercial applications is the **GU10** spotlight. The construction is similar to the capsule lamp except the capsule is set in a further glass case. This luminaire fitting can contain a **reflector** shield and a **diffuser** to soften the intense white light. Although these lamps get hot, the issue of grease causing the quartz to crack does not exist as the capsule is protected by a glass shield (see Figure 6.28).

Progress check 6.8

1 Describe briefly how an incandescent lamp works.
2 Describe why you should not handle a halogen lamp with bare hands.

Activity 6.12

1 Using the Internet, search manufacturer websites and download data sheets for a GU10 lamp. Justify which GU10 you would choose based on the data.
2 Investigate different household lamp applications (life span, power rating, lumens and cost) and make a comparison list for the most common types.

Key terms

GU10 – a bayonet cap fitting with two pins that are 10 mm apart. This type of cap is available in a range of different technologies, including LED lights.

Reflector – makes the lamp more efficient by reflecting the light that escapes from the back of the source, redirecting the light beams to the front. The angle of the reflective cone dictates the spread of the beam on the surface being lit.

Diffuser – a device that is added to the front of the bulb to break up and soften the light. The diffuser is constructed of a series of glass prisms or ground glass that reflect the light beam in different directions.

Fluorescent tubes – these are given the code MCF by manufacturers.

Discharge lighting

Discharge technology uses conductive gases instead of the solid metal filaments found in incandescent lights. The principle is based on producing enough charge on start-up to force current to pass over a contactor gap through a metallic gas. Once the current is flowing the arc is struck and the lamp stays on. There are different types of discharge lighting based on the gas contents of the bulb and the method used to strike the arc. The most common discharge lamp that you will come across as a building services engineer is the fluorescent tube.

Figure 6.29: Compact fluorescent light

Fluorescent tubes (low pressure mercury lamps)

With customer pressure for more ecologically friendly lighting, **fluorescent tubes** have seen something of a revival over the past few years with the availability of new compact fluorescent light fittings, as shown in Figure 6.29.

Fluorescent lights are slightly more efficient than their GLS equivalent with an efficacy of between 40 and 90 lumens per watt. A standard fluorescent tube is linear in design (like the linear tungsten halogen lamp). The inside of the tube contains a low pressure mercury vapour.

Other gases such as argon, neon or krypton are also in the glass tube. Each different gas will give the fluorescent tube a different colour. Neon will make a red light, mercury a blue light and argon a green light.

Because fluorescent tubes are so common, it is very important for you to understand how they work. If you understand their start-up stages, fault finding will become straightforward.

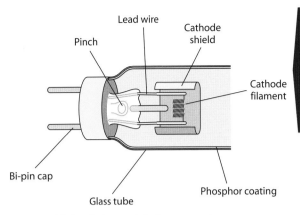

Figure 6.30: Detail of one end of a fluorescent tube

A fluorescent tube is made up of a starter circuit, the tube and a power factor correction circuit. The starter circuit requires a choke or ballast to provide the necessary high voltage to strike the tube and a set of contacts that allow the release of energy to the tube. When the power is turned on current flows through the choke and a large magnetic field starts to build up in the coils.

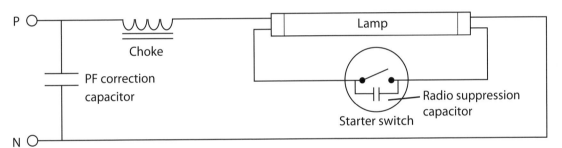

Figure 6.31: Glow-type starter circuit

To understand the starter circuit you need to consider the current path initially. The current passes through to one end of the lamp which heats up the cathode filament. Heating the tungsten filament excites the electrons creating an electron cloud at the cathode. The current then passes through to the starter switch. This switch is contained in a glass bulb and is made of bimetallic strips that heat up and bend towards each other. The reason current passes through from one contact to another in the starter switch is the partially conductive helium gas. A glow discharge happens around the bimetallic contacts. To complete the conductive path, the current passes through the switch contact and on to the other cathode at the far end of the tube before continuing on to the neutral.

So how does the magnetic energy stored in the coil actually get the tube to strike? Back at the starter switch, the partially conductive helium gas allows the current to pass from contact to contact within the starter bulb. As this occurs the bimetallic contacts heat up and bend to each other until they touch. The conductive helium gas causes a glow in the starter.

As they touch they go from being a high resistance path to a very low resistance path. As they fully touch, the bimetallic strips cool down rapidly. The rapid cooling makes the bimetallic strips spring apart very quickly which causes the magnetic field to collapse in the coil of the choke. If you remember from the electromagnetism section in *Chapter 1: Principles of electrical science*, a collapsing magnetic field will create a large back emf (voltage). This large voltage is dropped across the two cathodes which is enough to encourage the already excited electrons at the cathodes to bridge the gap and create a fully conductive path. As the discharge happens from one end of the tube to the other, the gases warm up and the path becomes less resistive. If there is less resistance, more current is able to flow from cathode to cathode. The choke therefore has a second function – one of a current limiter. If this whole process does not strike the tube, it will be repeated until it lights up – have you ever noticed fluorescent lights flickering?

Once the light has fully struck, visible light is created by the photons given off in the conductive tube reacting with the phosphorous coating inside the tube wall.

There are two other components worth mentioning: the power factor correction capacitor and the radio frequency suppression capacitor. The choke is essentially a large inductive coil. You will remember from page 240 that an inductor will make current lag the voltage. This is not a good thing as it could mean more current being drawn than is actually required by the lamp. A capacitor is therefore added in parallel to the circuit to correct the current and bring it back in phase with the supply voltage. The other component is the radio frequency suppression capacitor. This is required to suppress the spark given off as the two bimetallic strips spring apart. If this electromagnetic pulse were not suppressed it would interfere with other electronic equipment like TVs and radios (crackles).

High pressure mercury lamp

The construction of the **MBF** lamp is complex, making it expensive compared to other alternative technologies such as metal halide lamps.

The lamp consists of two glass envelopes, one inside the other. The outer envelope acts as a filter for the ultraviolet light the lamp produces. The outer envelope can also be lined with phosphor to give a better light and colour rendition. The inner smaller quartz envelope contains the high pressure mercury vapour where the arcing occurs as the gas discharges and strikes. This type of lamp is more efficient than most low pressure mercury lamps, with an efficacy in the region of 65 lumens per watt. As the discharge starts in the lamp on start-up it initially appears blue because of the ionising mercury, but after a while the light turns pure white. The bright white light means it is very suitable for large overhead lighting applications such as warehouses, sports fields, car parks or street lights. The colour rendition is also much better than the sodium light equivalent you are about to cover.

> **Key term**
>
> **MBF** – the classification term given to high pressure mercury lamps by manufacturers.

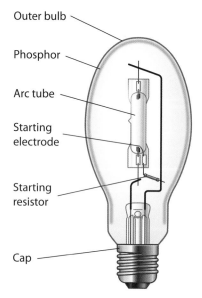

Outer bulb

Phosphor

Arc tube

Starting electrode

Starting resistor

Cap

Figure 6.32: High pressure mercury vapour lamp

Metal halide lamp

As mentioned previously, metal halide lamps are a more efficient version of the high pressure mercury lamp. By adding metal salts to the mercury vapour lamp it makes them a high output, efficient light source but smaller in size. The operating principle is the same for all discharge lighting. The arc is struck through a mixture of vaporising chemicals, including mercury, argon and metal halides. As the temperature and pressure increases the light is produced.

Low pressure sodium lamp

This lamp is commonly used in street lighting because of its long life (6,000 hours) and good efficacy (61–160 lumens per watt). However, the U-shaped glass tube contains an amount of solid sodium. Sodium is dangerous because it is hygroscopic. This means it particularly likes water and will not stop burning until it is no longer thirsty! If it gets onto skin it will cause very serious burns. For this reason, this type of lamp has to be very specially constructed. It is made with sodium-resistant seals and glass. The lamp also has to operate horizontally (or at a maximum of 20° out) so the sodium evenly distributes over the length of the U-shaped tube.

Figure 6.33: Metal halide lamp

When the discharge happens in the lamp it takes a while for the normal operating colour to appear. Initially on start-up the lamp appears red but after 6 to 8 minutes the lamp turns yellow. The yellow colour of the lamp means it gives very poor colour rendering as you will notice from street lights – the main application. This type of lamp is given the code SOX or SLI by the manufacturers.

High pressure sodium vapour lamps

As with all sodium lamps, high pressure sodium vapour lamps require special handling and disposal. The high pressure nature of this lamp requires a special **sintered** aluminium oxide arc tube within a hard glass outer envelope.

Figure 6.34: Low pressure sodium lamp

This lamp contains sodium, argon and xenon to help the starting process. An electronic starter is required to initiate the discharge. Once the lamp strikes it takes about 5 minutes for the colour to grow into a warm white. The warm nature of the light makes it perfect for a whole range of applications: food halls, reception areas, factories, floodlighting and warehouses. Areas where maintenance could be an issue are also suited to high pressure sodium lamps as they have an average life span of 6,000 hours, so they don't need to be changed often.

Figure 6.35: High pressure sodium lamp

Progress check 6.9

1 What is the main advantage of a metal halide lamp over a high pressure mercury lamp?
2 What is the code given to high pressure mercury lamps by manufacturers?

Key term

Sintered – glass specially manufactured from powdered glass that is moulded and set by firing.

Activity 6.13

Write a risk assessment for changing a high pressure sodium floodlight in a high bay warehouse.

Chapter 6

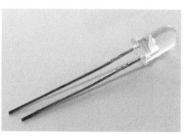

Figure 6.36: LED light

LED lighting

LED technology is revolutionising the lighting industry. LEDs (light-emitting diodes) are becoming cheaper to manufacture and more solutions are being developed as straight replacements for many of the established light systems. LEDs come in a wide range of packages and forms but are all based on semiconductor electronic devices (the operation of an LED will be covered fully later on page 328).

The main advantage of LED lighting is the reduced running cost, with only a fraction of the current required compared to any other equivalent light source. LEDs also have by far the longest life of all the light technologies available commercially. The only limitation is the availability of cheap mass-produced lamps. Manufacturers are working very hard to bring out new products every week that appear and act like GLS or tungsten halogen lamps. Many domestic households are changing low voltage or mains-powered downlighters to LED technology because of the LED performance. Within the new breed of LED lamps that copy domestic lamps, many different effects can be achieved through the clever design of diffusers and reflector plates. One difficulty that manufacturers are still working hard on is the ability to dim the LED in a domestic situation. Because the operating current is so small, to vary this current requires complicated and expensive electronics.

Case study

A major hotel chain has recently changed all their head office lighting to LED technology. The change has been made to take advantage of the longer life span of the LED compared to the mix of older GLS, fluorescent and halogen lights. The saving in power is reported to be over 800 000 kWh each year together with substantial maintenance savings and fewer bulbs being changed.

Lamp type	Advantages	Disadvantages
GLS	• Cheap • No control gear required • Fairly good colour rendering • Can operate in any position • Lots of colours and styles	• Rated life only 1,000 hrs • Poor efficacy (14 lm/w)
Tungsten halogen	• Cheap • No control gear required • Good colour rendering • Good power range	• Efficacy of 20 lm/w • Cannot be handled without cleaning • Can get very hot
Low pressure mercury	• Cheap • Long life • Better efficacy than GLS, tungsten halogen and high pressure mercury (60 lm/w)	• Special disposal required • Contains mercury • Several failure points in circuit • Requires starting gear • Requires power factor correction
High pressure mercury	• Expensive	• Special disposal required • Contains mercury

Metal halide	• Better efficacy than high pressure mercury (64–84 lm/w, high pressure mercury is 34–54 lm/w) • Better colour rendering than sodium lights • Good for large overhead lighting such as sports grounds	
Low pressure sodium	• Long life • Good for street lighting • Good efficacy of 100–175 lm/w	• Sodium is hygroscopic • Special disposal required • Very poor colour rendering
High pressure sodium	• Long life • Good efficacy of 67–120 lm/w • Good for a wide range of commercial/industrial applications	• Sodium is hygroscopic • Special disposal required • Only fair colour rendering
Light-emitting diodes (LEDs)	• Very good efficacy of 60–100 lm/w • Growing range of fittings and styles • Very low current • Good colour rendering	• Still relatively expensive • Difficult to dim • Technology still developing

Table 6.3: Advantages and disadvantages of light technologies

Activity 6.14

Find a plug-in lamp, safely turn it off at the socket, unplug it using gloves if it is hot (or wait 10 minutes for it to cool off) and then take the bulb out either by unscrewing or push/twist and release. Find a data sheet for this particular lamp on the Internet and make your own notes on the lamp, including power rating, lumens, energy rating and the type of fitting. Create your own data sheet to give to a customer explaining the main features.

Progress check 6.10

1 Describe luminous intensity.
2 What is the formula for illumination directly under a light source?
3 What is the formula for illumination at a point to the side of the source?
4 Draw the circuit for a low pressure mercury lamp and label all the parts.
5 Describe how a fluorescent luminaire starts.
6 Name five different luminaire types.
7 Describe the main advantages and disadvantages of each.
8 Describe an application for each type of light system.

Working practice 6.1

A large college has noticed that over the past two years their electricity bills have risen faster than expected and more than the advertised rates. They call in a specialist to investigate where all the electricity is being used as the onsite maintenance team has not been able to solve the problem. The electrical contractors notice there is a very high volume of fluorescent lights. All of the fluorescent tubes are changed when they blow by the maintenance team and the number of lights and plug-in electrical equipment has not changed significantly over the past few years.

1 What could be the problem?
2 How could this be proved?
3 What could be the solution?

LO3   # THREE-PHASE SYSTEMS

You are now familiar with the idea that the three-phase system is three similar sine waves that have been created by armatures that are 120° apart and spinning within a magnetic field. You have also seen previously that the three-phases of voltage created can be connected up in different configurations at the transformer. How these phases are connected will determine if the system is a delta or star system.

Star network calculations

With a star connected system there is only one current. Looking at Figure 6.37, you can see by following the current passing along the line from the left, that it reaches the corner, turns down into the phase and then carries on to the star point. This star point is connected to the neutral and to earth. So, the line current is actually the same value as the phase current that passes through the phase coil on the transformer. This means there is only one current, giving the formula:

$I_L = I_P$

where I_L is the line current travelling along the line and I_p is the phase current travelling along the phase of the transformer.

If you look at Figure 6.37 again, you can see that between any line and the neutral you will get a voltage level. This voltage is across a single-phase of the transformer and is called the phase voltage. However, there is another option. If you go between any two lines you will get a different voltage. The voltage is the combined effect of two voltages but they are 120° apart or out of phase. This voltage is bigger than the single-phase voltage by a value of √3 and is called the line voltage because it is between two lines.

To calculate the line voltage, use the formula:

$V_L = \sqrt{3} \times V_p$

Worked example

What is the line voltage if the single-phase voltage is 230 V?

$V_L = \sqrt{3} \times V_p$

$V_L = \sqrt{3} \times 230$

$V_L = 400 \text{ v}$

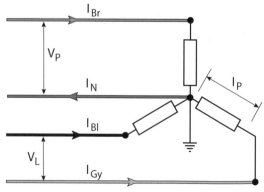

Figure 6.37: Star network connection

Power in a star configured network

In the d.c. world power is found by the formula:

$P = V \times I$

In the a.c. world power has to take into account coils, capacitors and hence power factor, giving the formula:

$P = V \times I \times \cos \varnothing$

In the a.c. three-phase world, this formula develops further with the addition of the root 3.

$P = \sqrt{3} \times V_L \times I_L \times \cos \varnothing$

Remember, three-phase power is still basically found by $V \times I$ but you need to consider how efficient the circuit is (how close in phase the voltage and current are). You also need to consider the power of the other two phases even though they are not all simply added together – hence, the use of $\sqrt{3}$. Notice also that the voltages and current are the line values.

Worked example

A star connected network has a line current of 30 A and a phase voltage of 250 V. Find the following values by calculation if the power factor is 0.8:

1 the phase current

2 the line voltage

3 the power.

1 Remember that in a star network there is only one current, so use the formula:

$I_L = I_P$

$I_P = I_L$

$I_P = 30$ A

2 $V_L = \sqrt{3} \times V_P$

$V_L = \sqrt{3} \times 250$

$V_L = 1.73 \times 250 = 433$ V

3 $P = \sqrt{3} \times V_L \times I_L \times \cos \varnothing$

$P = \sqrt{3} \times 433 \times 30 \times 0.8$

$P = 18$ kW

Calculation of the neutral current in a three-phase star network

As mentioned earlier, a balanced star network will have no current flowing in the neutral. However, this is not always the case as a fully balanced load on each phase is not always possible. If the loads are not fully balanced, there will be a current flowing down the neutral. This current will cause the neutral to heat up and should be minimised by careful design. The easiest way to find a neutral current is by creating a

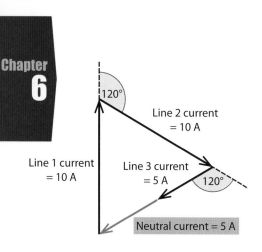

Figure 6.38: Phasor diagram showing resultant neutral current on a three-phase unbalanced star network

simple phasor diagram. Each line current is drawn to scale and 120° apart. To make the phasor diagram easier to work with, the currents can lead on from each other as shown in Figure 6.38. Draw line current 1 to scale, measure 120° and then start the next line current. When the third and final line current is drawn, the gap between the start point and end point represents the current flowing in the neutral. This can be found by simply measuring and using the scale, as shown in Figure 6.38.

Activity 6.15

1   A star connected network has a line voltage of 6.35 kV and a line current of 100 A. Find the following values by calculation if the power factor is 0.7:

   • the phase current

   • the phase voltage

   • the power.

2   A star connected network has a line voltage of 65 kV and a phase current of 600 A. Find the following values by calculation if the power factor angle is 32°:

   • the phase current

   • the phase voltage

   • the power.

   Now find the values by calculation if the power factor changes to 0.7.

3   An unbalanced star connected network has the following line currents: line 1 = 13 A, line 2 = 13 A and line 3 = 5 A.

   What is the current flowing in the neutral? Show by phasor diagram using an appropriate scale.

Delta network calculations

A tip – you only need to remember one set of formulae, the star or the delta. Delta and star formulae are opposite from each other.

Delta systems: one voltage and two currents.

Star systems: one current and two voltages.

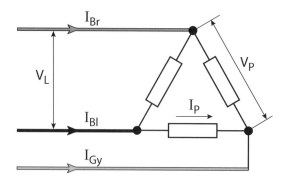

Figure 6.39: Delta connection

Looking at Figure 6.39, you can see there is no centrally connected point or neutral in a delta system. This means the only way you can get a voltage is between two lines, so there is only one voltage and line voltage must equal phase voltage. This gives the formula:

$V_L = V_P$

You can also see by following the current passing along the line from the left, it reaches the corner, turns down into the phase but this time the line current can split down either phase. This means the actual current passing down a phase is different from the line current.

This is shown by the formula:

$I_L = \sqrt{3} \times I_P$

where I_L is the line current travelling along the line and I_p is the phase current travelling along one of the phases of the transformer coil.

Worked example

What is the line current if the phase current is 30 A?

$I_L = \sqrt{3} \times I_P$

$I_L = \sqrt{3} \times 30$

$I_L = 51.96 \text{ A}$

Power in a delta configured network

To find the power in a delta connected network you use exactly the same formula as the star connected network.

$P = \sqrt{3} \times V_L \times I_L \times \cos \varnothing$

Notice again that the voltage and current are the line values. (However, it doesn't matter about the line voltage as it is the same as the phase voltage.)

Worked example

A delta connected network has a line current of 50 A and a phase voltage of 412 V. Find the following values by calculation if the power factor is 0.75:

1 the phase current

2 the line voltage

3 the power.

1 Remember that in a delta network there are two currents, so use the formula:

$I_L = \sqrt{3} \times I_P$

$I_p = \dfrac{I_L}{\sqrt{3}}$

$I_p = \dfrac{50}{1.73} = 28.9 \text{ A}$

2 $V_L = V_P$

$V_L = 412 \text{ V}$

3 $P = \sqrt{3} \times V_L \times I_L \times \cos \varnothing$

$P = \sqrt{3} \times 412 \times 50 \times 0.75$

$P = 26.76 \text{ kW}$

Progress check 6.11

1 How many different currents and voltages are there in a star network?

2 How many different currents and voltages are there in a delta network?

3 What is the formula for power in both star and delta networks?

Activity 6.16

1 A delta connected network has a power factor angle of 25° and a line voltage of 750 V. If the line current is 100 A, find the following by calculation:

a) the phase voltage

b) the phase current

b) the power.

2 A delta network has a phase voltage of 400 V, a phase current of 95 A and a power factor angle of 12°. Calculate:

a) the line voltage

b) the line current

c) the power.

Three-phase supplies and loads

Until now you have only used the voltages and currents given. If you were to look at this a little deeper you will need to consider using Ohm's law to find unknown currents in three-phase transformer coils (or loads). Consider Figure 6.39 on page 274 for a delta connection. The actual phases are shown as resistors. They are coils but will have resistance and inductive reactance – they are after all just long pieces of coiled-up wire in a transformer coil. From earlier work in the chapter you will remember resistance and reactance can be combined to give an overall effect called impedance (page 242). This impedance can be used in the Ohm's law formula by simply replacing the R with a Z.

Worked example 1

A delta connected supply has three equal loads of 40 Ω (for this example assume the load to be pure resistance) and a phase voltage of 400 V. Calculate the following values:

a) the line current

b) the phase current

c) the power.

For this example you only need to consider the load as a resistance so by using Ohm's law you can find the first missing value, the phase current. (Note that sometimes you have to answer the questions out of order at Level 3.)

$$I_L = \sqrt{3} \times I_P$$

$$I_P = \frac{I_L}{\sqrt{3}}$$

But you do not know I_P or I_L so you must start with Ohm's law. If you treat one of the phase legs as a simple Ohm's law problem to solve:

$$I_P = \frac{V_P}{R}$$

$$I_P = \frac{400}{40}$$

$$I_P = 10 \text{ A}$$

continued

Now the rest can be solved.

$I_L = \sqrt{3} \times I_P$

$I_L = \sqrt{3} \times 10$

$I_L = 17.32 \text{ A}$

And the last part, power, can be found.

$P = \sqrt{3} \times V_L \times I_L \times \cos \varnothing$

$P = \sqrt{3} \times 400 \times 17.32 \times 1$

$P = 12 \text{ kW}$

Notice that because the load is just a pure resistor, the power factor is unity, or 1. This means the current is completely in phase with the voltage and the circuit is at its most efficient.

What about if, as mentioned before, we have a coil with impedance?

Worked example 2

A delta circuit has three equal coils of resistance, 40 Ω, and inductive reactance of 10 Ω with a line voltage of 400 V. Calculate:

a) the power factor

b) the phase current

c) the line current

d) the power.

The best way to solve problems like this one is to keep it simple, write down everything you know and draw a simple labelled diagram, as can be seen below. This will enable you to identify the missing things. As an engineer you will probably be helped by this visual representation of the problem.

So the first step is to just consider what is going on in one of the phases. Draw the circuit out, label it up and then look for anything obvious to work out – remember, you do not necessarily need to answer the question in order!

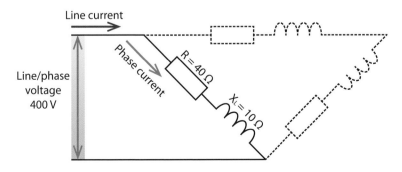

continued

Just concentrating on one phase, you will be able to work out the phase current. So, to work out the impedance of one phase you need to use the impedance formula from page 243, using Pythagoras' theorem.

$$Z = \sqrt{R^2 + X_L^2}$$

$$Z = \sqrt{40^2 + 10^2}$$

$$Z = 41.23 \ \Omega$$

The current in one of the phase legs can now be found using Ohm's law.

$$I_P = \frac{V_P}{Z}$$

$$I_P = \frac{400}{41.23}$$

$$I_P = 9.7 \ A$$

The line current can now be found.

$$I_L = \sqrt{3} \times I_P$$

$$I_L = \sqrt{3} \times 9.7$$

$$I_L = 16.8 \ A$$

Power factor, cos Ø, can also be found from the impedance triangle (see page 253).

$$\cos Ø = \frac{Adjacent}{Hypotenuse}$$

$$\cos Ø = \frac{R}{Z}$$

R is the resistance of the coil and Z is the overall impedance of the coil, giving:

$$\cos Ø = \frac{40}{41.23}$$

$$\cos Ø = 0.97$$

The last part of the problem can now be solved – power.

$$P = \sqrt{3} \times V_L \times I_L \times \cos Ø$$

$$P = \sqrt{3} \times 400 \times 16.8 \times 0.97$$

$$P = 11.29 \ kW$$

This approach can be used if the three-phase system is delta or star. To solve the problem, simplify into parts, draw and label the circuit, then identify what is missing. If all the transformer coil loads are identical, only one leg needs to be worked on until the missing parts of the puzzle are solved. The only decision required is – do you use star or delta equations?

Activity 6.17

1 A star network has three identical phase loads. One phase has a resistance of 55 Ω and an inductor of 38 mH operating at 50 Hz. If the line voltage is 175 V, calculate:

   a) the inductive reactance

   b) the impedance

   c) the power factor

   d) the line current

   e) the phase current

   f) the phase voltage

   g) the power.

2 A delta network has three identical phase loads. One phase has a resistance of 15 Ω and an inductor of 12.7 mH operating at 60 Hz. If the line voltage is 200 V, calculate:

   a) the inductive reactance

   b) the impedance

   c) the power factor

   d) the line current

   e) the phase current

   f) the phase voltage

   g) the power.

ELECTRICAL MACHINES

LO4

In this section you will build on your Level 2 knowledge of machines and you will learn the principles of a range of machines. As well as learning how to cause movement using d.c., a.c. single-phase and a.c. three-phase, you will cover how to control motors, including start, stop and speed control. The motor principle is simple but there are a number of things you can alter in a motor design that make it suitable for different applications. Some motors need to move very slowly but have great power, such as the Kings Cross escalator or the Covent Garden deep level lift. Other motors need great speed but less power, such as a high speed milling machine or lathe. There are a number of different types of a.c. and d.c. motors, each with a range of applications.

The main motor topics covered include:

- d.c. machines
  - series motor
  - shunt motor
  - compound motor
  - separately excited
- three-phase a.c. machines
  - cage induction motor
  - wound rotor
- single-phase a.c. machines
  - split phase induction motor

- o capacitor start induction motor
- o capacitor start, capacitor run induction motor
- o universal motor
- o shaded pole motor
- o synchronous motor (both single-phase and three-phase)
- start/stop
- speed control.

d.c. machines

The basic principle of causing movement in motors of any kind is magnetic attraction and opposition. Imagine bringing two magnets together and waiting in anticipation for the reaction. Are they going to jump apart or slam together? Imagine if you were able to switch this reaction on and off or control in some way the speed of that reaction. This is possible. You have covered transformers and know that the most efficient transformers are the ones that can be turned into magnets and then turned back into un-magnetised silicon steel (low hysteresis magnetic curve, Chapter 1, page 45). If you were able to do this to one set of the motor magnets then you could also turn the opposition/attraction on and off.

The idea of varying one or more of the motor magnets by making them electromagnets is the principle of all d.c. motors. By passing current down a conductor, you create a magnetic field. If this conductor is long enough, you can create quite a large magnetic field. It is also possible to wind a conductor around an existing magnet and when current passes through the conductor this will in turn make the permanent magnet even stronger. So we have proved that magnets and magnetic fields can be induced. We have also proved that you can increase magnetism to a permanent magnet by wrapping a current-carrying conductor around it. Let's look at combining all of these things and give you the correct terminology.

The construction of a simple d.c. motor consists of a conductor that is bent into a loop. This loop enables you to connect up the d.c. supply with one leg connected to the +ve and the other leg connected to the –ve of the battery or d.c. source. A switch is added in series with the supply so the power can be turned on and off. This loop is now called the **armature**.

> **Key term**
>
> **Armature** – the moving loop or loops of conductor that rotate within a magnetic field.

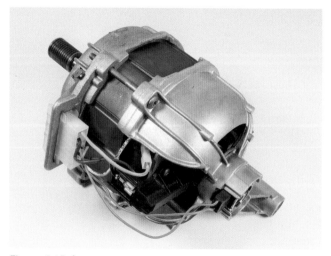

Figure 6.40: d.c. motor

Figure 6.41: d.c. motor armature

From the formula 'forceful Bill' (page 35), you will remember that one of the things you can change to make movement more powerful is the length of the conductor. So, if you have a much longer conductor by creating more loops, you will create a bigger, stronger motor. In reality, the armature will be made of a great number of conductor loops. When current passes through this armature, the first magnetic field will be set up for your motor reaction. All that is required now is to create a second magnetic field for the armature to react to. The second magnetic field is created by placing magnetic pole pairs in a casing that surrounds the armature. Whereas the armature will rotate on a spindle, the outer magnetic field is static. The outer case that surrounds the rotor is called the **yoke**. The yoke houses the magnetic pole pairs securely. As mentioned before, a good way to make the magnets even stronger and get a bigger motor reaction is to electrify the permanent magnets. A conductor is wound around the permanent magnets and these windings are called **field windings**. So there is an armature that requires power and a field winding that requires power. A single d.c. source is required and it is the way in which this is connected to the field winding and armature winding that changes the motor's characteristics.

Before the operation of a series d.c. motor is explored further, consider the fact that, if the armature is spinning, how is the d.c. supply kept connected to it? This is achieved through a simple but clever process called **commutation**.

Key terms

Yoke – the metal case that forms the outer support for the moving motor components, such as the armature in a d.c. motor.

Field windings – the insulated conductors wound around the magnetic pole pairs housed in the yoke of a d.c. motor.

Commutation – literally means to interchange or exchange. In electrical terms it means to change the direction of current.

Commutator – attached to the armature and spins with it. It is made of copper segments that are insulated from each other. For smooth rotation and better motor characteristics, more segments and loops are added.

Chapter 6

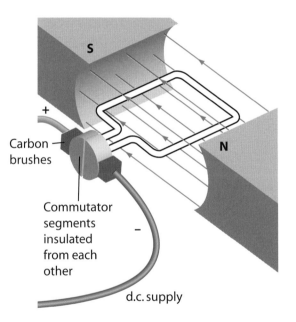

Figure 6.42: Single loop commutator

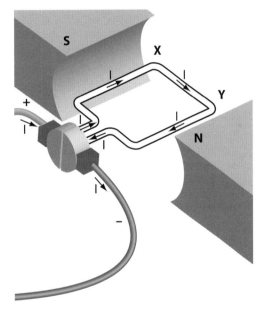

Figure 6.43: Single loop with d.c. flowing

The **commutator** is designed to be part of the armature and each separate copper segment pair is connected to a leg of the same loop. As the armature spins, so do the commutator segments. If there are many loops, the commutator also has many segment pairs to match the +ve and −ve legs of the loop.

So how does the commutator work? It is not totally obvious on first inspection. You have already seen in the introduction to machines (page 29) how a current-carrying conductor will react to the magnetic flux of a permanent magnet when brought close.

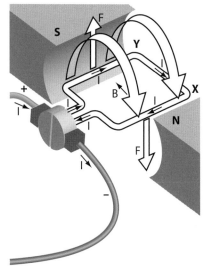

Figure 6.44: Single loop rotated 180 degrees

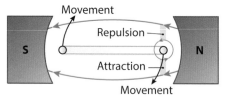

Figure 6.45: Single loop starts moving

Safe working

Carbon brushes that do not have a wear and tear sensor need to be regularly checked and maintained to avoid metal rubbing against metal.

Progress check 6.12

1 Describe how a commutator works.

2 Why is carbon a good material for a motor brush?

3 What is the outer case that holds the pole pairs called?

As can be seen in Figure 6.44, the permanent magnet flux lines are flowing left to right over the top and underneath the armature loop. The current is flowing clockwise around the loop. If you just concentrate on the right-hand side of the loop and use the 'corkscrew rule', you can work out the direction of the induced magnetic field around the conductor as anticlockwise. Now put the two magnetic fields together and you can see on the top of the conductor the flux lines are travelling in the same direction – alike = repulsion. So the top of the right-hand side of the loop will feel repulsion. Imagine you are at the bottom of the conductor now and take a fresh look at the magnetic flux lines. The permanent magnetic field is still flowing left to right but the flux lines around the conductor are in the opposite direction now – unlike = attraction. The same process is happening on the other side of the loop, so the loop will start to spin.

However, as the loop moves from the horizontal position to vertical, the situation changes. If the supply stayed flowing in the same direction as the loop carried on rotating past vertical, the forces would be reversed and the loop would stop and change direction! This is not good for a motor as all you will have is an armature that is continuously changing direction. This is where the commutation comes in. The supply is connected to the moving commutator and the armature loops via static carbon brushes. The carbon brushes maintain contact with the commutator by a spring-loaded housing pushing them onto the copper commutator. As the commutator rotates, it slides against the brushes but electrical contact is maintained. Now imagine the loop is approaching the vertical position – this is the point at which ideally you want the supply to change direction. The loop has just enough momentum to keep moving and this pushes the commutator round past the insulated strip between the two commutator segments. The supply effectively switches direction and the flux lines around the conductor reverse – the forces are now the same as the first phase when the rotation started.

The carbon brushes have several advantages. Carbon has a negative temperature coefficient. This means as the brushes heat up with friction, the resistance decreases and they perform better from an electrical view point. Carbon also acts as a self-lubricant allowing the commutator to slide easily.

The d.c. series motor

As mentioned previously, the way in which the field and armature windings are connected to the supply determines the motor characteristics. The first configuration can be seen in Figure 6.46. The field winding is in series with the armature. The current flowing through the field and armature conductors is the same. This means when the motor is starting and the start-up current is very high, the magnetic forces will be very strong. Because of the strong magnetic forces, you can imagine the motor will have a lot of torque at start-up. As the motor accelerates to full speed the current drawn by the armature and field windings will reduce, as will the torque. To help understand the series motor, you need to look at the graph in Figure 6.47. Put your finger on the bottom axis and slide it along the line from right to left as if you were reducing the current I_a.

Now look up the graph and you will notice the torque is small but the speed is high. If you slide your finger from left to right (as if you were increasing the current when the motor starts up), you will notice the speed is slower but the torque is at its peak.

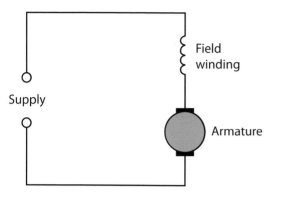

Figure 6.46: Series motor

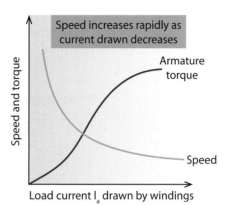

Figure 6.47: d.c. series motor operating characteristics

A danger with a series-connected motor can be seen by looking at this graph. If this motor were connected up to do work via a belt drive, what would happen if the belt broke suddenly? The torque would reduce suddenly and the motor would speed up. The motor would actually speed up until the point where it destroyed itself! Imagine a tug of war between two very powerful teams. The rope snaps and the teams fly apart at full speed – this is what happens if you connect a series motor up to a load by a belt drive.

Because of the operating characteristics, a series d.c. motor is good for low speed and high torque applications such as cranes, hoists and conveyor belts. A series d.c. motor is not good for loads that frequently change or that need to switch on/off a lot (a water pump that has to switch on and off constantly to maintain a water tank level).

Speed control of a d.c. series motor is achieved by varying the strength of the magnetic field. A variable resistor is connected across the field winding and this diverts some of the current, as can be seen in Figure 6.48.

Safe working

Avoid connecting up series d.c. motors to loads with belt drives – a snapped belt could lead to total motor destruction!

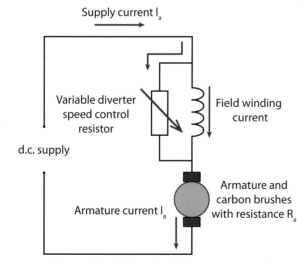

Figure 6.48: d.c. motor series motor speed control

> **Key term**
>
> **d.c. faceplate starter** – an external starting mechanism used to start large d.c. motors.

The start-up current, however, is very large and this requires some external control. A set of external resistors is connected to a mechanical starting plate also known as a **d.c. faceplate starter**.

A handle is moved across this starting plate to change the total resistance in the circuit. As the mechanical lever handle is rotated, resistance is gradually cut out of the circuit, as can be seen in Figure 6.49. The external resistors allow the power to be gradually increased until the motor is up to speed.

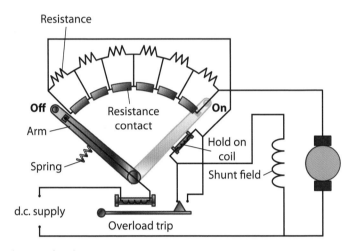

Figure 6.49: d.c. motor faceplate starter

Reversing a d.c. series motor is a case of simply changing the supply polarity to either the armature or the field winding. This has the effect of reversing one of the interacting magnetic fields and hence the direction the armature spins.

One other point for you to consider is the back emf that is generated in the armature. As a conductor loop spins, the other loops that are also in the armature will have voltage induced in them with a current flowing in the opposite direction. This back emf is the difference between the supply voltage and the volt drop in the armature (volt drop is just because the armature has resistance, R_a). You can calculate the armature volt drop using the formula:

Armature volt drop $= I_a \times R_a$

where I_a is the current flowing around the series circuit and through the armature and R_a is the resistance of the armature windings.

The back emf can be found using the formula:

$E = V$ – armature volt drop

where E is the back emf and V is the supply voltage.

The d.c. shunt motor

Although 'shunt' is an old-school term, it actually means parallel and is still used in industry. Parallel means exactly that – the field winding is connected in parallel across the armature.

> **Progress check 6.13**
>
> 1 How do you control the speed of a d.c. series motor?
> 2 Why should a series d.c. motor never be connected up to a load?
> 3 What is one method of starting a large d.c. motor?

The supply current now has two paths to flow down and the total current can be found by the formula:

$$I_T = I_a + I_f$$

where I_T is the total circuit current, I_a is the current flowing in the armature winding and I_f is the current flowing in the field winding.

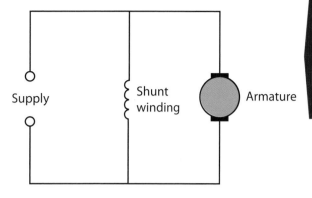

Figure 6.50: Shunt d.c. motor

One advantage of connecting the field winding in parallel is ease of control. If the load is suddenly removed by a belt drive breaking, only the armature current will decrease. The field current will remain the same and the motor will not self-destruct. This can be seen in the motor characteristics curve, Figure 6.51. The shunt motor is sometimes called a 'constant speed motor' as it has much better speed control than the series motor. The torque only drops off at very high speeds when the back emf becomes too large and the motor reaches a stall. The ability of the shunt motor to keep a constant speed over a range of torque means it is good for applications that require accurate torque and speed, such as small machine tools.

The shunt motor is started by the d.c. faceplate starter as per the series motor. Speed control is via a variable resistor in series with the field winding, as per Figure 6.52.

Progress check 6.14

1 What does 'shunt' actually mean?
2 What is a shunt motor also known as?
3 How is speed control achieved on a shunt motor?

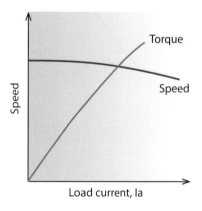

Figure 6.51: Motor characteristics curve

The d.c. compound motor

The series motor has very good torque at low speeds. The shunt motor has very good speed regulation. So why not combine the two configurations into one motor and have the best characteristics of both types? This is where the compound motor comes in. There are two main types of compound motors with slightly different wiring configurations. The main types are the short shunt and the long shunt compound motor.

- **Short shunt d.c. compound motor:** has both a series and shunt field winding. The series winding is connected to the parallel network of the armature and shunt winding, as can be seen in Figure 6.53.
- **Long shunt d.c. compound motor:** also has a series and shunt field winding. The series winding is in series with the armature and both of these are parallel to the shunt winding. This can be seen in Figure 6.53.

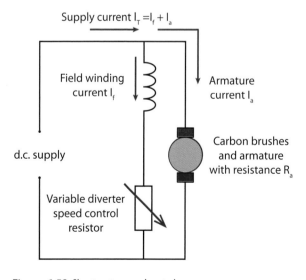

Figure 6.52: Shunt motor speed control

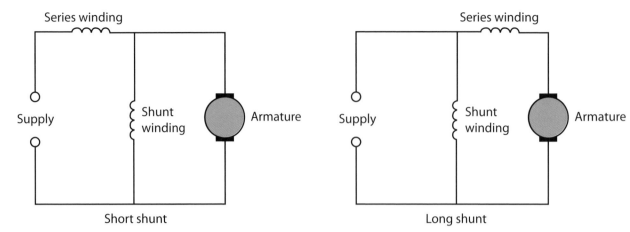

Figure 6.53: Compound motors

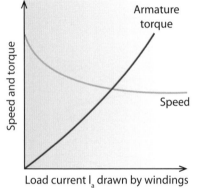

Figure 6.54: d.c. compound motor
operating characteristics

Both types of compound motor have constant speed over the full range of torque, as can be seen in the operating characteristic in Figure 6.54.

The d.c. separately excited motor

The last d.c. motor for you to consider is the separately excited motor. This is so called because the control circuit for the field winding is completely separate to the armature supply. The field current passes through the field winding and creates a magnetic flux. This induced magnetic field then reacts with the armature magnetic field that has been created by a separate d.c. supply.

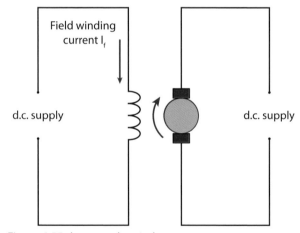

Figure 6.55: d.c. separately excited motor

Three-phase a.c. machines

In the introduction to machines you learnt about how an a.c. supply can be generated by rotating an armature within a magnetic field. You also learnt that to create a three-phase or poly-phase electrical source all that is required is to have three evenly spaced coils rotating in the magnetic field. For generation to occur, physical movement needs to go into the machine. For a motor, electrical energy needs to go in to create physical motion. The generator and the motor are essentially the same basic machine. The easiest of the a.c. machines to understand is the three-phase induction motor. So you will be looking at this first. As with the d.c. motors, there are a number of different types of a.c. motors for you to consider. The correct terminology will be introduced in this section, as well as some necessary calculations that you are expected to know as a building services engineer.

Cage induction motor

One of the most common a.c. motors you will come across is the cage induction motor. This is also known as the 'squirrel cage' induction motor because if taken to pieces it looks like a small animal's exercise wheel.

The construction of a cage induction motor can vary but the main type has copper bars connected together by a shorting ring at either end. The bars are not completely straight but at an angle (rifled). Imagine squeezing a bar of soap. If your fingers are right in the centre of the soap when you squeeze, the chances are you would just squash the soap and it would go nowhere. However, if you squeeze the soap from one end, where the soap is starting to slope, the soap will shoot out of your hands. Having the bars at an angle helps the initial movement and also makes the motor quieter in operation.

The metal shorting rings are important to complete the magnetic circuit generated within an induction motor. Remember, as well as adding structure and strength to the cage motor, the copper (or aluminium) bars are simply conductors ready and waiting to conduct current if it is induced.

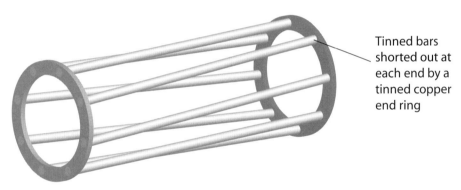

Tinned bars shorted out at each end by a tinned copper end ring

Figure 6.56: Squirrel cage rotor

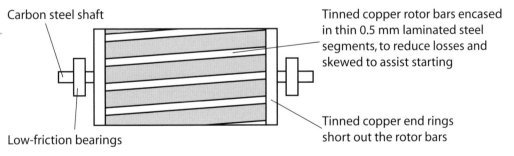

Carbon steel shaft

Tinned copper rotor bars encased in thin 0.5 mm laminated steel segments, to reduce losses and skewed to assist starting

Low-friction bearings

Tinned copper end rings short out the rotor bars

Figure 6.57: Cage fitted to shaft and motor

Inside the rotor cage are hundreds of very thin silicon steel pressed disks. These disks are insulated from each other with a lacquer to stop eddy current from being generated. The steel disks also help the conductor bars keep the skew angle that assists the start-up, as mentioned before. The efficiency of the motor is increased by having high-quality bearings at either end of the rotor shaft for it to spin on.

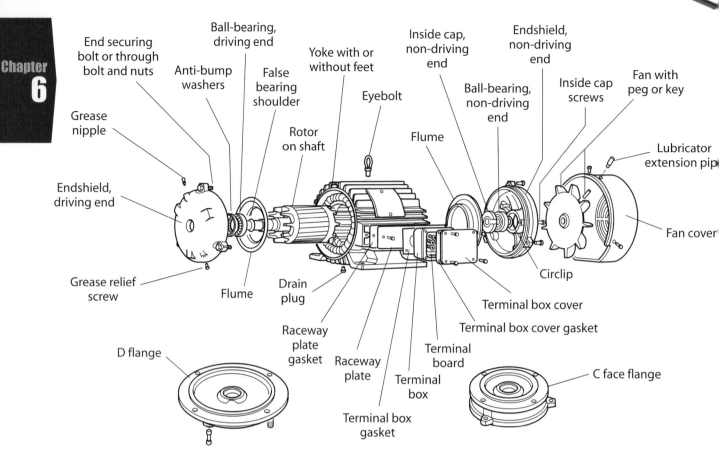

Figure 6.58: Construction of a three-phase squirrel cage induction motor

Three-phase rotor movement

Three-phase motors are much easier to start than single-phase motors. Each of the three-phase windings are placed around the inside of the yoke evenly spaced at 120°. As each phase reaches a peak current, the magnetic field will also reach a peak, one after the other. Because the three-phases are evenly spread around the outside of the rotor, the growing magnetic field will appear to be rotating around the outer case of the motor. You have already found that a moving magnetic field creates a current in any conductor it passes near. This rotating magnetic field passes through the stationary rotor. As the field cuts the rotor, it cuts the rotor bars. This in turn will induce an emf and a second current in the rotor bars. If a current is induced in a conductor bar, it will also produce a magnetic field. The magnetic flux lines will form around the conductor bars and it is this field that will react with the outer rotating three-phase magnetic field. The magnetic reaction between the rotor and the stator can be seen in Figure 6.59. Remember, the rotor is not actually connected to a supply – the current that produces the rotor magnetic field is induced.

Another way to imagine the rotating magnetic field is to consider the analogy of a 'lazy river ride'. The wave moves around a circular pool. Imagine now putting a beach ball on the wave. The beach ball would be on the wave and continue to move as long as the wave moved. From Figure 6.60, the ball could be the rotor, sitting on the magnetic wave. The lazy river wave is the magnetic wave moving around the stator.

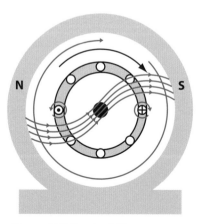

Figure 6.59: Rotating field

Figure 6.60: Three-phase magnetic wave analogy to a wave pool or lazy river

The beach ball is like the rotor – it is moved around by the wave

Waves

Moving magnetic field is just like a wave on a 'lazy river'

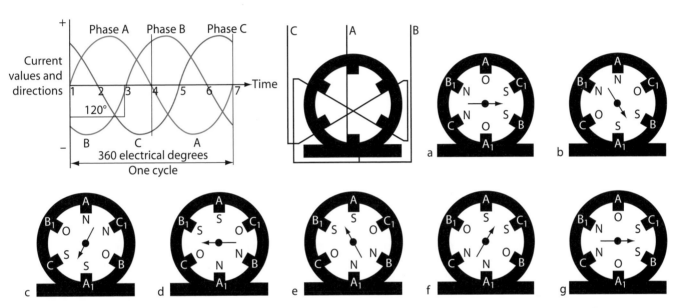

Figure 6.61: Three-phase machine

Three-phase motor speed and torque

The speed of the motor is very important as it dictates exactly what it can be used for. The torque of the motor is also important for the same reason. If a motor that is capable of high speed cannot shift heavy loads, it will be useless as a crane or escalator.

To help understand torque, imagine a car towing another car at a set speed. If the car being towed is slightly slower than the pulling car, there will be a certain amount of tension in the tow rope. If the two cars are moving at exactly the same speed, there will be no tension in the rope and no torque.

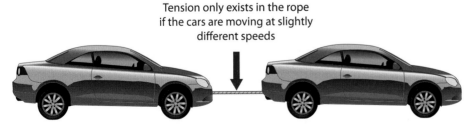

Tension only exists in the rope if the cars are moving at slightly different speeds

Figure 6.62: Towing a car and torque analogy

Key terms

Rotor speed (N$_r$) – the speed at which the rotor actually spins – it is always less than the synchronous speed.

Synchronous speed (N$_s$) – the speed at which the three-phase magnetic field moves around the stator.

From this analogy you can see that for there to be torque in a three-phase motor, there needs to be a difference between the **rotor speed (N$_r$)** and the rotating magnetic field in the stator – otherwise known as the **synchronous speed (N$_s$)**.

The speed at which the three-phase magnetic field rotates around the stator is limited by the supply frequency and the number of magnetic pole pairs that exist in the stator. This can be shown by the formula:

$$N_S = \frac{\text{Frequency}}{\text{Number of magnetic pole pairs}}$$

Worked example

A three-phase cage induction motor is connected to a 50 Hz supply. The motor has four pole pairs. What is the synchronous speed of the motor?

$$N_S = \frac{\text{Frequency}}{\text{Number of magnetic pole pairs}}$$

$$N_S = \frac{50}{4}$$

$N_S = 12.5$ revolutions per second

Note: because frequency is measured in cycles per second, synchronous speed is also measured in revolutions per second.

You have now seen that there is a slight difference between the synchronous speed and the rotor speed. The difference is called the slip. As a magnetic field is rotating around the outer stator, it cuts the rotor bars but only if there is a difference in speed between the two. If there is no difference between the two, no conductor bars will be cut, no current will be induced and no magnetic reactions are possible.

The slip difference between the synchronous and rotor speed is expressed as a percentage and can be found by the formula:

$$\text{Percentage slip} = \frac{N_s - N_r}{N_s} \times 100$$

Worked example

A three-phase motor has a rotor speed of 12 revs per second and is connected to a 50 Hz supply.

What is the synchronous speed and the slip if the motor has four pole pairs?

$$N_S = \frac{\text{Frequency}}{\text{Number of magnetic pole pairs}}$$

$$N_S = \frac{50}{4}$$

$N_S = 12.5$ revolutions per second

continued

To work out the slip, use the formula:

$$\text{Percentage slip} = \frac{N_s - N_r}{N_s} \times 100$$

$$\text{Percentage slip} = \frac{12.5 - 12}{12.5} \times 100$$

$$\text{Percentage slip} = 4\%$$

This means the rotor is moving 4% slower than the synchronous speed of the motor.

What happens if you need to find the rotor speed when you have the synchronous speed and the slip? The formula will have to be rearranged to make N_r the subject. If you feel confident, have a go at rearranging the formula and check to make sure you get:

$$N_r = N_s (1 - S)$$

Activity 6.18

1. Find the synchronous speed of a four pole motor that is connected to a 50 Hz supply.
2. For the motor in Question 1, what is the synchronous speed in revolutions per minute?
3. A six pole motor has a slip of 5% when connected to a 60 Hz supply. Calculate the synchronous speed of the three-phase motor and the rotor speed in revolutions per second.

Progress check 6.15

1. What is rotor speed?
2. What is synchronous speed and how do you work it out?
3. What is slip and how do you work it out?

Wound rotor

An alternative type of induction motor is the wound rotor. The construction is very similar to the standard squirrel cage motor but the solid copper or aluminium conductor bars are replaced with heavy gauge conductor windings. These are held in place by laminated silicon steel disks in the same way as the squirrel cage motor.

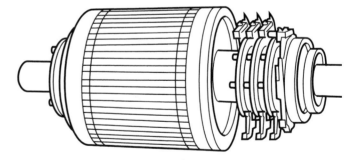

The rotor windings are connected in either a star or delta configuration but the ends are terminated on external slip rings. The advantage of this is the

Figure 6.63: Wound rotor motor assembly

slip rings can now connect to external resistors to vary the current in the rotor on start-up. The external resistance allows better start-up control and gives higher torque. When the motor is starting, all the resistance is in the circuit, and as the rotor picks up to operating speed the resistors are cut out. A disadvantage of this type of motor is maintenance of the slip ring and brushes.

As with all three-phase motors, direction of rotation can be changed by reversing any two phase connections.

Key term

Pulsates – a pulsating magnetic field occurs in a single-phase motor, whereas a three-phase motor has a rotating magnetic field.

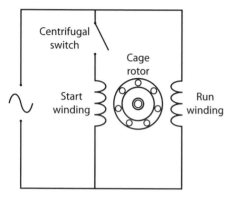

Figure 6.64: Split phase induction single-phase motor

Single-phase a.c. machines

Split phase induction motor

The three-phase motor has a huge advantage because it starts easily. All the rotor has to do is get pushed around by the moving magnetic field caused by the three-phase rotation in the stator. In comparison, a.c. single-phase has a difficulty because the magnetic field generated by a single-phase if applied to the same set-up as the three-phase motor would simply shake. A single-phase magnetic field **pulsates** – unlike a three-phase magnetic field which rotates. However, if the motor was given a little push to get it started, it would move and continue to rotate.

In reality this is not going to happen, so you need to get a little bit smarter – you need to trick the motor into starting. One method is to simulate multiple phases. You have seen in a.c. single-phase theory how an inductor makes current lag behind voltage (page 240). If you were to have a start winding with a certain inductance, this would create a lag. If you then had another winding with greater inductance, the lag in that winding would be greater. In other words, by having separate windings with different inductive properties, you can create something that appears like two phase – this is called a split phase induction motor. The two windings are called the run and the start winding because that is what they actually do – start the motor and run the motor. If you look at Figure 6.64, you will notice it is drawn with the start and run windings on opposite sides of the rotor for ease of explanation.

Because the start winding is made of a thinner gauge copper it has a higher resistance and lower inductive reactance. Because the start winding has a higher resistance it will burn out if it is left in the circuit for any length of time. For this reason the start circuit is connected via a centrifugal switch. The centrifugal switch operates and cuts out the start winding once the rotor reaches approximately 75 per cent of its top speed. The run winding is made of heavier gauge wire and can therefore stay in circuit continuously. Because the run winding is larger, the resistance is lower and the inductive reactance is higher. The high reactance causes the run current to be out of phase with the start current. The split phases can be as much as 30° out of phase, as can be seen in the phasor and wave form diagrams (Figure 6.65).

To change the direction of rotation for a single-phase motor, reverse the start or run winding but not both.

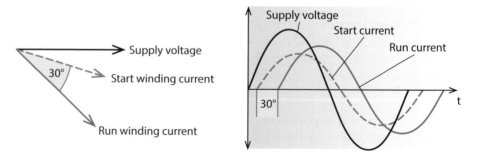

Figure 6.65: Phasor and wave form diagram of a split phase induction motor

The split phase induction motor is particularly suited to domestic white goods such as washer/dryers.

Capacitor start induction motor

The key to getting a good start with a single-phase motor is the split between the start and the run currents. To make the start even better, a capacitor can be added to the start circuit of a split phase motor. By adding the capacitor the current is made to slightly lead the voltage. This increases the split between the start and the run current and gives the motor much more of a kick on start-up. The circuit diagram and the phasor/wave form diagram can be seen in Figures 6.66 and 6.67.

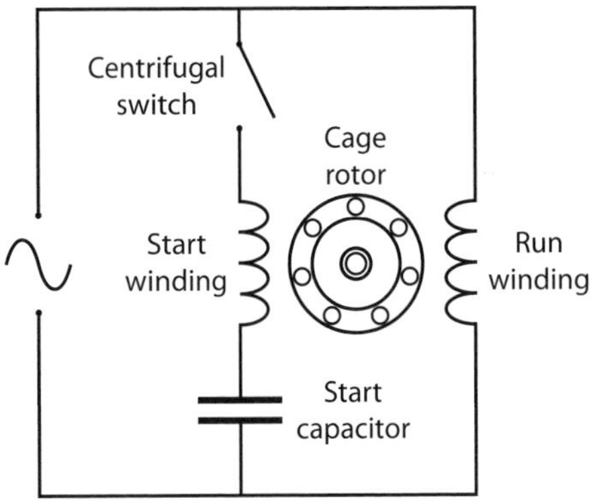

Figure 6.66: Capacitor start, induction run single-phase motor

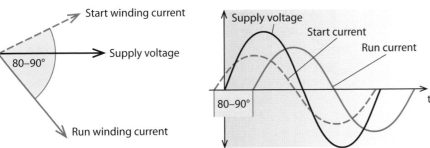

Figure 6.67: Phasor and wave form diagram of a capacitor start, induction run motor

The capacitor start induction motor is particularly good for all applications that require a larger torque on start-up, such as a compressor. The construction of a typical capacitor start induction motor can be seen in the assembly drawing in Figure 6.68 (overleaf). As can be seen, the capacitor is a cylinder externally fixed to the outside case of the yoke. In the next motor discussed a second capacitor is added. This will also be fixed to the outside of the yoke and is therefore a very good way of identifying different types of a.c. single-phase motors.

Capacitor start, capacitor run induction motor

Although the split phase motor is very common, it is not the most efficient. You have seen how the most efficient a.c. circuits have very good power factors (page 253). For an a.c. circuit this means making sure the supply voltage and current are in phase or as close as possible by neutralising any of the effects caused by reactive components. The split phase motor does not have a very good power factor when running at full speed. This is because the run winding has such a large inductive reactance. One way to remedy this is to introduce capacitance or power factor correction. This is done by fitting a capacitor in series with the start winding that is not cut out of the circuit when the rotor reaches 75 per cent of its top speed. This also means the start winding stays in the circuit. When the motor starts, the start current slightly leads the voltage supply, as can be seen in the phasor and wave form diagram in Figure 6.69 (overleaf).

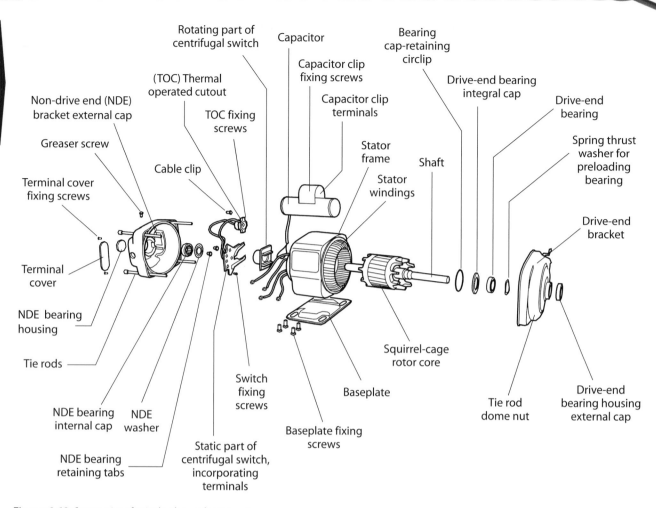

Figure 6.68: Construction of a single-phase induction motor

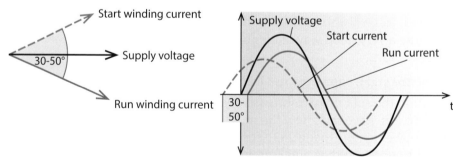

Figure 6.69: Phasor and wave form diagram of a capacitor start, capacitor run induction motor

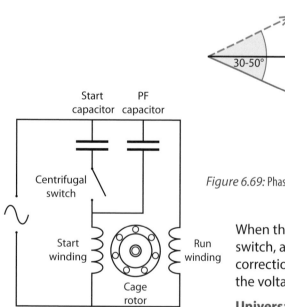

Figure 6.70: Capacitor start, capacitor run single-phase induction motor

When the larger capacitor is cut out of the circuit by the centrifugal switch, a capacitor is still in the circuit. This acts as the power factor correction capacitor and brings the running current more in line with the voltage. The circuit can be seen in Figure 6.70.

Universal motor

A universal motor is capable of operating with an a.c. or a d.c. supply connected. Its construction is basically the same as a d.c. series motor with the field windings in series with the armature. As with a d.c. series motor the universal motor also has a commutator and brushes. Previously you saw that to change direction of any d.c. motor you had to reverse the field or the armature windings. If you changed both, the motor would keep going in the same direction. This is where the advantage of a.c. comes in because a.c. by definition alternates. To fully understand the universal motor, you need to consider the positive part of the a.c. wave form and then the negative part of the wave form. As can be seen in Figure 6.71, the first positive half cycle of the a.c. supply passes down through the field windings and also the armature, resulting in magnetic fields and movement. This is exactly the same principle as the d.c. motor. When the second half of the cycle happens and the current direction is reversed, this occurs in both the field and the armature. Because this is synchronised and happens at exactly the same time, the armature keeps spinning in the same direction, as can be seen in Figure 6.71.

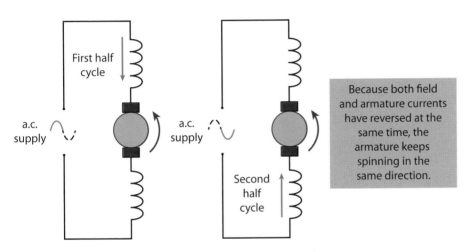

Figure 6.71: Universal motor – the positive and negative stages of the cycle

The universal motor benefits from d.c. and a.c. advantages. This motor is capable of high running speeds as it is not limited by the supply frequency like normal a.c. motors. Speeds of up to 12 000 revolutions per minute can be achieved. The universal motor is also good at high start-up torques, making it suitable for household machinery such as washing machines, food mixers/blenders or anything that is used intermittently. The low cost mixed with the highest power output for its size makes this motor a very popular motor in domestic applications.

Disadvantages of the universal motor are mainly down to the maintenance required due to friction at the commutator and brushes. They can also run too fast so they either need electronic speed control or, as is often the case, a fan blade is fitted to the armature shaft to introduce a load or drag resistance. When used with a.c. supplies, there is a high risk of eddy currents. For this reason the outer yoke is made of insulated laminations in a similar way to the transformer core. The constant switching of a.c. will cause the brushes and commutator to wear out quite quickly.

Shaded pole motor

The shaded pole motor is a very common and cheap a.c. single-phase motor. This motor is cheap because of its simplicity. Complex start winding arrangements are not required as it has only one winding.

Like other induction motors it also has a cage rotor and a stator winding. The stator winding creates a pulsating magnetic field that cuts across the rotor bars. If the magnetic field created by the stator was completely even, the rotor would simply shake. However, the poles are constructed in such a way that areas of magnetic shade are created in one section of each pole. The shade creates an uneven magnetic field on the stator poles. This difference is enough to start the motor turning – again, imagine squeezing a bar of soap and watching it shoot off. With such a small difference in magnetic strength across the face of the poles, the motor is not very strong when starting so is unsuitable for high torque starting applications. The shaded pole motor is good for household fans or other low-cost applications around the house.

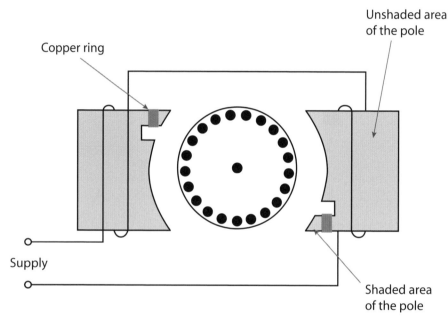

Figure 6.72: Shaded pole motor

Synchronous motor

Another form of motor that can be single-phase or three-phase is the synchronous motor. This motor does not rely on slip difference between the rotor and stator to produce torque, making it different to other a.c. machines. Instead, the rotor is actually made to run at the same speed as the rotating stator magnetic field. This is done by effectively magnetising the rotor by an external supply.

On a three-phase synchronous motor the rotor is made into a pole magnet by connecting it to a d.c. supply through slip rings. Small single-phase synchronous motors (as used in clocks) are started via mechanical means. Once the motors are started they lock into the supply frequency and keep running unless the power is cut or the motor stalls due to too much load.

Synchronous motors are very good at tracking the frequency supplied to them but very poor at starting, so different methods need to be used, including variable frequency or externally connected rotor resistors.

Constant speed applications such as pumps and fans use synchronous motors because of their accuracy. Synchronous motors also have the ability to run with a lagging, unity or leading power factor because extra current can be fed to the rotor. This means they can actually be very efficient with no 'watt-less power'.

Methods of starting and controlling machines

Several factors need to be considered when trying to control a motor of any kind. Certain applications will mean extra precautions are required. Some of these precautions will have implications for the cost, operation and maintenance of the machine.

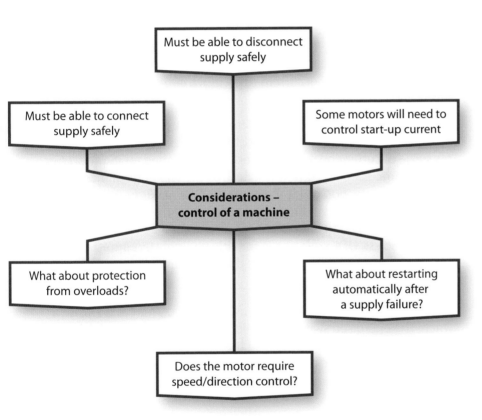

Figure 6.73: Factors to consider when thinking about the control of a machine

All machines need to have some functional method of supply connection and disconnection. You need to be able to turn the machine on and off but also safely disconnect the machine for maintenance purposes. Some motors will need to be reversed, such as a drill that needs to screw and unscrew. Most machines require speed control, such as the spin speed of a washing machine or the speed at which a lathe turns. One very important consideration is the loss of supply. After a power failure, all machines will stop. If an industrial machine stops, you need to be able to put in a safety

Progress check 6.16

1   Define the capacitor's function in a capacitor start induction motor.

2   What does the second capacitor do in a capacitor start, capacitor run induction motor?

Chapter 6

Key terms

Under-voltage protection – automatically disconnects the supply when it reaches approximately 80% of its nominal value. Contactors are used to trip the supply – a manual start button must be pushed to re-energise the contactor again.

Contactor – an electromechanical switch that uses a small current to energise a solenoid. When this solenoid operates, a switch is moved allowing the larger currents to pass through to the load.

Start-up current – the surge current required to overcome resistance and inertia for a motor to start turning. This can be as much as 600% of the final running full load current.

procedure to stop the machine automatically restarting, such as a lathe. This process is called **under-voltage protection**.

Control of potentially damaging currents also requires control on some larger motors. Start-up currents can be large in high torque motors. This current may need to be gradually allowed in to the motor circuitry via an external control circuit. Most motors can be destroyed by large currents that can occur in an overload situation. For example, such an overload can occur if string becomes tangled in the rotating brush of a vacuum cleaner. When you have vacuumed you may have noticed the cleaner 'racing' sometimes as things get caught up in the mechanism. As this happens, more current is drawn by the supply. If this extra current is not enough to make the 13 A plug fuse blow, it will make the motor winding heat up over a period of time. This heating will eventually destroy the windings. All motors rated 0.37 kW must have overload protection within their control circuits.

Direct on-line starter (DOL)

You would imagine that to start a motor you simply need to put a switch in the circuit. However, if a large current is required to start and run a motor, this could be a little scary – large arcing. A safer and more sensible approach is to use a small current to control a large current. This is how a **contactor** works.

Induction motors can have very different **start-up currents** compared to the actual run currents – typically six to ten times larger start-up currents. This surge of current on start-up needs to be planned for in the starter circuit. A large start-up current will destroy the windings of most motors so an overload protection coil is built into the DOL circuitry. An overload can be caused by a motor starting with too much load or too much resistance in the bearings. Overload conditions can also occur if a phase is lost, causing the motor stress. DOL starters are suitable for light start industrial motors up to around 5–7.5 kW but no more than this generally.

Starting a motor using DOL

As mentioned before, a DOL starter uses a small current to energise a switch. The energised switch then allows the three-phases of the supply to pass through to the three-phases of the motor load. Imagine a barrier at a passport control – checks are made, permission is given, the barrier comes up and people are allowed through.

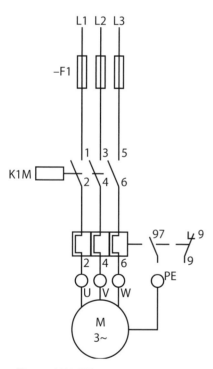

Figure 6.74: DOL starter

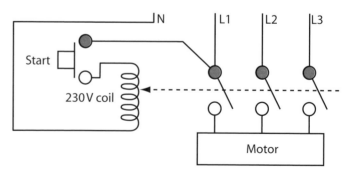

Figure 6.75: Three switch relay with normally open start button

Looking at Figure 6.75, you can see the start button is connected to one line of the supply and it is in series with the coil. As soon as the start button is closed a circuit is made.

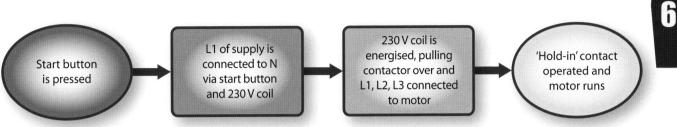

Figure 6.76: The DOL start process

The start button once pressed also operates a 'hold-in' contact, as can be seen in Figure 6.77 below. Once the motor has started the start button can be released, as it no longer has anything to do with the circuit.

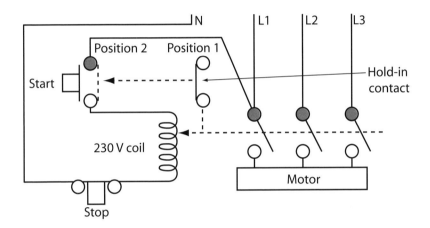

Figure 6.77: Hold-in contact

The 'hold-in' contact is responsible for keeping the main contactors energised allowing the supply to get through to the motor. For the motor to be switched off, the hold-in contact must be de-energised. A stop button is placed in series with the start button and the 230 V coil. The stop button is 'normally closed'. If this is pressed the control circuit is broken and the hold-in contact is de-energised. By removing the energy from the 230 V coil, the three-phase contactor switches return to their normally open position and the motor stops, as can be seen from Figure 6.78.

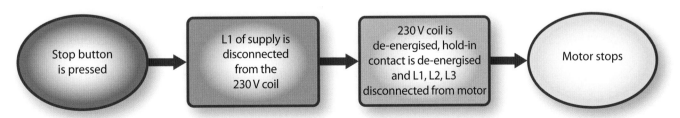

Figure 6.78: DOL stop process

Key terms

Remote stop/start buttons – these are placed in convenient positions for the operators of machines. They may serve as safety stop buttons or possibly because the environment where the machine is operating is not suitable for an operator, such as underground/out-of-reach machines or very hot rooms. Extra remote stop buttons are wired in series with existing ones.

Stop buttons – these are normally closed and wired in series with the start button on the control circuit.

Start buttons – these are normally open and wired in parallel to each other but in series with the stop buttons.

Starting and stopping a motor needs to be a convenient and safe process. For this reason there must be multiple buttons for most industrial applications, otherwise known as **remote stop/start buttons**. Emergency **stop buttons**, for instance, may need to be placed around the workshop so if something goes wrong the nearest person has access to stop the machine. To break power to a circuit, stop buttons are placed in series. If any stop button is pressed the circuit power is cut instantly – it doesn't matter where the circuit is broken, as long as it is broken. The important part to stopping a machine is to make sure it cannot be restarted automatically. If the power was lost to a heavy duty machine and the engineer began to investigate, it could be disastrous if the machine started up again automatically if the power returned.

There needs to be a manual reset process for some motor applications. The hold-in contact acts as the safety device, breaking the circuit and disconnecting the contactor. To restart the motor, the whole start process has to happen with the **start button** being pressed again.

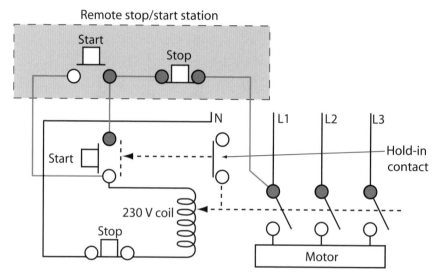

Figure 6.79: Remote stop/start control

One other method of controlling a motor is via an 'inch button'. This operates in the same way as the start button but additional circuitry is used to prevent the hold-in contact energising. This means the motor only operates when the inch button is pressed. As soon as it is released, the motor stops. With large machine applications it can be important to rotate the motor small amounts in a very controlled way.

Star delta

For relatively small three-phase motors DOL starters will be enough but larger three-phase motors require a little more help. Imagine starting a car. Although it is possible to get a car moving in second gear, if the car is heavy this can be very difficult. Normally you would expect to start in first gear and then when you have enough momentum, change to second gear. This is the principle of a star delta starter.

As with cars you can have a manual gear box or an automatic. Hand-operated star delta starters rely on the windings of the three-phase motor being available in a terminal box for control. Either end of each phase needs to be available for connection. This means six connection points can be onward connected and switched as required, as shown in Figure 6.80.

As the name suggests, a star delta starter is initially in the star configuration (first gear). This gives a reduced voltage and current on start-up to allow the motor to start turning gradually.

Once the motor is up to a constant speed, a switch is made to delta configuration allowing the full line voltage and load current through to give maximum speed. The motor windings and the two-stage starting connections can be seen in Figure 6.81.

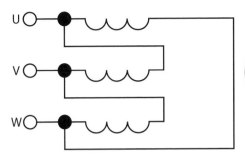

Figure 6.80: Motor windings with six connections

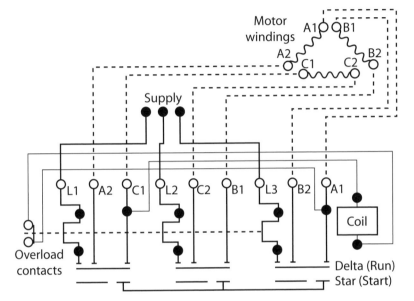

Figure 6.81: Hand-operated star delta connections

The difficulty with a manual handle is its reliance on human intervention and a quick movement. One solution to this is to use the automatic version, as can be seen in Figure 6.82.

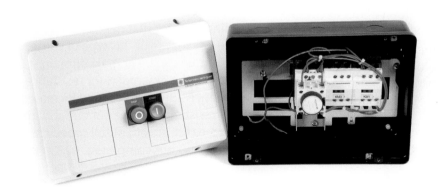

Figure 6.82: Star delta starter

Once pressed, the automatic star delta starter moves the contacts into star position allowing the motor to start with 230 V across each winding.

When the button is released the motor stays in star configuration but an electronic timer circuit is operated. When the timer triggers on 'timeout' the changeover contact switches from star to delta and gives the full 400 V to the motor windings. Once the motor is at full speed, the delta contactor coil stays energised via its own hold-on contact until such a time as the stop button is pushed. The full diagram can be seen in Figure 6.83.

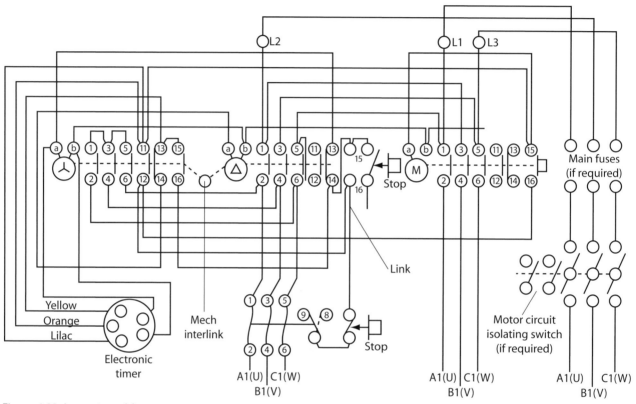

Figure 6.83: Automatic star delta contactor starter

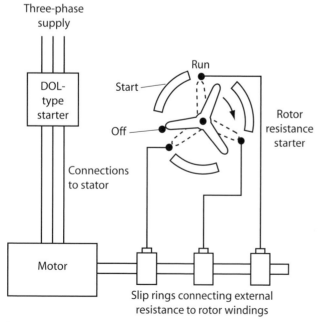

Figure 6.84: Rotor resistance starter

Rotor resistance

Wound rotor induction motors are configured in such a way that the rotor is connected to the supply via external resistors and slip rings. The rotor resistance starter is suitable for starting a motor that needs to start on full load. The advantage of this type of starter is that the torque can be at maximum on start-up by having the external rotor resistance at a maximum. As the motor speed increases the rotor resistance can be cut out until the rotor reaches full speed. At full speed the rotor resistance is cut out completely.

Soft starter

To start a motor in a very controlled way requires electronics in the form of **solid state** technology. Solid state soft starters have several advantages over older technology such as contactor control. Soft starters can control the starting voltage and current very accurately to the level required for the application and therefore limit the potentially damaging inrush of current that can occur.

Components such as transistors are layered into the silicon semiconductor in a way that allows a very high concentration of components per area.

A larger starting current is required to give the rotor the energy to overcome inertia and start moving. However, too much current in the form of an inrush must be avoided to prevent the need to over-engineer a motor circuit design. The electronic circuit restricts the start-up current to approximately 200 per cent of the full load current. This compares to 600 per cent of the full load current for a DOL starter. The solid state starter is able to start at a predetermined voltage and current level, and when the motor has run up to the required speed the full voltage level can be released to the motor. The control circuit can also stop instantly or ramp down the voltage so a controlled soft stop is also achieved.

Variable frequency and inverter

Variable frequency drive or VFD is a particularly good way of controlling motor speed where a continuous range of speed is required rather than simply speed steps. The speed of an induction motor can be controlled by several factors. Speed is indirectly proportional to the number of pole pairs – pole pairs increase, speed goes down in steps. Motor speed is directly proportional to the frequency of the supply voltage to the stator – frequency goes up, rotor speed increases. If you remember, synchronous speed is given by the formula:

$$n_S = \frac{f}{\text{Pole pairs}}$$

It is not practical to change the speed of a motor by varying the pole pairs. However, the supply frequency can be controlled electronically to give an infinite range of speed as required.

Programmable logic control

A programmable logic controller, or PLC, is a digital computer used to automate and control electrical machinery and motors. PLCs were designed to replace the earlier relay technology. Industrial control computers have several advantages over relays because they can be designed to operate in very harsh conditions and the logic can be held in programmable non-volatile memory. The control sequence is written in a special computer logic language that is easy to change and update as required. The hydraulic control signals to the motor are given in response to inputs received by the PLC computer. For complex industrial applications, PLCs have an advantage over any other form of control – changes can be simulated, trialled and debugged before release to the machines that are to be controlled. See *Chapter 3: Electrical installations technology*, page 126, for more information on these systems.

See *Chapter 3: Electrical installations technology*, page 126, for more information on these systems.

Key term

Solid state – electronic components that are made from solid material such as can be found in an integrated chip.

Chapter 6

Progress check 6.17

1. What method is used to start a wound rotor a.c. motor?
2. What is a simple way to explain star delta starters?
3. What does no volt protection do?

Chapter 6

LO5

ELECTRICAL DEVICES

Components

Solenoids, contactors and relays were all covered extensively in *Chapter 1: Principles of elecrical science,* including operating principles and applications. Please re-read the section on electromagnetism before checking your knowledge.

Progress check 6.19

1 Describe the two methods by which an MCB operates when a fault occurs.

2 Describe how the trip coil is activated in an RCD.

3 Describe four different types of protective device and the different applications they are suitable for.

Progress check 6.18

1 How do you identify the polarity of a solenoid?

2 Explain briefly how a DOL contactor starts a three-phase motor.

Overcurrent protective devices

The full range of protective devices has been covered to the required level in *Chapter 3: Electrical installations technology*, pages 145 and 149, and in *Chapter 7: Electrical systems design*, pages 372 –375. Please read these sections before checking your knowledge.

LO6

ELECTRICAL HEATING SYSTEMS AND CONTROL

Electric space heaters

Storage heaters

A traditional storage heater uses heating elements that are embedded inside special firebricks. These bricks are particularly dense and have very high thermal mass. The idea is to store the heat energy from the element in the bricks and then release the heat slowly as required. The hot bricks are insulated and contained within a metal case that is generally fitted on the wall. The bricks are made of clay, olivine, chrome and magnesite which makes them very heavy.

When considering how to heat a property, cost efficiency is very important. Storage heaters have the advantage of using cheap 'off-peak' night-time electricity. The 'off-peak' electricity is used to heat the element and the heat energy is stored in the bricks. There are several types of storage heaters commercially available, including:

- automatic charge control that measures the room temperature and only draws the electricity required
- combination storage heaters that have a built-in convection heater to take over from the heater element or supplement if the weather turns cold
- fan-assisted storage heaters that use a thermostat control to deliver the heat only when required.

The cheaper energy comes from the need for power stations to run 24 hours a day. However, most households and businesses do not require much electricity between midnight and 7 a.m. This energy is still produced and needs to be used rather than stopping and starting the electricity generation process each day. This system is called Economy 7 and traditionally has been a third to a quarter of the cost of the day-time rate.

Safe working

Take great care when lifting storage heaters as the dense bricks make them very heavy. Generally, the bricks should be taken out and moved separately in manageable loads.

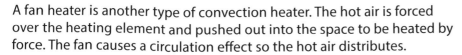

Some household functions can be timed to happen at night to save money, for example, water heating, dishwasher, washing machines and tumble dryers.

Economy 7 systems do however require separate metering to measure usage against 'on-peak' and 'off-peak' charging tariffs.

Convection heaters

Direct heating by passing a current through a resistive element will heat the immediate area in front of it. However, this is not the most efficient way of heating a larger area. A better way of heating a space is by circulating air convection currents through the heating appliance and over a heating element. As the air passes over the element, it heats up and rises. The rising hot air pushes through the colder air, leaves the heater vents and moves around the room. A typical household convection heater can be seen in Figure 6.85. A convection heater will have a range of controls, from element heater settings to thermostatic control to monitor the room temperature.

Figure 6.85: Convection heater

A fan heater is another type of convection heater. The hot air is forced over the heating element and pushed out into the space to be heated by force. The fan causes a circulation effect so the hot air distributes.

Radiant heaters

There are many other types of heaters, including the radiant heater. Radiant heaters come in a range of shapes and sizes as well as technologies. The main difference between the action of a convection heater and a radiant heater is the application. A convection heater will heat a space by causing a thermal current to flow in the room. A radiant heater will heat whatever object is placed directly in front of it. A radiant heater will not raise the temperature of a room unless it is left on for a long time. Imagine a camp fire. If you are standing by the camp fire you are warm. If you move away from the fire you immediately become cold because of the short range of the heat.

Figure 6.86: Oil-filled radiator

Domestic situations often require a little bit of extra heat. This comes in the form of a plug-in electric fire or an oil-filled radiator, as can be seen in Figure 6.86. Some smaller rooms like a bathroom or kitchen may have a tubular-type radiant heater, as can be seen in Figure 6.87, or an infrared heater, as can be seen in Figure 6.88.

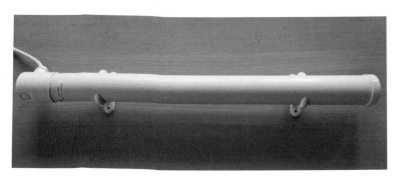

Figure 6.87: Tubular heater

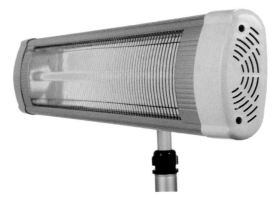

Figure 6.88: Infrared heater

Underfloor heating

The two main types of underfloor heating used in modern installations are electrical heating element or fluid system. If room cooling is required then a pipe and fluid or hydronic system is more appropriate. The electric element system is going to be considered here.

Underfloor heating is a very efficient method of heating a space as it has the advantage of rising heat. The element heats the floor and the floor acts as a large low-temperature radiant heater. An electrical element that produces heat through the resistive properties can be laid under a floor in a mat before the main floor is installed. Underfloor heating applications can also be used in non-domestic situations including landing pads, driveways and football pitches (hydronic systems are also used in these applications).

A further advantage of underfloor heating can be gained from the **thermal mass** properties of the material the element is laid beneath. Typically, underfloor heating is installed under stone or ceramic tile floors. These normally have a cold feel to them. The stone or tiles can be heated during off-peak times (making use of cheaper electricity) and then give off the heat slowly. When the room is hot, the stone can absorb some of the heat and as the room cools down, the heat can be given back to the room by the stone. This leads to a much more efficient system of heating as the thermostat only turns on the electricity to the underfloor element when the pre-determined temperature has been reached.

Unlike the fluid-based hydronic systems, electric underfloor heating can only be used to heat rooms, not cool them. A flexible heating element is made up of pre-formed cable mats fixed to a mesh and a carbon film. The mats are very thin in profile so can be laid under a range of floor finishes.

> **Key term**
>
> *Thermal mass* – relates to the thermal properties (heating and cooling) of a building material to store heat and give it out when required – cool when it's hot outside, hot when it's cool outside.

Figure 6.89: Underfloor heating

The mat is installed as per the manufacturer's instructions for each system and type of floor. Some require different insulation layers or screeds to be used in conjunction with expanded steel, as can be seen in Figure 6.89 (although this is a hydronic system!). This will allow for expansion movement with temperature and avoid cracking the expensive floor.

A typical mat used on a household tiled kitchen floor will be rated up to about 450 W per m^2 and comes in a range of sizes, starting at about 0.5 m wide rising to several meters long. Systems can be linked together from some manufacturers, but if larger areas of floor require underfloor heating you might require extra design help from the manufacturer.

Electrical water heating systems

The two main methods used to heat water are storage and instantaneous. The type of system installed will depend on the quantity of water required and how the water is used. For small amounts of hot water required randomly, an instantaneous system might be best. If large amounts of water are required on a continuous basis, then a storage system might be better suited.

Immersion heaters

Storage water heaters are particularly suited to applications where large amounts of water are required. The water is heated and kept in an insulated tank ready for any demands made on it. An immersion heater has one or more heating elements that are fed by a heat-proof flexible cable from its own double pole isolated and protected circuit. A dual element is used where Economy 7 supplies are fitted. This makes use of the cheaper 'off-peak' night-time electricity supply. The water is heated up at a fraction of the cost at night and then used first thing in the morning for showers, etc. The dual element water heater can be seen in Figure 6.90.

The tank of water can also be heated by a boiler and the immersion element can become the back-up or hot water booster as more hot water is required.

Larger cistern-type storage heaters may be appropriate for larger houses. The advantage of a cistern is that multiple users can take hot water at the same time without affecting the flow rate drastically. The cistern storage tank can be seen in Figure 6.91.

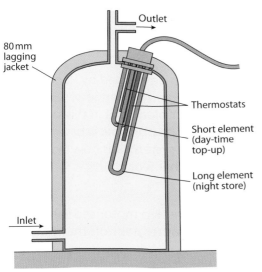

Figure 6.90: Dual element immersion heater, hot water

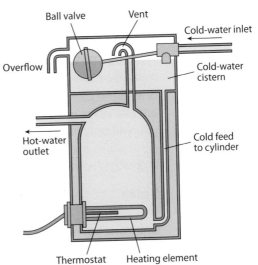

Figure 6.91: Cistern-type storage

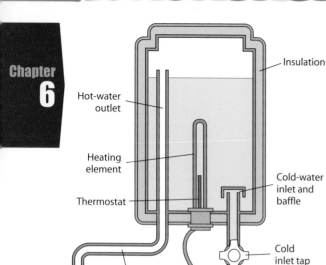

Insulation

Hot-water
outlet

Heating
element

Thermostat

Cold-water
inlet and
baffle

Cold
inlet tap

Swivel outlet

Figure 6.92: Non-pressure water heater

Another common water heater is the type found in toilets or over commercial sinks. This type is called the non-pressure storage heater. Only a small amount of water needs to be heated up at any one time. Generally used for hand washing or for washing cups and saucers, it can be seen in Figure 6.92.

Instantaneous heaters

If hot water is required in short bursts like a shower, it is probably not a good idea to heat up a tank of water. All that is required is a small amount of water to be heated up as and when called for. The water heating system best suited to this type of application is the instantaneous water heater, as can be seen in Figure 6.93.

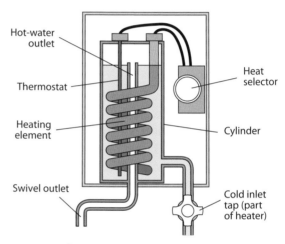

Hot-water
outlet

Thermostat

Heating
element

Heat
selector

Cylinder

Swivel outlet

Cold inlet
tap (part
of heater)

Figure 6.93: Instantaneous water heater

An electric shower relies on heating mains pressure cold water very quickly. This requires considerable power and some modern showers are rated at 10 kW or more and require a specially designed and isolated circuit with RCD protection. The temperature of the water is directly controlled by a dial on the body of the heater where the flow rate is also controlled.

Methods of control

Heating control circuits

A number of heating control systems have been developed by Honeywell, one of the pioneer companies in heating technology. A range of systems are available, including the Sundial C, S, W and Y plans, but in this section you will only cover two: the S plan and Y plan. The principle is based on controlling the water and heating through one system. It is how the system is configured that makes the control system either S or Y.

S plan

This system can be seen in Figure 6.94. There are two motorised **valves**. One zone valve controls the hot water and the other controls the heating.

Key term

Valve – a valve is like a diode. Water must flow through in the direction of the arrow on the side of the valve when it has been activated.

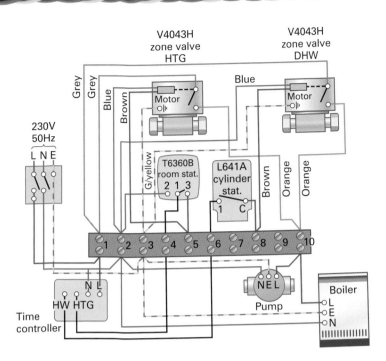

Figure 6.94: Wiring diagram for an 'S' plan temperature control system

A zone will have a programmable thermostat to monitor and control temperature. If a zone thermostat switches or there is a demand for heat, the relevant zone valve will operate. Just before the valve reaches fully open, the auxiliary switch activates the pump for the system and the boiler. This means that the water is heated and pumped around the system on demand.

Generally there are room thermostats and cylinder thermostats that sense water and room temperature. If more heating zones are required these can be fitted with more thermostats to make the heating more efficient. When both the thermostats' required temperatures are reached, i.e. the water or the room heating level is reached, the valves are closed, the boiler turns off and the pump shuts down. It is generally considered best practice to have a time switch to control the water and a programmable thermostat to control each heating zone.

Y plan

The Sundial Y plan system is designed to provide independent temperature control of both heating and domestic hot-water circuits in fully pumped central heating installations. The wiring diagram for a Y plan can be seen in Figure 6.95.

The system consists of a three-port mid-position valve. This means the control system can provide independent switching of water and heating circuits. The thermostats and the mid-position diverter valve work together to ensure the pump and boiler only come on when they are needed. This level of control ensures energy is not wasted by overheating the water or radiators.

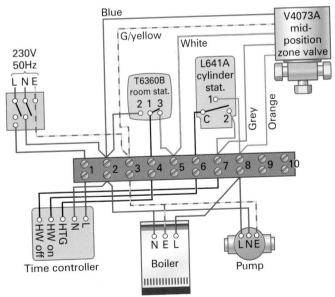

Figure 6.95: Wiring diagram for a 'Y' plan temperature control system

Thermostats

Thermostatic controls are a traditional method of controlling heating. A thermostat is a device that switches when a certain pre-determined temperature is reached. Bimetallic strips made of two metals with different expansion coefficients are bonded together. As the temperature rises, the two metal strips expand at different rates. If the bonded metals expand at different rates, the net effect is they will bend one way or the other. This action will make or break a contact and therefore act as a switch, sending a signal to the heating system.

Time switches and programmable controls

In most domestic households, heating and hot water are required at pre-determined times. Typically, you need hot water for showers first thing in the morning and for baths and washing dishes later on in the evening. However, during the day, hot water and heating may not be required if everyone is out. Time switches come in a number of types, with the latest digital programmable timers allowing a wide range of daily settings. Individual days can have unique time settings to reflect the requirements of the house. This is one area where the rising fuel costs for heating can be managed. Careful monitoring of heating and hot water times can save money and resources and should be regularly reviewed.

> **Progress check 6.20**
>
> 1 Describe what the motorised valves are used for on a Y plan.
> 2 How many motorised valves are there on a Y plan? Describe what they do.
> 3 Describe the difference between convection and radiant heaters.
> 4 Describe how a thermostat operates.

LO7

ELECTRONIC COMPONENTS AND APPLICATIONS

The role of a building services engineer and electrician has become a lot more complex as consumer technology and requirements have developed and changed. The point at which an electrician stops and a specialist takes over has become less defined. Because of this you are expected to be more skilled in a wider range of disciplines. To become fully competent in electronics takes many years of training and experience but a limited amount of knowledge can help identify specific areas when finding faults or particular hazards with equipment. Control systems have a much greater range of capabilities now due to electronics and you can expect to deal with a wide range of them. Identification of basic electronic components will help you to understand some systems sufficiently to be able to install, fault find and rectify some circuits such as alarm or heating control systems.

Electronic components

Resistors

Resistors do what they say – they resist. Resistance is present in any conductor and more so in insulators. Electrons will pass through any conductor with the right amount of encouragement in the form of voltage pressure. You have seen in *Chapter 1: Principles of electrical science* how resistance can be calculated based on known conductive properties of a material (resistivity), and the length and cross-sectional area of the conductor. Resistors can also be made into known values. All electronic applications need components that react in known ways so that circuits can be made and to ensure their performance is accurate. Resistors come in many sizes and ratings, based on their ability to handle power. They are quite easy to recognise as there are only a few basic types. Resistors can be variable with an operating resistive range or they can be fixed.

Resistor construction

There are various types of resistor construction but the main types include wire wound or resistive materials such as carbon or metal oxides. The wire wound resistors were the original purpose-made resistors.

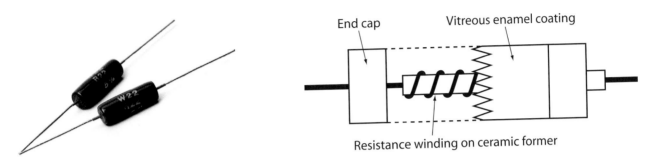

Figure 6.96: Wire wound resistors

A conductive material such as brass with a known resistivity is wound around an insulated ceramic core. The cross-sectional area and the length of the brass wire determine the actual resistance of the resistor. With current passing through the wire wound resistor in continuous operation, an important consideration is the resistor's ability to handle the heat. For this reason the power rating must be determined before the type of resistor is chosen.

Other types of resistor commonly used are carbon or metal oxide resistors. A layer of conductive material is coated on an insulated core. The amount of the conductive material determines the resistor value. Improvements in manufacturing over the past 30 years have meant tolerance levels have become very accurate for very little cost. However, it would be impractical to make every single value of resistor that may be required, so manufacturers make a range of set values or **preferred value resistors** with a range of tolerances. If other values are required other than the resistors available, combinations are made by putting resistors in series or parallel.

Identification of resistor values

The need for building services engineers to recognise a wide range of components and values is becoming increasingly important as more control systems become sophisticated. Resistors have two main methods of identification: written values on the resistor body or a colour code. Because the space available on even the largest high-powered resistor is small, a written value still has to be relatively small or abbreviated. In *Chapter 1: Principles of electrical science*, you saw how 000s can be replaced with letters, as shown in Table 6.4.

If you need to pick a resistor for an application it would be very difficult to read the 1 GΩ if all the nine zeros were on the body. It is much easier to read the G. A more common method for small resistors is to use the standard internationally recognised colour code. It is important when trying to read a

> **Key term**
>
> *Preferred value resistors* – a range of resistors that cover a majority of engineering needs.

Actual resistance	Equivalent resistance using engineering notation symbols
0.000001 Ω	1 μΩ
0.001 Ω	1 mΩ
1 Ω	1 Ω
1 000 Ω	1 kΩ
1 000 000 Ω	1 MΩ
1 000 000 000 Ω	1 GΩ

Table 6.4: Abbreviated resistance figures

resistor value to make sure the resistor is facing in the right direction. Most colour code resistors have a set of bands grouped at one end of the resistor, as can be seen in the example of the carbon resistors in Figure 6.97.

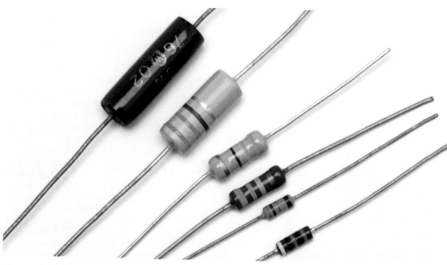

Figure 6.97: Metal oxide and carbon-composition resistors

At the other end of the resistor there is a separated colour band or a space with no colour band. This end shows the tolerance of the resistor. No band indicates the resistor has a tolerance of +/– 20%. Other single bands and the tolerance levels are shown in Figure 6.98.

Once you have placed the resistor the correct way for reading, the next step is to decode the colours into real resistance values.

The first two bands in a four-band resistor give the first two digits. The third band tells you how many zeros need to be added to the resistor digits.

The colour code values can be seen in Table 6.5 below.

Tolerance colour code

Band colour	±%
Brown	1
Red	2
Gold	5
Silver	10
None	20

Figure 6.98: Tolerance colour code

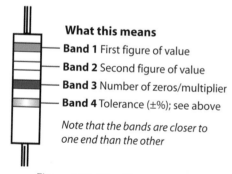

What this means

Band 1 First figure of value
Band 2 Second figure of value
Band 3 Number of zeros/multiplier
Band 4 Tolerance (±%); see above

Note that the bands are closer to one end than the other

Figure 6.99: What this means

Colour	Word association to help memorise	Value
Black	Black	0
Brown	Birds	1
Red	Running	2
Orange	Over	3
Yellow	Your	4
Green	Garden	5
Blue	Biting	6
Violet	Violently	7
Grey	Grey	8
White	Worms	9

Table 6.5: Colour codes

Worked example 1

What is the value of a resistor with four bands: orange, orange, red and a tolerance band of gold?

Orange = 3

Orange = 3

Red = 00

Gold tolerance band = +/– 5%

So the resistor is 3 300 Ω +/– 5% so it can be in the range 3 135–3 465 Ω.

Worked example 2

What is the value of a resistor with four bands: brown, black, violet and no tolerance band?

Brown = 1

Black = 0

Violet = 0,000,000

No tolerance band = +/– 20%

So the resistor is 100 000 000 Ω (or 100 MΩ) +/– 20% so it can be in the range 80 000 000 – 120 000 000 Ω or 80 MΩ to 120 MΩ.

Further examples of colour coding can be seen in Figure 6.100.

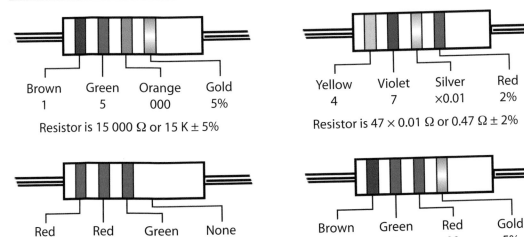

Brown	Green	Orange	Gold
1	5	000	5%

Resistor is 15 000 Ω or 15 K ± 5%

Yellow	Violet	Silver	Red
4	7	×0.01	2%

Resistor is 47 × 0.01 Ω or 0.47 Ω ± 2%

Red	Red	Green	None
2	2	00000	20%

Resistor is 2 200 000 Ω or 2.2 M ± 20%

Brown	Green	Red	Gold
1	5	00	5%

Resistor is 1 500 Ω or 1.5 K ± 5%

Figure 6.100: Examples of colour coding

As mentioned earlier, another method of identifying resistors is simply to write the value on the resistor body. This is only possible if the resistor is big enough, such as a high-powered resistor. Earlier in *Chapter 1: Principles of electrical science*, you covered the use of prefix symbols. A prefix code allows the number to be shortened. By dividing or multiplying a number by a thousand, million or billion and then replacing the 000s with a letter, the number can be put in its shortest possible space-saving format. There is also a code letter to show the resistor tolerance. Examples of resistor codes can be seen in Tables 6.6 and 6.7.

0.1 Ω	is coded	R10
0.22 Ω	is coded	R22
1.0 Ω	is coded	1R0
3.3 Ω	is coded	3R3
15 Ω	is coded	15R
390 Ω	is coded	390R
1.8 Ω	is coded	1R8
47 Ω	is coded	47R
820 kΩ	is coded	820K
2.7 MΩ	is coded	2M7

Table 6.6: Examples of resistance value codes

F	=	±1%
G	=	±2%
J	=	±5%
K	=	±10%
M	=	±20%
N	=	±30%

Table 6.7: Codes for common tolerance values

Activity 6.19

1 Gather a selection of resistors and use the colour code to work out the actual values and tolerance levels. Put the results in a table of your own design for the task.

2 Choose and set an ohmmeter to the correct range for your first resistor. Measure the resistor and decide if it is within the permitted range. Complete this for all the selected resistors and present your findings in a table of results.

Variable resistors

If you look at, for instance, a computer motherboard, you will notice a selection of variable resistors. Variable resistors can be tuned to a required value to make the circuit behave how you need it to. The construction of a variable resistor depends on the application it is required for. If the resistor needs to be used continuously, such as a slide control on a piece of equipment, then a linear or rotary track variable resistor is used, as seen in Figures 6.101, 6.102 and 6.103.

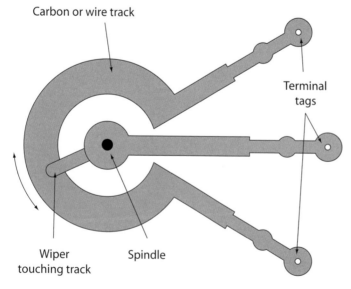

Carbon or wire track

Terminal tags

Wiper touching track

Spindle

Figure 6.101: Layout of internal track of a rotary variable resistor

Figure 6.102: Variable resistors/potentiometer and pre-set resistor

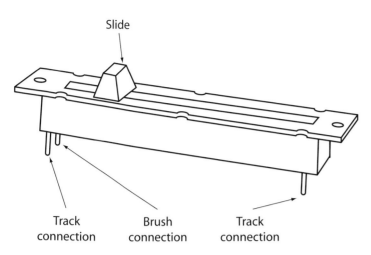

Figure 6.103: Linear variable resistor

If the variable resistor is required to fine tune a circuit when the circuit is being set up, a **pre-set resistor** will be used. A pre-set resistor generally has a tuneable screw at one end. The circuit characteristics are monitored as the pre-set resistor is tuned. Once the circuit is tuned and works, the circuit board is put back in its case and not touched again until maintenance is required.

The principle of a variable resistor is straightforward. The whole track of the resistor has a set value. Imagine connecting to the top or bottom of the resistor track, in this case, the top. As well as having the top and the bottom connection of the resistor, now imagine having a third connection available somewhere in the middle. This third connection is able to slide up and down the resistor track. This middle sliding contact taps into the resistor making a good electrical connection.

If you put the current into the top of the resistor, the current can come out of the bottom of the resistor or it can come out of the sliding connector. If the current is taken from the sliding connector, it only passes through part of the resistor track and therefore only experiences part of the total resistance of the track. The amount of resistance the current experiences will depend on how far the slider is, up or down, on the resistor track.

Looking at Figure 6.104 you will notice the voltage taken between the wiper and the 0 V rail will vary according to the position of the wiper. The resistance is spread evenly along the length of the resistor so this means the voltage will be directly proportional to the wiper's position on the slider. If the wiper is only 10 per cent up from the bottom 0 V rail, the output voltage will be 10 per cent of the supply. If the wiper is half way, the output voltage will be half the supply voltage. Potentiometers can be found in sound equipment such as mixing desks, as can be seen in Figure 6.105.

> **Key term**
>
> *Pre-set resistor* – used to tune a circuit. Some circuits will require a level of adjustment to work in a particular way. A pre-set resistor will be tuned and then probably not touched again in its lifetime.

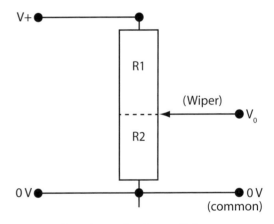

Figure 6.104: Circuit diagram for voltage applied across potentiometer

Figure 6.105: Potentiometers are used in mixing desks to control sound levels

Chapter 6

Light dependent resistors (LDR)

Another form of variable resistor is the **light dependent resistor** or **LDR**. As its name suggests, the value of resistance varies according to the amount of light that falls on it.

Key term

Light dependent resistors (LDRs) – also known as photoresistors. They react to light by either increasing resistance or decreasing resistance when light intensity varies – this process is called photoconductivity.

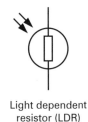

Light dependent resistor (LDR)

Figure 6.106: Light dependent resistor

Figure 6.107: LDR schematic symbol

Figure 6.108: LDRs are used in automatic lights

LDRs are commonly found in security lighting circuits or street lights. In fact LDRs can be found in any electrical application that needs to operate when it gets dark.

LDRs are made of semiconductor material that reacts to light falling on it. If the light level is high enough, the semiconductor material will absorb photons and this will pass enough energy to free up some electrons. You will recall that moving free electrons enable current flow. So, as more free electrons are made available by more light and photons being absorbed by the semiconductor, the resistance decreases. Most cheap LDRs are made of cadmium sulphide.

Thermistors

The last type of variable resistor is the thermistor. This component varies with temperature. Because of their sensitivity to heat, thermistors are particularly useful for measuring temperature and acting as switches for applications that are temperature critical. These applications can be classed as protection or control. Control systems that use thermistors include climate control applications that might be found in cars.

Figure 6.109: A car's climate control system uses thermistors to detect temperature

An example of a thermistor being used for protection is in a vacuum cleaner motor. When you are vacuuming, things can get caught up around the rotating brushes. When this happens the motor starts to race as it tries to draw more current. If the current being drawn is just below the fuse rating it will not blow but the windings of the motor will heat up. Prolonged heating will eventually burn out the motor windings so protection needs to be added to the circuitry. Because the motor winding is difficult to reach, this protection needs to be built in, automatic and, ideally, self-repairing.

The thermistor is wound into the motor winding so it can sense when a pre-set temperature is reached. This pre-set temperature will cut out the thermistor and protect the winding before expensive damage happens. When the motor windings have cooled down sufficiently the thermistor cools, resets itself and the vacuuming can continue. You will also find this application in a hairdryer. A hairdryer motor is subjected to particularly high temperatures and stress for such a small motor. As the hairdryer is moved close to the head, the motor can get stressed. The back flow of air caused by blocking the path with a head will cause the motor to work harder and the windings to overheat. Hairdryers will often simply cut out without warning and then start again.

Thermistors can have a positive or negative temperature coefficient which means resistance can increase (+ve) or decrease (–ve) with temperature. Positive temperature coefficient thermistors (PTC) and negative temperature coefficient thermistors (NTC), each have their own specific applications and uses. Thermistors can be identified by a simple colour code, as shown in Table 6.8.

Thermocouples

Although not strictly speaking a variable resistor, it is worth considering one more electrical component in this section. When two different metals are joined to produce a thermocouple and then exposed to heat, a small voltage will be generated at the junction – suggesting a resistance has been created. This reaction is called the thermoelectric effect or the Seebeck effect (named after the physicist that discovered it). The voltage generated has a direct relationship to the temperature to which the metals are exposed. Different metals and alloys joined together will cause different ranges of voltages to be generated when exposed to different temperatures. For this reason a thermocouple is a very cheap way of converting a temperature reading into an electrical reading. Thermocouples are used to measure temperatures in very hot applications such as turbines, diesel engines, kilns, furnaces and other industrial applications.

As a building services engineer, you will also come across thermocouples in ovens or heaters where they are used as a fail-safe device sensing when a pilot light goes out. Thermocouples are cheap, can measure a wide range of temperatures and do not require any other input as they are self-powered with no moving parts. An example of a thermocouple is shown in Figure 6.112.

Figure 6.110: Thermistor

Temperature-sensitive
resistor (thermistor)

Figure 6.111: Thermistor schematic symbol

Colour	Resistance
Red	3 000 Ω
Orange	5 000 Ω
Yellow	10 000 Ω
Green	30 000 Ω
Violet	100 000 Ω

Table 6.8: Colour coding for rated resistance of thermistor

Figure 6.112: Thermocouple

Progress check 6.21

1 Write down the resistor colour codes and associated values.
2 Write down the colour code for the tolerance bands and the associated values.
3 Describe briefly in your own words another method of identifying a resistor without colours, using an example.
4 Describe briefly the applications of different types of variable resistors.

Key terms

Charged – when a capacitor is filled with electrical charge.
Discharged – when a capacitor has released its stored charge.

Activity 6.20

1 A resistor reads: red, orange, green with a tolerance band of brown. What is the resistor value and acceptable range?
2 A resistor reads: yellow, violet, orange, brown. What is the resistor and the operating range?
3 A resistor body is marked 3M4M. What is the actual value?

Capacitors

A capacitor is a very common electronic component that exists in almost all electronic circuits. Capacitance as an effect is also present in every standard electrical circuit. In fact capacitance is all around you.

The basic capacitor that you will have seen is a device for storing **charge**. It consists of two metal plates that are separated by an insulated material. This insulator (known as the dielectric) can be made of polystyrene, oil-impregnated paper, or even just air, as long as it stops current flowing between the plates. If a d.c. supply such as a battery is connected across the plates, the capacitor will charge up to the same voltage level as the supply. This occurs because, in the battery and the conductors connecting the metal plates of the capacitor, there are a large number of free, negatively charged electrons. Before the capacitor is connected to the circuit, the plates have no charge – said to be **discharged**.

As the supply is connected, the free electrons in the circuit will be attracted to the positive part of the circuit. Imagine pouring water onto an uneven floor – the water will always try and get to the lowest point before settling – this is the same for electrons. The electrons will leave the metal plate of the capacitor and the conductor connecting it and flow to the positive plate of the battery. The electrons will also at the same time leave the negative plate of the battery and flow to the other capacitor plate. The electrons are attracted to the uncharged capacitor plate as it is effectively more positive than the –ve battery plate. The electrons will flow until the charge across the capacitor plates is identical to the charge difference (and potential difference) of the battery source.

If an ammeter was to be placed in the circuit, and a switch activated to complete the circuit, a current would be seen. Initially the current would be seen and, as the capacitor started to reach the same voltage difference as the supply, the current would tail off. Once the capacitor reaches the same voltage as the supply, the current would stop flowing altogether. Strictly speaking this is not a circuit as there is a circuit block in the way – the capacitor dielectric insulator. When a capacitor is used in a d.c. circuit it is actually a block as no current can pass between the insulated plates.

Because the plates charge to the same level as the d.c. supply connected to it, a capacitor can act as a charge store, just like a battery. If the capacitor, once charged, is taken out of the circuit, you could use it to power a load until it discharged. One recommended method of discharging a large capacitor is to put a light bulb across the terminals. A light bulb placed across a capacitor's +/− terminals will glow brightly and as the electrons flow through the bulb to the opposite plate they will give off their energy in the form of light. As the electrons reach equilibrium with both directions appearing to be the same charge level (effectively discharged), the electrons stop flowing and the light goes out. The fact that the light has become bright proves that the capacitor has actually stored energy. Large capacitors can also be discharged by special high-powered discharge resistors connected across the terminals. These large capacitors can be found where large banks of fluorescent lights exist, for example in the Dartford Tunnel.

A charged capacitor with one plate negatively charged and the other positive will have an electric field between the plates. This has similarities to the magnetic field that exists between poles of two opposite magnets. In the same way as a magnet, a capacitor will have imaginary magnetic flux lines that cannot be seen but the effect can be seen.

The capacitor's ability to store energy is directly proportional to the size of the plates, the distance between them, the p.d. (voltage supply) applied and the nature of the insulator. The capacitance value given to a specific capacitor is the relationship between the charge stored and the voltage applied to the capacitor plates. For this reason a capacitor component will have a fixed constant value of capacitance and this is measured in farads (F).

Capacitance can be found using the formula:

$$C = \frac{Q}{V}$$

where C is capacitance in farads (F), Q is charge in coulombs (C) and V is voltage in volts (V) across the plates.

Figure 6.113: Large capacitor and discharge resistor

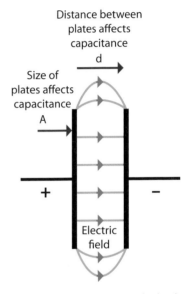

Figure 6.114: Capacitor strength related to plates

Worked example 1

What is the capacitance of a capacitor storing a charge of 320 μC when connected to a 5 V supply?

$$C = \frac{Q}{V}$$

$$C = \frac{320 \times 10^{-6}}{5}$$

$$C = 64 \ \mu F$$

You have already seen charge is related to current flow and the time in Chapter 1, so using the formula:

$$Q = I \times t$$

and then combining with the formula for capacitance, a further equation becomes available:

$$C = \frac{Q}{V} = \frac{I \times t}{V}$$

Worked example 2

A steady current of 10 A flows into a previously uncharged capacitor for 7 seconds when the potential difference between the plates is 10 000 V. What is the capacitance of the capacitor?

$$C = \frac{I \times t}{V}$$

$$C = \frac{10 \times 7}{10\ 000}$$

$$C = 0.007\ \text{F or } 7\ \text{mF}$$

Energy stored in a capacitor

If a capacitor is accidentally shorted out between the terminals, a large 'crack' will be heard and a flash seen as the electrons are attracted to the positive plate and the capacitor is discharged. This flash/crack proves there is energy stored in the capacitor. Imagine the capacitor is now charged up again by applying a p.d. across the plates. The average voltage over the charge cycle is V divided by 2 because the voltage starts at 0 V and then reaches the supply voltage level. So, the average voltage must be half the supply voltage by definition. You will also remember from previous work the formula:

$$P = V \times I$$

So the power required to charge the capacitor becomes:

$$P = \frac{V}{2} \times I$$

Also, from the section on energy, you will remember:

Energy = Power × time

or:

$$W = P \times t$$

So putting both formulae together gives:

$$W = \frac{V}{2} \times I \times t$$

Also, $I \times t$ can be replaced by C and correspondingly, $V \times C = Q$, so the formula to find the energy stored in a capacitor becomes:

$$W = \frac{CV^2}{2}$$

where W is the energy stored in the capacitor's electric field, V is the p.d. applied across the capacitor plates and C is the capacitance of the capacitor.

Worked example

What is the energy stored in a 64 µF capacitor when a 1 000 V d.c. supply is connected across the plates?

$$W = \frac{CV^2}{2}$$

$$W = \frac{64 \times 10^{-6} \times 1\ 000^2}{2}$$

$$W = 32\ \text{C}$$

Capacitors in parallel

A capacitor size is directly related to the size of the plates. Imagine you have connected two capacitors in parallel. If you look end on you will see the two plates next to each other. Effectively you have doubled the size of the plates exposed to the oncoming electrons by putting them next to each other. To find the total capacitance of capacitors in parallel with each other, add them together, as can be seen in Figure 6.115.

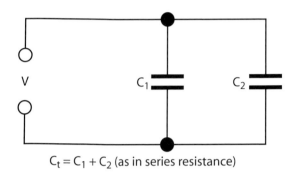

$C_t = C_1 + C_2$ (as in series resistance)

Figure 6.115: Capacitors connected in parallel

Worked example

If four identical 64 µF capacitors are connected in parallel, what is the total capacitance of the circuit?

$64 \times 10^{-6} + 64 \times 10^{-6} + 64 \times 10^{-6} + 64 \times 10^{-6} = 256 \times 10^{-6}$ or 256 µF

Strangely enough, this is the opposite process to adding resistors!

Capacitors in series

As hinted previously, the method of combining capacitors is simply the reverse of how you deal with resistors. So for capacitors in series you add the values as fractions:

$$\frac{1}{C_t} = \frac{1}{C_1} + \frac{1}{C_2}$$

As found when working out resistors in parallel, there is a shorthand version for adding two capacitors in series using the formula:

$$C_t = \frac{C_1 \times C_2}{C_1 + C_2}$$

Worked example

Two identical capacitors of 64 µF are connected in series. What is the total capacitance of the circuit?

$$C_t = \frac{C_1 \times C_2}{C_1 + C_2}$$

$$C_t = \frac{(64 \times 10^{-6}) \times (64 \times 10^{-6})}{64 \times 10^{-6} + 64 \times 10^{-6}}$$

$$C_t = 32 \times 10^{-6}$$

Figure 6.116: Capacitor showing polarity

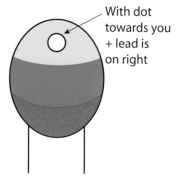

With dot
towards you
+ lead is
on right

Figure 6.117: Tantalum capacitor

 Safe working

Electrolytic capacitors are generally polarity-sensitive components that must be checked before connecting. Wrong polarity can be dangerous.

As you can see, capacitors in series follow the same rules as resistors in parallel. When two identical capacitors are placed in series the resultant value is half of one of them.

Types of capacitors

A large range of capacitors are available on the market for a wide range of applications. There are large industrial power factor capacitors used on fluorescent lighting banks, down to the smallest electrolytic capacitors found in computer circuits. Some capacitors are constructed to work in a d.c. circuit only and therefore are polarity sensitive. An electrolytic capacitor, as shown in Figure 6.116, has the polarity marked on the metal can. The shorter leg with the lighter stripe indicates the negative connection.

Other capacitor types, such as the tantalum capacitor, will have a mark of some kind to show the positive and negative connection. This can be seen in Figure 6.117.

As with resistors, capacitors are available as a fixed or variable value. The fixed value capacitors can be polarity sensitive as with most electrolytic capacitors. If an electrolytic polarity-sensitive capacitor is connected up the wrong way round it will fail and this can be quite dangerous.

The aluminium electrolytic capacitor construction can be seen in Figure 6.118, which shows one that has failed. The unrolled aluminium plates can be seen. One side is coated in an insulating oxide and the plates are separated by electrolyte-soaked paper spacers. With this capacitor, the negative terminal or anode has been connected to the supply the wrong way round. Most electrolytic capacitors are polarised and must have the lower voltage connected to the –ve terminal.

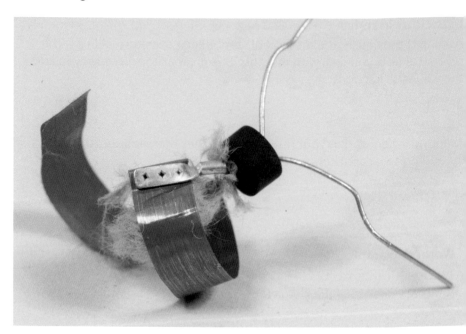

Figure 6.118: Exploded capacitor

The advantage of electrolytic capacitors compared to other types is the size to capacitance ratio. Much bigger values of capacitance can be achieved in small packages. Table 6.9 shows types of capacitor.

Capacitor type	Description
Mica capacitors 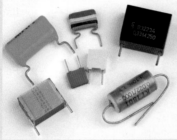	Because of the stable nature of this type of capacitor it is slightly more expensive than its plastic film equivalent. The stability of a mica capacitor is exactly why it is suitable for sensitive, accurate equipment working at very high frequencies. Even operating at high temperatures the mica capacitor can operate within +/– 1 per cent of its label value.
Film capacitors 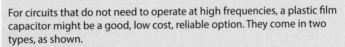 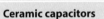	For circuits that do not need to operate at high frequencies, a plastic film capacitor might be a good, low cost, reliable option. They come in two types, as shown. Film capacitors are made from layers of metal electrodes covered in a dielectric insulating film. The film can be made of a number of different materials, each having unique electrical properties. There are: • polypropylene • polyester • polyethylene • polystyrene • polycarbonate. This type of capacitor is generally referred to as a plastic film capacitor. (Paper film capacitors can still be found although that technology is very old now.)
Ceramic capacitors 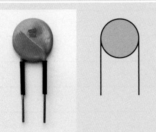	Ceramic capacitors are particularly suited to moderately high frequency applications. They are a low cost option compared to other equivalent capacitors. They are constructed with layers of metal plates separated by layers of ceramic disks. This capacitor is sometimes also referred to as a disk capacitor because of its shape. This type of resistor is also increasingly being used to replace electrolytic aluminium and tantalum capacitors because of the cost and reliability. The construction of some capacitors gives rise to a certain amount of inductance but the ceramic capacitor does not suffer from this, giving it another advantage over equivalent capacitors.
Variable tuning capacitors 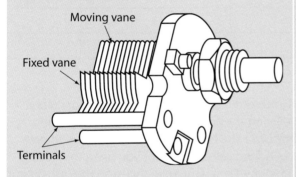	There are many reasons why you might need a variable capacitor. When working on a.c. single-phase theory previously, you found that capacitive reactance goes down when the frequency increases. Capacitive reactance also decreases if capacitance is increased (and if frequency reduces, reactance increases). For this reason variable capacitors are good for electronic tuning and filtering applications. As with resistors, variable capacitors are constructed in different ways depending on the application they are used for. Tuning capacitors are variable capacitors that are in constant use, for example a frequency tuner on a radio. This type of variable capacitor has two sets of plates – one that moves and one that is fixed. The plates interleave and when turned the amount the plates overlap also varies without actually ever touching. The dielectric insulation between the plates is air and ceramic.
Pre-set capacitors 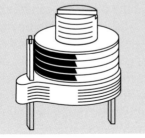	Similar to the pre-set resistor and sometimes called a trimming capacitor. The adjustment is via a small correctly rated screwdriver. The pre-set capacitor is much finer with a reduced tuning range. The diagram shows a pre-set trimming capacitor. This type of variable capacitor might be found on a circuit board that needs an element of tuning to function correctly, such as a filter circuit. As with a pre-set resistor, the only reason to touch this capacitor is initial circuit set-up, maintenance or servicing. Apart from these reasons, a pre-set capacitor will not need to be touched.

Table 6.9: Types of capacitor

Identification of capacitors

Capacitors are small components like resistors and therefore it is logical they should also have a similar colour and number code. Larger capacitors can have codes and smaller capacitors can have colours. The standard capacitor colour codes can be seen by referring to Figure 6.119 and using Table 6.10 below.

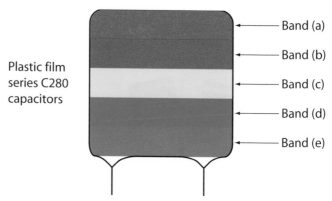

Plastic film series C280 capacitors

Band (a)
Band (b)
Band (c)
Band (d)
Band (e)

Figure 6.119: Capacitor colour bands

Colour	1st digit	2nd digit	3rd digit	Tolerance band	Maximum voltage
Black		0	None	20%	
Brown	1	1	1		100 V
Red	2	2	2		250 V
Orange	3	3	3		
Yellow	4	4	4		400 V
Green	5	5	5	5%	
Blue	6	6	6		630 V
Violet	7	7	7		
Grey	8	8	8		
White	9	9	9	10%	

Table 6.10: Standard capacitor colour coding

As you will notice the capacitor colour code is similar to the resistor colour code, with the first two digits giving the actual capacitor digits. The colour bands are read from top to bottom – the bottom of the capacitor is where the legs are. The third colour band indicates the factor you need to multiply the two digits by. The fourth colour will give the tolerance and the fifth colour band indicates the maximum voltage the capacitor can be connected to. Because capacitors are very small in value they tend to be measured in pico or nano farads – a capacitor of 1 F is almost unheard of and certainly not used in standard electronics. The capacitor colour code assumes all capacitors are starting in the pico farad range. If the multiplier band is black, this means the capacitor is actually a pico farad capacitor. If the third band is, for instance, orange, this means the capacitor needs to be multiplied by 1 000 (moved up three positions) and turned into a 'nano' farad capacitor.

Worked example

A five-band polyester capacitor reads from top to bottom: blue, white, orange, black, red. What is the capacitor?

Blue	= 6
White	= 9
Orange	= × 1 000
Black	= +/−20%
Red	= 250 V

The capacitor is a 69 nF, 250 V capacitor, + or −20%.

Most capacitors have a little more space than resistors for written codes. As with resistors, engineering notation and prefixes are used to give the capacitor value. If a capacitor code works out to be 2.2 pF, this can be written as 2p2F or simply 2p2. This saves even more space as the capacitor component is obvious to the trained eye and there is no need to use the farad symbol.

Capacitor applications

Capacitors can be used for a.c. and d.c. applications. Some of the general uses include:

- to store energy and give it back to a circuit, as used in car 'power caps' on car hi-fi systems
- to smooth rough d.c. wave forms after a.c. to d.c. rectification
- acting as filters with their ability to be tuned in radio technology
- to correct over inductive circuits to make them more efficient in the form of power factor correction for fluorescent lighting or inductive motor plants
- radio frequency suppression – radio frequency electromagnetic radiation caused by high speed switching can be suppressed by capacitors, as seen in a fluorescent starter circuit
- timing circuits; they often make use of a capacitor's predictable charge and discharge relationship.

Capacitor as a timing device

If connected in series with a known resistor value, a capacitor will charge and discharge in a very predictable way. Imagine a d.c. voltage supply connected across the series resistor and capacitor and a switch is flicked to connect the circuit. The instant the switch is flicked on, the electrons will be attracted to the more positive uncharged capacitor plate. The electrons will meet a certain amount of resistance as they pass through the series resistor. The initial rush of electrons slows down as the capacitor approaches the same p.d. as the supply voltage. All of this charging activity takes time and the time is referred to as the time constant of the circuit. This time constant can be calculated using the formula:

$$\tau = C \times R$$

where τ is the time, C is the circuit capacitance and R is the circuit resistance.

Initially when the capacitor starts to charge, it charges rapidly, as can be seen by the curve in the charge and discharge diagrams in Figures 6.120 and 6.121.

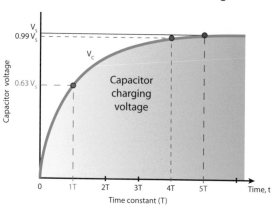

Figure 6.120: Increase in voltage as a capacitor charges

Figure 6.121: Decrease in voltage as a capacitor discharges

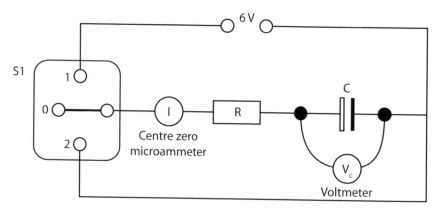

Figure 6.122: A typical charge and discharge circuit

If this rapid acceleration were allowed to continue, the capacitor would reach full charge in a set time. This time is referred to as τ. Because the flow of electrons slows down as the capacitor approaches full charge, the slope of the charge curve levels out. The capacitor actually never gets to the full charge value of the supply because of small amounts of resistance in the circuit. At approximately 5 × τ, the capacitor is about 99 per cent of the full supply value.

The same is true for the reverse process called capacitor discharge. If the circuit is now turned into a discharge circuit by 'shorting out' the capacitor and resistor, the discharge process would take approximately 5 τ to empty the capacitor of charge. The charge and discharge circuit can be seen in Figure 6.122.

> **Worked example**
>
> Refer to Figure 6.122. A capacitor of 24 µF is connected in series to a 68 kΩ resistor. What is the time constant of the circuit and how long will it take to reach 99% of the supply voltage level?
>
> $\tau = C \times R$
>
> $\tau = 24 \times 10^{-6} \times 68 \times 10^{3}$
>
> $\tau = 1.632 \text{ s}$
>
> For the capacitor to reach approximately 99% of full charge level:
>
> 99% full charge $= 5 \times \tau$
>
> Charge $= 5 \times 1.632 \text{ s} = 8.16 \text{ s}$

Activity 6.21

Set up the circuit as shown in Figure 6.122 and time the capacitor charge cycle approximately. A d.c. extra low voltage supply is required that has an ammeter and voltmeter display. You are looking for when the ammeter slows down as it approaches 0 amps flowing. The voltmeter should start briefly at 0 V and then rise to the supply setting.

Semiconductors

As the name suggests, semiconductors conduct under some conditions and not under others. This becomes particularly useful if you can control the 'on' and 'off' states like a switch by simply changing the voltage level applied to them.

Semiconductors are made out of pure silicon crystals that are formed into a lattice structure, as can be seen in Figure 6.123.

The crystal structure of a semiconductor means that whichever way you cut it the properties will look the same. The chemical and electrical properties of any atom depend on the number of electrons in the outer shell, otherwise known as the **valence shell**.

As can be seen from Figure 6.123, each atom is sharing its four electrons with four other neighbours. When combined in this way the atoms bond strongly to form a crystal. Carbon, silicon and germanium are known as tetravalent atoms because of the four bonds. The actual bonds between neighbouring atoms are called covalent bonds. When carbon is formed in this way it becomes diamond. The closer the covalent bond becomes, the more difficult it is to convince an electron to break free and current flow to happen. If an electron is forced free by enough pressure from the outer valence shell it will also leave a hole or space for another electron to jump in to. In a pure semiconductor, otherwise known as an intrinsic semiconductor, free electrons are few and far between, as are the holes. To increase the possibility of electron flow, other elements can be added to the semiconductor. This process is called **doping**.

Impurities that can be added have two options – they can either bring with them more electrons or more holes. An impurity such as arsenic or antimony will bring more electrons and is referred to as a pentavalent impurity because it has five electrons in its outer shell. This will mean there will be one free electron to act as a charge carrier. When silicon is doped with a pentavalent impurity, the overall effect is to make the silicon more negative due to the extra electron charge carrier. This material is referred to as an n-type doped semiconductor.

An impurity such as aluminium, with three valence electrons and hence a spare hole, can be added to silicon also. As electrons jump into the free holes, it is as if the holes are moving. Imagine a car moving off in a traffic jam and the car behind moves into that space. If this continues, the space is effectively moving backwards! Semiconductors that are doped in this way are more positive and are referred to as p-type doped semiconductors. A p-type impurity with three valent electrons is a material such as indium or gallium.

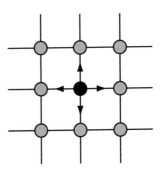

Figure 6.123: Lattice structure of semiconducting material (silicon or germanium)

Key terms

Valence shell – the outer shell of an atom that determines how good or bad a conductor the material will be.

Doping – when another impurity is added to the semiconductor. Doping can make the semiconductor more positive by adding holes or more negative by adding electrons.

Diodes

Diodes are created by joining a p-type and n-type semiconductor material together. This means that on one side there are free holes and on the other free electrons. When the diode is not connected to a battery supply, there is a region between the p- and n-type materials where a certain amount of recombination has happened. This is where holes and electrons have moved across the barrier and joined again. The region between the p- and n-type material where recombination occurs is called the depletion layer, as can be seen in Figure 6.124.

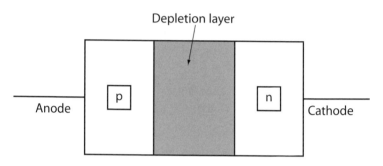

Figure 6.124: p-n junction

Once this recombination of electrons and holes has happened there is no possibility for current to flow, as no free electrons or holes exist. If, however, a battery is connected with the −ve of the supply to the n-type material and the +ve of the supply to the p-type material, current flow will occur as the electrons and holes are attracted by the battery plates. The voltage level required to start the flow is only a few tenths of a volt. If the battery connection is reversed, the non-conductive depletion layer between the n- and p-type material will increase. This increased barrier will stop any chance of current flow – effectively the diode will switch off. When a diode is connected to allow current flow it is said to be in forward bias mode. When the diode is reversed it is said to be in reverse bias mode. If a diode is in reverse mode, no current will flow unless a very large voltage pressure is applied. A large voltage applied to a diode in this way will destroy the diode. The symbol and examples of diodes can be seen in Figure 6.125.

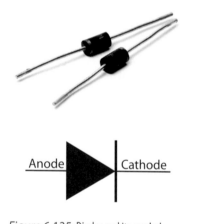

Figure 6.125: Diodes and its symbol

Zener diode

One type of diode that can operate in reverse bias mode is the zener diode (sometimes referred to as a power diode). This component is found in voltage regulator circuits because of its ability to provide a constant voltage level. One unique property of the zener diode is its ability to actually work in reverse bias mode without completely destroying itself. By very careful doping the zener diode can break down in a controlled way. Once the reverse bias breakdown voltage level is reached – otherwise known as the zener voltage – current will flow. Zener diodes can be manufactured with various pre-determined breakdown voltage levels. An example of a voltage regulator circuit with a zener diode can be seen in Figure 6.126.

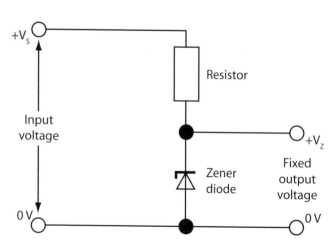

Figure 6.126: A simple voltage regulator circuit

Light-emitting diodes

Developments in technology and manufacturing techniques have seen a rapid rise in the demand for LED technology in lighting applications. The big advantage of LED technology is ongoing reduced running costs. LEDs do not use much energy for the light output they provide.

LEDs are semiconductors and have a p-n junction. When enough voltage is applied in a forward bias mode, current will flow. As the electrons flow across the junction, light energy is released. Generally LEDs cannot take a very large voltage and are typically rated at 3–4 volts. Anything over this will blow the diode. The connection of a diode is very important as reverse bias will damage the diode if the applied voltage is large enough. If a small voltage is attached across a reverse bias diode it simply will not work. The anode has to be connected to the positive rail of a d.c. supply and the cathode to the 0 V rail. The symbol and lead identification diagram can be seen in Figure 6.127.

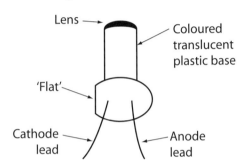
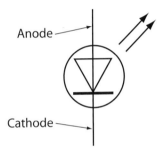

Figure 6.127: Light-emitting diode

Diacs

A diode for alternating current, otherwise known as a diac, is a semiconductor device that conducts current in both directions but only under certain voltage level conditions called the breakdown voltage. The breakdown principle works in the same way as the zener diode breakdown voltage.

When the voltage level of approximately 30 V is reached, current can pass through. The diac is bidirectional, unlike the zener diode, so current can pass in both directions as long as the trigger voltage level is reached. Until the breakdown voltage is reached the diac acts like an open circuit or open switch. A typical application of a diac is as a controlling switch for other semiconductor components such as a triac or thyristor.

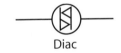

Figure 6.128: Diac symbol

Triacs

A triac is a semiconductor bidirectional switch that can be triggered. The triac has three connections: MT1, MT2 and a gate. The gate is the trigger when a signal of a certain level is reached. The pre-determined trigger level can be negative as well as positive. When triggered, the triac will allow current to pass in both directions as per the diac. Typical applications for triacs include the dimmer circuit that can be seen in the circuit diagram in Figure 6.129.

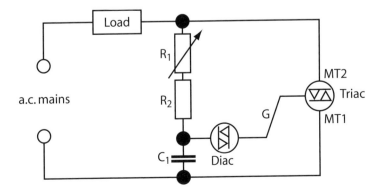

Figure 6.129: GLS lamp dimmer circuit

The charge rate of the capacitor is varied by the potentiometer in the tuning circuit. The charging capacitor controls when the diac triggers and turns on the triac. By changing the triac 'on-time' this varies the amount of supply sine wave getting through to the GLS lamp. If the average power getting to the lamp is less, the lamp is dimmer.

Thyristors

The thyristor or silicon controlled rectifier (SCR) is a semiconductor component made up of four layers of alternate p- and n-type material that act as a very high-speed switch. The SCR can be seen in Figures 6.130, 6.131 and 6.132.

Figure 6.130: Thyristor

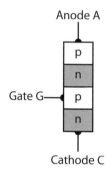

Figure 6.131: Construction of a thyristor

Figure 6.132: Thyristor circuit symbol

The thyristor is very similar to a diode except it has an additional connection called a gate. The gate is a trigger to switch the diode on and off. Normally the thyristor is in the off position but, once the trigger voltage has been reached, the thyristor can open and pass current through in one direction only. The thyristor will remain on as long as there is enough holding current at the gate. If the gate current reduces below the threshold or the SCR is put in reverse bias, the device will close the gate and the thyristor will switch off. An advantage of the SCR is its ability to use a very small current to control a very large current – similar to older technology, contactor control.

Rectifiers

Diodes have the ability to conduct current in one direction. This is particularly useful for turning alternating current, a.c., into direct current, d.c. If a single diode is connected in series with a load and then connected to an a.c. supply, only half of the sine wave can get through to the load, as can be seen in Figure 6.133.

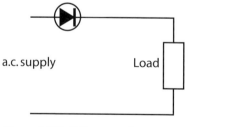

a.c. supply   Load

Figure 6.133: Half wave rectifier

The positive part of the wave will pass through the diode but the negative half of the sine wave will be blocked. This type of rectification is called half wave rectification. The disadvantage of half wave rectification is that half the wave is lost – this is very inefficient.

Diodes – full wave rectification

The majority of electronic circuits work on d.c. only so it is therefore essential to be able to convert a.c. to d.c. for most electronic devices found in the home and workplace. To rectify a sine wave fully into d.c., more diodes are required. One method uses two diodes, as can be seen in Figure 6.134.

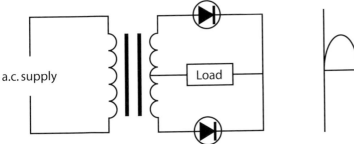
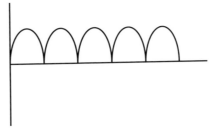

Figure 6.134: Full wave rectifier

The load is centre tapped onto the transformer coil. The positive part of the cycle passes around one half (the upper part) of the circuit that is made up of one diode and the load. The positive half cycle is, however, blocked by the other diode (the lower part of circuit) as it is in reverse bias. When the negative half of the sine wave arrives, the diodes reverse roles. The second diode now makes the circuit with the load and the original (upper) diode is reverse biased. This effectively turns both parts of the sine wave positive – full wave rectification.

Diodes – full wave bridge rectifier

A more traditional full wave rectifier that is available in an integrated form is the bridge rectifier. This consists of four diodes connected as shown in the schematic diagram in Figure 6.135.

The integrated components have four terminals, as can be seen in Figure 6.136.

The four diodes are connected in such a way that positive and negative parts of the sine wave are blocked and only allowed to pass through the load in one direction. This can be seen from Figure 6.137 where first the positive half cycle and then the negative half cycle pass through the diode bridge.

Bridge rectifier

Figure 6.135: Bridge rectifier schematic symbol

Figure 6.136: Bridge rectifiers

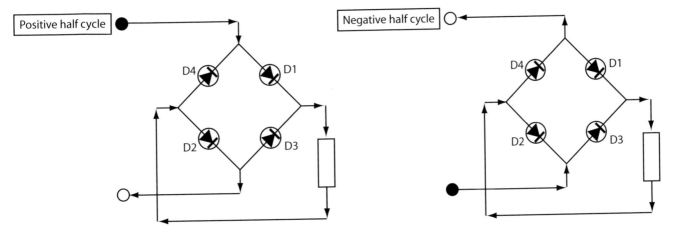

Figure 6.137: Full wave bridge rectifier

At this stage the sine wave has been fully rectified. This means it no longer has a positive and negative component – it is all positive. The signal still looks like a series of bumps and is not very useful. To make the new rectified signal into a smooth d.c. constant value the bumps need to be ironed out. This is achieved by a process called smoothing.

Bridge rectifier smoothing

Capacitors have an ability to store charge. Capacitors also take time to charge and discharge. This effect can be used in wave form smoothing. By connecting a capacitor across the output of a bridge rectifier, the charging and discharging action that occurs effectively smooths out the ripple of the wave form. This is sufficient for relatively small levels but larger currents require a different approach. If an inductor is now placed in series with the output of the circuit, a back emf is generated as the current changes. The back emf will help maintain a steady current. The best method to achieve a smooth consistent d.c. value is to combine capacitors and inductors, as can be seen in Figure 6.138.

This type of circuit is called a filter circuit because it makes use of the capacitor and inductor to filter out the ripple and create a relatively smooth d.c. supply for electronic applications. The before and after signal can be seen in the wave form diagram in Figure 6.139.

High-powered d.c. supplies

Although single-phase 230 V rectification is commonplace, there is also a need for high power d.c. supplies as can be found on the railway networks. Three-phase rectification is one method of creating a high powered d.c. supply. This method uses six diodes connected into a bridge configuration, as shown in Figure 6.140.

Transistors

Another form of component that uses semiconductor technology is the transistor. A transistor can be used as a simple switch or as an amplifier of signals. As with all semiconductors a transistor has layers of doped material that are involved in the transfer of electrons and holes between the transistor regions. A transistor is made up of three layers that can either be

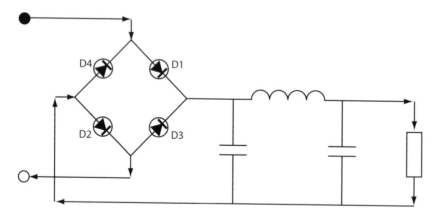

Figure 6.138: Capacitor input filter

Figure 6.139: Wave form for a capacitor input filter

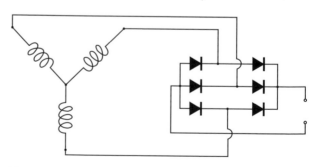

Figure 6.140: Three-phase bridge rectifier

two n-type and one p-type or the other way round. Either way, a transistor will have two p-n regions where electron and hole transfer can happen. A typical configuration of a pnp transistor can be seen in Figure 6.141.

It can be seen from Figure 6.142 that the central connection is labelled base, the top is the collector and the bottom is the emitter. The arrow in the circuit symbol shows the direction of conventional current flows. If the doping is changed, a pnp transistor is created.

Figure 6.141: Transistors

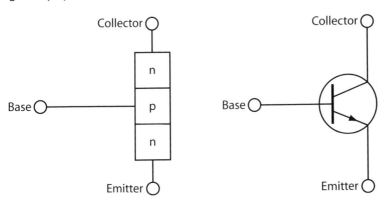

Figure 6.142: npn transistor and its associated circuit symbols

There are two main families of transistor: the bipolar junction transistor (BJT) and the field effect transistor (FET). Each has its own unique properties that makes it suitable to specific applications.

The BJT is capable of providing very high **gain** and is therefore found in amplifiers and high-speed digital circuits.

The FET is particularly suited to high-density, low power consumption digital integrated circuits such as digital watches, timers and calculators.

Operation
The transistor operates like two back-to-back diodes with a controlling tap. The control (tap) for the flow of electrons and current flow is managed by the base current. The base emitter must be forward biased for current to flow anywhere. Charge carriers can be holes or electrons within an npn or pnp transistor. In a p-type material the majority of carriers are holes. In an n-type material the majority of charge carriers are electrons. An electrical input to one pair of the three leads can control the output from another pair. In a pnp transistor this will mean the emitter base connections will control the output from the collector emitter connections.

npn transistor
Looking at Figure 6.144 (overleaf), you can see that battery B1 has created a voltage drop across the base emitter junction. This has the effect of forward biasing the p-n junction. Holes will enter the thin base and electrons will be attracted to them from the emitter – the start of current flow. A small amount of electron-hole recombination will occur in the base but, because it is only lightly doped, a considerable amount of electrons will be left in the base. The base collector junction is actually reverse biased due to battery B2. In reverse bias conditions, the collector is more positive than the base. Electrons that are sitting in the base are therefore attracted to the positive plate of the battery, B2, and they leave

Key term

Gain – used to describe the range an amplifier can turn up to.

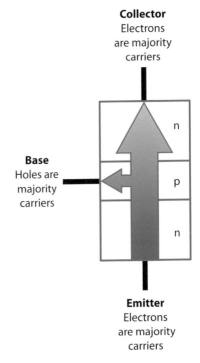

Figure 6.143: Electron flow in an npn transistor

the base area into the collector. The actual amount of electrons and hence current that flows through from the emitter to the collector is controlled by the conductivity of the base. The base conductivity is controlled by the amount of holes and hence the base current. The flow of electrons in an npn transistor can be seen in Figure 6.143 on the previous page.

Transistor applications

A transistor can be used as an amplifier as the base can control the level of current flowing through the emitter collector like a tap. A small control current in the base can be represented by a larger version of itself flowing through the emitter/collector. If the base current increases, the emitter/collector also increases. If an npn transistor is connected to a supply and is forward biased with a voltage of approximately 0.6 V, as shown in Figure 6.144, a base current of 0.5 mA will cause a collector current of 50 mA to flow.

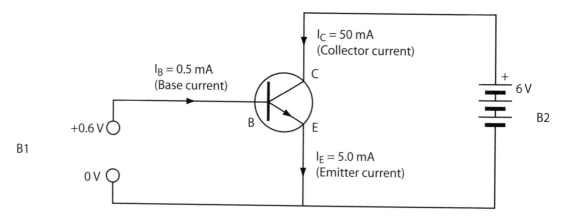

Figure 6.144: Current amplifier

This relationship is called the gain of the transistor and is given the symbol h_{fe}. This amplification ratio can be shown by the formula:

$$h_{fe} = \frac{I_C}{I_B}$$

$$h_{fe} = \frac{50 \text{ mA}}{0.5 \text{ mA}} = \text{A gain of } 100$$

If the base current changes so will the collector current by the same ratio.

A transistor can also be used as a switch. For a transistor to be on, a voltage of at least 0.6 V must be applied across the base emitter junction. If the transistor is on, the collector/emitter resistance is very low and there will be 0 V dropped across it. If the resistor is off, all the supply voltage will be dropped across the collector/emitter. By supplying a forward bias and then removing the forward bias, a logical switch is created.

Transistors can be manufactured in minute proportions and integrated into silicon chips. Thousands upon thousands of these logical switches can be combined in one small chip to create logic gate circuits that are found on all electronic circuitry.

Inverter

The opposite of a rectifier is an inverter. The purpose of an inverter is to change d.c. back into a.c. for equipment that requires a.c. Commercially available equipment is typically found in cars where there can be a need to power equipment from a car battery supply. There are also much larger

Progress check 6.22

1 Describe how an a.c. supply can be converted to d.c. using electronic components.

2 Name three applications for capacitors.

3 Describe how to work out a resistor value from a colour code by creating an example.

applications that need to have a d.c. supply converted back into a.c. Some large sites such as big department stores or computer centres are often backed up by uninterrupted power supplies. These emergency power systems might have battery rooms that provide a constant d.c. supply. This is of no use unless it can be converted to a.c. via an inverter. These back-up emergency systems are an alternative solution to back-up generators as they can be fairly instantaneous in operation, thereby avoiding data loss.

> **Safe working**
>
> Lead acid battery back-up units produce hydrogen which is highly explosive when mixed with air – great care must be taken when working with this type of supply.

Function of electronic components in electrical systems

There are many areas that a fully qualified electrician can specialise in. Some specialist areas have a mix of electrical and electronic/control work. These areas include security/fire alarm systems, telemetry/telephony and electrical heating to name a few.

Security alarm systems

There are plenty of specialist security companies but there will always be an occasion where you may get involved with this area due to the need for electrical supplies to cameras, alarm panels, detectors, sounders, etc. Alarm systems, including intruder and fire alarms, will be covered in more depth in *Chapter 7: Electrical systems design*, pages 363–364.

Telephony

Telecommunications has always been fairly closely linked with the work of a building services engineer or electrician. When an electrical installation is underway it will often be the electrician that is also contracted to lay in the data and communications cables at the same time. The various cables used in data and telecommunications have been covered in previous chapters. As well as installing the cables, you may be involved in the termination of data points. Telephony and data are in many instances the same with voice over Internet protocol (VoIP) – all you install are data points and the equipment that is plugged in does the conversion. Installation work will mainly involve infrastructure in the form of structured cabling systems and equipment racking. Data and telephony cabling is extra low voltage and has to be installed in separate containment or insulated to the highest voltage level if combined. Increasingly, fibre is available domestically and this will be supplied by a telecoms company as it requires special equipment to install, test and commission.

There is a rapid growth in electrical equipment being controlled over IP devices such as iPhones turning on pumps, opening electric gates and closing curtains. All of this electronic equipment will be installed by a trained electrician.

Motor control

Motor control and the components used have been covered earlier in this chapter on pages 296–302.

Heating control

Heating controls have been covered extensively earlier on in the chapter. Revisit this section on pages 307–309 and check your knowledge.

Light dimmers

Dimmers and the use of diacs and triacs has been covered extensively on pages 328–329. Revisit this section and check your knowledge.

> **Progress check 6.23**
>
> 1 Describe three areas where a building services engineer might require some knowledge of electronic components.
> 2 Explain the function of diacs in a dimmer circuit.
> 3 What is the function of a triac in a lighting control?

Knowledge check

1 The current in a coil changes from 5.4 A to 2 A in 0.04 seconds and induces a voltage of 20 V. What is the inductance of the coil?

a   25 H
b   0.023 H
c   0.24 H
d   2.5 H

2 A circuit consists of two series resistors of 5 Ω and 3 Ω. What is the power in the 3 Ω resistor if the circuit current is 3 A?

a   36 W
b   9 W
c   15 W
d   1 W

3 A series network is made up of a capacitor of reactance 50 Ω, resistor of 40 Ω and inductor of 80 Ω. What is the power factor of the circuit?

a   0.6
b   0.7
c   0.99
d   0.8

4 A luminaire producing a luminous intensity of 1 000 cd is installed 4 m above a surface. What is the illuminance on the surface directly beneath?

a   62.5 lux
b   250 lux
c   250 lm
d   62.5 cd

5 In a four wire balanced three-phase system, the line current is 40 A. What is the neutral current?

a   40 A
b   60 A
c   0 A
d   1 A

6 The supply is connected to the rotor of a three-phase wound rotor induction motor by what method?

a   Direct wires
b   Slip rings and brushes
c   Commutator and brushes
d   Stator

7 A current-carrying conductor is placed at right angles to a permanent magnetic field. The movement that follows can be described as what?

a   The generator effect
b   The rectifier effect
c   The alternator effect
d   The motor effect

8 A four pole three-phase induction motor with 4% slip is connected to a 400 V, 50 Hz supply. What is the rotor speed?

a   24 revs/s
b   12 revs/s
c   24.4 revs/s
d   200 rev/s

9 What method is used to start a d.c. motor?

a   A starter capacitor
b   A rectifier
c   A faceplate starter
d   Star delta starter

10 The no-volt circuit in a DOL starter is designed to do what?

a   Prevent starting when the voltage exceeds supply
b   Prevent starting after loss of supply
c   Prevent fault currents damaging the armature
d   Allow starting at lower voltage levels

11 The centre part of a solenoid is known as what?

a   The armature
b   The relay
c   The former
d   The contactor

12 An infrared heater is an example of what type of heater?

a   Underfloor
b   Convection
c   Radiant
d   Storage

13 A Y plan heating system consists of what components?

a   Two-port motorised valve
b   Three-port mid-position valve
c   Three-phase contactor control
d   Two-phase contactor control

14 If the capacitors used in the networks in the diagram below are all 3 µF, which network will give an overall capacitance of 4 µF?

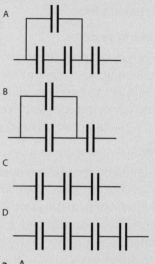

a   A
b   B
c   C
d   D

15 A residual current device will operate under what conditions?

a   A neutral to live fault
b   Phase to phase fault
c   CPC to MET fault
d   Phase to CPC fault

Electrical systems design

Chapter
7

This chapter covers:

- how to interpret information needed for electrical design
- the principles of designing electrical systems
- how to design complex electrical systems
- how to plan work schedules for electrical installations.

Introduction

Good electrical design is essential to ensure that a new or modified installation is safe and able to do the job it is intended for. It is also vital to the efficient running of a building engineering services project: the more accurate the information available, the less stoppage time there will be for the resolution of queries.

The job of the designer is not just to mark positions of switches and lights on blank scale drawings, but to calculate the sizes of the cable needed to feed the installation and its final circuits, to match the right wiring system to the environment, to make sure that the protective devices will operate within the required disconnection times, and to protect the installation and its users from fire and shock.

A construction project such as the Olympic Park in London needed thousands of people from a wide variety of trades and professions for its design, build and commission. Even smaller projects will require the carefully coordinated input from multiple trades. This needs to be planned so that the right people are on site at any given moment and the work is carried out in a logical order.

How do we arrive at our design decisions? How do we know if the installation can be wired in twin-and-earth or needs to be protected in steel conduit and trunking? When do we install low smoke-emitting cable or even fireproof cables? How can we coordinate the complex threads of a construction project?

INTERPRETING DESIGN INFORMATION

The first stage of an electrical design project is to ask questions. What does the client want, what regulatory issues does this raise, what sort of building will the electrical system serve, what will be the most appropriate wiring system and installation method, and are there any sustainable installation considerations?

Criteria for selecting electrical systems

The Electricity at Work Regulations describe an electrical system as a 'system in which all the electrical equipment is, or may be, electrically connected to a common source of electrical energy, and includes such source and such equipment.' In other words, all the parts of the electrical installation we are to design.

There are many different types of electrical system and each has its own attributes. Therefore it is an important first step to establish the type of electrical system the client wants. There are a number of criteria which will form the basis for that decision.

IET regulations

All electrical design must conform to the BS 7671:2008 IET Requirements for Electrical Installations. Any customer requirement must be weighed against BS 7671:2008 and, where there is a clash, the customer must be made aware that their requirements may have to be modified to bring them in line with BS 7671:2008.

Suitability of system

The electrical system installed in premises must be suitable for:

- the purpose of the installation, e.g. industrial, alarm system, domestic
- the building and environment – is it a harsh industrial premises, or a house or office block where the wiring system will not be at risk of mechanical damage?

Customer needs

The customer is first and foremost. This is the person, or group of people, paying for the work and looking for a finished product which will meet their requirements. There are a number of ways in which communication with the customer is set up and the customer represented.

- Direct contact – for smaller and medium-sized work the customers themselves may contact and speak with the electrical company directly through a phone call, followed by a meeting (often on site) and an invitation to put in a price for the work.
- Architect – for larger jobs the main contact between the electrical contractor and the customer may well be via an architect. This is the professional who has designed the new building (see Figure 7.1). (*Chapter 5: Understand how to communicate with others within building services engineering* describes the role of the architect in more detail.)
- Main (or principal) contractor – some larger jobs are given to a main contractor, usually a building company, who will then employ sub-contracting companies for specialist roles (see Figure 7.2). This means that the electrical work will be undertaken by an electrical sub-contractor. (*Chapter 5: Understand how to communicate with others within building services engineering* describes the role of the principal contractor in more detail.)

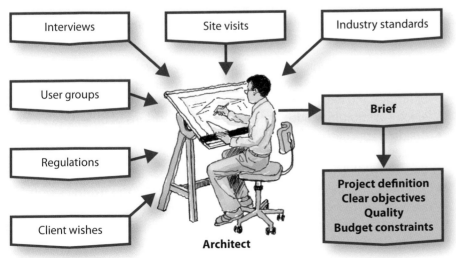

Figure 7.1: An architect and the process surrounding design

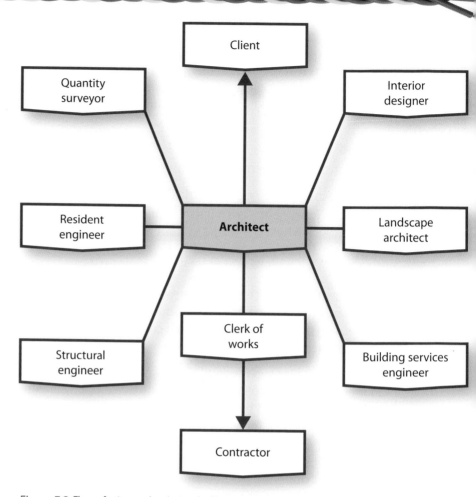

Figure 7.2: The professions and trades involved in a construction project

Building layout and features

The building structure is important to electrical design because it determines where cables can be routed and what sort of fixings will be needed for equipment and accessories. There may also be parts of the building where fixings and cable routing cannot take place. Fire walls, for example, will either require specialist fixings or may not be fixed to at all.

Positioning of equipment may well be affected by the building structure. There may be weight limitations or planning restrictions which preclude additions to external features. This sort of information is mainly provided by:

- scaled drawings which show positions and an overall view. These will include drawings marked up for other trades to show positions of fittings and fixtures, pipe runs, etc. It is important that positions of electrical equipment do not clash with the positions of other equipment
- building specifications which give information about the building itself, particularly the materials and the finishes used, which is important to the electrical designer in terms of fixings and cable routes
- electrical specifications which give details such as the fixing heights for accessories, fixing methods, types of wiring systems and equipment to be used, as well as stating any restrictions on cable routing or fixing in the building.

Sustainable design

Sustainable designs are design methods intended to reduce impact on the natural environment. The designer must take a number of factors into consideration. It is no use simply deciding to install renewable and sustainable technologies and materials if they are not suitable for the situation. For example, a wind turbine is of no use in a location where there is little wind. The main considerations for sustainable design are outlined below.

- Carbon reduction – the UK government has set a target of a 37 per cent reduction by 2014 and therefore CO_2 reduction must be taken as seriously by the electrical industry as by any other.
- Energy efficiency – low energy lighting, heat pumps and solar heating are all ways of reducing both energy consumption and cost, and have been available for a long time. Sensors and timers ensure that lights only operate when there is someone in the room.
- Environmental impact:
  o Micro-generation reduces the energy demand from sources with a high **carbon footprint**.
  o Physical appearance – planning regulations are intended to reduce negative impact from equipment and services added on to premises. The designer must take these restrictions into account when making their design decisions.
- Cost-effectiveness – many renewable, low carbon technologies can be expensive to install. However, in the long term, they can prove more cost efficient than conventional technologies.
- Increased efficiency – many renewable systems are very efficient in terms of input and output. A solar panel used for power generation has no moving parts and requires no mechanical input whatsoever. Heat pumps need only a small amount of electricity to operate but produce most of their output by using natural heat sources.
- Increased comfort – the use of natural products for construction work is considered to be healthy for the occupants of a building. Certainly reduced use of toxic chemicals in paints and adhesives, etc. will lower the amount of harmful fumes present in the atmosphere.

> **Key term**
>
> *Carbon footprint* – the amount of CO_2 produced by a technology or process is its carbon footprint.

Figure 7.3: A low energy light bulb

> **Case study**
>
> Recycling reduces cost and protects resources by utilising materials that have already been created, either reforming them or simply using them in their original form. The construction of the RFL building in the USA made extensive use of recycled materials. This included scrap gas pipes which were converted into the columns for the structure.

> **Progress check 7.1**
>
> 1 What must the electrical system installed in premises be suitable for?
> 2 What is the role of an architect?
> 3 What are building specifications?

Positioning requirements

The components of an electrical system will not be the only equipment and fittings to be installed in a building. In a factory, for example, there will be machines, pipe and ductwork and in a kitchen there will be cupboards, a sink and so on. The position of electrical equipment and services is important for a number of reasons, including:

- aesthetics
- mechanical stress
- regulatory requirements.

Customer preference

Every effort should be made to provide the customer with the electrical installation they require. This includes fitting equipment and accessories where the customer wants them. This may clash with the designer's view of what is convenient or aesthetically pleasing but if it does not contravene any regulations, the customer's decision is final.

Working practice 7.1

A young and inexperienced electrician carried out the electrical installation for a recently renovated kitchen in a farmhouse. When asked for their preferred positions for the isolating switches for the hob and oven, the customer answered that they should be installed in the cupboard beside the oven. The electrician replied that this was not a good idea as the switches would not be immediately accessible in case of emergency. The customer produced the brochure and pointed out that there were no switches visible in the photograph. The electrician argued that the photograph was for a 'fairy tale' kitchen and did not show real life – the switches had to go on the wall. An argument followed and in the end the electrician reluctantly complied with the customer's wishes and installed the isolator in the cupboard.

1 Was this the correct position for the isolating switches?

2 How could the electrician have handled the situation differently?

3 Was it right for the electrician to call the customer's view of the kitchen a 'fairy tale'?

Regulations

Both BS 7671:2008 and Part P of the Building Regulations include specific regulations about the positioning of electrical equipment and accessories.

- BS 7671:2008 states that the positioning of electrical equipment and services must not:
  - cause heat damage or fire to surrounding materials (Chapter 42)
  - be inaccessible for maintenance and repair (including cable joints) (Chapter 51)
  - have any adverse effects on other electrical and non-electrical services (Chapter 52).
- Part P of the Building Regulations:
  - states that domestic electrical accessories should be within reach of anyone who has to use them and recommends heights between 450 mm and 1200 mm from finished floor level
  - gives guidance regarding accessories in kitchens where there are problems with water in sinks and heat from hobs and ovens.

Figure 7.4 shows the minimum distances for electrical equipment in a kitchen.

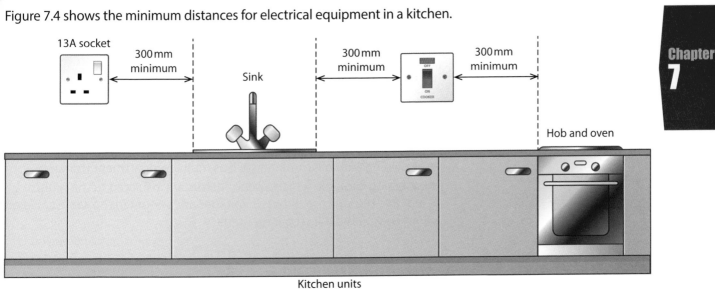

Figure 7.4: Minimum distances between electrical accessories and hazardous features in a kitchen

Clearances and space

Electrical equipment will need not only the space it occupies but room for it to be operated and also repaired and maintained. Covers may have to be removed and specialist tools brought in. On the other hand, electrical services must not obstruct access to other service equipment.

Disabled access

First and foremost, no electrical equipment should obstruct disabled access or cause any sort of hazard. As for the electrical system itself:

- switches and sockets should be installed at a height which is accessible to a disabled person, for example someone in a wheelchair
- doors will need to be fitted with easy-to-operate buttons for automatic opening, again positioned at a height and position which make them accessible to everyone
- disabled toilets need to be fitted with emergency pull cords or buttons.

System performance

Electrical equipment, cables and accessories should, of course, be positioned so that there are no factors which will adversely affect their performance. Excessive heat increases the resistance of cable. Increased resistance means lower current-carrying capacity. It can also weaken the sheath and insulation, as will extreme cold.

BS 7671:2008 requirement 528.3.1 states that a wiring system should not be installed in the vicinity of anything that produces smoke, heat or fumes that could cause it damage. If it isn't possible to avoid this type of environment, then the cable must be suitably protected.

Dusty atmospheres will also affect the performance of electrical equipment, clogging moving parts or blocking ventilation slots. All electrical enclosures and equipment should conform to an international protection (IP) rating. Only those with the appropriate rating should be installed in locations in which their performance might be undermined by **external influences**.

> **Key term**
>
> **External influences** – described by BS 7671:2008 as anything from outside which will affect the design and performance of an electrical installation.

Proximity to other services

A magnetic field forms around any cable which is carrying current (see Figure 7.5). The magnetic field around low voltage (230 V and 400 V) cables can seriously disrupt and reduce the performance of data and extra low voltage circuits that come into contact with it. The designer must allow for this by:

- running low voltage, extra low voltage and data circuits through separate routes
- providing suitable trunking such as multi-compartment trunking
- selecting cables in which all conductors are insulated to the highest voltage rating if the different voltages are present in the same cable or if it is impossible to separate them.

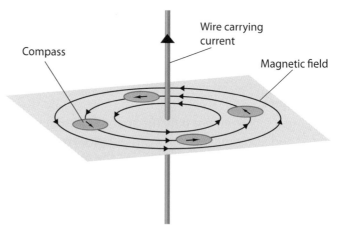

Figure 7.5: A magnetic field around a conductor

The Electromagnetic Compatibility (EMC) Regulations 2007 are intended to protect electronic, data and communication equipment from the kind of electromagnetic interference caused by mains electrical services.

Measurements from design plans

To ensure that all parts of an electrical system are positioned correctly, whether to avoid interfering with other services or for customer use, measurements need to be taken from the main design plans.

Scale

A layout drawing shows positions of features of a building and its equipment. Most layout drawings are drawn to scale. Scale is a way of reducing the real-life dimensions to manageable proportions. So, for example, a 10 m wall in real life can be drawn as a 100 mm line on a drawing.

Scale also enables you to take measurements off the drawing and calculate the exact position of an item of equipment or accessory. It is usually expressed as a ratio – for example, 1:100. This means that 1 mm on the drawing represents 100 mm in real life.

Remember:

- the left-hand number is the drawing measurement
- the right-hand number is the real-life measurement.

Progress check 7.2

1 What does BS 7671:2008 say generally about positioning electrical equipment?

2 What effect does excess heat have on a cable?

3 What are EMC regulations?

Worked example

The wall on a layout drawing is 60 mm long. How long is the actual wall if the drawing scale is 1:50?

The wall on the drawing is 50 times smaller than the real wall, so the calculation will be:

60 mm × 50 = 3 000 mm or 3 m.

Activity 7.1

Calculate the actual positions of these items of electrical equipment in metres from the following drawing measurements.

1   A 13 A socket outlet 45 mm from the edge of a doorframe – scale 1:100.

2   A row of fluorescent luminaires run parallel with a wall. The line of lights is 12.5 mm from the wall and the scale is 1:40.

3   Each light is shown as 3 mm apart – what will be the actual distance between the lights?

4   A cable run from a distribution board to a sub board is shown as 689 mm on a drawing. If the scale is 1:20, how much cable should be ordered? (Allow four metres for connection at either end.)

5   A row of six 13 A sockets is installed along the wall of an office. The wall is shown as 266.67 mm and the scale is 1:15. How long is the wall and how could the socket outlets be evenly spaced along the wall? The two end sockets must be 0.25 m from each end of the wall.

Area

Area is the flat space taken up by an object or feature or the face of an object. Area calculations may be needed to work out the location of a large item of equipment, or lighting calculations when a certain level of lighting is needed for a certain area, for example a workbench. Area is expressed as m^2.

- Area of a four-sided space = width × length.

- If the four-sided object is a square then it is the product of two sides.

- Area of a triangle = $\dfrac{height \times base}{2}$ or height × (0.5 × base).

- Area of a disc or circle is πr^2 or $\dfrac{\pi d^2}{4}$

Activity 7.2

Find the area of:

1   a playing field 126 m × 205 m

2   a triangle with a height of 465 mm and base width of 23 cm

3   a helicopter circular landing pad with a diameter of 11 000 mm.

Volume

The volume of an object is the amount of space inside the object. Volume may have to be calculated when designing a heating system. The amount of air in the room and the changes required to that air will need to be worked out. Volume is expressed as m^3.

- For a box-type object, the formula = width × height × length.
- For a cylinder, the formula = cross-section area × length.

Activity 7.3

1 A room is 3 m wide, 4.5 m long and 4 m high. What is its volume?
2 What is the capacity of a fuel tank that is 3.2 m long and has a diameter of 2 500 mm?
3 What is the volume of a container that is 9 275 mm × 3.2 m × 4 m?

Weight

Weight will be an issue because of structural considerations. There may be a weight limit on the floor and this must be taken into consideration when selecting the equipment. Weight of equipment is expressed in kilograms (kg).

Conversion of measurements

1 m = 1 000 mm

Table 7.1 shows specific measurement conversions.

Measurement type	How to convert
Linear	To convert m to mm multiply by 1 000 (10^3) To convert mm to m divide by 1 000 (10^3)
Area	To convert m^2 to mm^2 multiply by 1 000 000 (10^6) To convert mm^2 to m^2 divide by 1 000 000 (10^6)
Volume	To convert m^3 to mm^3 multiply by 1 000 000 000 (10^9) To convert mm^3 to m^3 divide by 1 000 000 000 (10^9)

Table 7.1: Measurement conversions

Information needed for electrical systems

Due to the dangerous nature of electricity and the complexity of the systems it powers, the designer will need a considerable amount of technical and regulatory information. Table 7.2 describes some of the main sources of information for electrical design.

Information source	What it tells the designer
Manufacturer's technical instructions	This can be in the form of drawings, circuit diagrams, schematics, exploded views and step-by-step instructions.
Building Regulations	The Building Regulations are divided into 14 parts: Part A. Structure Part B. Fire safety Part C. Site preparation and resistance to contaminants and moisture Part D. Toxic substances Part E. Resistance to the passage of sound Part F. Ventilation Part G. Hygiene Part H. Drainage and waste disposal Part J. Combustion appliances and fuel storage systems

	Part K. Protection from falling, collision and impact Part L. Conservation of fuel and power Part M. Access to and use of buildings Part N. Glazing – safety in relation to impact, opening and cleaning Part P. Electrical safety – dwellings* *This is the part electricians will be most interested in. Part P takes the main requirements of BS 7671 and applies them to the domestic installation.
British Standards (BS)	The British Standards Institution (BSI) is an organisation which sets quality standards for technology, construction, manufacturer and management.
European Standards (EN)	The European equivalent of British Standards.
Industry standards	Generally accepted requirements followed by the members of an industry.
IET regulations	The Institute of Electrical Technology (IET) is a professional body that encourages the study of engineering and technology. They consist of a network of experts and professionals who are considered to be the main authority on electrical safety and good practice. The IET publishes the BS 7671:2008 Requirements for Electrical Installations.
Verbal and written feedback from the customer	Ultimately all work is carried out for the benefit of the customer, so the customer is always the first source for information about a project.

Table 7.2: Sources of information for electrical design

Special locations

Part 7 of BS 7671:2008 refers to what it terms 'special locations' and applies existing regulations to these types of installation. A special location is an installation which is hazardous or an extreme environment. Special locations can be loosely grouped into a number of different types, each with their own set of hazards and requirements.

Progress check 7.3

1 What is scale when used on a layout drawing?
2 How many mm^3 in 1 mm^3?
3 What are British Standards?

Presence of water	Main regulatory requirements
• Bathrooms and showers • Swimming pools and fountains • Saunas	• Divided into zones, with zone 0 being within the actual water source, e.g. bath • No 13 A socket outlets allowed • ELV equipment within the zones nearest the source of water. The power supply must be SELV, with the transformer and mains supply outside the closest zones • Equipment should be resistant to water ingress • Additional protection by RCDs is required

Table 7.3: The main regulatory considerations for installations in wet environments

Harsh environments	Main regulatory requirements
• Construction and demolition sites • Agricultural and horticultural • Caravan parks • Marinas	• Risk of damage to cables and equipment • Harsh weather conditions • Cables and equipment should be protected against mechanical damage, corrosion, and damage from livestock and rodents • Underground supplies should be buried deep enough to remain undisturbed • Overhead cables run at a minimum of 6 m where there is vehicles activity, 3.5 m in other locations • Additional protection by RCDs is required

Table 7.4: The main regulatory considerations for installations in harsh environments

Figure 7.6: A marina is an installation in a harsh environment

Temporary installations	Main regulatory requirements
• Exhibitions, shows and stands • Fairgrounds and amusement parks	• Shock and fire risk to public must be kept to a minimum • Fairground equipment must be ELV with transformers out of reach of public • No joints made in cables • Signs, lamps or exhibits to be controlled by an emergency switch • Additional protection by RCDs is required

Table 7.5: The main regulatory considerations for temporary installations

Figure 7.7: A fairground is an example of a temporary installation

Mobile	Main regulatory requirements
• Mobile or transportable units • Caravans and motor caravans	• Maximum voltages 230 V/400 V a.c. • Protection against vibration is a consideration so cables should preferably be a flexible type and all cable route holes should be fitted with grommets • Display information notices detailing the installation • Additional protection by RCDs is required

Table 7.6: The main regulatory considerations for mobile installations

Restricted movement	Main regulatory requirements
• Conducting locations • Operating and maintenance gangways	• Access restricted to skilled persons • Enough room must be provided for safe working and opening electrical equipment coves and doors • Must be an emergency exit • Protection of the conducting location is provided by use of SELV supplies with the transformer outside the area

Table 7.7: The main regulatory considerations for installations with restricted movement

Others	Main regulatory requirements
• Medical locations	• Dedicated supplies • IT earthing system with functional earth • Emergency back-up supply needed which restores supply within 0.5 seconds
• Solar photovoltaic power supply systems	• Voltage present as long as there is light • Equipment exposed to weather so needs to be able to withstand extremes • Protection by double insulation is preferred for the d.c. side; overcurrent protection for a.c. side
• Floor and ceiling heating systems	• Elements must be installed at a minimum of 10 mm from flammable material • Maximum temperature of any part that can be touched is 35°C • Location must be recorded to protect it from damage during any future works

Table 7.8: The main regulatory considerations for other special locations

Progress check 7.4

1 What is a special location?
2 What sort of earth system is used for a medical location?
3 What are the hazards for the electrical installation in a visiting fairground?

Tender content

Prospective contractors for medium to large construction projects have to go through the competitive tendering process before the contract can be awarded to the successful company. Each company which wants the contract has to construct and enter a **bid**, which includes the price they will charge and how the work will be carried out. (For more details about tenders, estimates, etc. see *Chapter 5: Understand how to communicate with others within building services engineering*.)

Key term

Bid – an offer of a price for a contract, similar to a bid made for an object at an auction.

Working practice 7.2

A large office block and computer centre was to be fitted with a new air conditioning system. The work involved installation of heavy pipework, armoured cables and cable trays through the ceiling voids above each open-plan area. To achieve this, most of the ceiling tiles would have to be removed and mobile scaffolding erected around and over the desk islands. The work simply could not be carried out with the staff in place. However, the centre could not be closed down during normal working hours for strategic reasons. Tenders were invited from a number of building engineering services companies. The customer employed its own design and project management department and two project managers were given the job of judging the tenders. Interviews were set up with three different contractors who each presented their proposals. The tender accepted was not the cheapest but it was the most suitable for the situation. The proposal was to close the office on each floor at 5 p.m. promptly each Friday evening. Once the office was empty of staff, the work would commence with the erection of scaffolding. The ceiling tiles could then be removed to clear the way for the engineers. Work was to be carried on round the clock until the early hours of Monday morning, so that the area could be cleaned up and the contractors long gone by the time the office staff returned at 9 a.m. The price for this proposal was, in fact, the highest but the work plan was the most suitable.

1 Why couldn't the work be carried out during normal working hours?

2 Why were three contractors interviewed?

3 Why wasn't the cheapest tender accepted?

Day work

There are two ways of pricing labour for a construction.

- Set price – a fixed price which is agreed before the work commences.
- Day work – payment based on the number of hours actually spent on the work.

Changes to work

Construction projects do not always go according to plan. Problems arise, for example it may not be possible to install a system or its components exactly as proposed in the design, or the customer may change their mind. Any change will have both a time and a cost implication. *Chapter 5: Understand how to communicate with others within building services engineering* describes the variation order which is used to control changes to planned work. An example of a simple change process is shown in Figure 7.8.

Legal declaration

A legal declaration is a judgement made on an individual issue.

Penalty clauses

A penalty clause is a safeguard intended to keep the project on track. A common penalty clause is a financial penalty imposed on a contractor for each day they overrun the agreed completion time for a project.

Schedule of work

A schedule of work is a list of the work required for a new construction project and is issued to contractors as part of an invitation to tender. The schedule breaks the project down into a table of tasks from which materials can be itemised and labour costs calculated. It does not include price, location, sizes, etc.

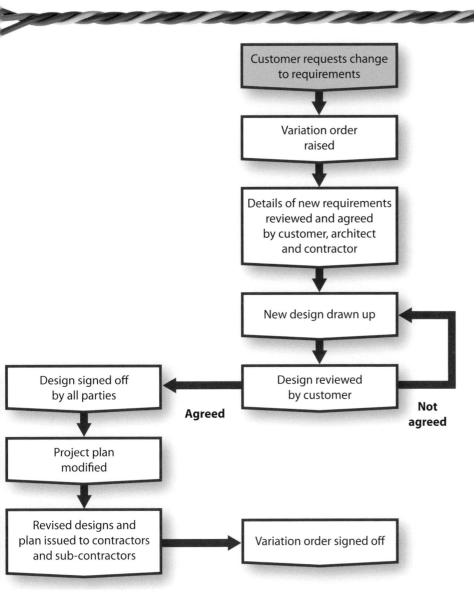

Figure 7.8: Flow chart for a change request

Quotations

A quotation is given by a contractor as part of their tender. It is a statement of how much the work will cost. When a quotation is accepted it becomes legally binding. A set of documents are needed in order to prepare a quotation. These are called 'tender documents' and include:

- a full set of all plans and drawings
- specification documents
- a schedule of work
- details of any materials the customer will be providing
- details of any work to be sub-contracted out.

ISO9000

ISO9000 is an international standard for the relationship between a contractor and their customer. Being registered as an ISO9000 company is a way of showing that you have a formal management system in place. The main requirements of ISO9000 are outlined on the next page.

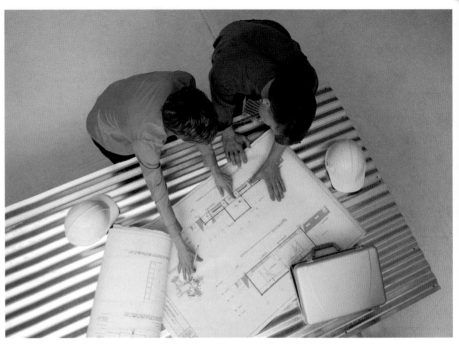

Figure 7.9: A drawing being reviewed

- The contractor must have a formal management system in which it is clear exactly who is responsible for what.
- A formal relationship must be established between the contractor and the customer, usually in the form of a contractual agreement. This agreement states exactly what the contractor will provide and what the customer expects to receive.
- All design must be formally reviewed by the contractor and customer and signed off by both parties as the current, approved version. If the design is changed the new version must be reviewed (see Figure 7.9) and signed off and all previous versions removed.
- Only current, approved documentation must be used. All documents should be given a version number and issued only to those who need them. A list should be kept of everyone who has the document. This means that everyone on the list can be issued with the replacement if a document is superseded by a newer version.
- A formal change control process must be set up so that all changes to the original agreement can be tracked and handled without dispute between the customer and contractor (see page 351 for a change control flow chart).
- Evidence of training should be available to prove that all personnel are qualified to carry out the work they are doing.
- When work is completed, a formal agreement must be signed to prove that the customer is satisfied that the work is done. This sign-off may also include any warranty and maintenance agreements.

An ISO9000 registered company will be audited regularly by the awarding body to make sure it is meeting the requirements of the standard.

Sources for costing electrical equipment

A quotation for electrical work is made up of costs for labour and materials. Labour costs are set by the company. The main sources for materials' pricing information are listed and described below.

- Manufacturers – some equipment will have to be constructed to order and purchased from a manufacturer. An example of this is an electrical control panel which will need to be custom-made for a particular job. Prices will need to be negotiated because it is sometimes expensive to buy directly from the manufacturer.
- Merchants and independent suppliers – the main source of materials for the electrical industry. These companies allow the contractor to open an **account** and negotiate discounts on large orders. The staff are often a good source of technical knowledge for the products they sell.
- Catalogues and brochures – issued both by manufacturers and suppliers, catalogues are useful for researching what is available, along with technical information about the product. There is usually a price list included, although not always, in the main body of the publication.
- Internet – increasingly suppliers and manufacturers are putting their products, prices and other information on the Internet. Ordering can also be carried out online.

> **Key term**
>
> **Account** – having an account with a supplier means that instead of paying for each order when it is made, the bill is settled at regular intervals, e.g. monthly or quarterly.

> **Progress check 7.5**
>
> 1 What is a quotation?
> 2 What does ISO9000 require with regards to design?
> 3 When would you need to go to a manufacturer for a price?

Requirements

We will look at the main points for pricing a domestic installation as an example of the requirements for putting together a tender for a specific electrical job.

Materials and components

These are the 'parts' used for an electrical job. This is a domestic electrical installation so these materials will include the following, as outlined in Table 7.9.

Component	Type	Purpose
Cable	Sheathed cable	Sheathed, multi-core cable such as twin-and-earth and three-core-and-earth for power and lighting circuits. Typical sizes are: • 1.5 mm^2 • 2.5 mm^2 • 4.0 mm^2 • 6.0 mm^2 Sheathed single cables used as meter tails (usually 16.0 mm^2)
	Singles	Equipotential bonding, minimum cross-sectional area of 6.0 mm^2
	Coaxial and other data cables	Television aerial, telephone extension and Internet supplies
Accessories	13 A sockets	Plastic or brushed metal finish, single and double. Installed flush in the wall
	Light switches	Plastic or brushed metal finish, one-way, two-way and intermediate, also one or multi-ganged types. Dimmer switches may also be required. Installed flush in the wall
	Light fittings	From basic pendants and batten holders to spotlights and decorative fittings Also extra-low voltage, flush spotlights and concealed lighting

Table 7.9: Main components for electrical jobs

Continued ▼

Component	Type	Purpose
Accessories (continued)	Isolators	Spur outlets, cooker switches, 45 A pull-cord type for showers. Will be plastic or brushed metal finish. Fitted flush. All isolators must be double-pole
	TV and other data outlets	Plastic or brushed metal finish, fitted flush in the wall
	Junction boxes	May not be needed for a new installation but required for modifying circuits if the installation is an addition to an existing one. Typical variations are: • 30 A three-terminal • 20 A four- and six-terminal • 5 A four-terminal
Protective devices	Consumer unit	The main connection point and source of the final circuits in the installation Consumer units are available in different sizes, depending on how many final circuits it will feed
	Circuit breakers	Provide overcurrent protection. Available in different types and ratings, depending on the circuit they feed. Typical ratings used in a domestic installation are: • 6 A – lighting • 16 A – small water heaters • 20 A – water heaters • 32 A – ring final circuit, shower and cooker circuits • 45 A – larger cooker
	RCD	Usually a 30 mA rating. Provides additional shock protection
Sundries	Metal back boxes	Back boxes for switches, 13 A sockets, etc. Cut into the wall and available in different depths. Typically: • 16 mm for light switches • 25 mm for 13 A sockets Plasterboard fitting types also available (sometimes called 'dry-liner' boxes)
	Earth bonding connectors	Used to connect equipotential bonding conductors to exposed metalwork such as water and gas pipes
	Insulation tape	General purpose tape with insulation properties
	Fixings	Screws, wall plugs, cavity fixings, etc.
	Cable clips	For fixing cables. Not often required in a domestic installation other than securing cables on the side of joists under floors, etc.
	Cable capping or oval conduit	Used to protect and secure cables installed in wall chases

Table 7.9: Main components for electrical jobs

Labour

Labour is the time spent by the electricians actually carrying out the work. There are two ways of pricing this.

- Pay by the hour – the electrician is paid for each hour they spend on the job (this type of pricing is best for smaller jobs).
- Fixed price – the hours are estimated before the work starts and a price is agreed.

Remember, a domestic installation will be carried out in two stages.

- First fix – before ceilings and floors are boarded and walls plastered – cables are installed and back boxes are cut into the walls.
- Second fix – after floors and ceilings are boarded and walls are plastered – final connections and testing.

Progress check 7.6

1 List three typical protective device ratings for a domestic electrical system.

2 What are labour costs?

3 What are the two stages of an electrical installation in a new domestic property?

THE PRINCIPLES OF DESIGNING ELECTRICAL SYSTEMS

LO2

You have won the tendering exercise and the project is yours. You have the schedule of works, a building specification, an outline electrical specification and a set of drawings. The next stage is the technical design itself. We have seen already how the design must meet the customer's criteria, must take other services and building features into account and that there may be special locations within the installation. Now it is time to select the wiring systems, protective devices, lighting and power systems for the job.

Regulatory requirements

We have already considered the way regulatory requirements impact on the initial stages of the electrical design process. The main source of regulatory information for the electrician is BS 7671:2008 IET Requirements for Electrical Installations, which are, of course, the electrical wiring regulations. These are supplemented by the *IET On-Site Guide* and various IET Guidance Notes. Some of the main regulatory points are shown in Table 7.10.

Subject	Where to find this	Main points
Selection and erection	Part 5 and Appendix 4	• All equipment installed must comply with British and European Standards (BS EN) 511.1 • Wiring systems must take into account any external influences that may affect them – 512.2 • Voltage drop must be taken into account when selecting the cables – Appendix 4 • Cables feeding rotating equipment with a high starting current must be able to carry that high current – 552.1.1
Associated equipment and enclosures	Parts 4 and 5	• The equipment used in an electrical installation must be right for the job – 512.2.1 • All termination and equipment must be accessible for repair and maintenance – 513.1 • If the door of an enclosure can be opened without use of a tool or key then the live parts inside must be inaccessible behind a cover or barrier – 412.2.2.3
Isolation and switching	Part 5	• Isolators are off-load switches used to completely isolate part of a circuit or an item of equipment from the supply – 537.2.1.1 • An isolator must disconnect all the live conductors (live includes the neutral) – 537.2.2.1 • Under no circumstances must an isolator switch the earth conductor – 537.2 • If an isolator is separate from the equipment it switches, there must be a means of locking it in the off position – 537.2.1.5 • Switches are functional devices and part of the normal operation of a circuit, for example a light switch. Switches should be in the line conductor and not the neutral – 537.5
Protection against fire and flammable/explosive atmospheres	Part 4	• Operation of electrical equipment must not cause fire – 421.1.2 • If fixed electrical equipment emits heat as part of its normal operation it must be far enough away from any flammable materials so that it does not start a fire – 421.1.2 • Lamps must be installed so that they are at the following distances from flammable materials: • up to 100 W – 0.5 m • between 100 W and 300 W – 0.8 m • between 300 W and 500 W – 1 m – 422.3.1 and 422.4.2 • Electrical arcs caused by the operation of an item of electrical equipment must be contained inside an enclosure – 421.1.3

Table 7.10: Main points from the regulatory requirements

Continued ▼

Subject	Where to find this	Main points
Protection against shock	Parts 4 and 5	• Basic protection must be provided to prevent contact with live parts – 411.1 • All exposed conductive parts of premises must be connected together, and to earth – 411.3.1.1 • The earth fault loop impedance must not exceed the maximum stated in Tables 41.2 to 41.4 • Additional protection is provided by an RCD – 411.3.3
Special locations	Part 7	• See the tables on pages 347 to 349
Segregation	Part 5	• Circuits of different voltages should, if possible, be physically separated, or segregated, from each other – 528.1 • Where it is impossible to segregate different voltages, such as the cores of a multi-core cable, then the insulation of the lower voltage conductor must be rated to the highest voltage present – 528.1

Table 7.10: Main points from the regulatory requirements

Selection of electrical systems

The selection of an electrical system depends on a number of factors, the most important of which are described below.

External influences

The environment in which the electrical system is installed can sometimes carry a heavy risk of water or mechanical damage, or even tampering. This means that enclosures and equipment must be selected to be suitable for the environment and provide the right level of protection from these influences. IP ratings are an indication of the type and level of protection this equipment will have.

The codes are shown in Table 7.11. The first digit shows how protected the system is against solid objects, such as body parts, and the second digit shows how protected the system is against water. Sometimes an X can be found in place of one of these digits, which means that the test has not been carried out or it is not applicable.

Earthing system

There are three main earthing systems. These are defined by the route from the main earth terminal at the supply point of the installation, back to the star point of the sub-station transformer. These are looked at in detail in *Chapter 3: Electrical installation technology*. As a reminder, there are three main earthing systems in use.

- TT – terra-terra – installation is connected to earth via an electrode driven into the earth itself adjacent to the premises.

- TN-S – terra neutral separated – installation is connected to the sub-station earth via a separate conductor (usually the armouring of the supply cable).

- TN-C-S – terra-neutral combined-separate – installation earth connected to the supply neutral.

Supply voltage and frequency

The supply voltage, and system, must be established. Table 7.12 shows the main voltages used for UK electrical installations. UK supply frequency is 50 Hz.

First digit	Mechanical protection	Second digit	Water ingress protection
0	No protection	0	No protection
1	Protected against solid objects larger than 50 mm, such as accidental touch by back of hand	1	Protected against vertically falling drops of water, such as dripping pipes or condensation
2	Protected against solid objects larger than 12 mm, such as fingers or similar	2	Protected against dripping water when the system is tilted by 15°
3	Protected against solid objects larger than 2.5 mm, such as tools or thick wires	3	Protected against spraying water when the system is tilted by 60°
4	Protected against solid objects larger than 1 mm, such as most wires or nails	4	Protected against water splashed from any direction
5	Protected against dust, unless it enters in large quantities	5	Protected against low pressure jets of water from any direction
6	Fully protected against dust and all contact	6	Protected against strong, pressurised jets of water from any direction
n/a	n/a	7	Protected against taking in water at a harmful level during a 30-minute period when immersed in between 15 cm and 1 m of water
n/a	n/a	8	Protected against continuous immersion at depths of greater than 1 m

Table 7.11: IP codes and their meanings

Voltage	Usage	Notes
230 V	Single-phase supply for domestic and small to medium-sized installations	Standard voltage for UK power and lighting circuits
400 V	Three-phase supplies for industrial and large installations	Used either as 400 V three-phase to feed three-phase machinery or to feed power and lighting in a large premises – with the loads balanced over all three phases
Extra-low voltage (ELV) ≤50 V	Actual voltage required by many items of electrical equipment, particularly electronic equipment and those fed by SELV and PELV systems	Transformer integrated into equipment itself, its plug or fitted in-line in its power lead. This often includes a rectifier to convert the a.c. supply to d.c. for the load
Micro-generated supplies	Electrical supplies produced by micro-generation	This will be d.c. and needs to be transformed to 230 V and changed to a.c. using an inverter
High voltage (HV) ≥ 1000 V	Supplies to local distribution sub-stations. Direct supplies to large premises and also HV equipment in large industrial complexes	HV supplies to premises are often configured as a ring main, which provides built-in redundancy for maintaining supplies even if part of the supply system is broken

Table 7.12: The main voltages used for UK electrical installations

Working practice 7.3

A large computing centre which acted as a strategic node for a complex network received a complete buildings service upgrade. This included industrial scale air conditioning and environmental control and the installation of a 1000 V ring to feed each section of the computer centre. The ring took the form of buried armoured cables fed from two main transformers. The cables were single-core, aluminium conductors. The reason the complex was fed from two identical transformers and the supply was in the form of a ring was built-in redundancy. Because the computer centre was critical to the operation of a large international corporation, it simply could not be allowed to shut down through power failure of any sort. The ring supply meant that if one transformer failed or part of the supply was broken, or needed to be isolated for maintenance, a supply route remained in place.

1   Why are single-core conductors used?
2   What is built-in redundancy?

Progress check 7.7

1   What is the requirement for an isolator if it is situated remotely from the equipment it controls?
2   What is a TN-S system?
3   What is the UK frequency?

Special locations

Details on special locations can be found on pages 347 to 349 of this chapter.

Factors affecting selection of electrical systems

An installation will be made up of a number of different systems – lighting, power, alarms, etc. These all have individual characteristics and factors that govern their selection, for example the use they will be put to, the environment, the supply and external influences. In this section we will look at these factors and their impact on the design process.

Lighting systems

Chapter 6: Principles of electrical science examines lighting in detail, particularly the calculations that need to be carried out to ensure that there is sufficient light available to create a safe and healthy environment.

Lighting design factors

- Light levels
- Colour rendition
- Load
- Switching
- Thermal effects

Power systems (final circuits)

A final circuit is the circuit fed by an individual protective device in a distribution board or consumer unit. This can be a ring final circuit feeding 13 A socket outlets, a lighting circuit or the supply to an item of equipment or system such as control circuitry or emergency alarms. Considerations to be made when designing a final circuit are outlined below.

- Size of the load – this will determine how much current flows and this, in turn, determines the size of the cable.
- Rating of protective device – this must be able to allow enough current through to operate the load but not enough to damage the cable.
- Type of protective device – will the load demand a high starting current, or will it draw the same current at all times? What is the

prospective fault current for the circuit? Different types of protective device afford different types of protection. Some fuses and circuit breakers are designed to withstand high fault currents in the moments before they operate.

- Type of cable – although this usually depends on the wiring system, cable type might also be decided by its use. For example, control circuitry may need multi-core cable with the capacity to bring both the main electrical supply to the load and contain extra cores for the controls.

- Switching – will the load require functional switching? A lighting circuit certainly does, whereas items of electrical equipment such as a cooker or a heater are fitted with their own functional switching and will require only an isolator.

- Location of the load – is the load outside or indoors? Will it be subjected to harsh mechanical stress or extremes of weather? Are there issues with accessibility for operation and maintenance?

- Are there any smoke-emission or heat-protection issues? If the wiring is to be run in a public place then it should have a low smoke-emitting sheath and insulation. For a fire alarm system it will have to be fireproof. Is the equipment itself a fire or heat hazard?

Distribution systems (sub-mains)

A sub-main is a separate supply for part of premises taken from the main supply point. This can be single- or three-phase and will feed a distribution board. In an industrial installation this is often an armoured cable. The main consideration for a sub-main supply cable will be the size of the load. In this case, maximum demand has to be calculated but diversity also taken into consideration.

Sub-main design factors

- Three-phase, single-phase or HV
- Load
- Diversity
- Switchgear/electrical panel
- Cables and cable routes
- Containment and cable support systems

Environmental control/building management systems

Nowadays, the building management system (BMS) is increasingly common in new shopping malls and other large construction projects. A building management system consists basically of a central processing unit which can control:

- lighting
- heating
- humidity
- air conditioning
- opening and locking doors
- security
- back-up power
- alarms.

The heart of a BMS is a computer system. This is programmed with pre-set requirements for the building's environment and security.

1 Information is received by the control processor from various sensors. These can be anything from temperature and humidity to light levels, and even occupancy of various rooms and areas.

2 The computer system uses the information to calculate the optimum levels of light, heat and so forth and then sends instructions to outstations.

3 These outstations are also pre-programmed and use the information they receive from the central processor to set necessary levels of control. The outstations are linked together so that information can be shared between them.

The **communications protocol** used for a BMS is called BACnet, which stands for Building Automation and Control networks. BACnet provides communication between all the components of a BMS. It is able to exchange all information, regardless of the particular building service they perform.

Key term

Communications protocol – a system of digital messages, formats and rules used for exchanging messages between computing systems.

LON – Local Operation Network, a networking platform used to join together otherwise separate components of a BMS system so that they can exchange information and instructions.

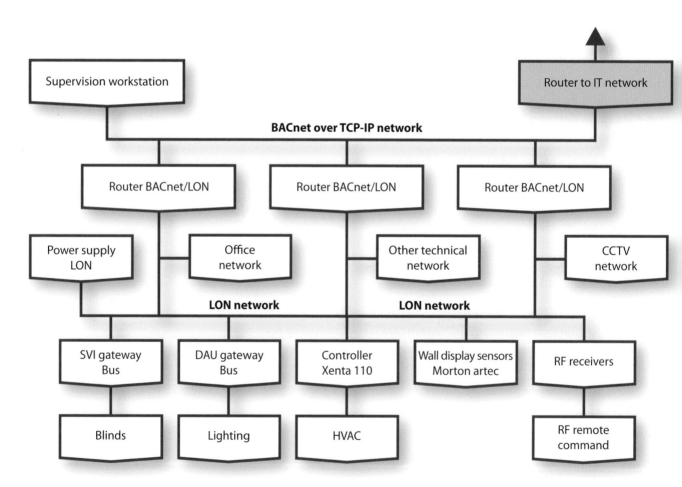

Figure 7.10: A basic building management system

BMS design factors

- Scope of BMS control
- Location for central processing station
- Provision of mains supply
- Locations and positioning of sensors
- Routing and containment for control wiring
- Separation of wiring from mains supplies
- Commissioning
- Handover and training

Fire alarm systems

The purpose of a fire alarm system is to detect smoke and heat from a small flame, alert the occupants of the building to the presence of fire and, in some cases, fight the fire by activating a sprinkler or inert gas system.

Fire alarm installations are usually divided up into zones. Each zone will have its own sensors and alarm points and information is routed between these zones and the central fire alarm panel.

Many modern fire alarm systems are equipped with addressable heads. Each sensor is programmed with a unique address that will not only indicate which zone the fire is in, but pinpoint the exact location. Existing fire alarm systems can be upgraded to an addressable head type with a minimum of modification.

A fire alarm system is typically made up of the components described below.

- Detector – automatically changes the system to alarm state. Typical detectors are:
  o photoelectric smoke and heat detectors
  o ionisation smoke detectors
  o heat detectors (see Figure 7.11)
  o in-duct smoke detectors
  o alarm buttons
  o sprinkler water flow sensors.

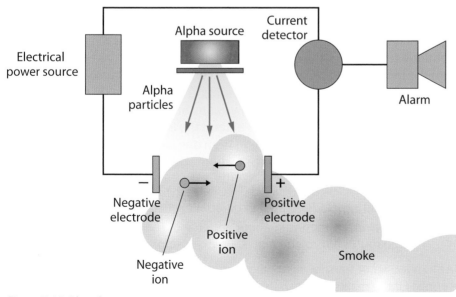

Figure 7.11: A heat detector

- Alarms – lets the occupants know that there is a fire and that they should evacuate. They can also be fitted to remote locations to alert the emergency services. Alarms can be:
  - o horns
  - o strobe lights
  - o chimes
  - o bells.
- Control panel – containing programming and operating electronics. Supplied by a standard radial circuit, it is the point at which information about the fire is displayed and the central processing unit of the fire alarm system. It must be accessible.
- Batteries and charger – usually 6 V and wired in series to make up 24 V d.c. This is back-up power for the system so that if power is lost to the building, fire protection isn't.
- Electromagnetic door holders (floor- or wall-mounted) – in case of alarm, the magnet is de-energised, allowing the door to swing shut.

A resistor is connected across the line after the final device. When this resistance is seen by the control panel, normal status is maintained. If the resistance increases, it means that an open circuit has developed, and the panel alerts maintenance control. The alphanumeric display will read something like 'Open Circuit – Zone One'.

Fire alarm design factors
- Segregation of premises into zones
- Location for control panel
- Provision of mains supply
- Locations and positioning of sensors and alarms
- Routing and containment for control wiring
- Separation of wiring from mains supplies
- Fireproof cabling, routing, fixings, support systems and containment
- Commissioning
- Handover and training

Emergency lighting

Emergency lights are self-contained units wired into the mains supply but which are also fitted with battery back-up. This means that they will continue to operate if the mains supply fails. The batteries are either fitted into the individual lights or installed as a central power source to which the lights are connected. Emergency lighting batteries are on permanent charge, usually from the mains supply. Many modern emergency lights operate on a low-voltage charge to extend the life of their batteries. The batteries should keep the light going from one to three hours. A typical battery type is lead-calcium. Lamps include energy-efficient types such as:

- halogen incandescent with xenon filaments
- light-emitting diodes (LEDs) in parabolic reflector (PAR) lamps
- LED pathways – mounted about a foot and a half off the ground along escape routes to provide a guide-path in smoky conditions.

A changeover or transfer switching system will bring in the back-up supplies if the mains supply fails. This could be simply switching to battery

power or bringing a UPS and/or generator online. There are two families of emergency lighting.

- Maintained emergency lighting – illuminated by the mains supply during normal conditions and by back-up battery supplies during an emergency.
- Non-maintained emergency lighting – not illuminated until the mains supply fails.

Emergency lighting should be manufactured and installed according to BS 5266 and BS 5588.

Emergency lighting design factors

- Positioning of emergency lighting
- Maintained or non-maintained
- Lighting levels
- Location of battery source – if a central back-up type
- Connections to UPS and generator
- Fireproof cabling, routing, fixings, support systems and containment
- Commissioning

Unlawful-entry alarms (burglar alarms)

Burglar alarms can be found in banks, shops and homes. Many different types of systems are available. A burglar alarm is an electric circuit. When the circuit is opened or closed, the flow of electricity in the circuit is changed and the alarm is sounded. This means there are two basic systems.

- Closed-circuit – the electric circuit is normally closed, for example when an infrared beam is unbroken, or a window or door is shut. If the window or door is opened, or the beam interrupted, the circuit is broken and the alarm triggered.
- Open-circuit – the circuit is normally open. It will be closed when a window or door is opened and the alarm is triggered when the circuit is made.

Motion sensors use either radar or infrared to detect movement. Radar sensors will operate an alarm if the radar beam is broken by an object passing through it. The photo sensor detects infrared energy given off by an intruder's body.

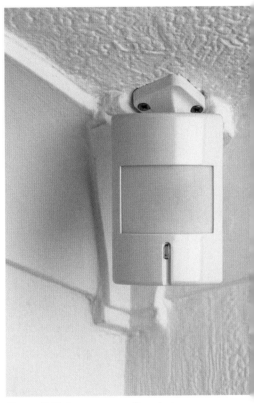

Figure 7.12: A room protected by radar beams

Intruder alarm design factors

- Location of main panel
- Type of system: closed- or open-circuit
- Positioning of sensors
- Type of alarm
- Routing cables
- Mains supply
- Segregation of mains and signal cables

Closed-circuit TV

Closed-circuit television, or CCTV, is a security system that uses cameras as a means to observe an area or room, etc. The cameras themselves may

be fixed or motorised so that they can be remotely moved to focus on a particular point or event. A CCTV system consists of:

- a camera
- a lens
- a monitor – often just a simple television set (usually just a black and white set), but increasingly a PC or laptop is used
- cables – wireless versions are also available
- image-recording equipment – either tape or digital.

Most modern systems incorporate more than one camera. However, a single monitor and recording device can still be used. The monitor screen is split into separate images, each displaying an image from one of the cameras. Alternatively, the image from one camera can be expanded to fill the whole screen.

Multiplexers are used to process information from multiple cameras. A multiplexer code-marks each image from each camera. This allows images from all the cameras to be recorded onto one tape or hard drive. The multiplexer then uses these code marks to play back the recorded picture from the camera that you wish to view.

CCTV design factors

- Location and positioning of cameras
- High-level mounting for cameras
- Cable routing
- Wireless or wired
- Mains supply for equipment
- Segregation of mains and signal cables
- Location of monitoring station
- Commissioning
- Handover and training

Communication and data transmission systems

Data transmission is the movement of electronic data (a digital bit stream) from one point to another. Conductors for the data can be:

- copper wires
- optical fibres
- wireless communication channels.

The data is transformed into a signal, such as:

- electrical voltage
- radio waves
- microwaves
- infrared signals.

Messages are represented by a sequence of pulses by means of either:

- line code (baseband transmission) – 'digital-over-digital' transmission, in the form of a sequence of electrical pulses or light pulses and used in wired local area networks such as Ethernet, and in optical fibre communication

- continuously varying wave forms (passband transmission) – digital-to-analogue conversion carried out by modem equipment. Examples of analogue signals are telephone calls and video.

There are two types of transmission.

- Asynchronous transmission – this method of transmission is used when data is sent intermittently as opposed to in a solid stream.
- Synchronous transmission – a continual stream of data sent between the two nodes. The data transfer rate is quicker, although more errors can occur.

Communication system design factors

- Cable routing
- Cable containment and support systems
- Segregation between mains and data cabling
- Location of patch panels
- Mains supplies to routing equipment
- Commissioning and handover

Progress check 7.8

1 What is a BMS outstation?
2 What is a maintained emergency lighting system?
3 What is a CCTV multiplexer?

Maximum demand and diversity

Maximum demand is the current demand for an installation. This information is needed by the supplier for selection of their supply cable and main fuse rating. It is also as part of the inspection and testing documentation. However, diversity is allowed for an installation. Diversity is based on the assumption that not all the electrical equipment will be operating at once and therefore an allowance can be made when calculating maximum demand. BS 7671:2008 and the *IET On-Site Guide* recommend the following diversity calculations (Table 7.13 is based on Table A2 in the *IET On-Site Guide*).

Type of final circuit	Domestic installations	Small shops, stores, offices and business premises	Small hotels, boarding houses, guest houses, etc.
Lighting	66% of total current demand	90% of total current demand	75% of total current demand
Heating and power, other than cooking appliances and storage radiators	100% of demand up to 10 A plus 50% of any current demand over 10 A	100% of demand for largest appliance plus 75% of current demand for all other appliances	100% of demand for largest appliance plus 80% of current demand for second largest plus 60% of all other appliances
Cooking appliances	10 A plus 30% of remaining current demand. Add on 5 A if there is a 13 A socket in the cooker isolator	100% current demand of largest appliance plus 80% of second largest appliance plus 60% of remaining appliances	100% current demand of largest appliance plus 80% of second largest appliance plus 60% of remaining appliances
Electric motors	Not applicable	100% of demand of largest motor plus 80% of second largest plus 60% of remaining motors	100% of demand of largest motor plus 50% of remaining motors

Table 7.13: Diversity calculations

Continued ▼

Type of final circuit	Domestic installations	Small shops, stores, offices and business premises	Small hotels, boarding houses, guest houses, etc.
Instantaneous water heaters	100% of demand of largest appliance plus 100% of second largest plus 25% of remaining appliances	100% of demand of largest appliance plus 100% of second largest plus 25% of remaining appliances	100% of demand of largest appliance plus 100% of second largest plus 25% of remaining appliances
Thermostatically controlled water heaters	No diversity allowed	No diversity allowed	No diversity allowed
Floor warming	No diversity allowed	No diversity allowed	No diversity allowed
Storage space heating	No diversity allowed	No diversity allowed	No diversity allowed
Standard domestic-type power and lighting circuits	100% of demand for largest circuit plus 40% of demand for every other circuit	100% of demand for largest circuit plus 50% of demand for every other circuit	Not applicable
Socket outlets (other than in the row above)	100% of demand of largest utilisation point plus 40% of demand for every other point	100% of demand of largest utilisation point plus 70% of demand for every other point	100% of demand of largest utilisation point plus 75% of demand for every other point in the main rooms (dining rooms, etc.) plus 40% of demand for every other point

Table 7.13: Diversity calculations

> **Worked example**
>
> What is the maximum demand for three domestic lighting circuits of loads 950 W, 1000 W and 1200 W respectively?
>
> **Step 1**
>
> Calculate current for each circuit.
>
> Circuit 1 $= I = \dfrac{P}{U} \; \dfrac{950}{230} = 4.13$ A
>
> Circuit 2 $= 4.35$ A
>
> Circuit 3 $= 5.22$ A
>
> **Step 2**
>
> Check diversity allowance for the load – in this case 66% of total current.
>
> **Step 3**
>
> Calculate diversity.
>
> Total current $= 4.13 + 4.35 + 5.22 = 13.7$ A
>
> 66% of 13.7 $= \dfrac{66}{100} \times 13.7 = 0.66 \times 13.7 = 9.04$ A

Activity 7.4

Calculate the allowable diversity of:

1 a 45 A cooking appliance in a house – the isolator incorporates a 13 A socket
2 a guest house with three instantaneous water heaters: 3 kW, 2.5 kW and 1 kW
3 a small shop with four electric heaters: 5 kW, 3 kW and two at 2.5 kW.

Suitability of wiring systems

Chapter 4: Installation of wiring systems and enclosures looks at wiring systems in detail. From a design point of view we need to consider the following points, which will influence the choice of wiring system to be used.

- Constructional features of the building – for example, will the walls be plastered or finished face brickwork? Is the building of metal construction or reinforced concrete? Is there a way to hide cables and other components, or will all components and cables need to be run on the surface?
- Applications – what will the installation be used for? Does it require fireproof or moisture-resistant equipment? Is it an industrial, agricultural or domestic installation?
- Advantages – the advantage of a particular wiring system must be weighed up in terms of effectiveness, cost, speed of installation and ease.
- Limitations – what are the limitations of a particular wiring system? Will it fulfil the requirements of the job? What about the long term, for example would it be better to install trunking rather than conduit, to allow for future expansion?
- Environmental factors – the surrounding environment must be considered.
  - o Ambient temperature – is it extremely hot or cold?
  - o Moisture – is it a humid atmosphere?
  - o Corrosive substances – are there corrosive chemicals in the atmosphere that might damage the wiring system?
  - o UV – sunlight which contains ultraviolet light can have a detrimental effect on some cables' insulation and sheaths.
  - o Animals – particularly in agricultural installations, in terms of rodents that may damage the wiring and livestock that may be harmed by it.
  - o Mechanical stress and vibration – is the environment a physically harmful one in which the wiring might be damaged?
  - o Aesthetic considerations – the installation must be neat and discreet and fit in with the decor and ambience of the environment.
  - o Exposure to elements – is the system an outside one and will it be able to withstand rain and other weather?

There are a number of wiring systems available to the designer. Table 7.14 lists the main ones and their applications. They are described in detail in *Chapter 3: Electrical installations technology* and *Chapter 4: Installation of wiring systems and enclosures*.

Wiring system	Application
Cable tray and ladder racking	Cable support systems are used mainly in industrial installations or in the service areas of commercial and public buildings.
Cable trunking	Containment system that allows for the routing of multiple circuits. Removable lid means it can be used as a route for additional circuits or as an existing route for rewires. Available in various sizes and available in metal or plastic. *Chapter 3: Electrical installation technology* describes how to calculate the maximum number of cables allowable for various sizes of trunking.
Conduit	A pipe system into which cables can be run and which provides routing and mechanical protection. Available in various sizes and available in metal or plastic. *Chapter 3: Electrical installation technology* describes how to calculate the maximum number of cables allowable for various sizes of conduit.
PVC/PVC	Often known as twin-and-earth, it is the basic cable used for domestic and many commercial and public buildings.
SWA	Steel wire armoured cable is a versatile heavy-duty cable with a layer of armouring between its inner and outer sheath. It is often used for underground supplies and sub-mains.
MICC and MICV	Mineral-insulated cable has excellent fire-resistant qualities and is ideal for fire alarm and emergency systems.
FP200	Rapidly replacing mineral-insulated cable as fireproof cabling. It is simpler and cheaper to install and needs only a simple compression gland for termination.
PVC single core	The most basic type of electrical cable, singles have only one layer of insulation and no sheath and are usually routed through conduit or trunking systems.
Thermoplastic insulation	Insulation and sheathing used for most conventional wiring systems.
Thermosetting insulation	Used for low and high voltage applications. It can tolerate higher temperatures and can be used safely with conductor temperatures of up to 90°C. LSF (low smoke and fume) thermosetting cable is available.

Table 7.14: Main wiring systems and their applications

Progress check 7.9

1 What is meant by 'aesthetic considerations'?
2 What is ladder racking?
3 When would you use thermosetting insulation?

Current-carrying capacity of cables

In *Chapter 3: Electrical installation technology* we looked at the volt drop and current-carrying capacity tables in Appendix 4 of BS 7671:2008 and carried out basic cable selection calculations. Now we must also consider other factors such as ambient temperature and the effect of different types of protective device on the calculation.

The basics

The total amount of voltage dropped by a cable is calculated using the formula:

$$\text{Total volt drop} = \frac{\text{Length of run} \times \text{current} \times \text{mVd/A/m}}{1\,000}$$

where:

- total volt drop is the amount of voltage dropped by the cable you select for your circuit
- length of run is the length of cable feeding the circuit (in metres)
- current is the current taken by the load
- mVd/A/m is the amount of voltage dropped for each amp of current over each metre of its run
- the answer must be divided by 1 000 because the volt drop for the cable is stated in millivolts.

The tables

The maximum current-carrying capacities of various cables are listed in Appendix 4 of BS 7671:2008. You will see from these tables that the amount of current a cable can carry depends on:

- the type of cable
- its installation or reference method – these are described in another set of tables also in Appendix 4.

The amount of voltage each of these cables will drop per amp per metre is also listed in these tables in Appendix 4.

Remember: these volt drop values are given as mV.

De-rating factors

The type, length and installation method all affect the amount of current a cable can carry. However, there are other factors which need to be included in our calculation. These are listed in Table 7.15.

De-rating factor code	Description	Where these are found in BS 7671:2008
Ca	Ambient temperature	Table 4B1 (Appendix 4)
Cc	Buried circuit – factor will always be 0.9	5.1.1 (Appendix 4)
Cd	Factor applied for the depth the circuit is buried	Table 4B4 (Appendix 4)
Cf	Circuits protected by rewireable fuses. The factor is always 0.75	Regulation 433.1.101 (Part 4)
Cg	Grouping factor, applied when a number of cables are grouped together	Table 4C1 (Appendix 4)
Ci	Applied to circuits run through thermal insulation	Table 52.2 and Regulation 523.9 (Part 5)
Cs	Thermal **resistivity** of soil	Table 4B3 (Appendix 4)

Table 7.15: Factors which affect the amount of current a cable can carry

> **Key term**
>
> *Resistivity* – the amount of resistance possessed by a material. Its symbol is ρ and it is measured in W/m.

The currents

The final piece in the cable selection jigsaw is the type of current. These are shown in Table 7.16.

Current code	What it measures
Ib	Design current, the amount of current taken by the load.
In	The protective device rating for the circuit.
I_2	Current-carrying capacity required of the cable once factors are applied.
It	The tabulated current. When I_2 is calculated you will look in Appendix 4 for a cable which will carry this current. It is unlikely that you will find this exact value so it must be the next largest value – which is It.

Table 7.16: Types of current

- The first calculation is to find Ib for the circuit.

$$Ib = \frac{P}{U} \quad \text{(power consumed by the load)} \\ \text{(supply voltage)}$$

- Once Ib has been calculated then the rating of the protective device (In) must be chosen. So, if Ib = 14 A, In will be a 16 A circuit breaker.

Calculate I_2 using the formula:

$$I_2 = \frac{In}{Ca \times Cc \times Cd \times Cf \times Ci \times Cs}$$

It follows that It ≥ I_2.

The formula shown includes all the rating factors. It is unlikely that they will all apply to a given situation so only those that do should be used.

Worked example

An armoured cable is run underground (buried direct) at a depth of 0.5 m to feed a single-phase electrical installation in a garden shed. The load consists of two 100 W lights, a 3 kW electric fire and three 13 A socket outlets which will be used for electric hand tools, the total power of which is 1 kW. The length of run is 18 m. We will assume a thermal resistivity of 3 Km/W.

Step 1

Calculate design current Ib.

Total power

200 + 3 000 + 1 000 = 4 200 W

$$Ib = \frac{P}{U} \quad \frac{4\,200}{230} = 18.26 \text{ A}$$

Step 2

Select protective device rating In.

Ib = 18.26 A, therefore protective device will be a 20 A circuit breaker.

In = 20 A

Continued

Step 3

Find relevant rating factors.

Cc   buried circuit

Cd   factor applied for the depth the circuit is buried

Cs   thermal resistivity of soil

Values for our circuit will be:

Cc = 0.9

Cd = 1 m depth factor is 0.97

Cs = 3 KmW factor is 0.9

Step 4

Apply the factors and calculate I_2.

$$I_2 = \frac{In}{Cc \times Cd \times Cs} \frac{20}{0.9 \times 0.97 \times 0.9} = 25.46 \text{ A}$$

Step 5

Find matching cable in Appendix 4 tables. Remember It $\geq I_2$.

1 Table 4D4 – current-carrying capacity for multi-core armoured with normal 70° thermosetting insulation (copper conductors).

2 Look on top row of columns for appropriate reference method – reference method D 'direct in ground…'.

3 Select '1 two-core cable, single-phase…'.

4 The first cable of this type to take 25.46 A is 2.5 mm², which will carry 29 A.

Step 6

1 Using Table 4D4B, find the mVd/A/m for the selected cable.

2 Select column 'Two-core cable, single phase a.c.'.

3 The mVd/A/m for 1.5 mm² cable of this type is 18 mVd/A/m.

Step 7

Carry out volt drop calculation. Ib is used as the current.

$$\text{Total volt drop} = \frac{\text{Length of run} \times \text{current} \times \text{mVd/A/m}}{1\ 000}$$

$$\text{Total volt drop} = \frac{18 \times 18.26 \text{ A} \times 18}{1\ 000} = 5.92 \text{ V}$$

This is acceptable, so 2.5 mm² cable can be used for this job.

Activity 7.5

Using Appendix 4 of BS 7671:2008, carry out cable selection calculations for the two circuits described below. Remember to include the appropriate de-rating factors.

1 A single-phase 5 kW heater is to be fed by twin-and-earth cable (copper conductors). The cable passes through holes in the joists with four other cables. The length of run is 12 m.

2 A circuit of four 500 W spotlights is fed using steel conduit and single cables. The maximum length of run is 22 m. There is another radial circuit in the conduit.

Protective devices

Chapter 3: Electrical installations technology looks at protective devices in general. This section focuses on breaking capacities and discrimination for protective devices. However, let's remind ourselves of the basics.

- Circuit breaker – single- and triple-pole, automatic switches that provide protection against overcurrent.
- Fuse – used to provide overcurrent protection by physically melting and cutting off the current path.
- Residual current device (RCD) – an electromagnetic device designed to protect against electric shock.
- RCBO – provides both overcurrent and shock protection and incorporates the components of both an RCD and circuit breaker.

There are a number of currents associated with protective devices. Table 7.17 provides a key to these.

Current code	What it measures
I_2	The current that ensures a protective device's effective operation
Ia	Current that causes operation of a protective device within the time specified
Icn	The rated short-circuit capacity of a protective device; in other words, the maximum fault current the device can operate under but will probably be destroyed
Ics	Rated in-service short-circuit capacity. The maximum fault current a protective device can operate under and be useable afterwards
$I\Delta n$	Residual operating current of a residual current device (RCD), e.g. 30 mA
Ipf	Prospective fault current

Table 7.17: Current codes

Disconnection time

Although the operation of a protective device seems instantaneous to us, it does take a certain amount of time to work. This short time-span is called the disconnection time. Obviously, the shorter the disconnection time, the better.

When it comes to disconnection times for earth faults, BS 7671:2008 splits protective earthing systems into two types.

- TT – earth connection for an installation is provided by an earth electrode driven into the ground adjacent to the installation itself.
- TN – where the earth is connected to the star point of the supply sub-station transformer, i.e. TN-S and TN-C-S.

Table 41.1 BS 7671:2008 gives maximum disconnect times for TT and TN circuits that feed loads of less than 32 A. For single-phase 230 V a.c. circuits they are:

- TT – 0.4 seconds
- TN – 0.2 seconds.

For a 400 V three-phase circuit these are:

- TT – 0.07 seconds
- TN – 0.2 seconds.

There are a set of regulations referring to disconnection times.

- 411.3.2.1 – requires that a device operates within its required disconnection time.
- 411.3.2.3 – TN-protected circuits feeding loads of more than 32 A and also distribution circuits may be allowed a disconnection time of 5 seconds.
- 411.3.2.4 – TT-protected circuits feeding loads of more than 32 A and also distribution circuits may be allowed a disconnection time of 1 second.
- 411.3.2.5 – systems in which the fault voltage is 50 V or less (e.g. the 110 V centre-tapped transformers used for tools on construction sites) need not rely on the disconnection of a protective device but on other means, such as the extra-low voltage itself.
- 411.3.2.6 – if disconnection time cannot be achieved, then supplementary equipotential bonding must be added to the installation.

Tables 41.2 to 41.4 show the maximum earth fault loop impedances for various protective devices and their ratings.

- Tables 41.2 and 41.3 apply to circuits which are allowed a 0.4-second disconnection time.
- Table 41.4 is for circuits allowed a 5-second disconnection time.

Used in conjunction with the graphs in Appendix 3 of BS 7671:2008, the actual disconnection times can be calculated.

Worked example

Protective device – 16 A BS88 fuse – system C (0.4-second disconnection time)

Step 1

Refer to table – in this case Table 41.4.

Step 2

Find maximum earth loop impedance value – 2.42 Ω.

Step 3

Calculate prospective fault current that should flow when this fuse operates.

$$Ipf = \frac{U}{Z} \frac{230}{2.42} = 95.04 \text{ A}$$

Continued

Step 4

1 Refer to the graphs and their inset tables in Appendix 3.

2 Figure 3A1 is the graph we need.

3 The table inset in the top right-hand of the page shows the disconnection time for a selection of fault currents for each protective device rating.

4 Select 16 A and cross refer to 95 A; you will see that the disconnection time should be 0.4 seconds.

Using the graph, the horizontal axis is the fault current which will flow in the event of a fault. The gradient is exponential. This means that:

- for the first set, starting at '1', each vertical line represents 1 A, so the lines represent 1 A, 2 A, etc. to 10 A
- between 10 A to 100 A, each vertical line represents 10 A, so the lines represent 10 A, 20 A, etc.
- between 100 A and 1 000 A, each vertical line represents 100 A, and so on.

The same applies to the time gradients on the vertical axis.

1 On Graph 3A1 find the vertical line for 95 A – it is difficult to be completely accurate but it is halfway between the 90 A and 100 A lines.

2 Follow this up until it meets the 16 A curve.

3 From the point where they meet, cross to the time axis and read off the time, which is, again, 0.4 seconds.

Discrimination of protective devices

It would not be practical for the main 100 A fuse to blow every time there was a fault in a domestic installation. Therefore, the protective device closest to the point of fault is the one that should operate. To achieve this, the lowest possible value protective device is installed at each point at which the cable size changes.

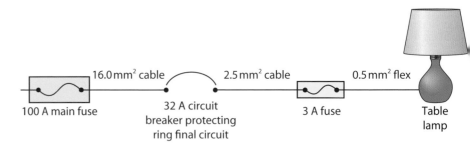

Figure 7.13: Discrimination

The time/current graphs in Appendix 3 of BS 7671:2008 show the characteristics between the protective devices. The tables inset into the graphs show the operating current for each type of device. These tables can be used to select protective devices which will not operate before the next down the line.

The graphs for fuses are different to those for circuit breakers. Also, there are a set of different operating times in the inset tables on the fuse graphs, but only one operating time for circuit breakers.

Circuit breaker graphs (for example Figure 3A4) have a vertical line for the first part of the overcurrent. This is because the magnetic element operates much faster than the thermal one and requires a much smaller fault current to work. The length of this line depends on the type of circuit breaker. For example, for a type B circuit breaker, the vertical line extends to just over 10 seconds. For the type D, it extends to only 3 seconds. This indicates that in the event of an overload, the bi-metal strip in a type D circuit breaker will heat and bend and operate the circuit breaker after 3 seconds, whereas in a type B this will take 10 seconds. (*Chapter 3: Electrical installations technology* lists the different types of circuit breaker and their uses.)

Breaking capacities of protective devices

The breaking capacity of a protective device is the amount of current it can safely carry and still operate correctly. Remember that when a fault occurs, a large fault current will flow into the circuit. The protective device must be able to withstand this high current. Nothing works instantaneously, so no matter how quickly the device appears to operate, it will still have to withstand this current for a certain amount of time (the disconnection time).

This maximum current is indicated on the device.

- Circuit breakers – the letter M and a number are used to indicate breaking current. For example, M4 means that it will withstand a fault current of 4 kA.
- Fuses – the breaking capacity is stamped onto the metal end cap or on the fuse body.

Thermal constraints

Requirement 434.5.2 states that the protective device must operate before the heat caused by the fault current damages the conductor. BS 7671:2008 gives a formula to calculate the maximum time a fault can be allowed to run through a conductor. The formula is

$$t = \frac{k^2 s^2}{I^2}$$

where:

- t is the maximum time fault current can run through a particular conductor
- k is a factor for the maximum temperature allowed for certain-sized conductors for a maximum of 5 seconds
- S is the cross-sectional area of conductor in mm^2
- I is the current that will flow down the conductor during a fault.

Worked example

A 6.0 mm^2 armoured cable feeds a single-phase, 230 V, 6.21 kW load. The circuit is protected by a type-B circuit breaker to BS EN 60898. What is the maximum time this cable can take the full load fault current?

Step 1

What is the protective device rating?

$$I = \frac{P}{U} \quad I = \frac{6210}{230} \quad I = 27 \text{ A, so the nearest protective device rating is 32 A.}$$

Step 2

Find the maximum earth fault loop impedance for circuits protected by this protective device. Remember, these values can be found in BS 7671:2008 in Tables 41.2 to 41.4.

Table 41.3 refers to circuit breakers which match the one we are looking for. Type B is the first row.

The maximum earth fault loop impedance (Zs) for a 32 A circuit breaker to BS EN 60898 is 1.44 Ω.

Step 3

Calculate the fault current, the I in our thermal constraint formula.

$$\text{Fault current } I = \frac{U}{Z} \quad I = \frac{230}{1.44} \quad I = 159.72 \text{ A}$$

Step 4

1 Look up the k factor. Remember, this is in BS 7671:2008, Table 43.1.

2 The type of insulation was not mentioned; therefore assume it to be 70°C thermoplastic (this is normal cable insulation).

3 Conductor csa is 6.0 mm^2 – refer to column 1 of the 'Conductor cross-sectional area' row.

4 Our cable is ≤300 mm$^{2.}$

5 For a copper conductor the k factor is 115.

Step 5

Carry out the calculation.

$$t = \frac{k^2 S^2}{I^2} \quad t = \frac{115^2 \times 6^2}{159.72^2} \quad t = 18.66 \text{ seconds}$$

Progress check 7.10

1 What is discrimination?

2 What is the disconnection time for a TN circuit feeding an a.c. circuit of less than 32 A?

3 What is meant by thermal constraint?

Activity 7.6

Calculate the maximum time a cable can carry fault currents in the following circuits.

1 A 9 A circuit protected by a Type C BS EN 60898 circuit breaker. Cables are 1.5 mm^2 with 90°C thermosetting insulation.

2 A 28 A circuit protected by a BS 3036 fuse (0.4-second disconnection time), using a mineral-insulated cable with a 70°C thermoplastic sheath. Conductor size = 1.5 mm^2.

3 A 58 A circuit protected by a 63 A BS88-2 type system E (bolted) fuse (5-second disconnection time). Cable size = 16.0 mm^2 with aluminium conductors and 70°C thermoplastic sheath.

COMPLEX ELECTRICAL SYSTEMS

Electrical supplies are often used for more than simple power and lighting. Industrial installations, for example, may include high voltage and complex control systems such as programmable systems. Intake positions often include panels and heavy duty switchgear. There may also be back-up power.

Poly-phase

This is looked at in detail in *Chapters 1* and *6: Principles of electrical science* and *Chapter 3: Electrical installations technology*. As a reminder, poly-phase, normally three-phase, serves two main purposes.

- Distribution – the supply from the sub-station transformer is 400 V three-phase. Each dwelling on the supply route is then fed from one of the phases and the common neutral.
- Industrial equipment – electrical equipment such as industrial motors use the relationship between the phases to operate.

Working practice 7.4

A small curtain producer and installer, based in Hertfordshire, was experiencing problems with its ageing heating system. The heating finally broke down and several of its elements burned out. A new heater unit was needed. Electric heating is always expensive to run, but there was no gas supply to the factory building and getting one installed was extremely costly. The factory owner did his research and found a heating system that offered some level of economy. The heater chosen was a three-phase, industrial storage heater. The unit was three metres tall and one and a half metres wide. The heating element banks were spread over the three phases. The storage heating principle was the secret of its economy. The heater was switched on at night when a cheap **tariff** was available. The elements were set into specially designed thermal bricks which held heat for many hours. This meant that the heating supply could switch off early in the morning when the full-cost tariff came back into play. The bricks then gave off their stored heat throughout the day.

1. What is the principle of storage heating?
2. What is the advantage of a three-phase heater?

Key term

Tariff – the rate at which electricity is charged to the customer.

Commercial

Although the term 'commercial' implies business premises, it can also be used as a general description of a place that is open for public use. Commercial electrical installations can, therefore, include those in:

- shops
- offices
- pubs
- cinemas
- churches.

The main design focus when planning for a commercial installation is people. These types of building are designed around their customers or visitors. Therefore, while all installations must be safe for those who work in, or visit, them, a commercial installation is different from an industrial one in so far as there are no machines. Instead, there are customers or visitors, and the staff who serve, or look after, them. The main design issues are outlined in Table 7.18.

Design issue	Design considerations
Fire safety	Low smoke-emission insulation on cables, e.g. XPFE cables. Fire safety – an alarm system will be needed. This will also require emergency lighting. Intruder alarms will be necessary so decisions will need to be made about open or closed circuit types and whether CCTV will be required.
Lighting	Lighting is an important issue. Good lighting design will achieve a number of aims, including: • health and safety – correct lighting levels are vital for commercial premises • suitable colour renditions, particularly in a shop where it can set off the products to their best advantage. A harsh, white light is not conducive to well-being.
Data	Many commercial installations rely on computing systems, whether in an office or for the tills in a shop. This means the installation of data networks, which raises issues of circuit segregation, IT earth systems and the need for back-up supplies.
Building management systems	Larger commercial installations benefit from building management systems which can control environment and security, and optimise the use of energy.
Test and inspection	Commercial installations need to be inspected and tested regularly. Guidance Note 3 gives the recommended frequency for test and inspection. Electrical equipment will also need to be regularly tested.
Flexibility	Office installations need flexibility so a false floor system incorporating underfloor bus-bar trunking and the use of power tiles is a consideration.

Table 7.18: Design issues and considerations for commercial installations

Industrial

For the most part, industrial installations are those for premises where some type of manufacturing, processing or repair-related activity is carried out. The term also covers complexes such as large warehouses and docks. Examples of industrial installations include:

- factories
- vehicle repair workshops
- general engineering workshops
- food processing plants.

The main design concern in relation to industrial electrical systems is mechanical damage both from physical wear-and-tear and from the presence of corrosive substances, extreme heat or cold, etc. Wiring systems tend to be protected within containment and heavy duty enclosures. Supplies are usually three-phase – sometimes high voltage is made available. The main supply intake will be in the form of high current switchgear and include:

- main switch – multiple-pole-and-neutral switch which provides connection point to the supply
- bus-bar – means of distributing the supply to the switchgear components

- fuse-switches – high current-rated fuses physically switched in and out of circuit by the fuse-switch operation
- main distribution board – usually supplies other sub-distribution boards via sub-main cables (usually steel wire armoured cable).

In modern industrial installations, this switchgear is contained within an electrical control panel. These panels not only contain the components of the supply intake but also monitoring devices for energy use, current balancing, and fault and performance monitoring, including:

- voltmeters – connected in parallel with the circuit. In a very large commercial or heavy industrial installation, the supply side may be 11 000 V (11 kV), 33 000 V (33 kV) or higher. One of the purposes of the voltmeter is to make sure that all three phases are connected and working. A recording version will show if there have been any:
  - o surges – when the voltage is much higher than it should be. This can be caused by a lightning strike. Surge protectors can be fitted into a sensitive installation
  - o dips – when the voltage is much lower than it should be
- ammeter – this shows how much current is being used. For a three-phase supply, the current should be balanced, as far as possible, across all three phases. The ammeter will show if one phase has to work harder than the others. If there is too much difference, current will flow in the neutral which reduces the efficiency of the supply and increases the cost. Ammeters are normally connected in series. However, clip-on ammeters can be used instead. These use the magnetic field around a conductor to read the current. In electrical panels, the clip-on jaws are replaced by a permanent coil wound round the conductors.

There are many types of control panel. Electrical panels are mainly used for:

- the main supply intake into large buildings and premises
- controls for automated systems
- control points for monitoring and alarm systems
- back-up and emergency supply controls.

Figure 7.14: A typical electrical control panel

The main two groups are:

- form 1 panels – there is no segregation (physical separation) between the panel's bus-bars and the rest of the equipment in the panel
- form 2 panels – bus-bars are segregated from the rest of the panel by being run in a separate compartment.

Containment and cable routing in an industrial installation is usually by:

- trunking – metal channelling used as the main cable routes around an area
- conduit – used to route circuits from trunking to their switching and loads, e.g. lighting circuits
- bus-bar trunking – comes pre-fitted with copper bus-bars which can carry a high current. Take-off boxes are fitted to the trunking wherever power is needed. When installed overhead, bus-bar trunking provides a flexible supply source for machinery
- lighting trunking – trunking specifically designed to bear the weight of lighting and provide a fixing and installation grid in large open areas
- armoured cables – used as sub-main cables as well as feeds to industrial equipment
- tri-rated cable – used in control panels for both low and high voltage
- emergency stop buttons – installed to protect the users of machinery in the event of an accident. These buttons can be easily operated to shut off an individual machine or a complete area.

Protective devices should be selected to cope with the high current demand from equipment at start-up – type-B circuit breakers and BS88 fuses, for example. There is also the problem of highly reactive loads from large numbers of electric motors and fluorescent luminaires. Corrective capacitor banks will be required to counteract the lagging power factor of highly inductive loads.

Fluorescent lighting is problematic in the industrial environment due to stroboscopic effect. Fluorescent tubes flash at a rate of 100 times/second due to the fact that the sine wave for their a.c. supply crosses the zero line twice in every cycle. This can create the illusion that rotating machinery is stationary and result in accidents. Ways to overcome this problem are:

- a non-flickering light source such as an incandescent lamp installed near rotating parts to break up the stroboscopic effect
- twin or multiple-tubed fittings which have leading and lagging power factors present and cause the tubes to flicker at different rates, thereby cancelling out the stroboscopic effect.

Programmable systems, or PLC, are a common form of industrial process control. *Chapter 3: Electrical installations technology* describes these systems but basically they consist of a central processing point which is fed information from sensors and other inputs. The processor responds by sending out instructions based on their programming. Installing PLC systems raises the issue of data and control wiring which needs to be routed separately from mains voltage supplies.

Industrial machinery invariably incorporates some type of electric motor. Motor control is, therefore, a major part of an industrial electrical system. *Chapter 6: Principles of electrical science* describes electric motor control principles in detail.

Agricultural

BS 7671:2008 considers an agricultural installation to be a special location (see page 347 of this chapter). Modern agricultural installations often include environmental control, temperature, humidity, and feed and processing systems. Feed systems can be automated so that food is delivered straight to livestock. Some farms incorporate their produce-grading equipment based on the industrial conveyor belt system. Temperature and humidity are important for livestock and for storage of grain and other produce. The main design issues and consideration for agricultural installations are shown in Table 7.19.

Design issue	Design consideration
Mechanical stresses	Containment or armoured cable needs to be used.
Fire	Thermal effects from lighting and heating are a serious hazard due to wooden construction of many farm buildings, presence of flammable materials, such as straw and hay, and chemicals such as fertiliser and fuel storage for agricultural machinery.
Moisture	Containment and enclosures should be resistant to dust and moisture ingress. Containment should ideally be plastic conduit which offers mechanical protection but will not corrode in damp or humid conditions.
Livestock protection	Animals are very susceptible to electric shock so electrical protection is very important in an agricultural installation. This is both mechanical protection for the cables and components of the system and also earth bonding, overcurrent and shock protection. BS 7671:2008 and page 347 of this chapter give details of the type of electrical protection needed for agricultural installations.
Protection against rodent damage	Containment must be used to prevent rodents chewing through cables and causing electrical faults.
Industrial-type processing systems	Modern farms are equipped with automatic feeding systems for their livestock. These will require timing and process control. Also, vegetable and fruit farms often have their own produce processing and packing set-ups. Working on the same principle as industrial production and processing lines, these systems can benefit from PLC-type controls.

Table 7.19: The main design issues and considerations for agricultural installations

Interpret sources of information for the connection of complex electrical systems

As with all design, complex systems require sources of information and must also be designed within the limitations of statutory regulations. These documents and information sources are detailed on pages 346 and 347 of this chapter and also in *Chapter 1: Principles of electrical science* and *Chapter 2: Health and safety in building services engineering*.

Assess connection methods for given situations

Chapter 3: Electrical installations technology describes connection and termination methods, comprising:

- standard clamped terminal
- post type
- crimped termination
- soldered
- push types.

For the type of complex electrical systems we have been discussing in this chapter, the following connection methods may also be used.

Compression joints

This means everything from crimp lugs to larger compression terminations. The correct-size termination lugs must be used and fitted and secured to the cable. Larger compression joints require special tools such as hydraulic crimpers to secure the joint.

Heat shrink

This is a type of insulation that is secured over a termination. When heated, the insulation tightens and seals the connection.

Resin compound joints

Resin and compound joints are used for underground cable jointing. The jointing kit is supplied in a pack which includes the body of the joint and the resin itself. This is sometimes separated into two parts and must be mixed before use. The enclosure is formed from two halves which are taped together around the actual terminations. Then the resin is poured into the enclosure through a filling hole. Once the resin has hardened, the finished joint is buried in the cable trench.

Catenary systems

Supplies can be routed overhead, supported by means of poles and brackets fitted to the outside walls of buildings. Specialist overhead cables are available. These are usually single-core with stranded aluminium conductors and an integral support, or catenary, wire. This wire takes the weight of the cable so that the conductors are not put under strain.

Twin-and-earth and even small armoured cables are sometimes run overhead. Although this is acceptable, they must be supported by a separate catenary wire.

The connection point for overhead cabling should not take any weight. The components of an overhead wiring system are as follows.

- Insulators – support the cables and keep the metalwork separate from the connection point. These can be made from ceramic material, glass or, more recently, plastic-based materials.
- Bracket – a U-shape bracket into which the insulator is fitted. The bracket is fixed to the side of a power pole or the wall of the building. The fixings must be secure because overhead cables exert a lot of weight, particularly when they are strained to their full tension.
- Straining vice – a tool designed to pull overhead cables to their full tension and hold them there while the cable is fitted to the insulator and terminated. If there are two or more cables in the run, care should be taken to make sure they run parallel to each other.
- Overhead cables – specialist cables with stranded aluminium conductors and an integral catenary wire. Low voltage types have black insulation; high voltage types have either green insulation or no insulation at all.
- Connectors – weather-resistance connector blocks in which the overhead cable is connected to the tails or supply cable for the building it feeds.

| 1. Road crossing accessible to vehicles | 2. Accessible to vehicles but not a road crossing | 3. Inaccessible to vehicles |

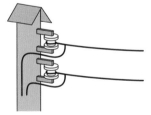

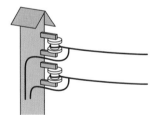

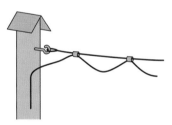

All methods of suspension
5.8 m minimum above ground

All methods of suspension
5.2 m minimum above ground

PVC cables supported by a
catenary wire
3.5 m minimum above ground

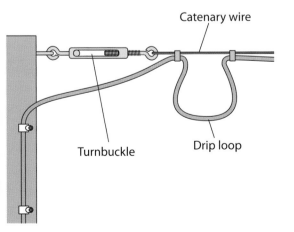

Catenary wire

Turnbuckle

Drip loop

Figure 7.15: An overhead cable installation

BS 7671:2008 gives minimum heights for some overhead cables. These are listed in the special locations table on page 347. The Building Regulations consider overhead wiring, particularly uninsulated overhead cables, to provide shock protection by 'placing out of reach'.

Buried

Buried supplies are usually armoured cables, which are designed to withstand harsh environments, such as burial in soil. BS 7671:2008 requires certain depths for underground cables running through special locations and details are in the special location tables in this chapter. The main point is that they should be deep enough not to suffer damage from vehicles or any activity that takes place on the surface.

Buried cable should be laid in a soft bed, sand for example, to prevent damage from sharp stones or other debris in the soil. A marking tape must be laid above the cable to warn anyone who is excavating the area of its presence.

In some locations, ducts are laid in the ground so that cables can be drawn in and out easily. A draw rope is usually available in the duct and when it is used to draw in new cables, a second rope should be pulled in at the same time. This can either be left in place as a new draw rope or used to pull the existing one back into the duct.

The location and routes of buried cables must be marked on drawings and updated as new cables are installed and old ones removed.

Procedures for proving that terminations and connections are sound

It is vital that all electrical terminations and connections are sound. To make sure that all terminations are sound, a number of tests must be carried out. This is part of the Schedule of Inspections carried out during an inspection and test routine (see *Chapter 8: Electrical installations: inspection, testing and commissioning*). The terminations must be:

- electrically safe – a loose or poor termination can cause heat, fire, shock or poor performance of equipment
- mechanically sound – not subject to mechanical damage and securely terminated.

Consequences of unsound termination

A number of problems result from poor termination of cables. Most obvious is that equipment and loads will not work properly. There may be an intermittent supply, or overheating. The problems described in more detail in *Chapter 9: Electrical installations: fault diagnosis and rectification* are:

- high resistance joints
- fire risk
- shock risk.

Erosion and corrosion

Care must be taken when connecting two different types of metal. Many metals are not compatible with each other and the point of connection between them can cause electrolytic effect. This happens particularly when the joint becomes damp. The two surfaces react to each other and one acts as a cathode, the other as an anode, exactly as in a battery. The anode then slowly breaks down as particles migrate to the cathode. The cathode is seen as the 'noble' metal and the anode is the 'less noble'. Some incompatible metal combinations are:

- copper and iron
- aluminium and steel
- copper and aluminium.

This effect is made worse by a loose connection between the two metals and also by water ingress, especially by salt water.

PLANNING WORK SCHEDULES

Once a contract has been awarded, the project needs to be carefully planned to ensure that the works are completed efficiently. Lines of communication must be set up between all the parties involved, with, as far as possible, a single point of contact for each one. Every trade will, quite naturally, see its own part in the project as highly important and while all the trades are necessary for successful completion, there can be clashes and difficulties arising from unclear communication and inadequate planning. Control of large construction works is called project management, with the project manager acting as the hub for the various threads that make the project work.

Effective working relationships

The structure of the team working on a construction project will vary from job to job, depending on which trades are required for the work. The specialist companies who carry out the work are called contractors. The main specialist trades and how their work affects the work of the electrician are described below.

- Gas fitter – installs gas pipes, boilers and other gas-fired equipment. Much of this equipment needs an electrical supply and some of it requires complex control circuitry. Gas pipes must be connected to the main installation earth.
- Plumber – installs water supplies and distribution throughout premises. Plumbers fit taps, sinks, baths and showers. The pipework requires an earth connection. If the shower is an electric version it will need an electrical supply. The positioning and types of electrical services in areas such as bathrooms are dealt with in Part 7 of BS 7671:2008 under special locations.
- Floor layer – some cable routes may be underfloor so cable routing needs to be carried out before the finished floor is in place.

Progress check 7.11

1 What is the difference between a Form 1 and a Form 2 control panel?
2 What is a catenary wire?
3 What causes electrolytic action at cable terminations?

LO 4

The trades whose work will also impinge on your own work as an electrician are:

- tiler
- carpenter (or joiner)
- plasterer
- painter and decorator.

Table 7.20 shows the main considerations for a job on which a number of different trades are working.

Consideration	Description
Coordination	Trades must work in a coordinated manner, each one performing their tasks in the right order and in such a way as to allow the other trades to get in and do their work.
Communication	Clear lines of communication must be established, often with one main point of contact for each trade.
Alterations to schedule	Alterations to the schedule must be notified and handled formally using a variation order (see *Chapter 5: Understand how to communicate with others within building services engineering*). An example of a formal change process is shown on page 351.
Consideration of customer requirements	The work is all about the customer and therefore the customer's requirements are of utmost importance and above all other considerations – with the exception of regulatory ones.
Negotiation	Negotiation is a vital part of any project. There are times when it is not possible to give a customer exactly what they want. In this case negotiation may be needed to work out a compromise. Negotiation is a way of getting the best for all parties involved. A good starting point is to ask for more than you anticipate getting. This way, you can afford to make concessions and still get most of what you originally wanted.
Timing	A project relies on timing. Each part of the project, each task, will be given a set amount of time to be completed. It is important, therefore, that: • timescales are realistic • each trade keeps to their timing • timings are coordinated using a project plan. If it seems that the timescales cannot be met then this has to be made known and the plan formally reviewed and altered.

Table 7.20: Considerations for a job on which a number of different trades are working

Stages of an electrical installation

Many electrical jobs, particularly domestic installations, are divided into two definite parts: first fix and second fix.

First fix

This is the initial work. In the case of a domestic installation this will be installation of the cables and the flush back boxes for accessories such as switches and sockets. This is carried out before floorboards and plasterboard are fixed so the cable can be run in easily. Once the first fix is completed, the electrician will leave the site to return for the second fix. The other trades need to be coordinated so that noggins are installed and plasterboard left off walls to enable cables to be run and the first fix completed.

Second fix

Once the plasterboard and floorboards are installed and the walls are plastered, the electrician returns to connect the sockets and switches and other accessories such as light fittings. The consumer unit and mains position will be installed and wired. Following this the installation is inspected and tested.

Difficulties

A construction project will seldom run perfectly. There may be conflict between the trades involved or there may be snags encountered which require changes in the plan.

Conflict (suppliers, trades, clients)

Chapter 5: Understand how to communicate with others within building services engineering described the problems, and resolution, of conflicts in the workplace.

Poor performance of staff

Good workmanship not only means a quality job but also a safe and functioning one. The staff involved must be properly trained and understand exactly what it is they should be doing. Companies will have a formal disciplinary process. This will usually consist of a number of stages, beginning with a verbal warning and leading to a written one and finally to dismissal if the poor performance or ill-discipline persists. (See *Chapter 5: Understand how to communicate with others within building services engineering* for more information on handling poor performance.)

Resource shortages

Materials and tools must be ordered and delivered on time. This requires planning so that work doesn't tail off due to shortages. Allowance must be made for difficulties in purchasing and delivery of specialist equipment, particularly if it has to be custom-built for a job. An example of this is electrical control panels which tend to be tailor-made for specific projects.

When tendering for work, a company should make sure they have enough trained staff and equipment to complete the work on time and to quality. It is not the customer's fault if you have not taken this into consideration and find yourself struggling to meet deadlines because your staff are spread too thin.

Quality of components and incomplete fittings

The cheapest is not always the most cost-effective and this applies to materials used for electrical installations as much as anything else. The components must be right for the job and of sufficient quality to do their job without breaking down. This is where the supplier's expertise can be useful. Cheap components might enable you to enter a competitive tender, but there may well be a cost implication later if the components fail and your company is bound by a warranty agreement to carry out repairs and replacement.

Damage

Construction work is very heavy duty and involves teams of people, heavy-duty tools and difficult conditions. Damage to structure and equipment is inevitable. Damage can be done to existing decor, fittings and property during rewires. It can also occur in new installations where someone from one trade accidentally damages work put in by another – the unwary electrical apprentice stepping onto freshly laid, wet concrete, for example.

Damage to a customer's possessions or property must be owned-up to and paid for. Damage to work done by another trade must be reported as quickly as possible. Always tell the truth and, if appropriate, offer to pay for any damages done.

Programme of work

Before work can start in earnest, a number of issues need to be resolved. The following are among those which have to be considered. Because construction projects are complex and involve many different trades, a plan must be drawn up to make sure the right people are on site at the right time. This means that the work can flow from start to finish without breaks. These plans can take many forms.

- PERT (programme evaluation and review technique) – a flow chart constructed from circles which represent tasks in a project (see Figure 7.16).

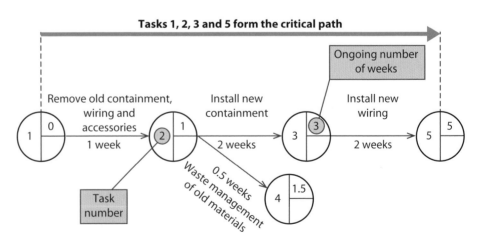

Figure 7.16: A simple PERT chart

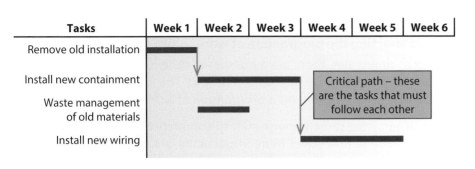

Figure 7.17: A Gantt chart

- Gantt chart – a bar or Gantt chart is a visual representation of a project. In its simplest form, it consists of a list of trades and tasks to be undertaken. These are represented on a timeline by bars similar to those on a graph. Figure 7.17 shows a typical Gantt chart layout.

Critical path analysis

Although many jobs can be carried out in parallel, there are some that cannot be started until the previous one is finished. A simple example of this is that the roof of a building cannot be constructed until the walls have been built. This procession of interdependent tasks is called the critical path.

Spreadsheets

A spreadsheet is a chart or table that shows costs against work. There are some that are generated on computer and have an in-built calculation programme. Spreadsheets are used for tracking costs, because it is important to know at all stages during a project how much money has been spent and, more importantly, how much remains.

Commissioning

Commissioning consists of the inspection and testing of the new installation and is described in detail in *Chapter 9: Electrical installations: fault diagnosis and rectification*.

Snagging

Snagging is slang for the final stage of a job when small faults must be put right. This might be anything from securing conduit box lids that may have been left off, to replacing broken or faulty fittings and accessories. Snagging is an important part of the project and must be considered when planning the work.

Sign-off

When the work is complete, inspected, tested, commissioned and ready for the client to use, the client signs off the work and officially takes it over. There may, of course, be a warranty period in which the contractor must return to repair any faults, or even a maintenance contract which means that the contractor will carry out both routine and preventative maintenance and possibly be on call-out in the event of a fault.

The sign-off is a formal document and proof that the customer is happy with the work and has agreed that it is complete.

Factors affecting the programme of work

As well as timescales, there are a number of other factors affecting the execution of an electrical installation project. The following must be considered when drawing up a programme of work.

- Staffing – are there enough staff to complete the work within the programme timescale? More importantly, are the staff you have available fully trained to do the work?
- Materials and ordering materials – ordering and delivery of materials can cause delays, particularly if the materials are specialist parts or those that need to be manufactured to order. Remember to allow enough time for material ordering and delivery.
- Storage – materials and tools are valuable and as such must be stored securely. Theft may be covered by insurance and may not incur direct financial loss but it will cause delays and inconvenience. Also, materials and tools will need to be protected from the elements.
- Design – although there will be drawings and other information for the main construction project, there may be a need for the electrical company to draw up its own set of designs before ordering and work can commence.
- First fix – most electrical installation work is divided into two parts: first and second fix. First fix usually takes place before floors and ceilings are boarded and walls plastered, second fix when these jobs have been completed. This separation of works needs to be considered in the programme.
- Timescale – the timescale proposed by any contract must take into account all the above issues and be realistic. There could well be financial penalties imposed on contractors who cause delays to the programme, so there is little to be gained by putting in an unfeasibly tight timescale just to win the contract.

Benchmarking

Benchmarking is a method of comparing your performance with a standard, for example a national average or values set by a company or manufacturer. It could be used for drawing up provisional programmes of work by using benchmarks as a guide to how long tasks may take.

Handover procedures

Once a job has been completed you cannot simply walk away and say the job is done. There needs to be a formal handover. There may be equipment or systems that require training to use, warranty and repair issues, as well as the need for a maintenance contract.

Walk-through with client and demonstration of systems and equipment

Show the client the job, the location of the distribution board or boards, the protective devices and how to reset them. If there are alarms or controls then the client needs to be made aware and some training put in place. The client should also be able to comment on whether they are satisfied with the work and if it is fit for purpose.

As stated before, if there is any equipment installed then the client or the personnel who are going to use the equipment or system need to be shown how to use it. This may take the form of a formal training course.

Manufacturers' instructions

Manufacturers' instructions must be handed over and secured in a place where they will be protected and can be made available if necessary. These can take the form of a handbook, a leaflet or even a collection of manuals and documents.

Warranty

A warranty is a guarantee. There will be a warranty agreement written into most contracts. This states the warranty period and lists the repairs and service it covers. This might include a call-out agreement so that the electrical company will attend to a breakdown immediately. It may only cover parts, or labour, or it may cover all work associated with a fault.

Agree final invoice

The contractor will, of course, need to be paid for the work. There may have been a fixed price or there may have been an agreement in which the electrical company is paid for the hours they spent on the work. This must be established clearly and agreed formally by all parties at the outset of the work. The final invoice may include a bill for extra work. This should be already known to the customer who would have been involved in a formal change process for the design.

> **Progress check 7.12**
>
> 1 How do plasterers interact with electricians?
>
> 2 What is the difference between a Gantt and a PERT chart?
>
> 3 What is snagging?

Knowledge check

1 The de-rating factor for a cable grouped with other cables is:

   a Ca
   b Cf
   c Cg
   d Ci

2 Design current is represented as:

   a Ib
   b In
   c It
   d I$_2$

3 The disconnection time for a 230 V, single-phase TT-protected circuit feeding a 16 A load is:

   a 0.2 seconds
   b 0.1 seconds
   c 0.4 seconds
   d 0.3 seconds

4 The maximum earth fault loop impedance (Zs) for a Type C, 25 A circuit breaker to BS 60898 is:

   a 1.84 Ω
   b 0.92 Ω
   c 0.46 Ω
   d 2.3 Ω

5 Where conductors carrying different voltages are run together:

   a all cables must be double insulated
   b all outer cable sheaths must be rated to the lowest voltage
   c all cables must be tri-rated
   d all insulation must be rated to the highest voltage

6 The regulations intended to protect electronic and data equipment from electromagnetic interference are:

   a Electromotive Compatibility
   b Electromagnetic Continuity
   c Electromagnetic Compatibility
   d Electro-electronic Continuity

7 Farms, construction sites and fairgrounds are:

   a statutory locations
   b separate locations
   c special locations
   d segregated locations

8 The transformer feeding SELV equipment in a bathroom should be located outside:

   a Zone 1
   b Zone 2
   c Zone 3
   d Zone 4

9 A quotation is:

   a a price entered by a contractor when tendering for work
   b a scheme of the tasks which make up a construction project
   c an arrangement between the contractor and the architect
   d a scheme for repair and maintenance of the new installation

10 A series of interdependent tasks running through a project are a:

   a PERT path
   b Gantt path
   c critical path
   d series path

Electrical installations: inspection, testing and commissioning

This chapter covers:

- requirements for commissioning electrical systems
- procedures for the inspection of electrical systems
- procedures for completing the testing of electrical systems
- requirements for documenting installed electrical systems
- inspecting electrical wiring sytems
- testing the safety of electrical systems.

Introduction

Inspection and testing applies as much to the simple alteration of a single circuit in a house as it does to a large installation, such as a new hospital or sports stadium. It is essential that all electrical systems and installations are inspected and tested. It should be verified that correct practices are being used during installation and installations must then be tested before they are handed over to the customer. This confirms that they are working properly and safe to use.

Electrical equipment and systems will also need to be commissioned, i.e. taken through their functions to make sure they work correctly and also to adjust settings to their operating values.

As well as the verification, inspection and testing of new work, existing installations in commercial, industrial and public premises must also be inspected and tested at regular intervals. This is to make sure that they have not deteriorated in any way and become dangerous.

Inspection and testing is a specialist role within the electrical industry. Only electricians who are qualified inspectors and test engineers are authorised to sign a test certificate. Signing this document places a great responsibility on the test electrician, who is vouching for the safety of the installation.

This chapter looks at the regulatory requirements for inspection and testing and the responsibilities of the test electrician, as well as taking you through the inspection, testing and commissioning process, step-by-step.

REQUIREMENTS FOR COMMISSIONING ELECTRICAL SYSTEMS

A new electrical system must be commissioned to confirm that it is safe, in good working order and ready for use. Commissioning consists of verification, inspection and test. Before we go any further, however, it is a good idea to clarify the three categories of inspection and test.

- Initial verification – commissioning inspection and test for new electrical installations, carried out during, and on completion of, the electrical work. This applies equally to the installation of a single new circuit and to a major construction project such as the Olympic Stadium.
- Minor Works Certificate – inspection and test carried out on a modification to an existing circuit.
- Periodic inspection and test – inspection and test to ascertain the condition of an existing installation.

Table 8.1 describes the components of commissioning a new electrical installation.

Type of inspection and test	Description	Documentation
Verification	Carried out during construction to ensure that wiring, enclosures and accessories are being correctly installed.	
Inspection	An examination of the completed electrical installation. This does not require test instruments.	Schedule of inspections
Test of electrical installations	A procedure, using test instruments, for establishing that a new electrical installation (from a single new circuit to a complete installation) is safe.	Electrical Installation Certificate Schedule of Test Results
Minor works	Carried out on completion of an extension or alteration to an existing circuit.	Minor Works Certificate

Table 8.1: Components needed when commissioning a new electrical installation

The regulatory requirements for commissioning of electrical systems

Electricity at Work Regulations, 1989

The overall regulations relating to electrical systems are the Electricity at Work Regulations, 1989. These regulations state that all connections should be sound, that equipment should be earthed, and that equipment within an electrical system should be suitable for its environment, kept in good condition and maintained regularly.

BS 7671:2008 IET Requirements for Electrical Installations

Part 6 of BS 7671:2008 is very clear about the need to inspect and test electrical installations. It describes the process, states the acceptable readings and findings, and lists the test procedures in the order in which they should be completed. Requirement 134.2.1 of BS 7671:2008 states that all new installations must be inspected and tested and that the purpose of these tests is to make sure they meet the requirements of BS 7671:2008.

IET On-Site Guide

The electrician's handbook, the *IET On-Site Guide*, is a stripped-down practical version of BS 7671. It uses plain language and clear diagrams to explain the main points of the requirements and how they apply to the work itself. There is also information in the Guide which does not appear in BS 7671, such as conduit and trunking capacities and factors that can be added to the maximum earth fault loop impedance (Zs) values. The *On-Site Guide* shows the main testing routines using a set of colour diagrams.

Part P of the Building Regulations

The Building Regulations 2013 cover all aspects of building work in domestic premises. Part P refers specifically to electrical installations. Sections 1 to 3 of these regulations require an electrical installation to be safe, and that all electrical installation work be tested and inspected. They also require:

- test certificates to be completed
- labels to be placed at earth bonding points, consumer units and RCDs
- instructions and log books and detailed drawings for electrical equipment to be kept and made available.

Chapter
8

It describes non-notifiable work. Non-notifiable works are:

- alterations or repairs to existing circuits
- modifications to an electrical installation that is not in a special location or on the regulation's list of notifiable work.

This work does not have to be carried out by a qualified electrician. However, it must be checked for safety. This can be carried out either by the electrician, if you engaged one for the job, or by hiring an electrician to carry out the check if you did the work yourself.

Notifiable work includes:

- the installation of a new circuit
- the replacement of a consumer unit
- addition or alteration to existing circuits in a special location.

The regulatory requirements for inspection of electrical systems

Inspection is the first stage of the commissioning routine. It consists of a set of checks, carried out with the power off but without the use of test instruments. The main regulatory requirements for the inspection of electrical systems are found in BS 7671:2008 and IET Guidance Note 3.

BS 7671:2008 IET Requirements for Electrical Installations

The electrician's 'bible', BS 7671:2008, states that all electrical systems must be inspected during installation and on completion. Part 2 defines inspection as an examination of an electrical installation, and one that uses most of the human senses. Part 6 of BS 7671:2008 states that inspection must be carried out while the supply is switched off. Part 6 lists:

- the inspections and tests to be carried out
- the documentation to be completed.

According to Part 6, inspection is carried out to verify that the electrical equipment that has been installed is not damaged or faulty, and complies with the relevant British Standards (as well as the equivalent international standards).

IET Guidance Note 3

The IET have published a set of guidance notes which cover all aspects of electrical work. Guidance Note 3 is specific to test and inspection and is a clear, readable, practical guide to carrying out all stages of the process.

Guidance Note 3 calls the inspection of new work 'initial inspection'. It states that inspection should be carried out at all stages of the installation work. It lists and describes the inspections to be carried out. The guidance note also contains an inspection checklist. This is an exhaustive list of every item that should be inspected, including:

- switchgear
- general wiring accessories
- lighting controls and points
- junction boxes and terminations
- cables
- conduit
- trunking.

These inspections are listed and described in detail later in this chapter.

The regulatory requirements for testing electrical systems

Testing follows inspection. This part of the procedure is carried out using test instruments and includes readings taken while the installation is live.

BS 7671:2008 IET Requirements for Electrical Installations

Part 2 of BS 7671:2008 defines testing as proving the effectiveness of an electrical installation, including the use of instrument readings. Test results must fall within the acceptable values given by BS 7671:2008. There are a number of tables in the requirements which give these minimum and maximum values.

- Maximum earth fault loop impedance (Zs) for various types of protective device – Tables 41.2 to 41.4.
- Maximum earth fault loop impedance (Zs) for non-delayed RCDs – Table 41.5.
- Part 6, Table 61 gives the minimum values of insulation resistance for circuits of various voltages.

IET Guidance Note 3

Guidance Note 3 lists the tests in the sequence in which they should be carried out and describes them in great detail. It also contains diagrams showing correct connections and test methods. It includes test routines for some of the rarer installations such as those with non-conducting walls and floors.

PROCEDURES FOR THE INSPECTION OF ELECTRICAL SYSTEMS

Inspection must be carried out by a competent person who has been trained and qualified in the process. Because of potential inconvenience and hazards to the customer, particularly if inspection is to be carried out in premises that are in use and occupied, careful preparation and good communication is needed. The extent and list of inspections to be undertaken will depend on the type of installation.

Procedures for preparing for inspection

You will not simply be able to walk into premises, switch off the power and start inspection. Equipment may be in use and you may not be able to switch it off for safety or commercial reasons, for example a respirator in a hospital. Another factor is that some areas of the premises to be inspected may be dangerous, or subject to restricted access. Inspection has to be organised and arranged in a considerate and formal way so that all parties know exactly what is going to happen, when and where, and how it will affect them.

Contact with client

The client is the person who has requested, and is paying for, the inspection. It is essential, therefore, that good relations are established and maintained with the client. The timing and logistics of the inspection must be discussed and agreed. Perhaps a formal timetable can be drawn up so that all parties know when you, as the inspector, will be working in each area. There must be clear agreement and understanding, ideally in the contract for the work, of the action to be taken if any faults are uncovered. Does the client want them repaired by the inspector or left for another electrical contractor to put right?

Progress check 8.1

1. What do the Electricity at Work Regulations say about the live parts of an electrical system?
2. Which part of the Building Regulations deals with the electrical installation in a domestic property?
3. Which set of guidelines could be described as the electrician's handbook?

LO2

Arrange isolation timings

It must be made clear to the client that power will need to be shut down, sometimes for long periods of time. Isolation of electrical supplies must be negotiated because it would be both wrong, and possibly disastrous, to switch off any supplies without warning. Sudden power shutdown could cause loss of data and work in an office equipped with computers, damage to industrial machinery and even pose a safety hazard.

Safe and orderly isolation of individual circuits may not be easy to achieve if the distribution board is not correctly labelled, or, as is often the case with older installations, not labelled at all. It can be easy to switch off the wrong circuit breaker. The client needs to know this and to prepare his or her staff accordingly.

Power shutdown must be undertaken using the safe isolation procedure which includes measures to stop anyone switching the power back on again. Fuses must be taken with you, circuit breakers locked-off and the distribution board and work area placed behind barriers.

Range and limitations of inspection

The extent of the inspection has to be agreed between the inspector and the client. Will it be a full inspection of all parts of the electrical installation? Are there any areas of the premises which the electrician cannot enter, and therefore cannot properly inspect? Is there sensitive equipment that cannot be easily isolated?

There may be alarm or control systems that require specialist inspection, or equipment and systems that are still under warranty and, therefore, cannot be dismantled without voiding that warranty.

Once any limitations are agreed, they are entered onto the inspection documentation.

Gather information (client, test results, certificates)

Any information about the installation will be extremely useful. The sort of information you will need includes:

- previous test documents and results
- fault reports
- wiring, circuit and layout drawings
- manufacturer's information
- details and information about any specialist systems such as alarms.

This will highlight any previous problems with the installation and anything you must consider or look out for during your own inspection. Wiring and circuit diagrams could help you to trace the full extent of circuits, which, in turn, could help ensure that the correct supply device is operated when attempting to isolate the circuit.

Prepare safe

All electrical work is hazardous. Electric shock, fire and burns are very present dangers, as well as the risks involved in working at height, in confined spaces and near machinery.

Some sobering statistics are provided by the HSE, the most recent being for 2011 to 2012.

- 173 workers killed at work.
- 111,000 other injuries to employees were reported under RIDDOR.
- 212,000 over-3-day absence injuries occurred.
- 27 million working days were lost due to work-related illness and workplace injury.
- Workplace injuries and ill-health (excluding cancer) cost society an estimated £13.4 billion in 2010/11.

The statistics below are for electricity-related work accidents for 2011 to 2012 reported under Electricity Safety, Quality and Continuity Regulations 2002.

	Fatal injuries	Non-fatal injuries	Near misses	Fires/explosions
2011/12	15	556	11,714	376

General health and safety when working in the electrical industry is described in *Chapter 2: Health and safety in building services engineering*. However, the following health and safety considerations apply to inspection and testing and you need to consider them at all times.

Great care must be taken when working on an existing installation. Distribution boards and consumer units are not always marked up with circuit identification so it is not always easy to isolate the circuits you need to work on. An added hazard is well-meaning people who might inadvertently restore power without realising that electricians are working on the circuit.

The safe isolation procedure must always be deployed when isolating a circuit or a distribution board, and only approved voltage indicators used to confirm that a circuit is dead.

A further hazard for electricians on inspection is the presence of back-up supplies such as generators which will automatically come online if there is a power failure. There are also back-feed hazards from highly inductive and capacitive circuits as well as from photovoltaic supplies which produce electricity whenever there is light and need to be properly isolated before work begins.

Safe working

During the safe isolation procedure, remember to remove fuses from the distribution board so no one can put them back, and to lock circuit breakers in the 'off' position.

Working practice 8.1

A periodic inspection was arranged for one of the office areas in a large UK research laboratory. The inspection was planned for the following weekend. A memo was sent out by email from the department manager to her staff to inform them that there would be a power shutdown for four hours on that date.

One member of her staff, however, an engineer named Dave, was off sick. When he came back to work he was inundated with emails and only read those he thought relevant to his job. Because of this, he did not read the shutdown memo. Neither did he read the shutdown notice that was pinned to the office notice board. Dave was very good at his job, but *his* job was his main focus and he paid little attention to the other activities going on in the department.

Continued ▼

Inspection day arrived and the electrician, Nathan, went straight to the service cupboard, unlocked the door, located the relevant distribution board and, using the board's circuit schedule, found the 100 A fuse that protected the sub-main supply to the office area. He pulled it out and laid it on top of the distribution board, then went into the office area to start the inspection. His first job was to remove the front cover of the office distribution board to confirm cable identification and connections.

Meanwhile Dave, who had come in to the office to catch up, was outraged to see his desktop computer suddenly shut down and most of the morning's work vanish. He was on his feet in a moment, angry and determined to rectify the problem. Thinking of himself as something of an electrician he went off to find out what was happening. On his way out, he saw that the service cupboard door was open. When he went inside he noticed the exposed distribution board and the fuse. He opened the board, retrieved the fuse and pushed it back into its carrier.

Fortunately Nathan noticed the office lights flicker back into life and jumped back from the distribution board he was working on without receiving a shock. But how could the near-accident have been avoided?

- List the precautions that should have been put into place to prevent this type of scenario occurring.

Fire and burns

Electrical shock can cause burns, some of which are severe and result in long-term damage. Short circuits are also a cause of burns because they produce an eruption of intense heat, and because the blast may scatter molten debris in all directions. Eye protection is a good idea when working live.

General injury

As well as the potential for electric shock, there is, of course, a risk of general injury. There are health and safety documents and guidelines for all aspects of working. It is essential that you are familiar with their main requirements. The right equipment must be used for the job, and be of a good quality. The appropriate PPE should always be available, and used.

Safe systems of work

There are working systems available which reduce the risk of accidents in the workplace and minimise the hazards you face when inspecting and testing.

- Risk assessment – should be drawn up to highlight hazards, the risks they pose and how they will be controlled. *Chapter 2: Health and safety in building services engineering* describes risk assessment in more detail.
- Permit to work – an authorisation for carrying out particularly hazardous work. This must be approved and signed by someone in authority and who understands the job and its associated risks. While all information entered onto the permit is relevant, the estimated finish time has particular importance. It can be used as a prompt for the person in charge to come looking for you and to check that you are all right if you overrun your time. A permit to work can only be used once.

Progress check 8.2

1 Apart from shock, what is another risk for an electrician working on a live circuit?

2 Why is a photovoltaic supply a hazard for an electrician carrying out an inspection and test?

3 What are safe systems of work for inspection and testing?

How human senses can be used during the inspection process

Among the best instruments for inspection are the human senses. We use these all the time, for example when crossing a road. Looking and listening for oncoming traffic is an instinctive action. So, before unpacking any test meters from your tool kit, use your eyes, ears and sense of smell and touch to detect any problems, evidence of bad practice or the telltale signs of a fault.

Sight

Looking at the installation and its components reveals a lot about that installation. The neatness of the work can be telling. If someone has taken the time to make sure everything is straight, level and neatly arranged then they probably took care over the connections and wiring. The characteristics of the installation will also be ascertained by looking: the type of earth system used, the wiring systems, whether it is a single- or three-phase supply, the presence of equipotential bonding.

Can you see any obvious problems or examples of poor installation and bad practice? Are the conductors identified using the correct colour scheme? Are the right accessories and fittings being used for the job? Is there labelling?

Touch

Touch is the second most used sense when carrying out inspection of an electrical system. The soundness of terminations can be checked using the tug-test, in other words pulling on the conductors to make sure they don't drop out of their terminals. Tracing individual cables is often done by feel. The soundness of fixings, lids and covers will be checked using your hands. Then there is touch itself – does anything feel hot; is there excessive vibration?

Smell

The sense of smell is perhaps not used as often during inspection, but it can tell us if a cable or component is running hot because it will detect the smell of burning. PVC gives off a particular odour when it gets too hot, as do the insulating chemicals on motor and transformer windings. This sense can also pick up on the risk of fire from hot-running electrical equipment and fittings such as the scorching smell given off when heaters and lamps are placed too close to flammable materials or surfaces.

Speech

You will need to talk to people who have installed or who are using the electrical system you are about to inspect. Power shutdowns will need to be agreed; moving staff or equipment out of a work area may have to be arranged. Premises may have to be closed to staff or public, or the work completed out-of-hours. Updates on the progress of the work and clarification of what it involves should be discussed. Finally, ask the client or person in charge of the building if they have noticed any problems with the installation.

Hearing

Hearing has two main uses during inspection.

- Listening to the customer or users of the installation, or electricians if commissioning a new system – they may need to tell you about problems they have experienced with the installation or use of the system.

- Listening for evidence of problems such as:
  - chattering – a coil is not holding in a set of contacts firmly
  - vibration – problems with the windings in a transformer or motor; problems with the mechanical components of an electric motor
  - crackling sounds – evidence of burning or a loose connection.

Justify choice of applicable items on an inspection checklist that apply in given situations

The checklist of inspections, which forms part of the inspection and test documentation, covers every conceivable type of electrical system. This means that not every inspection on the list will need to be carried out for every installation, for example 'presence of undervoltage protection'. Undervoltage protection is only needed if there are electric motor controls present, so will probably not be included in the inspection of a domestic or office installation.

Another factor is accessibility to certain components or a particular area of an installation. There may be hazardous areas, not accessible to the inspector, or specialist electrical systems such as fire and intruder alarms, which require specialist testing.

> **Progress check 8.3**
>
> 1 Give two inspections that can be carried out using touch.
> 2 Give two inspection checks that can be carried out using the sense of hearing.
> 3 Why would an inspector complete all the inspections listed on the Schedule of Inspections?

PROCEDURES FOR COMPLETING THE TESTING OF ELECTRICAL SYSTEMS

Once inspection has been completed, testing can begin. This is carried out using instruments which will give readings and indications. Checking these results against the acceptable values stated in BS 7671:2008 will confirm that the installation is safe and working correctly. Testing is in two parts.

- Dead tests – carried out with the power off.
- Live tests – carried out with the power on.

The result of the test and any relevant comments are entered onto the Schedule of Test Results.

Why regulatory tests are undertaken

While inspection is a valid and necessary part of the verification procedure, it can only give a limited view of what is going on in the electrical installation. A set of tests must be carried out to establish that cables and equipment are working correctly and safely and that the protective systems, such as circuit breakers and fault protection, will operate if needed.

Verify continuity of conductors (circuit protective, earthing, bonding, ring final)

The protective conductor is also known as the earth conductor and its purpose is to conduct any earth fault currents away from faulty equipment and down to earth. Because the protective conductor does not normally carry current it has to be tested to ensure there are no breaks, either from damage or loose connections.

Insulation resistance

Insulation resistance testing proves the integrity of the cable insulation. An acceptable reading will show that there are no short circuits between:

- line and neutral
- line and neutral and earth
- the phases in a three-phase system.

Polarity

Correct polarity means that all the conductors are connected into the correct terminals and the circuit is, essentially, the right way round and that switches and protective devices are connected into the line conductor and not the neutral. Many items of electrical equipment are not 'polarity-conscious', which means that they will operate normally whichever way round the line and neutral feeds are connected into them. However, it will cause electric motors to run in the opposite direction to the one intended. While this will not damage the motor itself, it can damage the equipment it operates.

Incorrect polarity can pose a shock hazard by making parts of an accessory, normally connected to neutral, into a line part. An example of this is the Edison screw lampholder. The threaded and more easily accessible part should be connected to neutral, while the less accessible connection point at the back of the lampholder should be connected to line.

Earth electrode resistance

Installations protected by a TT earthing system are connected to the mass of earth via an earth electrode. The resistance between the electrode and the earth itself must be measured to ensure that there is a good connection between the installation earthing system and the mass of earth.

Earth fault loop impedance

The purpose of the earth fault loop impedance test is to measure the impedance of the earth fault path (Zs) and ensure that it is within acceptable limits. If it is, then the earth fault loop path will provide a low impedance route for an earth fault current to get to the mass of earth, via the star point of the sub-station transformer. It will also provide a low impedance route for the prospective fault current to flow from the transformer to the point of fault, via the protective device which will operate and cut off the supply. This is a live test which means it is a test carried out with the power on. Because of this, care has to be taken as there is a risk of shock.

Prospective fault current

The prospective fault current is the current that will flow in the event of a fault (the current flowing from the sub-station transformer to the point of fault, as described in the earth fault loop impedance test section above). The reading should be taken at the installation supply point using a specialist instrument, or an earth fault loop impedance tester set to the prospective fault current setting.

Correct operation of RCDs

The RCD provides additional shock protection and is designed to operate in the event of an earth fault. BS 7671:2008 requires that many types of circuit

Figure 8.1: DOL starter

should include an RCD, particularly ring final circuits, showers and those in special locations (see *Chapter 7: Electrical systems design*). It is essential that the RCD operates correctly and at the minimum fault current, usually 30 mA, although 500 mA RCDs are installed in some situations. The RCD must not operate *below* its minimum fault current (known as 'nuisance-tripping') and this also has to be confirmed as part of the RCD test routine.

Functional testing

The functional test is simply a check to make sure everything works correctly and safely. Included in a functional test are:

- switchgear
- controls
- interlocks
- undervoltage protection – built into motor controls so that if there is a power failure the motor will not restart automatically when power is restored.

Phase rotation

In a three-phase circuit the three voltages must reach their sine wave peaks in a set order: L1, L2 and L3. If they are swapped over it will cause three-phase motors to run in the reverse direction, damage machinery and possibly electrical equipment. It also poses a safety risk. A phase rotation meter is used to indicate correct phase polarity and rotation, either by a set of lights or by LED display.

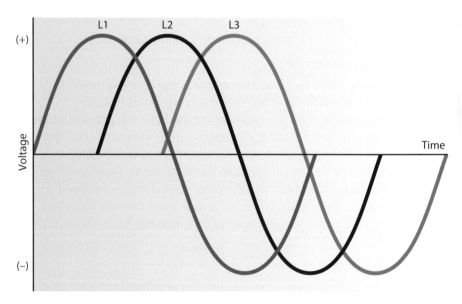

Figure 8.2: Three-phase sine wave showing the order of rotation

Verification of voltage drop

Voltage drop is an effect in a conductor similar to water pressure in a hosepipe. The longer the conductor, the lower the voltage measured at its furthest end. Because of this, cables must be selected to carry the required current all the way to the load without a voltage drop of more than 9 per cent.

Verification of voltage drop is not usually carried out as part of the test routine and the only practical ways of doing so are to:

- take voltage readings at the supply and load ends of the cable run
- check that the appropriate calculations were carried out during the design stage of the installation.

Confirmation of automatic disconnection of supply for earthing arrangements (TT, TN-S, TN-C-S)

Because fault protection is necessary to guard against electric shock, and therefore danger of injury or death, it is essential that the supply is automatically disconnected in the event of a fault. Confirmation of automatic disconnection is achieved by a number of tests. These are:

- continuity of protective conductors
- continuity of bonding conductors
- earth fault loop impedance test
- compliance with the maximum Zs values found in Tables 41.2 to 41.4 of BS 7671:2008.

Progress check 8.4

1 Why test for continuity of protective conductor?

2 Why carry out an insulation resistance test?

3 Why test for polarity?

Why regulatory tests are carried out in the exact order as specified

As part of the regulatory requirements for testing, the tests must be carried out in a certain order. This is because insulation resistance testing is invalid if the circuits are not whole and complete, which is why continuity tests are carried out first, and it would not be safe to switch on the power unless insulation resistance is checked before doing so.

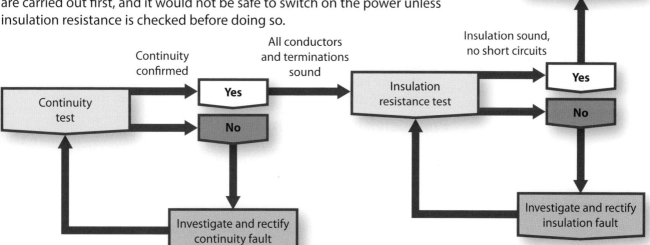

Figure 8.3: The order for testing

How to prepare for testing electrical systems

The main steps leading up to a test are similar to those for inspection (see pages 397–401). The main points are listed below.

- Safe system of work – care must be taken at all times when testing and a risk assessment should be drawn up to detail possible hazards and the control of potential risk. Permits to work should be raised to cover particularly hazardous tasks (see *Chapter 2: Health and safety in building services engineering* for more details).

- Safe isolation – because many tests have to be carried out while the circuit or supply is switched off, the safe isolation procedure must be followed to ensure that power is not inadvertently restored while work is in progress.
- Instrumentation fit for purpose – always use the right test instrument for a particular test. Make sure the setting is correct for the test. Test instruments must comply with the requirements of HSE GS38. Some of the main points are that:
  - o leads must be undamaged and clearly identified using different colours
  - o finger guards should be fitted to the probes to prevent electric shock
  - o probe tips must be between 2 mm and 4 mm across any surface
  - o in-line high rupture capacity (HRC) fuses must be fitted to voltage indicators
  - o instruments must be calibrated by an authorised calibration company once a year.
- Communication with clients – good relations should be established and maintained because power shutdowns have to be arranged, progress reported and any problems raised and discussed. Courtesy, honesty and clarity are vital.
- Range and limitations – the range of tests will depend on the type of installation and the systems it contains. There may also be limits to the tests that can be carried out, for example sensitive equipment in circuits which will be damaged by insulation resistance testing, fluorescent luminaire circuits for which line-to-neutral insulation resistance tests will not be possible.
- Implications for others – be aware that other people in the premises will be affected by testing. Power will be lost, access to certain areas may be restricted by barriers, obstacles such as access equipment may arise and there may be tripping hazards. Safe working practices are vital to prevent injury and careful negotiation must be undertaken to maintain good relationships with the customer and people in the vicinity.

Procedures for regulatory tests

The tests themselves are described later in this chapter, but it is important that the tests are carried out in the order specified by BS 7671:2008. They must also be carried out in the manner described in IET Guidance Note 3. The order of tests is:

1 continuity of protective conductors, main and supplementary bonding conductors

2 continuity of ring final circuits

3 insulation resistance

4 protection by SELV, PELV and electrical separation*

5 protection by barriers or enclosures*

6 insulation resistance of non-conducting floors and walls*

7 polarity

8 earth electrode resistance (TT systems only)

9 protection by automatic disconnection of supply

10 earth fault loop impedance

11 RCD test

12 prospective fault current

13 phase sequence (three-phase circuits and supplies only)

14 functional testing

15 voltage drop*.

*Not commonly carried out because they are not required or not generally in use as part of a fixed electrical system.

Implications of non-compliance of regulatory test results with regulatory values

Any test result that is not within the parameters set out by BS 7671:2008 is essentially a 'fail'. The result, and also comments about the problem and its severity, need to be entered into the Schedule of Test Results. Who actually repairs the fault is down to the contract drawn up for the work. However, once it is repaired the circuit must be re-tested. This includes a repeat of all the tests leading up to the discovery of the fault.

Shock

When electric current passes through a human body it results in electric shock. The minimum current that can cause injury and possibly death in a 230 V a.c. system is approximately 40 mA (0.04 A), and even lower if the path of the shock current takes it through the heart. Voltages of 230 V and above can result in serious burns due to tissue breakdown caused by dielectric effect.

The earthing system of an installation is designed to divert any fault current away from possible human contact and cause the operation of the protective device. To do this successfully, the impedance of the earth fault loop must be low enough to allow a clear path from the fault to earth, and from the supply transformer back to the point of the fault, via the protective device which should operate. To ensure that this happens, all parts of the earthing system must be securely connected and continuous. Earth conductors must also be the correct size. Earth fault loop impedance values are listed in Tables 41.2 to 41.4 of BS 7671:2008.

Fire

Overloading of cables, loose connections and high ambient temperature can cause fire. Visual inspection should reveal any evidence of fire hazard; continuity testing will show up high resistances which indicate a loose connection.

Heat-producing electrical equipment should also be inspected to make sure it does not pose a fire risk, that there are the required spaces around the equipment and that it doesn't touch any flammable surface or material; also that it is the correct equipment for the job.

Thermal imaging is sometimes used as part of the visual inspection to detect high resistance joints. This is not a common inspection method and not required by BS 7671:2008. However, it can be a useful tool in complex electrical systems or as an inspection method when systems cannot be switched off.

Chapter 8

Progress check 8.5

1 What can limit the extent of testing you can carry out in premises?

2 What are the first three tests to be carried out?

3 List three ways in which electrical wiring or equipment can pose a fire risk.

Burns

Burns, of course, result from fire. They will also be the result of electric shock and contact with heat-producing electrical equipment such as lamps. Again, visual inspection should reveal any burn risks in an installation.

Factors that affect insulation resistance values

Length

If an insulation resistance test is carried out on a long cable there is a chance it will give a reduced reading because there is more opportunity for current to flow between the cable cores. Some of this is due to naturally occurring holes in the insulation which allow electron flow. These holes will increase in size with time.

Parallel circuits

The reading will be lower when measuring resistance in a parallel circuit due to the effect of equivalent resistance in a parallel circuit (see *Chapter 6: Principles of electrical science*).

The requirements for the safe and correct use of instruments to be used for testing

Table 8.2 shows the test instruments used for the inspection and testing of electrical circuits. HSE Guidance Note GS38 applies to test instruments and describes what makes an electrical test instrument safe. See page 454 for the main points of GS38.

Instrument	Description
Voltage indicator 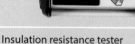	The voltage indicator is not intended to measure actual voltage but to show if a voltage is present. This is typically a lamp and two probes which are connected to line and neutral, or to line and earth. • Do not use a neon screwdriver for this task because they are unreliable and dangerous. • Do not use a cheap version. An electrician's life may depend on this piece of equipment!
Low reading ohmmeter 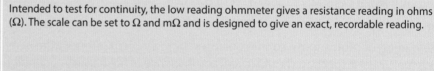	Intended to test for continuity, the low reading ohmmeter gives a resistance reading in ohms (Ω). The scale can be set to Ω and mΩ and is designed to give an exact, recordable reading.
Insulation resistance tester 	The scale on an insulation resistance tester is usually set to MΩ (megohms 1 MΩ = 1 million Ω). This is because the insulation resistance test is used to check that there are no short circuits or faults to earth caused by failed or damaged insulation. The lowest acceptable reading for most circuit types is 1 MΩ. Readings are taken between: • line and earth • neutral and earth • line and neutral. Insulation resistance testing is usually carried out at 500 V d.c. so care must be taken to ensure all sensitive equipment is unplugged or isolated before carrying out the test.

Instrument	Description
Earth electrode tester	This is a three- or four-terminal resistance meter designed purely to take earth electrode resistance measurements.
Earth fault loop impedance tester	The earth fault loop impedance tester measures the path from the point of an earth fault, along the earth conductor to the star-point (and then to earth) of the supply transformer, then back along the supply line to the point of fault. Only a low reading that complies with the maximum earth loop impedance stated in BS 7671:2008 is acceptable.
Prospective fault current tester	Usually incorporated into the earth loop fault impedance tester, the prospective fault current tester measures the maximum fault current that will flow in the event of a short circuit or earth fault. The reading will be high – typically thousands of amps (kA).
RCD tester	Any circuit protected by a residual current device (RCD) has to be tested. The sequence of tests that have to be carried out are: • 50% of the RCD operating current – RCD should not operate. This proves that the RCD will not 'nuisance-trip', in other words will not trip when there is no fault • 100% of the RCD operating current – RCD should trip within stated operating time • 150% of tripping current – RCD should trip more quickly than stated operating time.
Phase rotation meter	In a three-phase system, the connections should be such that the three phases will rise and fall in this order: L1 – L2 – L3. If this order is reversed in any way then three-phase motors will run in reverse and damage could be done to both mechanical and electrical equipment. The phase rotation meter will indicate correct, or incorrect, rotation using lights or an LED screen.
Plug-in socket tester	Plug-in testers display the correct operation of a 13 A socket. Combinations of green, red or similar indicator lamps (and sometimes a buzzer) give a quick confirmation of correct operation or presence of certain faults. Plug-in testers should be used only as indicators and not as a recordable test or fault-finding tool.
Plug-in test leads	Plug-in test leads can be as simple as a 13 A plug with its flex terminated into a connector block. They are intended as a method for taking readings at a 13 A socket without having to remove the socket's faceplate.

Table 8.2: Test instruments used for the inspection and testing of electrical circuits

---

Progress check 8.6

1 Which test instrument is used to confirm continuity of a conductor?

2 Which test instrument is used to confirm the soundness of cable insulation?

3 Which test instrument measures the impedance of the fault path from point of fault to star-point, then back to the point of fault?

Chapter
8

 LO4

REQUIREMENTS FOR DOCUMENTING INSTALLED ELECTRICAL SYSTEMS

Once complete, inspection and testing has to be signed off. This signifies that the installation is working correctly and is safe to use. Some of the documentation is completed during the inspection and test process. Other parts form the final approval and sign-off. Copies have to be distributed to various parties, such as:

- inspector
- client
- architect.

The purpose of certification documentation

Electrical Installation Certificate

Although the Electrical Installation Certificate is the final sign-off for the test and inspection, it is often completed first as it sits at the front of most inspection and test document sets.

Typical certificate headings are shown in Table 8.3.

Heading	Description
Installation and client details	Name of client and address of installation under test.
Extent of installation	Brief description of the installation and the part of the installation to be tested. If the building is completely new then the whole installation will be tested. If it is an addition or alteration to an existing installation, then only the new part will be tested.
Sign-offs for: • *design* • *construction* • *inspection and testing*	The signatures of the parties concerned. This shows approval and acceptance of the results of the inspection and test by the designer, the person responsible for installation of the electrical system, and the inspection and test electrician. There is also a field for the details of these signatories.
Next inspection	The date on which the installation will next be tested and inspected. This will be a periodic inspection.
Supply characteristics	The details of the installation supply are split into the following five headings. 1 Earthing arrangements – the earthing systems protecting the installation, e.g. TT, TN-C-S or TN-S, etc. 2 Number and type of live conductors – the supply system to the installation: single-phase, three-phase (three-phase only or three-phase and neutral), or two-phase. Also, whether it is a.c. or d.c. 3 Nature of supply parameters – this is: • the supply voltage, i.e. single-phase 230 V, three-phase 400 V • frequency (the normal mains frequency of UK is 50 Hz) • prospective fault current • Z_e, which is the external section of the earth fault loop path. 4 Supply protective device characteristics – the type of protective device found at the source of the installation supply and its rating. For most domestic installations this will be a type BS 1361 100 A fuse. 5 Other sources of supply – are there any other sources of supply? For example, a generator or renewable energy supply such as a PV array or a wind turbine.

Heading	Description
Means of earthing	For all systems, except TT, this will be the supplier's facility. For a TT system this will be an installation earth electrode.
Maximum demand	The calculated demand for the installation. Diversity must be taken into account when calculating maximum demand.
Details of installation earth electrode	If there is an installation earth electrode, the type, location and its resistance must be entered.
Main protective conductors	There are two main protective conductors. • The earthing conductor is the main conductor that connects the installation earthing system to the supply earthing system. • The main protective bonding conductors are those which connect the main services, e.g. water and gas, to the installation, and, ultimately, the supply earthing system.
Main switch or circuit breaker	Details of the main multi-pole switching device at the source of the installation. Needed are: • number of poles switched (double-pole for single-phase and three or four for three-phase depending on whether there is a neutral) • location – where the switch can be found • current rating – for a circuit breaker this will be its operating current; for a switch, it will be the maximum current it can safely switch • fuse rating or setting – if it is a fuse, or fused-switch, then what is its rating? • voltage rating – the operating voltage of the switching device.
Comments on existing installation	If the new installation work is an addition or alteration to an existing installation, then the condition of the existing installation must be described in this field.
Schedules	Which schedules will be included in this test documentation set? (Usually the Schedule of Inspections and Schedule of Test Results.)

Table 8.3: Typical certificate headings and their descriptions

Electrical Installation Condition Report

An inspection and test routine carried out on an existing installation is called a periodic inspection and test. The main report derived from this type of test and inspection is called an Electrical Installation Condition Report.

Table 8.4 shows only the headings on a Periodic Inspection Condition Report that are *different* to those on the Electrical Installation Certificate. The report is signed by the inspector and a copy retained by him or her and by the person who has ordered the inspection to be carried out.

Heading	Description
Reason for producing this report	A periodic inspection may have been ordered by the client because of a fault, the need to locate the fault, and also to ensure the rest of the installation is safe and working properly. It may also be that the period between inspections has expired.
Details of installation	This differs from the Electrical Installation Certificate in so far as there are tick boxes for the installation type, e.g. domestic, commercial, etc.

Table 8.4: Headings specific to a Periodic Inspection Condition Report

Continued ▼

Heading	Description
Extent and limitation of inspection and test	There may be limitations on the inspection due to a number of reasons. For example, a part of the installation may not be switched off, it may be physically inaccessible or it may be feeding equipment sensitive to conventional inspection and test. It may be covered by a warranty, or be a specialist system such as an alarm circuit, which requires its own test and inspection routines.
Summary of condition of installation	This is a general comment on the installation. It is summed up with the word 'satisfactory' or 'unsatisfactory'.
Recommendations	This applies if the summary comment for the previous field is 'unsatisfactory'. The recommendation includes a set of codes which indicate if repair work should be carried out immediately, or can be sorted out at a more convenient time. The codes are described in the next field.
Observations	This is a list of observations made on the condition of the installation. These are coded to show the seriousness of the situation and if and when repairs need to be carried out. The codes are: • C1 – danger present, immediate remedial action required • C2 – potentially dangerous, urgent remedial action required • C3 – improvement recommended. This is applied to older installations that may not have RCD protection.

Table 8.4: Headings specific to a Periodic Inspection Condition Report

Schedule of Inspection for Periodic Testing

The main headings for a typical Schedule of Inspections for an existing installation are described in Table 8.5.

Heading	Description
Distributor's supply intake equipment	The condition of the service head, the supply intake position and the tails.
Presence of adequate arrangements for other sources of supply (such as micro-generators)	Are there alternative sources of supply and if so are they correctly installed and adequate for the job?
Earthing bonding arrangements	Are the components of an earthing system in place and what condition are they in?
Consumer unit(s)/distribution boards	What condition are the consumer unit(s) and distribution boards in? Are they fitted with multi-pole switches? Are there RCDs? Is there mechanical protection for the cables at the point where they enter the board? Is all required labelling in place? For example: • presence of non-standard cable colours • RCD quarterly test notice.
Final circuits	This is a long list, which includes: • segregation of Band I and II circuits • provision of fire barriers where cables and containment pass through walls • correct conductor colours • the right wiring system for the installation • correctly rated protective devices.
Locations containing a bath or shower	Bathrooms and showers are divided into zones. Each zone is suitable only for certain types of electrical equipment and supply. Is the correct bonding in place? Does the equipment have the correct IP rating?
Special installations or locations	These are described in Part 7 of BS 7671:2008 and include construction sites, agricultural installations, fairgrounds, caravan parks and medical locations. These need to be given a separate report as there are extra considerations for these types of installations.

Table 8.5: Headings for a Schedule of Inspections

Minor Electrical Installation Works Certificate

Minor work is described by BS 7671:2008 as alterations to an existing circuit. Only the altered circuit has to be inspected and tested and a Minor Electrical Installation Works Certificate completed. This is a single-sided document which contains a basic set of headings, including:

- description of the minor works – the work that was actually carried out as well as the location and the date of the work
- installation details – what is the earthing system? What are the methods of fault protection for the installation and for the circuit you have been working on? There is also a field for any comments about the earth bonding and arrangements
- essential tests – the tests required are:
  - o continuity of protective conductor
  - o insulation resistances
  - o earth fault loop impedance
  - o polarity
  - o RCD operation if applicable (the Minor Works Certificate only requires the RCD to be tested at IΔn)
- declaration – signature and details of the qualified inspection and test electrician approving the work.

Schedule of Inspections (EIC and ECR versions)

The Schedule of Inspections lists the items in an installation to be inspected. The exact inspection that will take place is determined by the installation type. A domestic installation, for example, will not have a non-conducting location or incorporate earth-free equipotential bonding. Every box on the Schedule must be marked with either a tick or 'N/A' (if the inspection item is not applicable to the installation). The full set of inspections is listed and described later in this chapter.

The responsibilities of personnel in relation to the completion of the certification documentation

IET Guidance Note 3 (GN3) lists the people who have responsibility with regard to the inspection and testing of an installation. These people are involved in the preparation of what GN3 calls the 'specification'. This specification must be prepared before the installation work starts and must contain detailed design and other related information. It must contain enough data to enable the installation to be tested and commissioned. Three of the roles involved are:

- designer – carries out the planning and calculations for a new electrical installation and draws up the specification. The designer is required to sign the Electrical Installation Certificate
- installer – the person who carries out the actual installation work. The installer should be a competent person, someone fully trained. The installer is also required to sign the Electrical Installation Certificate
- tester – the competent person who undertakes the inspection and test of the new installation. This has to be carried out throughout the entire installation process. Only a fully qualified tester should sign the Electrical Installation Certificate.

Progress check 8.7

1 On an Electrical Installation Test Certificate, what is meant by 'supply characteristics'?
2 What is maximum demand?
3 What is an Electrical Installation Condition Report?

The regulatory requirements for documenting electrical systems

All electrical installations must be supported by documentation, such as distribution board schedules, as-fitted drawings, manufacturer's instructions and manuals, and inspection and testing certificates, and schedules of results and inspections.

IET Requirements for Electrical Installations

BS 7671:2008 requires that 'appropriate documentation' is provided for every electrical installation (Regulation 132.13). Appendix 6 gives examples of commissioning, inspection and testing certificates.

IET Guidance Note 3

GN3 states that the original paperwork of an electrical certificate, whether Electrical Installation, Minor Works or Condition Report, must be given to the customer or person requesting the test. The copy is kept by the test electrician. GN3 also describes two different types of Electrical Installation Certificate.

- Form 1 – a single-signature form used for a smaller installation which was designed, installed and tested by the same person.
- Form 2 – a multi-signature form for larger works where the roles were carried out by different people.

If someone is signing the certificate on behalf of a company they must include the name of the company alongside their signature.

For condition reports used for periodic inspection and testing, standard inspection schedules should be used only for installations with a maximum current demand of 100 A. If the electrical system demands more than 100 A, the test electrician will need to customise the schedule to suit that installation.

Recording

Information about the installation and results gathered while carrying out inspection and testing must be recorded. The actual readings given by test instruments must be entered into the relevant fields, even if they do not fall within the acceptable limits given by BS 7671:2008. The fact that the measurement does not conform to BS 7671:2008 must then be entered into the notes section of the form. All fields must be filled in, even if they are not relevant to the particular installation under test, in which case 'N/A' or something similar should be written.

Retention

All inspection and test documentation must be kept by the relevant parties and be available to anyone working on the electrical installation. It can have legal implications and may have to be produced in court if there is a dispute involving the electrical installation.

Progress check 8.8

1. What does GN3 mean by a design 'specification'?
2. What are Form 1 and Form 2 Electrical Installation Certificates?
3. A standard inspection schedule for a periodic test can be used for an installation with a current demand of up to how many amps?

INSPECTING ELECTRICAL WIRING SYSTEMS

LO5

The regulatory requirements and preparation for inspection are described earlier in this chapter.

Implement safe system of work for inspection of electrical systems

Safe systems of work are described earlier in this chapter and also in *Chapter 2: Health and safety in building services engineering.* As a reminder, however, the main systems for safe working are:

- risk assessments
- permits to work
- safe isolation.

Record inspection of electrical systems

Inspection is the first part of an inspection and test process and is, as the name suggests, a check on the condition of the electrical installation. BS 7671:2008 and GN3 list the following items to be included in the inspection.

Connection of conductors

You must make sure that conductors are correctly and securely connected into their terminals and connection points. There should be no copper showing, stranded conductors must be complete with no strands removed and terminal screws should be tight.

Identification of conductors

Conductors must be identified, both by the standard harmonised colour scheme and, if necessary, by numbered or lettered tags.

Single-phase:

- line – brown
- neutral – blue
- earth – green and yellow.

Three-phase:

- line 1 – brown
- line 2 – black
- line 3 – grey.

If an existing installation is wired using pre-harmonisation colours, they do not need to be replaced. However, a warning notice must be displayed at the distribution board. The pre-harmonisation colours are as follows:

Single-phase:

- line – red
- neutral – black
- earth – green, or green and yellow.

Three-phase:

- line 1 – red
- line 2 – yellow or white
- line 3 – blue.

Blue conductors are used as switch-wires in loop-in lighting circuits, and grey and black conductors are used as strappers in two-way circuits. These conductors must be identified as line conductors using brown sleeving or tape.

If a number of conductors of the same colour are used in a complex circuit (e.g. a control circuit), the conductors should be identified at each end using a number or lettered tag.

Routing of cables

Cables should not be run through areas where there is a risk of mechanical damage or where there are extremes of temperature, unless they are designed to withstand these conditions.

Ideally, cables in a loft space should not be laid on top of the joists but pass through by means of holes or slots.

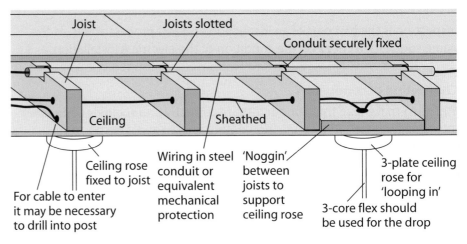

Figure 8.4: Cables running through joists

If run on the surface, cables should be clipped at intervals. The spaces between cable clips are shown in the *IET On-Site Guide* as follows.

- <9 mm diameter – horizontal: 250 mm vertical: 400 mm (2.5 mm^2 twin-and-earth is approximately 9 mm in diameter).
- 9 mm <diameter <15 mm – horizontal: 300 mm vertical: 400 mm.
- 15 mm <diameter <20 mm – horizontal: 350 mm vertical: 450 mm.
- 20 mm <diameter <40 mm – horizontal: 400 mm vertical: 550 mm.

There should be no clips on the bend of a cable and cable such as twin-and-earth must be fixed flat against the surface, not on its side. Cable bends should not be too tight as this puts enormous strain on the conductors inside the cable.

BS 7671:2008 lists things to be considered when re-routing a cable. These things are:

- presence of water or high humidity
- presence of foreign bodies such as dust
- presence of corrosive substances
- vibration.

Figure 8.5: The correct installation of cables

Selection of conductors

Chapter 2: Health and safety in building services engineering, page 131, describes the importance of selecting the correct size of cable for the current it has to carry. In some cases cable size is obvious while in others it may not be. For example, it can be difficult to tell how many cores are contained in a steel wire armoured cable, or their cross-sectional area. You may have to make enquiries or carry out the cable calculations described in Chapter 2, pages 149–153. A rule of thumb for cable sizes is:

- lighting circuits – 1.5 mm^2
- 20 A radial – 2.5 mm^2
- ring final circuits – 2.5 mm^2
- 30 A circuits such as a cooker – 6.0 mm^2.

BS 7671:2008 also warns against mixing metals that cause an electrolytic action when they are in contact. This effect results in an electrical charge between the two metals, with one acting as an anode and the other as a cathode. Molecules will migrate from the anode to the cathode and this causes deterioration and corrosion.

Activity 8.1

In the course of making additions to an existing lighting circuit, an electrician opened up a two-way switch at the bottom of a staircase. The lighting was wired using the loop-in method. The cable that runs between the two switches in this type of system is normally a three-core and earth (see Chapter 3), which gives the facility for two strappers and a common, needed to make the arrangement work. In this case a twin-and-earth had been run between the switches: brown as common, blue as one strapper, the CPC – identified and insulated using brown sleeving – as the other. Why is this wrong?

Connection of single-pole devices

Single-pole devices are those which connect and disconnect the line conductor only. Examples of single-pole switches are light switches and circuit breakers. Single-pole devices should NOT switch the neutral. If they do, the line supply to the light or other equipment may still be energised and pose a risk of electric shock even when the switch is in the off position. When inspecting an electrical installation, make sure that it really is the line conductors that are being switched by these devices.

Correct connection of accessories and equipment

The terminal points of electrical equipment are usually labelled to identify which conductors go where. The simplest identification is the L, N and E markings on a terminal.

Presence of fire barriers

Where a conduit, trunking or cable passes through a wall, the hole must be repaired around it. This is a requirement of the Building Regulations, Part B Fire Safety and is intended to prevent fire from spreading from one part of a building to another via an unrepaired hole. The filler must be of the same fire-resistant material as the wall. The inside of the trunking must also be filled with a fire-resistant material, although not one that will cause damage to the cables.

Figure 8.6: A distribution board door

Protection against electric shock

There are a number of ways of preventing electric shock. These must all be checked for effectiveness. The methods to be inspected are listed below.

- Basic protection – protection against shock under normal conditions. This would include insulation and enclosures.
- SELV and PELV – separated extra-low voltage systems in which the load is completely isolated from the supply via an isolating transformer.
- Double and reinforced insulation – is there any damage? The integrity of this insulation is your protection against shock, so it must be checked thoroughly to make sure it is not damaged or weakened in any way.
- Barriers, obstacles and placing out of reach – these are alternative methods of shock protection. Examples are as follows.
  - Placing out of reach – high voltage, insulated cables kept out of reach using tall poles or pylons. Out of reach means beyond arm's length, considered to be 2.5 m by BS 7671:2008.
  - Barriers – the door of a distribution board is an example of a barrier. Once this is open you can touch live parts inside the board.
- Automatic disconnection of supply – the parts of this system to be inspected will be:
  - earthing conductor
  - circuit protective conductors
  - protective bonding conductors
  - supplementary bonding conductors
  - functional earths and their monitoring devices
  - FELV – functional extra low voltage
  - protective devices such as circuit breakers and RCBOs.
- Locations – these are unusual specialist installations such as earth free and non-conducting locations. The conductive parts of these places must be bonded together to maintain them all at the same electrical potential, even if they are not connected to the installation earth.
- Additional protection – RCDs. Are they present and do they work when the test button is pressed?

Prevention of mutual detrimental influence

The electromagnetic field around a low or high voltage cable can interfere with the function of the other lower voltage or data-type circuits. When these circuits are installed in the same trunking they must be segregated into different compartments. Multi-compartment trunking has been available for many years and is an economical way of routing all cables through a single enclosure.

If it is unavoidable to mix data, communication or extra low voltage circuits with mains voltage circuits, then all the conductors must be insulated to the same standard as the cables carrying the highest voltage. An example of this is a multi-core cable in which data and low voltage conductors are contained within the same sheath.

Presence of appropriate switches and isolators

Although they both work by opening and closing a circuit, switches and isolators perform different functions within an installation.

A switch is designed to open and close under normal operating conditions, in other words while current is flowing. Light switches, time switches, thermostats and circuit breakers are examples of switches.

An isolator is designed to completely disconnect all live conductors (line and neutral) from the load. It is not designed to be operated while current is flowing. All items of electrical equipment must have an isolator. This is in addition to the equipment's own on–off switch, which does not isolate the supply but only stops the equipment itself from operating. BS 7671:2008 requires that isolators:

- are installed and designed so that they cannot be inadvertently switched on or off
- are lockable if remote from the equipment they control.

Example

The on–off switch on an item of electrical equipment does not isolate it from the supply but only stops the equipment itself from working. Let's take the example of a washing machine. You can operate the on–off switch on the front of the machine and it will no longer wash clothes. All the lights will be out and the machine is off. However, if you pull it out of its gap under the kitchen worktop and open the back, you will find that there is still a supply into the machine and that it is not only live at the main terminal but in several other parts of its circuitry as well. Only when the isolator is switched off will there be no supply whatsoever to the appliance. The isolator for a washing machine is usually the 13 A plug and socket.

Figure 8.7: The on–off switch on a washing machine does not isolate it from the supply

Presence of undervoltage devices

Undervoltage protection prevents injury caused by a sudden start-up of electrical equipment after a power shutdown. Typically, undervoltage protection is applied to electric motors which drive machinery. If the power is shut down for some reason, the motor starter must operate in such a way that when the power is restored, the motor will not restart automatically.

Labelling of protective devices, switches and terminals

Various parts of an electrical installation must be labelled and these labels have to be made from a durable material. A good-quality print should be used for the legend so that it does not become worn or faded and unreadable. All protective devices should be labelled to indicate which circuit they supply. Terminals must be labelled, for example 'L, N and E' for a single-phase connection and 'L1, L2, L3, N and E' for a three-phase connection. The requirements in Section 514 of BS 7671:2008 describe and show the wording for the labels required in an electrical installation.

Safety Electrical Connection – Do Not Remove

> **IMPORTANT**
> This installation should be periodically inspected and tested and a report on its condition obtained as prescribed in BS 7671 (formally the IET Wiring Regulations for Electrical Installations) published by the Institute of Electrical Engineers.
>
> Date of last inspection : ....
>
> Recommended date of next inspection : ....

Figure 8.8: Electrical installation labels must be durable and legible

Selection of equipment and protective measures appropriate to external influences

External influences are those which can affect electrical systems from the outside, such as temperature and the presence of water. Appendix 5 of BS 7671:2008 lists and describes these external influences. Each one is given a code, for example:

- AD4 – water splashes
- AB6 – extreme temperature and humidity.

Appendix 5 of BS 7671:2008 also shows the IP (ingress protection) codes for equipment suitable for the area. AD4 equipment, for example, must conform to IPX4. During inspection, confirm that the electrical equipment used is suitable for its environment.

Adequate access to switchgear and equipment

All electrical equipment and accessories should be accessible. This is so that repairs and maintenance can be carried out and switches operated and reset.

Presence of danger notices and warning signs

Examples of warning notices are those which warn of:

- alternative supplies present
- protective bonding
- non-standard conductor colours
- isolation of equipment voltages exceeding 230 V to earth.

These notices should be clearly displayed and of a standard type. They must be durable and not peel off. The legend should be in good-quality print so that the information remains legible.

Presence of diagrams and instructions

Diagrams and charts must be available and up to date. The following should be available.

- Circuit information showing the load, size of cable and type of wiring.
- Positions and purpose of switches and isolators.
- Information about any circuit that might be damaged by testing.
- Certificates and reports from previous inspections and tests.

Erection methods

All electrical equipment and accessories should be installed and wired according to Part 5 of BS 7671:2008. This part of the inspection checklist should include:

- suitability of accessories, equipment and wiring systems for the environment
- secure fixings
- lids on conduit boxes
- correct spacing of saddles and cable clips.

Sampling

When carrying out the inspection phase of a periodic inspection and test of an existing installation, it may be almost impossible to look at every single piece of electrical equipment or accessory in the installation. Because of this, sampling can be used. In other words, a percentage of accessories, etc. are inspected in detail and taken as a measure for the whole installation.

IET Guidance Note 3 recommends that 10 per cent of the components in a final circuit are sampled. For all other components for an installation, such as distribution boards and switchgear, it recommends 100 per cent but will accept 10 per cent to 25 per cent if necessary. No sampling is allowed when inspecting the earth protection system. Guidance Note 3 states that the sampling percentage should be based on:

- age of installation
- general condition
- environment
- effectiveness of ongoing maintenance if there is any
- time since last inspection and test
- installation size
- discussion with client
- quality of information about the installation.

TESTING THE SAFETY OF ELECTRICAL SYSTEMS

LO 6

As we have seen in the section of this chapter on the regulatory requirements for testing, the tests must be carried out in a certain order. This is because insulation resistance testing is invalid if the circuits are not whole and complete, which is why continuity tests are carried out first, and it would not be safe to switch on the power unless insulation resistance is checked before doing so.

Implement safe system of work for testing electrical systems

General health and safety considerations will be the same as for all electrical work. However, testing brings the added risk of electric shock, always a hazard for an electrician. This applies especially because some testing routines have to be carried out on live circuits. Live testing must only be undertaken by an experienced, competent person. The area around any exposed live parts must be closed off using barriers to protect members of the public who might be working in, or passing through, the vicinity.

Select the test instruments for regulatory tests

Instruments used for electrical tests are described earlier in this chapter, on pages 408 and 409.

Test electrical systems

This section describes each of the main tests in detail.

Continuity of protective conductor

The protective conductor is also known as the earth conductor and its purpose is to conduct any earth fault currents away from faulty equipment and down to earth. Because the protective conductor does not normally

---

Progress check 8.9

1. List two things that should be inspected for routing of cables.
2. Which conductor should single-pole devices be connected to?
3. What is AB6 the code for?

carry current it has to be tested to ensure there are no breaks, either from damage or loose connections.

Acceptable values depend on the cross-sectional area of the conductors under test. These can be calculated using Table B1 in Appendix B of IET Guidance Note 3 (GN3). This table gives the expected reading for $R_1 + R_2$ for various combinations of conductor sizes. The figures shown are mΩ/m (milliohms per metre). Once the length of the conductors is known, then the total resistance can be calculated.

Note that these figures are valid for an ambient temperature of 20°C. If the ambient temperature is higher or lower than 20°C, then the correction factors in Table B2 of GN3 must be applied.

Example

6.0 mm^2 line conductor with a 2.5 mm^2 cpc

Length of run = 12m

Ambient temperature = 20°C

According to Table B1, the expected $R_1 + R_2$ resistance will be 10.49 mΩ/m

$10.49 \times 12 \times 10^{-3} = 1.13$ Ω

If the temperature increases to 30°C then, according to Table B2, a correction factor of 1.04 must be applied:

1.13 Ω $\times 1.04 = 1.31$ Ω

There are two methods for testing the continuity of protective conductors.

Method 1

There are two reasons for using this method. One is purely practical. To confirm continuity of a conductor, a resistance reading has to be taken by connecting the instrument leads to each end of the conductor. However, in an electrical installation, the conductors are often many metres in length, which makes this direct connection impractical. Method 1 provides an easy way to overcome the problem.

The other reason is technical. You will see from Figure 8.9 that the line and CPC are connected together at the supply end, then the resistance is measured between them at the furthest point of the circuit, as shown in Figure 8.11. This provides a reading known as R_1 (line) + R_2 (CPC), in other words the total resistance of the two conductors. This can be used to calculate the full earth fault loop impedance (Zs) of the circuit.

$Zs = Ze + (R_1 + R_2)$

Ze is the external impedance of the earth fault loop path and can either be measured or obtained from the supply company.

If Method 1 is used for testing a lighting circuit, the furthest light can be used as the test point for the main CPC, but the switches may have to be tested individually because although the CPC is usually looped from light

Figure 8.9: Line and CPC connected together at the supply end of a circuit

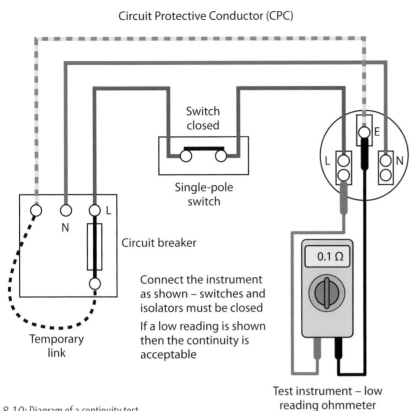

Circuit Protective Conductor (CPC)

Switch
closed

Single-pole
switch

L
N

Circuit breaker

Connect the instrument
as shown – switches and
isolators must be closed

If a low reading is shown
then the continuity is
acceptable

Temporary
link

0.1 Ω

E
L
N

Test instrument – low
reading ohmmeter

Figure 8.10: Diagram of a continuity test

fitting to light fitting, it is dropped down to each switch
as a spur. This means that the switches are not in the
main run of the CPC.

Method 2

As well as the CPCs in the installation, the continuity
of other earth conductors must be confirmed. This
is carried out using a low reading ohmmeter and
taking a reading from each end of the conductor.
A wander-lead may be needed due to the distance
between the test points. The wander-lead resistance
must be taken and then deducted from the reading.
Alternatively, the instrument should be zeroed with
the wander-lead connected.

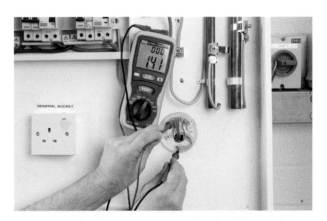

Figure 8.11: Carrying out test at furthest end of circuit by measuring
resistance between line and CPC

If conduits and trunking have been used as protective conductors, these
must be tested for continuity using Method 2. They should
also be inspected to make sure they are mechanically sound.

The reading obtained from Method 2 testing is R_2. Appendix B of IET
Guidance Note 3 gives the expected resistance per metre for conductors
of 1 mm² to 50 mm² cross-sectional area. From this table it can be worked
out that a 10 mm² copper bonding conductor, 12 m long, should have a
resistance of:

1.83 mΩ/m (0.00183 Ω/m) × 12 m = 22 mΩ (0.022 Ω).

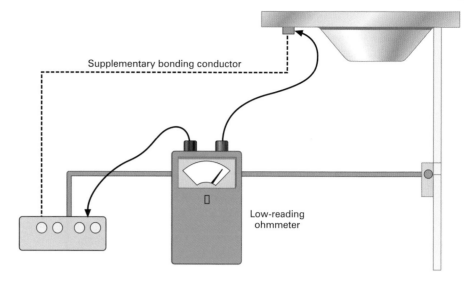

Figure 8.12: Continuity of earth bonding test

Ring final circuit tests

Because of its unique configuration, in so far as each conductor not only connects all the socket outlets on the circuit but also returns to the source, extra checks have to be made on the ring final circuit. It has to be confirmed that there actually is a ring and that its polarity is correct.

Step 1 – end-to-end resistance

To confirm there is no open circuit and to confirm the presence of a ring, follow the steps on the next page.

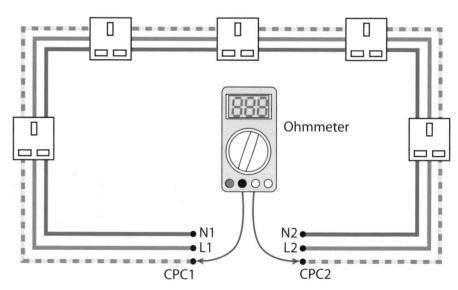

Figure 8.13: End-to-end test on a ring final circuit

Checklist

PPE	Tools and equipment	Consumables	Source information
No specific PPE needed. Standard workshop or construction site PPE may be required.	• Side cutters • Pliers • Insulation strippers • Screwdrivers • Low reading ohmmeter	• 2 × 30 A connectors	• BS 7671:2008 Part 6 • IET On-Site Guide • IET Guidance Note 3

1 Identify and separate phase, neutral and CPC ends.

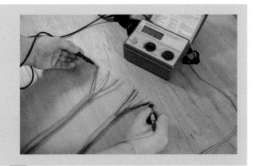

2 Use a low resistance ohmmeter to measure the resistance between each set of like cores.
- Line-to-line (r_1).
- Neutral-to-neutral (r_n).
- CPC-to-CPC (r_2). If this is smaller than the line and neutral conductors (in this case, a 1.5 mm^2 CPC in a 2.5 mm^2 twin-and-earth cable) the reading should be 1.67 times higher than that of the other conductors.

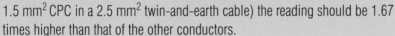

3 If all the conductors are the same size then the readings should be within 0.05 Ω of each other.

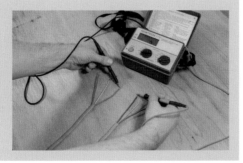

Figure 8.14: End-to-end ring final test

Step 2 – line-to-neutral resistance ($r_1 + r_n$)

The $r_1 + r_n$ test confirms that there is no open circuit on phase and neutral and also contributes towards the polarity check.

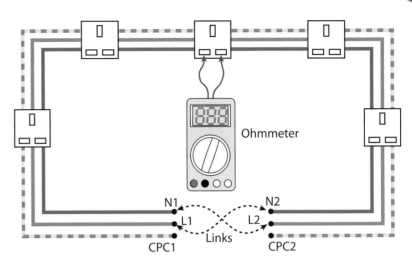

Figure 8.15: $r_1 + r_n$ test on a ring final circuit

Checklist

PPE	Tools and equipment	Consumables	Source information
• Hard hat • Protective footwear • Hi-vis jacket	• Cable cutters • Cable strippers • Combination pliers • Terminal and Philips screwdrivers • Low reading ohmmeter	• Connectors	• Wiring diagram • Circuit diagram • Test documentation

1 Connect together the opposing line and neutral of each end, or 'leg', of the ring.

2 Connect the test instrument to each socket in turn using a test plug.

3 Measure the resistance between line and neutral.
A similar, low reading should be obtained at each socket. This indicates line and neutral are both connected into the line and neutral.
They may not, however, be the right way round.

Figure 8.16: $r_1 + r_n$ test

The expected reading should be in the region of: $\dfrac{(r_1 + r_n)}{4}$

(r_1 and r_n are the end-to-end readings for line and neutral found in the previous test.)

Step 3 – line-to-CPC resistance ($r_1 + r_2$)

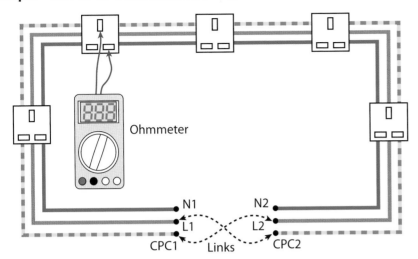

Figure 8.17: $r_1 + r_2$ test on a ring final circuit

Checklist

PPE	Tools and equipment	Consumables	Source information
• Hard hat • Protective footwear • Hi-vis jacket	• Cable cutters • Cable strippers • Combination pliers • Terminal and Philips screwdrivers • Low reading ohmmeter	• Connectors	• Wiring diagram • Circuit diagram • Test documentation

1 Connect the opposing line and CPC of each leg of the ring.

2 Connect the test instrument to each socket in turn using a test plug.

3 Measure the resistance between line and earth.
A similar, low reading should be obtained at each socket. This indicates that line and CPC are both connected correctly and confirms that line and neutral are also in their correct terminals.

Figure 8.18: $r_1 + r_2$ test

The expected reading should be in the region of: $\dfrac{(r_1 + r_2)}{4}$

(r_1 and r_2 are the end-to-end readings for line and CPC found in the first ring final circuit test.)

This reading can be used as the $(R_1 + R_2)$ in the Zs calculation.

Activity 8.2

1 While carrying out a $r_1 + r_n$ test on a ring final circuit, an electrician notices that the reading at one 13 A socket is much higher than the others. What could cause this?

2 The $r_1 + r_n$ test on another socket in this ring is fine. However, the resistance reading for the $r_1 + r_2$ test is so high it is off the scale. What is the probable cause for this?

Insulation resistance

Insulation resistance testing proves the integrity of the cable insulation. Tests are carried out at 500 V d.c. between line, neutral and earth. All covers and faceplates must be in place. The resistance is measured by an insulation resistance tester and displayed as megohms or MΩ. The minimum acceptable insulation resistance for a 230 V or 400 V circuit, according to BS 7671:2008, is 1 MΩ.

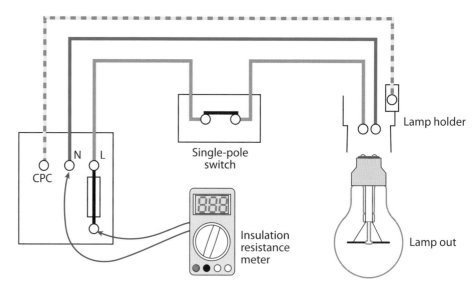

Figure 8.19: Insulation resistance test on a light

Care needs to be taken when carrying out this test because of the voltage used. Checks must be made prior to testing to ensure that:

- all appliances are unplugged
- all sensitive items of equipment, for example electronic equipment and dimmer switches, are disconnected from the circuit
- all lamps are removed from light fittings (these will give false readings because of their relatively low resistance between line and neutral)
- all accessories incorporating a neon indicator lamp are switched off or taken out of the circuit.

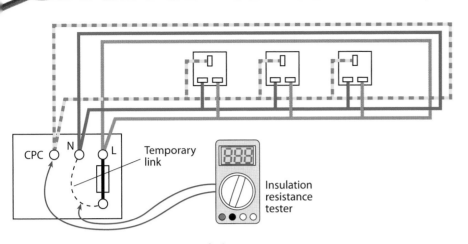

Figure 8.20: Insulation resistance test on a ring final circuit

If it is not possible to take an insulation resistance reading between line and neutral, for example in a lighting circuit feeding fluorescent luminaires, then only a line and neutral-to-earth test should be carried out. The reading must be taken with the line and neutral twisted together so that they can be simultaneously connected to one end of the test lead.

For speed when carrying out an insulation resistance test, connect line and neutral together so both can be tested to earth at the same time. A few seconds saved at each test will soon add up.

When carrying out an insulation resistance test on a three-phase circuit, readings must be taken between each phase (L1 to L2, etc.) and between each phase and earth, and each phase and neutral.

Polarity

Correct polarity means that all the conductors are connected into the correct terminals and the circuit is, essentially, the right way round, and that switches and protective devices are connected into the line conductor and not the neutral. Polarity can be checked:

- during inspection
- as part of other tests – for example the ring final $r_1 + r_n$ and $r_1 + r_2$ tests confirm ring final circuit polarity
- in a polarity test for the Edison screw lamp.

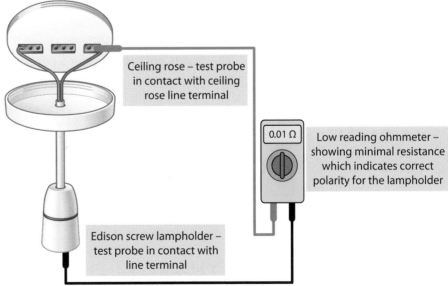

Figure 8.21: Polarity test on an Edison screw-type light fitting

Earth electrode test

Installations protected by a TT earthing system are connected to the mass of earth via an earth electrode. As part of inspection and testing, the resistance between the electrode and the earth itself must be measured. There are a number of methods for doing this. Method E1 is the one described here.

1 Isolate the supply.

2 Remove the earth conductor from the earth electrode.

3 Select a calibrated, four-terminal earth electrode tester. You will also need two temporary test earth spikes.

4 Connect instrument terminals C1 and P1 to the earth electrode.

5 Drive in the test spike T1 at a distance of at least 10 × earth electrode length, so if the electrode is 2 m long, T1 must be 20 m away.

6 Drive test spike T2 into the ground halfway between T1 and the electrode.

7 Connect terminal C2 to T1.

8 Connect terminal P2 to T2.

9 Take a reading.

10 Move T2 closer to the electrode by 10 per cent of the total distance between the electrode and T1 (if electrode to T1 is 20 m then T2 must be moved 2 m closer to the electrode).

11 Take a reading.

12 Move T2 back towards T1. This must be 10 per cent of the overall distance between the electrode and T1, measured from T2's original position at the start of the test.

13 Take a reading.

14 Find the mean of the three readings.

15 Calculate the maximum deviation between the mean reading and the original three readings.

16 Express as a percentage of the mean.

17 A percentage of under 5 per cent is acceptable.

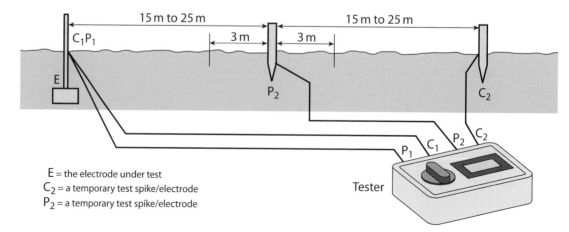

E = the electrode under test
C_2 = a temporary test spike/electrode
P_2 = a temporary test spike/electrode

Figure 8.22: Earth electrode test

Example

Using the procedure above, if the three readings were 5.5 Ω, 5.8 Ω and 5.3 Ω, the mean would be:

$$\frac{16.6}{3} = 5.53 \ \Omega$$

The maximum deviation is between the mean 5.53 Ω and 5.8 Ω = 0.27 Ω. 0.27 Ω as a percentage of the mean (5.53 Ω) = 4.88% which is just about acceptable and shows how close the readings need to be.

Safe working

The earth conductor can only be disconnected from the earth electrode if the supply is switched off. The supply must not be restored until the test has been completed and the earth conductor reconnected to the electrode.

Earth loop impedance test (Zs)

The purpose of the earth fault loop impedance test is to measure the impedance of the earth fault path (Zs) and ensure that it is within acceptable limits. This is a live test which means it is a test carried out with the power on. Because of this, care has to be taken as there is a risk of shock.

An earth fault loop impedance meter is used for this test. The instrument has three probes. The neutral probe is not actually used for testing but completes the instrument's own functional circuit. There is usually a plug-in lead supplied for testing ring final circuits. Because it puts a fault on the circuit being tested, an earth fault loop impedance tester can trip an RCD. Most modern versions of the instrument are fitted with an anti-trip setting.

Safe working

When connecting test probes to live terminals, connect them in the following order.

1 Earth
2 Neutral
3 Line

When removing them, do so in the reverse order:

1 Line
2 Neutral
3 Earth

This means that the most dangerous, current-carrying terminal will be the last to be connected to the test instrument and the first to be disconnected.

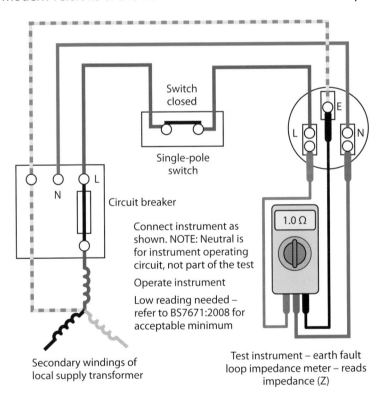

Switch closed

Single-pole switch

Circuit breaker

Connect instrument as shown. NOTE: Neutral is for instrument operating circuit, not part of the test

Operate instrument

Low reading needed – refer to BS7671:2008 for acceptable minimum

Secondary windings of local supply transformer

Test instrument – earth fault loop impedance meter – reads impedance (Z)

Figure 8.23: Earth fault loop impedance test

When carrying out an earth fault loop impedance test on a light fitting or any other equipment on which you need to hold the probes, get someone to hold the test instrument to free up your hands for safe testing.

Tables 41.2 to 41.4 of BS 7671:2008 state the maximum Zs values for various types of protective device at different ratings.

Activity 8.3

BS 7671:2008 states in Table 41.2 that the maximum earth fault loop impedance (Zs) for a 16 A Type B circuit breaker to BS EN 61009-1 is 2.87 Ω. However, if the earth fault loop impedance test returns a reading of 8.3 Ω, what action should be taken? Note: the Ze is within acceptable limits.

Measurement of external earth loop impedance (Ze)

This test measures the external part of the total earth fault loop path and along with the $R_1 + R_2$ reading (measured when carrying out the continuity of the protective conductor test) can be used to calculate Zs, the full earth fault loop path.

$$Zs = Ze + (R_1 + R_2)$$

To carry out the test, follow the steps below.

1 Set all protective devices to 'off'.

2 Disconnect the main incoming earth conductor from the main earthing terminal.

3 Test between the main incoming line conductor and the main incoming earth conductor.

4 Reconnect main earth conductor.

Safe working

The installation is not protected from shock fault while the main earth conductor is disconnected.

Prospective fault current (I_{pfc})

The prospective fault current is the current that will flow in the event of a fault. The reading should be taken at the supply using a specialist instrument or an earth fault loop impedance tester set to the prospective fault current setting.

There are two types of I_{pfc} meters available.

- Two-lead version – which means you have to carry out two sets of tests, one between line and neutral and one between line and earth.
- Three-lead version – which enables you to carry out both tests at once.

For three-phase circuits, the readings are taken between each phase to earth, each phase to neutral and then between each phase. The highest reading will be entered onto the test documentation as the prospective fault current.

I_{pfc} can also be obtained by asking the supply company, or by calculation.

$$I_{pfc} = \frac{U}{Zs}$$

Phase sequence

In a three-phase circuit the three voltages must reach their sine wave peaks in a set order: L1, L2 and L3. If they are reversed in some way it will cause three-phase motors to run in the reverse direction, damaging machinery and possibly electrical equipment. It also poses a safety risk. A phase rotation meter is used to indicate correct phase polarity and rotation, either by a set of lights or by LED display.

Functional

The final test is the functional test, which is simply a check to make sure everything works correctly and safely. Included in a functional test are:

- switchgear
- controls
- interlocks
- undervoltage protection – built into motor controls so that if there is a power failure the motor will not restart when power is restored

- RCDs:
  - test that the RCD itself is working by pressing the RCD's test button
  - use test instrument to test at:
    - 50 per cent rated current – device should not open
    - 100 per cent rated current – device should open within stated operating time
    - 5 × rated current – device should operate well within stated operating time
  - repeat tests with instrument set for 180° – tests will be carried out when the sine wave is at the opposite polarity to that of the first set of RCD tests.

Progress check 8.10

1 What is the Method 1 continuity test usually used for?

2 How do you carry out an insulation resistance test on a circuit feeding fluorescent luminaires?

3 What is Ze?

Record results of regulatory tests

Table 8.6 shows the main headings of a typical Schedule of Test Results form, to be filled in during the test routine.

Heading	What to enter
DB reference number	The individual number given to the distribution board which feeds the installation, or part of the installation, you are testing.
Location	A description of where to find the distribution board.
Zs at DB	The full earth fault loop impedance measured from the distribution board supply.
I_{pfc} at DB	The prospective fault current measured at the distribution board supply.
Correct supply polarity confirmed	Check that the supply cables are connected to the correct terminals, i.e. line to L and neutral to N. If the board is a three-phase distribution board then a phase rotation test should be carried out to confirm the polarity of the three line conductors: brown L1, black L2 and grey L3.
Details of circuits and/or equipment vulnerable to damage when testing	Are there any circuits that incorporate delicate equipment? An example would be a lighting circuit which includes a dimmer switch.
Details of test instruments used	The instruments used for the tests should be calibrated and marked with a serial number, which is entered into this field.
Tested by (signature, etc.)	The inspector's name, signature and dates.

Table 8.6: Headings of a Schedule of Test Results

Now we move to the main body of the form.

Heading	Description
Circuit number	Each circuit should be given a unique id number at the distribution board.
Circuit description	A brief description of the purpose of the circuit, e.g. ring final circuit, upstairs lights.
Overcurrent device BS (EN)	The BS (EN) number of the overcurrent device and its type. For example, 'Circuit breaker – BS EN 60898'.
Overcurrent device type	Protective devices are divided into types depending on what they are designed to protect. For example: • type B circuit breakers are used for ordinary domestic installations • type C or D are for circuits where there might be a current surge at switch on.
Overcurrent device rating	The rating of the circuit breaker is the current it is meant to operate at, for example 16 A, 32 A.
Overcurrent device breaking capacity	All protective devices must withstand the prospective fault current. For circuit breakers, breaking capacity is indicated by a code, for example M6. This means that it can withstand a fault current of 6 kA. The breaking capacity of fuses is usually written on the fuse itself.
Conductor reference method	The reference method is the code for the installation method, for example clipped direct to the surface is Method C, installed in conduit is Method B (BS 7671:2008 Appendix 4).
Live conductor size	The cross-sectional area of the line and neutral conductors, e.g. 2.5 mm^2.
CPC size	The cross-sectional area of the circuit protective conductor (earth), which can differ from that of the live conductors. For example, a twin-and-earth cable with 2.5 mm^2 live conductors would have a 1.5 mm^2 CPC.
Ring final circuit continuity r_1, r_n and r_2	The results of the end-to-end test carried out on a ring final circuit.
Ring final circuit continuity $R_1 + R_2$ (or R_2)	The results of the cross-connection test carried out between line (R_1) and CPC (R_2).
Insulation resistance	The results of the insulation resistance test. This must be entered as a figure. Some instruments will give a result such as '>5 MΩ'. Others show a symbol to denote that resistance is so high it is off the scale. Typically this is a '1' symbol. In this case enter the minimum allowable insulation resistance for the circuit type, e.g. 1 MΩ for a 230 V circuit, 0.5 MΩ for a SELV circuit.
Polarity	A 'Yes' or 'No' to be marked here to denote correct polarity throughout the circuit.
Zs	The result of the earth fault loop impedance test, carried out on each circuit.
RCD @ I∆n	The first of the RCD tests carried out at operating current. The operating time is entered into this field.
RCD @ 5 I∆n	The second RCD test carried out at five times operating current. Again, the operating time to be entered.
RCD test button operation	A 'Yes' or 'No', or 'Correct' or 'Incorrect', to be entered into this field.
Remarks	If there are any problems, unusual but acceptable readings, or parts of circuits that could not be tested, the details should be entered into this field.

Table 8.7: Further headings for a Schedule of Test Results

As referred to in Table 8.7 under the conductor reference method, installation methods are each given codes. Table 8.8 shows some of the 57 installation methods detailed in BS7671:2008, Table 4A2.

For a complete list, refer to BS7671:2008, 17th edition.

Installation method			
Number	*Examples*	*Description*	*Reference method to be used to determine current-carrying capacity*
13	TV / ICT / 13	Non-sheathed cables in skirting trunking	B
14	TV / ICT / 14	Multi-core cable in skirting trunking	B
15		Non-sheathed cables in conduit or single-core or multi-core cable in architrave	A
16		Non-sheathed cables in conduit or single-core or multi-core cable in window frames	A
20		Single-core or multi-core cables: • fixed on (chipped direct), or spaced less than 0.3 cable diameter from a wooden or masonry wall	C
21		Single-core or multi-core cables: • fixed directly under a wooden or masonry ceiling	B Higher than standard ambient temperatures may occur with this installation method.
22		Single-core or multi-core cables: • spaced from a ceiling	E, F or G Higher than standard ambient temperatures may occur with this installation method.

Table 8.8: Wire installation and reference methods

Verify compliance of regulatory test results

The test results must fall within the maximum and minimum values given by BS 7671:2008 such as the maximum earth fault loop impedance values in Part 4. The actual readings must be entered onto the documentation, even if they do not fall within these values. A note of this should then be entered into the notes section of the result schedules.

Commission electrical systems

Although commissioning is used to describe the verification, inspection and testing of a new installation, it is also used to describe the procedure for setting up and testing particular systems and equipment, for example:

- fire, intruder and other alarm systems
- electrical control panels
- boilers and other components of heating systems
- complex control systems such as those that operate industrial processes
- back-up power systems such as generators and uninterrupted power supplies.

Functionality

Functional testing is the main component of equipment and system commissioning and is described below.

Fitness for purpose

As well as commissioning the fixed wiring system, specialist electrical systems such as alarms or control circuitry and equipment must be tested for 'fitness for purpose' before they are handed over to the customer. Much of this will be detailed on commissioning schedules which are unique to the equipment or system.

Commissioning involves running the equipment or systems through all its functions. Inspections and tests are carried out, many of them being the standard routines described already in this chapter. However, there are other tests that are usually only undertaken as part of equipment commissioning procedures. Two of them are described below.

Touch current

If the 500 V d.c. insulation resistance test cannot be carried out, a touch current test can be used to test equipment insulation. Care has to be taken as this test is done while the equipment is switched on and operating. The test is carried out between the internal live parts of the equipment and its metalwork and insulation.

Acceptable reading: 3.5 mA, or 0.75 mA for appliances with heating elements.

Flash test

This is a high voltage insulation resistance test. For mains supply panels this can be carried out at 2 kV (2 000 V). The purpose is to check that the insulation is able to withstand this type of voltage.

Commissioning

All functions of the equipment or system are tested. This means that all buttons are pressed, alarms operated and isolators and fuse switches operated. This is done without the full load, then with the load connected. Included in a functionality test are:

- switchgear
- controls
- stop buttons

- interlocks – for example, panel doors cannot be opened until the power is switched off
- no-volt cut out – built into motor controls so that if there is a power failure the motor will not restart until the start button is pressed
- correct sequence or start-up and operation
- programming works correctly
- that all warning lights, on/off lamps, etc. work correctly.

Complete certification documentation

There will be specific commissioning documentation, such as checklists and schedules of settings, to be completed. These will be individual to a particular system or item of equipment. Completion of standard electrical certificates is described earlier in this chapter.

Progress check 8.11

1 What are two extra tests that may have to be carried out when commissioning an item of electrical equipment?

2 List three functionality checks carried out as part of commissioning equipment or a system.

3 What type of documentation is required for commissioning electrical equipment and systems?

Knowledge check

1 The M6 designation on a circuit breaker means that it:

  a is rated at 6 A

  b can withstand a fault current of 6 kA

  c is allowed a maximum earth fault loop impedance of 6 Ω

  d can carry 6 kV

2 If the Ze of an installation is 4 Ω and $R_1 + R_2$ is 0.6 Ω, what is Zs?

  a 2.4 Ω

  b 4.6 Ω

  c 6.67 Ω

  d 3.4 Ω

3 An end-to-continuity test is carried out on a ring final circuit wired in 2.5 mm twin-and-earth. r_1 and r_n are both 0.08 W. What would you expect the r_2 resistance to be?

  a 1.75 Ω

  b 0.134 Ω

  c 0.048 Ω

  d 20.85 Ω

4 If r_1 is 0.07 Ω and r_n is 0.08 Ω, the result of the $r_1 + r_n$ test should equal:

  a 0.6 Ω

  b 0.15 Ω

  c 0.075 Ω

  d 0.0375 Ω

5 What is the test voltage and acceptable reading for an insulation resistance test on a SELV circuit?

  a 500 V d.c., 0.5 MΩ

  b 250V d.c., 0.5 MΩ

  c 500 V d.c., 1 MΩ

  d 250 V d.c., 1 MΩ

6 An E1-type earth electrode resistance test requires:

  a a four-terminal earth electrode tester and two test stakes

  b a two-terminal earth electrode tester and a low reading ohmmeter

  c a four-terminal earth electrode tester and a wander-lead

  d a four-terminal earth electrode tester and an insulation resistance tester

7 When should verification take place for a new electrical installation?

  a At the beginning of the electrical installation process

  b At the end of the electrical installation process

  c At all stages of the installation process

  d It is not required for a new installation

8 A Minor Works Certificate is needed for:

  a a small installation with only a few circuits

  b swapping an old light fitting for a new one

  c a modification to an existing circuit

  d installation of a single new circuit

9 Which of the following would be carried out as part of an inspection?

  a Continuity of protective conductor, check for correct connection of cables, check that warning notices are in place

  b Check cable routes, check cable identification, check phase rotation is correct

  c Check that fittings and accessories are securely fixed, check if circuits are live, check for presence of notices warning that pre-harmonisation colours are present in the installation

  d Check for correct cable selection, check single-pole switches switch line only, check that remote isolators are lockable

10 RCDs should be tested at:

  a half $\times$ operating current, at 2 $\times$ operating current, 5 $\times$ operating current

  b half $\times$ operating current, at operating current, 5 $\times$ operating current

  c half $\times$ operating current, at 2 $\times$ operating current, at operating current

  d half $\times$ operating current, at operating current, operating current again

Electrical installations: fault diagnosis and rectification

This chapter covers:

- how electrical fault diagnosis is reported
- how electrical faults are diagnosed
- how to carry out fault diagnosis on electrical systems
- the fault rectification process.

Introduction

Faults will occur no matter how well an electrical installation is installed or maintained, and even if its components are of the highest standard. Therefore, it is important that you have the skills and knowledge to diagnose and repair faults. Not only does an electrical breakdown service provide a good source of income, it also tests and adds to your knowledge and understanding.

Fortunately, electrical faults rarely result in injury or major damage to property. Modern installations are protected by protective devices such as circuit breakers and residual current devices (RCDs). However, normal service will need to be restored, which is where you come in.

Locating, diagnosing and repairing electrical faults require patience, a logical approach and access to as much information as possible. It also requires good customer relations, because the customer is no doubt already frustrated at the loss of power, and, in the case of a business, possible loss of revenue.

Finding and diagnosing faults also bring their own hazards, from working in awkward and often confined spaces, to live working while checking for the presence of a voltage.

Once the fault is located and diagnosed, you need to decide if it is more cost-effective to replace the equipment or to rewire the complete installation. If a replacement part is ordered there may be a long wait for its delivery. If there is a wait, the faulty item or circuit will need to be made safe and put out of action.

REPORTING ELECTRICAL FAULT DIAGNOSIS

Procedures for recording information

Although most faults are reported verbally, formal fault report systems are sometimes used within large companies or organisations. The fault report process is outlined below.

1 Report raised when a fault is reported by a member of staff or public.

2 Report details the location and a description of the symptoms (or the cause if it is something obvious such as a blown lamp).

3 The fault is entered onto a database and may take its place in a waiting list. Some systems have a facility for giving a fault a priority rating, like a traffic light colour scheme.

4 The fault is allocated to an electrician.

5 Once it has been rectified the electrician will enter details of the action taken, either onto a paper-based form or, as is more likely these days, a computer-based form.

6 The fault will be closed and a message sent to the people who reported the fault and/or those affected.

Fault reports are useful because, while they may not give the fault's exact cause or nature, an experienced electrician might be able to diagnose the fault just by reading the report.

Fault reports are retained even after the problem has been rectified. One reason for this is because there may be a pattern, or a recurring fault. Such a fault can be cured more quickly if there is a record of a previous diagnosis and corrective action.

Codes used in Electrical Condition Report (BS 7671) for faults

Periodic inspection and testing will often reveal faults; some that need immediate attention, others that are not so urgent. Any faults or problems found are recorded in 'Section K: Observations' of the Periodic Inspection Condition Report (completed at the conclusion of a periodic inspection and test). Faults are coded as below.

- C1 – danger present – this should be rectified immediately.
- C2 – potentially dangerous – this should be rectified urgently.
- C3 – improvement recommended – the problem needs to be rectified but it is not urgent.

After being allocated a code, any fault is described and the need for further investigation indicated as 'Yes' or 'No'.

Implications of recorded information

As with all recorded information, there are implications – actions that may have to be taken, possible risks if action is not taken and difficulties that may arise if it is.

- Danger – the reported fault raises the possibility that there is danger to those using the installation if the fault is not diagnosed and repaired.
- Isolation – the faulty equipment, component or circuit will need to be isolated if repair cannot be carried out immediately.
- Action required – a decision needs to be made on what action to take; can the fault diagnosis take place immediately, can repair be effected, is the equipment or system under warranty?
- Recommendation – sometimes it is not immediate action that is needed, but a recommendation for a solution to the problem. This can then be considered by the customer.

Progress check 9.1

1 Why keep a fault report after a fault has been diagnosed and repaired?
2 What are the three fault codes used in a Periodic Test Condition Report?
3 Give two implications of recorded information regarding a fault.

HOW ELECTRICAL FAULTS ARE DIAGNOSED   LO2

The first stage in the fault diagnosis process is to gather information. This might be information from the client or person responsible for the area in which the fault occurred. You will need to select the correct test

instruments and understand what the instrument readings are telling you. Once you have found the fault, you will need to be able to make a decision as to how to rectify the problem. Above all you must work safely, following procedures and being aware of the hazards of this type of work.

Safe working procedures for completion of fault diagnosis

Safety procedures should be observed when carrying out fault diagnosis and repair. *Chapter 2: Health and safety in building services engineering* describes these procedures in detail but, as a reminder, here are the main ones.

- Effective communication with others in the work area – this is particularly important when fault finding because you will often be working in someone else's workplace or home. The customer needs to be made aware of what you are intending to do at each stage of the fault diagnosis and repair process, especially if it will mean switching off electrical supplies, or working in an area where other people are present.
- Use of barriers – isolate work areas so that people will be made aware that hazardous work is in progress and will think twice before entering the area.
- Positioning of notices – warning notices must be used to prevent people entering an area made hazardous by fault diagnosis and repair work.
- Safe isolation – follow the approved safe isolation procedure for individual circuits and for distribution boards (see pages 84–85).

Precautions to be taken in relation to hazards of fault diagnosis (risk to electrician)

All electrical work presents hazards. *Chapter 2: Health and safety in building services engineering* describes many of these in detail. There are, however, particular hazards associated with fault finding.

Electric shock

Because you may be examining and testing existing circuits and systems, it is important that these are switched off and safely isolated from the supply. A part of fault finding is often to switch circuits and equipment on and off. This can lead to circuits being left on by mistake or confusion over which protective device protects and powers which circuit.

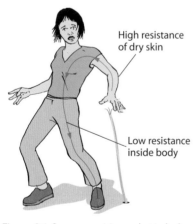

High resistance of dry skin

Low resistance inside body

Figure 9.1: Someone receiving an electric shock

Always use the safe isolation procedure if working on electrical circuits and equipment, and always test any potentially live parts with an approved voltage indicator before starting work.

Live working

Although live working is not recommended, the Electricity at Work Regulations accept that there are cases where this will have to take place. Fault finding is one of those. For example, you may need to take current or voltage readings to check for the presence of a supply.

All test instruments must comply with HSE Guidance Note GS38 (see page 454). Cheap or low-quality test instruments are dangerous because they can be unreliable and develop dangerous faults.

Only an experienced electrician should carry out live working unsupervised and even then, only carry out live tests if it is absolutely safe to do so. The live parts must be secure and accessible. Trying to push the probes of a voltmeter onto the live terminals of a light switch while it is hanging off the wall by its cables is certainly *not* safe. Probing the live terminals of a ceiling rose, which is securely fixed to the ceiling and has a set of shrouded terminals, is much safer.

If tests are to be carried out on a working circuit, any people who use the supply must be warned that testing is taking place.

Lone working

As an electrician you will often be sent out on your own to investigate an electrical fault. If you have an accident when working alone there will be no one to help you. It is a good idea to let someone know if you are going to be working alone. In extreme cases a permit to work should be raised. One of the critical entries onto a permit to work is the completion time for the job. If you are not back by the completion time, it acts as a prompt for the competent person in charge to come and look for you to make sure you are all right.

Hazardous areas

Electrical fault finding may take you into all kinds of places. Some of these areas may be hazardous. Hazards include:

- presence of harmful or toxic chemicals
- presence of machinery
- presence of toxic substances
- confined spaces.

Fault finding in a particularly hazardous area will require a risk assessment and a permit to work. When you do enter the area, make sure you wear the necessary PPE.

Presence of batteries (e.g. lead acid cells, connecting cells)

Many installations are equipped with batteries which provide back-up power or d.c. voltage for specialist items of equipment and systems. An example of this is the uninterrupted power supply (see Chapter 1, page 86). Remember, batteries, and their associated wiring, are always live, until they are drained of their energy. They can also give off toxic fumes, so make sure that any battery areas are well-ventilated.

Back-up supplies

Generators, uninterrupted power supplies and photovoltaic arrays are all methods of back-up power and are sources of energy that will flow even when the main supply is switched off. For more details on the hazards posed by this equipment, see *Chapter 2: Health and safety in building services engineering*.

Fibre-optic cable

Fibre-optic cable has clear glass or plastic conductors. Information packets are transmitted as light impulses, fired into the cable by a laser.

Figure 9.2: Factories can be hazardous places when you are trying to diagnose an electrical fault

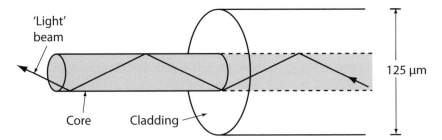

Figure 9.3: A fibre-optic cable

There are two hazards connected with fibre-optic cable.

- The conductor ends are thin and extremely sharp.
- The light pulses are extremely bright so you should never look directly into a fibre-optic conductor.

The cable itself is relatively fragile. If you have to replace it after finding a fault, be careful not to form any tight bends because this will damage, or even break, the conductors.

Functional earth

Computer systems use the earth conductor as part of their normal function. This means that the earth conductor will carry a certain amount of current. Therefore, care must be taken when fault finding in an area where there is computing equipment because, although the current carried by the earth is low, there is still a shock hazard.

Inductive circuits

One of the electromagnetic effects of an inductive circuit is back emf. This is a second voltage set up in opposition to the supply voltage by the fluctuating magnetic field in the iron core (see page 87). A hazard is present when an inductive circuit is switched off because, as the iron core's magnetic field collapses, it induces a current into the windings with a resulting shock hazard.

Capacitive circuits

Capacitors are designed to charge when connected to an electrical supply, then discharge once disconnected. This poses a hazard because, although the electrical supply might be switched off, any capacitors connected to the circuit will discharge current once power is lost.

Fault-finding hazards – equipment and other electrical systems

As well as posing a threat to your own safety, fault finding can also lead to damage to electric equipment and systems. While this is not as important as human life, this sort of damage can cost your customer a great deal of money and harm their business if they have to shut down all or part of their operation.

Electrostatic discharge

Electrostatic discharge (ESD) occurs when electricity flows suddenly between two objects. It is particularly harmful to electronic components. ESD has two main causes.

- Static electricity – takes place when two materials are rubbed together. Rubbing a balloon against a woollen sweater, removing some types of

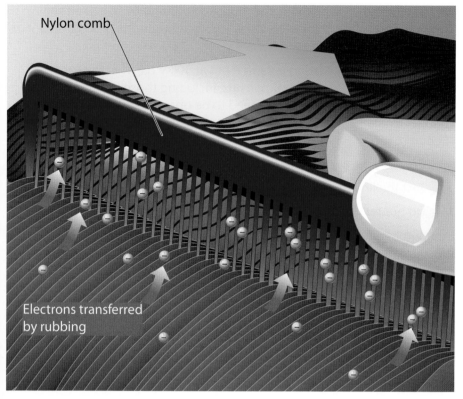

Nylon comb

Electrons transferred
by rubbing

Figure 9.4: Static electricity

plastic packaging or untangling clothes from a tumble dryer can all
cause static electricity.

- Electrostatic induction – if you place an electrically charged object
  near to a conductive object which is isolated from earth, the charged
  object will set up an electrostatic field. This will redistribute the
  electrical charges which exist on the surface of the other object. This
  effect can damage the components in an electronic circuit.

A further problem with ESD is sparking, which can cause fire or explosion
in certain atmospheres, such as those filled with coal dust or flour dust, or
those rich in oxygen.

To counteract the effect of ESD when working on or near electronic
components, you are required to wear a wrist band, connected to earth.
This prevents an ESD build-up by constantly discharging any static
electricity to earth.

IT equipment

Most working or public premises contain computers and other items of
IT equipment. Care has to be taken when testing circuits supplying or
associated with computing systems.

If power needs to be shut down as part of your fault-finding work, it is
essential to warn computer users before you do so. This is so that work can
be saved and systems shut down correctly. Computer users and supervisors
must be warned about possible power loss anyway. You may not need
to shut down an actual computer circuit, but it can be difficult to identify
individual circuits at a distribution board (for example, if the circuit breakers

Figure 9.5: ESD wrist band

are not labelled). It is possible, therefore, to accidentally switch off the power supply to the computers while trying to locate the protective device you need for *your* circuit.

Insulation resistance testing is carried out at 500 V d.c. This voltage can cause damage to electronic circuits, including those in computing equipment. The following points need to be observed in order to safely conduct insulation resistance tests on any circuit which contains IT equipment.

1   Connect the line and neutral conductors together.

2   Test the line-and-neutral combination to earth.

3   Do not test between line and neutral.

Computer systems use the earth conductors as part of their normal function. This means that the earth conductor will carry a certain amount of current. Care must be taken therefore when fault finding in an area where there is computing equipment because although the current carried by the earth is low, there is still a shock hazard.

Activity 9.1

You are called to a large office area to attend to a lighting fault affecting one of the office area circuits. The office is filled with desks and IT equipment. The distribution board is not labelled. The lighting circuit no longer works when the switch is operated. You are met by the office manager. What discussion would you have with her before you start work?

High frequency circuits

All cables and other electrical components have an electromagnetic field build-up around them when they are live. There is also a certain amount of capacitance. At the normal supply frequency of 50 Hz this is not much of a problem because the capacitance will be relatively low. However, as the frequency increases, the capacitive field becomes stronger. This has the potential to damage or cause malfunctioning of electronic components. It can also distort readings from test instruments.

Progress check 9.2

1   What is a functional earth?

2   Why are capacitive circuits potentially harmful?

3   What precautions can be taken against electronic discharge when working with electronic circuits and components?

The logical stages of fault diagnosis

Fault finding requires a logical, step-by-step approach. Some faults will be obvious and found quickly. Others will require investigation. There are a number of steps and fault procedures that can be carried out, some of which could save you a lot of work and time. These are described in this section.

Table 9.1 shows the main stages of the fault diagnosis process.

Action	Description
Identify symptoms	What happened? What form does the fault take? Is it intermittent? Has all power been lost?
Gather information	Talk to those affected; collect documentation if it is available, e.g. previous test documentation, manufacturer's instructions and diagrams.
Analyse the evidence	Carry out a visual inspection of the fault or circuit; study documentation and other information in relation to the fault.
Check the supply	Is supply available for the affected circuit? Has the supply been lost for the whole installation?
Check protective devices	Are protective devices the correct type and rating? Have they operated?
Isolate and test	Carry out safe isolation procedure before testing the circuit or equipment.
Interpret and test	Carry out appropriate tests and inspections in order to diagnose the fault.
Rectify the fault	Replace, repair or rewire as appropriate.
Retest, including functional tests	Carry out appropriate inspection and test, and complete certificate and schedule of results.
Restore supply	Reset controls and recommission equipment or systems if necessary.

Table 9.1: The fault diagnosis process

Step 1: Identify the symptoms

Even if the person who reported the fault is non-technical, with no electrical knowledge at all (described by BS 7671:2008 as an 'ordinary person'), they will be able to give you helpful information about the fault. The types of questions you need to ask are listed below.

- What is the problem?
- When did it happen?
- What actually happened?
- Did you see anything unusual such as a flash?
- Did you smell burning?
- Does a particular item of electrical equipment cause the protective device to operate?
- Has this happened before?

Step 2: Collect and analyse data

As well as the general information provided by the person reporting the fault, more detailed technical information could prove invaluable.

Fault report

Faults can be reported in a number of different ways. It may simply be a phone call in the middle of the night to a call-out electrician with the request for help, along with sketchy details delivered by someone affected by the fault. It could be a printout or job card given to an electrician at the start of the day. The report or card will contain the information about the fault.

In larger premises such as factories and hospitals, there will be a system for reporting and recording faults as well as for allocating the fault to an electrician.

When the fault is cleared a card may have to be signed and filled in with details of how long the job took, the nature of the fault and how, and if, it was repaired. These records are normally given to the client and anyone else who needs to be aware of the fault, its rectification and any other information. These would include:

- client
- contractor
- building control office
- certificating bodies.

These records will be added to the history and documentation for the installation and may help other electricians who have to attend similar faults on that circuit or system.

Location and nature of supply

It will be necessary to find the distribution board and the actual protective device for the circuit itself because you will probably have to switch the supply off at some stage. The distribution board also provides a starting point for testing. Continuity and insulation resistance tests can be carried out from the distribution board and the fault pinned down to a particular circuit.

It is important to obtain information about the supply itself when fault finding. Ask yourself these questions.

- Is the supply single-phase or three-phase, or even extra-low voltage?
- Where does the main supply enter the premises?
- Which earthing system is being used?
- What type of protective devices are in use? For example, some circuit breakers are designed to take a high current for a short while and are appropriate for motor circuits but will not be appropriate for other types of circuit where high current will be very damaging even if it flows for a short time.

Drawings and diagrams

Drawings and diagrams will be a great help in fault finding – in fact, for complex circuits such as control circuits, they are vital. Layout drawings will show the location of accessories and equipment, and circuit and wiring diagrams will show how complex circuits are wired. Being able to trace a cable route could prove invaluable if trying to find a fault in a cable itself. Chapter 5 describes the different types of diagram and drawing and their uses in detail.

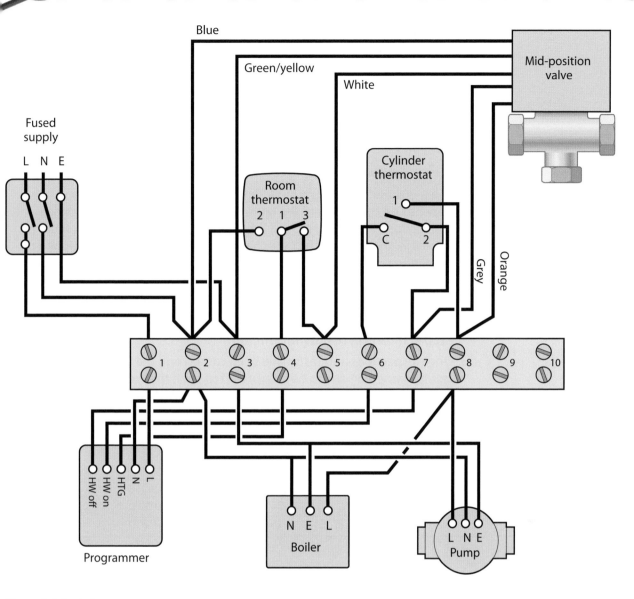

Figure 9.6: Circuit diagram

Manufacturer's instructions

For equipment and systems such as fire alarms, a set of manufacturer's instructions describe how the equipment or systems actually work and may well include a list of common faults and their causes.

Maintenance and test records

Test records will show if there have been previous problems with the equipment or circuit. They will also show details of any work carried out, including replacement of parts.

Installation Condition Reports are records of regular tests carried out on existing electrical installations (see pages 411 and 412).

Step 3: Checking and testing

Sometimes all it takes to find an electrical fault is a visual inspection. So, before starting any instrument tests, take a good look at the installation.

Activity 9.2

As you prepare to find the fault in a three-phase circuit in a factory, you study the Condition Report from the last periodic inspection and test of the premises. In Section K of the report you notice the code 'C2'. What does this alert you to?

Look for:

- signs of wear and tear
- mechanical damage to cables or accessories
- signs of overheating or burning
- corrosion.

Also listen for strange noises from equipment, etc. Is something overheating? Do cables or enclosures feel unusually hot? Can you smell burning?

Step 4: Fault finding using test instruments

There is no set first step in fault finding. How you start the process depends on the nature of the fault. You might need to probe terminals and connections while they are still live in order to establish whether there is a voltage. You may, on the other hand, decide to carry out a continuity or insulation resistance test, which means that the power to the affected area will need to be shut down. When shutting the power down in an area it is vital to follow the safe isolation procedure and to warn the occupants in the area that their power and lighting will be switched off.

Step 5: Interpret results/information

Once inspections and tests have been carried out, the information and results gathered have to be interpreted to see if they reveal the presence and nature of the fault. Table 9.2 gives a general guide to results and what they tell you.

Type of fault	Typical cause	Symptom	Test and instrument to be used	Result
Open circuit	Cable dropped out of terminal or broken conductor	Loss of power to all or part of a circuit	• Continuity test using low reading ohmmeter • Voltage indicator	• High resistance reading • No voltage indicated
Short circuit	Direct contact between line and neutral or between phases in a three-phase system	Circuit breaker or fuse operates	Insulation resistance tester between line and neutral (and between lines in a three-phase supply)	Low reading
Earth fault	Direct contact between live conductor and earthed part of a circuit	Circuit breaker, fuse or RCD operates	Insulation resistance test between live conductors and earth	Low reading
Overload	Current demand from the load is too high for the cable	Circuit breaker or fuse operates	Clamp-on ammeter test on line conductor	Current reading higher than rating of protective device and current-carrying capacity of cable
Faulty RCD	Fault on RCD itself, which causes constant tripping even when there is no live-to-earth fault	Nuisance-tripping	RCD test at 50% fault current	RCD will operate at a fault current 50% its rated operating current
Incorrect polarity	Conductors connected to the wrong terminals, e.g. line-to-neutral and vice versa	Short circuit, incorrect operation of equipment, e.g. electric motor running in reverse	• Polarity test • Phase rotation test	• High resistance reading • Incorrect phase connection indicated

Table 9.2: Possible faults with causes and symptoms

Working practice 9.1

Jim's friend, Carl, rang him late on a Sunday evening. Carl knew Jim was an electrician and could help. Carl had been decorating one of his bedrooms and had taken down the light fitting to paint the ceiling. He said that he was sure he reconnected the light correctly, but when he switched it back on there was a loud bang, a flash and the circuit breaker operated. When Jim asked him to think carefully about how he reconnected the light he said that he connected all the brown wires together and then all the blue wires together. As soon as Jim heard this, he knew what Carl had done.

What was his mistake?

Progress check 9.3

1 List three questions you would ask a customer when gathering information about an electrical fault.
2 What useful information can be gathered from maintenance and test records?
3 Which protective device will only operate if there is an earth fault?

The safe and correct use of instruments used for fault diagnosis

If the fault cannot be located simply by inspecting the installation or circuit, and sometimes this is possible, the circuit will need to be tested. Electrical test instruments are required to do this. We will now look at the main test instruments and how they are used for fault finding.

Continuity tester

Continuity means continuous. When a continuity test is carried out on an electrical circuit or part of a circuit it is to check that a conductor or circuit is continuous, in other words that there are no breaks or open circuits which prevent current from flowing. This type of test is carried out when power has failed because of a fault.

A low reading ohmmeter is used for continuity tests and it works by feeding a small current into the circuit at a set voltage, then calculating the resistance. The low reading ohmmeter will give you an exact reading, which can also be useful when testing a component or part of a circuit of known resistance.

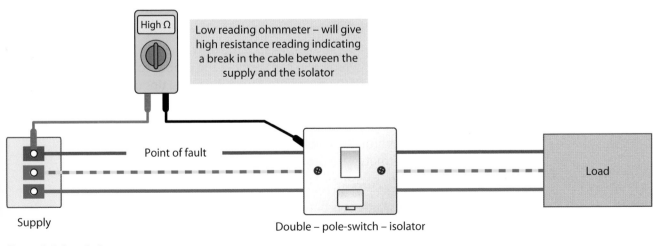

Figure 9.7: A continuity test

Activity 9.3

You are testing a newly installed ring final circuit, which has been wired in singles. The end-to-end tests results are:

L to L = 0.06 Ω
N-to-N = 0.06 Ω
E-to-E = 0.06 Ω

You move on to the $r_1 + r_n$ test and $r_1 + r_2$ tests. The results for most of the 13 A sockets are acceptable. However, one socket gives the following readings.

$r_1 + r_n$ = 0.03 Ω
$r_1 + r_2$ = open circuit

What is wrong with this 13 A socket?

(You may need to refer to the ring final circuit testing section in *Chapter 8: Electrical installations: inspection, testing and commissioning* before attempting this problem.)

Figure 9.8: An insulation resistance test

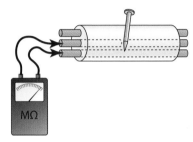

Figure 9.9: A voltmeter

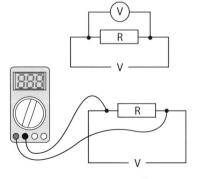

Figure 9.10: Measuring voltage

Insulation resistance test

The insulation resistance test is applied if a short circuit or earth fault has occurred. An insulation resistance tester should be used on its 500 V d.c. and MΩ (mega ohm – million ohms) settings. It does not give an exact reading but an indication that the insulation of the circuit cables is intact or damaged. A short circuit due to damaged insulation will be shown as a low reading or even as a zero. A healthy circuit will show as a '>200 MΩ' or as a '1' symbol.

Because insulation resistance testing is carried out at 500 V d.c., all sensitive equipment should be unplugged or isolated from the circuit. There is a risk of damage to electronic equipment at this voltage.

Voltage indication and measurement

Taking voltage readings has to be carried out as a live test, which makes it a dangerous procedure. Voltage measurement can pinpoint where an electrical supply has failed within a circuit. It can indicate if the voltage is fluctuating or is too low or too high. It is also used to confirm that a circuit is safely isolated prior to being worked on.

If you need to remove switches or sockets, then switch off the supply prior to removing them. This is because the problem might be a poor connection which will spring loose and possibly inflict an electric shock when you pull the plate off the wall.

Voltage is measured by connecting the instrument in parallel across the circuit. In other words, you can connect the probes to line and neutral or line and earth. Be careful not to touch the metal tips of the probe together while testing as this will cause a short circuit.

- A voltage indicator is not intended to give exact voltage readings but to confirm if voltage is present.
- A voltmeter will provide an exact reading.

Measuring current

Measuring current is difficult because current readings are taken in series within the circuit. This means that you have to open a circuit and connect the ammeter between the two open points – this is time-consuming and potentially hazardous.

Clamp-on ammeters avoid the need for opening a circuit and consist of a set of jaws which are clamped together around the cable. The ammeter contains a current transformer which is activated by the magnetic field set-up when current flows through the conductor.

Current measurements are often necessary when fault finding in complex circuits. In these types of circuit multiple components operate at various currents. It is also sometimes necessary to measure the current taken by an electric motor at start-up and run and then compare them with the manufacturer's values.

Monitoring voltage and current

If the problem with a circuit or electrical system is intermittent, or it seems to be related to the electrical supply rather than a short circuit or continuity problem, it may be necessary to record the voltage and/or current over a period of time. Recording voltmeters and ammeters will provide evidence of surges, or periods of low voltage or current.

Earth fault loop impedance (EFLI) tester

A live test, measurement of earth fault loop impedance (Zs), is vital to establish that the earth fault protection of an installation will work correctly. If the impedance of this path is within acceptable parameters, sufficient fault current will flow when there is a fault and the protective device will operate within its safe disconnection time. BS 7671:2008 contains a set of tables which give the maximum Zs value for each type and rating of protective device. If the measurement taken exceeds that value then further investigation of the fault protection system is needed.

Like any live test, this is a hazardous procedure. Most EFLI testers are equipped with both plug-in leads for testing 13 A socket circuits and probes for other circuits. The probes must conform to GN38 and be insulated with finger guards and retractable tips. Most testers are fitted with a strap which enables you to suspend the instrument from your neck and leave your hands free to work with the probes. If you are struggling to use the instrument safely, then ask someone to help you.

Prospective fault current tester

This instrument is used to measure the amount of fault current that will flow in the event of a short circuit or earth fault. It can be used at any point in the circuit, although the test is usually carried out at the intake position.

Prospective fault current test is often a function on an earth fault loop impedance tester. The connections made between the instrument and the supply depends on the type of fault current being measured (live-to-neutral short circuit or live-to-earth fault). It also depends on the instrument itself.

This is a live test, and therefore hazardous. Make sure the leads and insulated probes are undamaged and that you study and completely understand the instructions before carrying out the test. Prospective fault current measurements for three-phase systems are particularly dangerous as the voltage between phases is 400 V and you will be working with exposed terminals in order to obtain your reading.

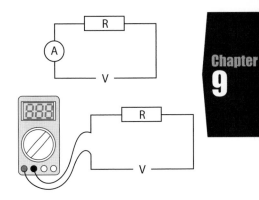

Figure 9.11: An ammeter in a circuit

Figure 9.12: A clamp-on ammeter

Figure 9.13: An RCD tester

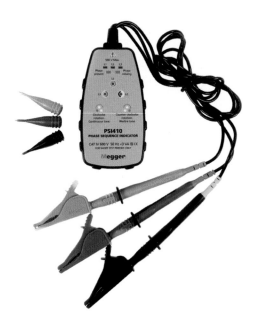

Figure 9.14: A phase rotation instrument

RCD tester

The residual current device (RCD) is intended to shut off the supply in the event of an earth fault. The device should work within its stated tripping time. If a fault occurs and the RCD does not operate correctly, or does not operate at all, it should be tested using a specialist test instrument. Many earth fault loop impedance testers are fitted with an RCD test function.

1 Press the test button on the RCD itself – the RCD should operate. If it doesn't then the RCD is faulty.

2 Carry out an RCD test using an RCD tester.

 • Test at half fault current – the RCD should not work. If it does then there is a problem and the device will operate at less than the intended fault current. This can cause 'nuisance-tripping', which means that the RCD will shut down the supply even when there is no fault.
 • Test at tripping current – the RCD should operate within its stated tripping time.
 • Test at five times tripping current – RCD should operate well within its tripping time.

If the RCD fails any of these tests it should be replaced and retested. If the new RCD fails, an earth fault loop impedance test must be carried out to establish if the earthing system and, in particular, the fault path is sound.

If the result exceeds the maximum values set out in BS 7671:2008 then a number of actions must be carried out, including replacement of the earth bonding conductors with cables of larger cross-sectional area.

Phase rotation

The voltage and currents in a three-phase system should rise and fall in a particular order: line 1 – line 2 – line 3. If the connections to a three-phase supply are incorrect, then this order will be disrupted on the load side of the reversed connections. A three-phase electric motor shows up this type of problem because it will run in the opposite direction to the one intended. Apart from this, however, it is very difficult to tell if the phase sequence is correct without using a phase rotation meter. The test instrument is connected to the three line terminals at the load or on the customer side of the supply point. Correct phase rotation is indicated using lights or an LED display.

HSE Guidance Note GS38

HSE Guidance Note GS38 lays down requirements for safe test instruments. These include the need to inspect an instrument before use. Look out for damaged leads, plugs and casing. If there is any sign of damage the instrument must not be used. GS38 also states the minimum safety features of test equipment, for example:

 • test probes must be insulated and fitted with finger guards; the probe tip must be no longer than 4 mm
 • test leads must be insulated and colour-coded
 • leads must be permanently connected to the probes with no conductor showing whatsoever.

Progress check 9.4

1 Which test instrument is used to carry out a continuity test?
2 How could you safely use an insulation resistance tester to find an earth fault in a circuit containing sensitive equipment?
3 What is the maximum length for the tip of a test probe on a test instrument?

Causes of electrical faults

Before you can begin the fault-finding process, it is vital to understand what an electrical fault actually is. Faults can result in the complete shutdown of a circuit or item of equipment, intermittent or partial failure, or catastrophic short circuits or shock. All these, however, are caused by a small number of actual failures within an electrical system.

High resistance

Electrical conductors have a low resistance (see *Chapter 1: Principles of electrical science*). Low resistance means that current can flow easily. If the resistance increases, less current can flow and heat will be generated. A high resistance fault is one where this has occurred. The reduced current flow can cause malfunctions while the high temperature poses a fire risk. Common causes of high resistance are:

- loose connections
- stranded cables terminated with strands cut away (sometimes done to make it easier to fit into a terminal). This is bad and unsafe practice because it reduces the cross-sectional area of the cable and this means that it can carry less current
- damaged conductor
- damaged winding or heating element.

Insulation failure

The purpose of insulation is to prevent both shock and short circuit by creating a barrier between you and a live conductor and between live conductors. BS 7671:2008 states that basic insulation should be provided in such a way that it can only be removed by destroying it, for example stripping it away with a cable stripper or knife. Insulation must, therefore, be sound. This is checked by the insulation resistance test which measures the resistance between two insulated conductors or parts of an electrical system.

Overcurrent

Overcurrent is described by BS 7671:2008 as a current that is more than the rated value of a conductor. In other words, a cable is carrying too much current for its size. There are two main causes of overcurrent, a short circuit or overload, both of which are described below.

Short circuit

When a circuit is working normally, the current travels along the conductors to the load. The load has a set resistance which will only allow a certain amount of current to flow. The cable is sized according to how much current will be flowing through the load, e.g. a 10 A load will require 1.5 mm² cable and a 30 A load will need a 6.0 mm² cable. If the line and

Figure 9.15: A burned-out connector

Figure 9.16: A burned-out terminal block

> **Key term**
>
> *Cable rating* – conductors will carry a certain amount of current. This is their rating and it depends mainly on the cross-sectional area of the conductor.

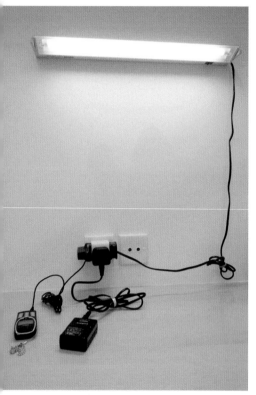

Figure 9.17: An overloaded 13 A socket

neutral or line and earth come into accidental direct contact, the load is bypassed and a very high current will flow. This is far beyond the current-carrying capacity of the cable and will cause a massive release of heat and explosive energy. The same will happen if a live conductor comes into direct contact with earth.

Overload

A short circuit will, of course, cause an overload. Damage can also occur when the current taken by the load itself is higher than the rating of the cable, for example (and this is an extreme example!) a 6 kW load fed by a 1.5 mm² cable.

The current taken by a 6 kW load is $\frac{6\,000}{230}$ = 26 A (approximately).

A 1.5 mm² cable will carry approximately 16 A. The cable will, therefore, become hot and this could cause a fire. The cable selected must be large enough to carry the load current. It must also be backed up by a protective device, rated to operate if the cable carries too much current.

A jammed electric motor can also cause an overload. The motor can no longer rotate but continues to call for more current until the motor windings overheat and burn out. Motor starters are fitted with a protective device called an overload cut-out and this is designed to operate and cut off the supply if the current exceeds the overload setting.

Open circuit

An open circuit occurs when a connection fails and the conductor is no longer connected to its supply or its load. It can be caused by a broken conductor or a faulty item of equipment, such as a malfunctioning switch which will not close.

Earth fault

Although a direct connection between a live conductor and earth will cause an overcurrent, it also poses a shock risk. The earth beneath our feet acts as a negative terminal for our mains electrical supply. This makes it the point to which electrical current naturally flows and a path is provided for the returning current along the neutral conductor, to the star point of the local sub-station transformer which is, in turn, connected to earth via an electrode. The amount of current is governed by the size of the load (see above).

An earth fault provides a low impedance path between line and earth. As a result, a high fault current will flow from the sub-station transformer to the fault point. The point of fault may be the metallic case of an item of electrical equipment or an appliance. This is now live and anyone touching it will provide another path to earth through their body and receive an electric shock.

In most modern electrical installations, additional protection against shock is provided by a residual current device (RCD). This is an automatic switch designed to operate and cut off the supply if there is an earth fault. Chapter 8, page 433, describes RCDs in more detail. RCDs must be tested regularly to confirm that they are in working order. If an RCD fails to operate it can result in an electric shock. (Page 454 describes the RCD tests in detail.)

Activity 9.4

An area of a farm is illuminated by a halogen floodlight, operated by a waterproof metal switch. The light is supplied from a 6 A circuit breaker via an RCD. Someone has reported receiving a mild electric shock from the switch, although it still works. The switch is old and looks a little corroded. What is going on?

Figure 9.18: Lightning

Transient voltages

A transient voltage is a momentary high voltage in a circuit and can cause a lot of damage and even result in fire. Transient voltages can be caused by:

- lightning
- high voltage (HV) switching faults
- faults on HV power lines.

To protect against the effects of transient voltages, surge protectors are fitted into HV supplies and can also be connected to sensitive equipment. They work by diverting the extra current to earth, or by blocking the current within themselves.

Progress check 9.5

1   State two faults that cause a high resistance joint or termination.
2   What is overcurrent?
3   What could cause a power failure in part of a circuit?

Typical faults

Symptom	Possible cause	Remedy
No power to any part of a circuit or installation	Protective device has operated due to: • short circuit • overload • earth fault	Insulation resistance test and repair or replacement of faulty cable or component
No power to part of a circuit or its load	• Broken conductor • Broken connection • Loose connection • Local fuse blown	Continuity test and/or throughput test, then repair or replacement of faulty part If the local fuse has blown then an insulation resistance test should be carried out on the load side of the circuit – but only if it is safe to do so. See page 408, Table 8.2
Intermittent power	• Loose connection • Broken or partially broken conductor that is still allowing some current to flow	Continuity test and repair or replacement of faulty part
Burning smell	• Loose connection • Windings or coil burning out • Incorrect-sized cable or flex • Incorrect component installed • Damage from a power surge – often caused by lightning strikes	Visual check and/or continuity test A substitution test might also be appropriate
Flickering fluorescent tube	• Faulty tube • Faulty starter • Faulty choke	Replace the faulty part • Faulty tubes usually strike but flicker while they are on • Faulty starter causes the tube to try to strike but with no success • If the choke is faulty the ends will glow but the tube will not strike

Table 9.3: Typical faults and how to fix them

9

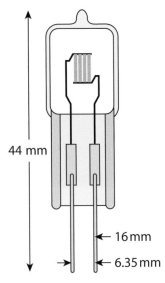

44 mm

← 16 mm

← 6.35 mm

Figure 9.19: A lamp filament

Safe working

Lamps will become extremely hot while they are on, so never touch a lamp straight after it has failed. Also, lamps will implode if broken which means that the jagged edges of the fragments from the opposite side of the lamp will fly towards you if a lamp shatters. Finally, some lamps contain chemicals which are irritating to the skin if handled.

Figure 9.20: A screw-it connector

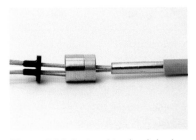

Figure 9.21: A mineral-insulated gland

Common fault locations

There are certain parts of an electrical installation where faults are more likely to occur than others. They are often the best places to start when trying to track down and identify an electrical fault.

Lamps

One of the most common faults found in electrical installations is lamp failure. Lamps cannot be repaired so the only remedy is to replace them. Some lamps such as fluorescent tubes can become less efficient as they get older. The ends tend to blacken and the light output decreases. In large installations where there are many hundreds of lamps, regular mass replacement is part of the maintenance schedule.

Lamps can also 'blow' suddenly. This is caused by a burnt-out filament which eventually ruptures. The result is a very small, but quite violent explosion. When this happens, a short circuit can occur between the line and neutral sides of the filament connections. The protective device will operate in these cases.

Terminations

The point where a cable is connected to an accessory or piece of electrical equipment is called the termination. This can be in the form of:

- a connector block
- post-type terminal
- push-fit terminal
- soldered termination.

The two main problems that occur at a termination are:

- broken conductor – the cable snaps off and causes a loss of supply, short circuit or earth fault
- loose conductor – the connection point becomes hot because of arcing between the loose conductor and the terminal. This also causes momentary dips in, or loss of current to, the load. For example, a lamp might flicker if there is a loose termination in its circuit.

Cable joints

Similar to terminations, cable joints are points at which:

- additions are made to an existing circuit, for example a spur added to a ring final circuit using a junction box
- a cable is extended, for example because an accessory or distribution board is replaced or moved and the cables will no longer reach.

Cable joints should be contained in a junction box or, if made using a connector block or through-crimps, contained within some sort of enclosure. You may find screw-it type joints in older installations. These must be replaced if found. All cable joints must be accessible, though not necessarily easily accessible. A cable joint can be located under a floor or in an attic space. They should never be sealed into a stud wall.

Mineral-insulated cable glands

The termination glands fitted to mineral-insulated cables are designed to protect the highly absorbent insulation from moisture ingress. Seals should be firmly crimped in place, but they can work loose, for example if they are subjected to constant vibration.

Another fault can occur if the conductors are twisted while the seal is being fitted to the terminal pot. Because the conductors are not covered with insulation within the cable and inside the gland, any contact between them will cause a short circuit. See page 134 for a detailed description of mineral-insulated cable.

Coils and windings

Contactors and relays work using an electromagnetic coil. If the coil is damaged it can overheat and burn out. The conductors used in windings are often insulated with a chemical lacquer. If this insulation becomes hot it will melt and cause a short circuit between the turns of the windings. The windings in motors and transformers can also burn out in the same way.

Implications of unsatisfactory readings or diagnosed faults

Once a fault is found it may simply be a case of carrying out a repair and restoring normal service. On the other hand, there might be warranty or contractual implications which means that you will either be obliged under the terms of those agreements to carry out the repair free of charge, or you will not be able to work on the faulty equipment because it will void the warranty.

If an unsatisfactory reading is taken during an inspection and test it will be recorded on the test documentation. A set of fault codes is shown on page 441 of this chapter. Again, decisions will need to be taken at that point as to whether the test electrician repairs the fault or whether it will be carried out by a specialist company or under a different contract. If the fault is severe and poses a hazard to the users of the system then it must be repaired as quickly as possible. The implications of fault diagnosis and repair are described in more detail on page 465.

CARRYING OUT FAULT DIAGNOSIS ON ELECTRICAL SYSTEMS

LO4

There is no point simply rushing in without any plan or structure to open enclosures and carrying out the first test that comes into your head. You will need to take the actual fault diagnosis process step-by-step. The actions you take will depend on the fault. If a circuit breaker has operated then you need to identify the affected circuit and begin there; if there are obvious signs of damage to a fitting or accessory then that would be your starting point. Also, a step-by-step approach helps to clear your mind and think through what the fault might actually be.

Use a logical approach for locating faults on electrical systems

Check protective devices

If there is a loss of power, the best place to start is at the protective device. This will tell you if there is any current at all in the circuit. If the power loss affects only part of the circuit, the protective device will be on. If it affects all of it, the protective device may well have operated.

Progress check 9.6

1 State two faults that can occur at a terminal.

2 What action should be taken if screw-it connectors are found?

3 State one fault that can occur in a mineral-insulated cable termination.

You will also be able to identify the rating and type of device. It may be that the wrong type of circuit breaker has been installed, for example the load is a motor which requires a high current at start-up. The breaker is a type which does not allow for this and operates the moment the current rises above its rating. On the other hand, you may see that the circuit breaker or fuse is not rated for the load and that the circuit needs upgrading. Remember, this may mean replacement of the cable as well.

You may discover that an RCD is faulty by pressing its test button. The problem may be nuisance-tripping. An RCD test will be needed, at 50 per cent of its tripping current.

Check for supply voltage – throughput test

Even if the protective device is in the on position, it may still mean that voltage has been lost on its supply side.

There may also be power at the supply end of the circuit, but somewhere along its length the circuit has been opened and current is not able to reach the load. This fault can be traced using the throughput method.

Starting at the part of the circuit which has lost its supply, work back with a voltage indicator until voltage is found. If the reason for the loss of power is not obvious, then further tests such as continuity tests can be carried out from that point (with the supply turned off).

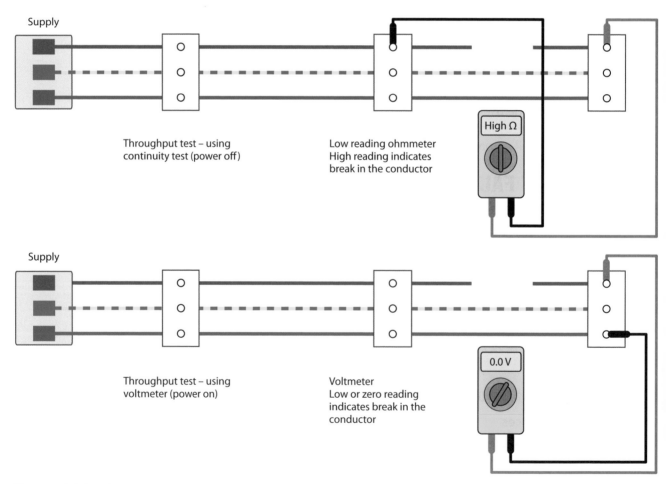

Supply

Throughput test – using
continuity test (power off)

Low reading ohmmeter
High reading indicates
break in the conductor

High Ω

Supply

Throughput test – using
voltmeter (power on)

Voltmeter
Low or zero reading
indicates break in the
conductor

0.0 V

Figure 9.22: A throughput test

Check loads and equipment

If there is a short circuit in a ring final circuit into which a number of loads have been plugged, the first step would be to unplug all the loads and switch the protective device back on.

- If the device *doesn't* operate, then it is highly likely that the problem is with one of the items of equipment and not the ring final circuit.
- If the device *does* operate, then the problem is with the ring final circuit itself.

The same method can be used for most other circuits.

In the case of radial circuits the same effect can be achieved by switching off the isolator for the load.

It will be possible to carry out tests on some equipment but not others. Equipment repair is often a specialism and a repair engineer may well have to be called in.

Half-split test method

Removing a cover or an accessory plate (such as a switch or socket front) may reveal the fault. On the other hand, removing every socket outlet or light fitting in a large circuit could be extremely time-consuming. In this case, the circuit needs to be broken down into sections. Disconnect the circuit at a certain point to isolate a specific area of the circuit. Test this area using:

- continuity test if the circuit, or part of it, doesn't work
- insulation resistance test if the fault has operated a protective device.

If the fault is not found, isolate another section of the circuit, and so on until the fault is located. The separated section can be further broken down until the fault itself is pinpointed.

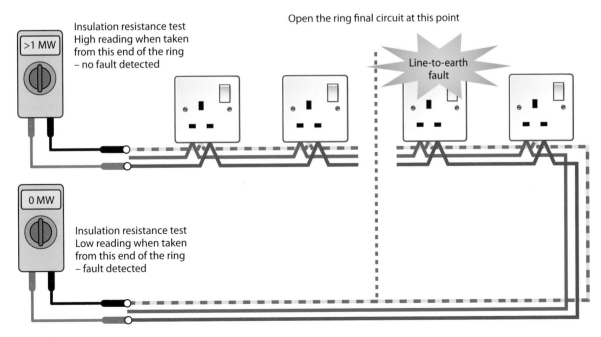

Figure 9.23: Split-circuit test

Typical faults are:

- broken or disconnected conductor
- failed part
- failed insulation leading to short circuit and overcurrent
- burned out windings, coil or element.

Long lead method

If the fault is an open circuit or suspected loss of continuity due to a loose connection, it is a good idea to use a Method 2 type continuity test on the faulty conductor. To do this you will probably need an extension or 'fly' lead.

1   Connect one of the instrument (a low reading ohmmeter) leads to one end of the faulty conductor.

2   Connect the other instrument lead to a fly lead.

3   The test instrument should either be nulled with the fly lead in place, or the resistance of the fly lead taken so it can be subtracted from the final reading.

4   The fly lead is connected to the other end of the faulty conductor.

5   A continuity test is taken using a low reading ohmmeter.

6   A high reading will indicate a loss of continuity.

Other methods

- Substitution – sometimes it isn't possible to tell if a component is faulty simply by looking at it. If this is the case the part should be swapped for a new or functioning one.

  o If that new part works then the fault is with the old component.
  o If the new part does not work then further investigation is needed.

  Substitution is often used when diagnosing lighting faults and should be the first test carried out before removing covers and switching off the supply.

- Equipment self-diagnostics – some equipment incorporates readouts or is fitted with meters that indicate normal and abnormal readings. Details of the normal readings should be obtained and compared with the reading on the equipment indicators. More complex systems may have alarms to indicate that there is a fault while others can be connected to a computer which will provide diagnostic information. This type of fault diagnosis often requires specialist training and knowledge.

Record throughout

Make a record of all readings and findings as you go. This will help with the fault diagnosis process, particularly if the fault is complex. The readings you take will also be entered onto the fault diagnosis documentation.

Document fault diagnosis

The final diagnosis of the fault must be recorded. This is described in more detail in *Chapter 8: Electrical installations: inspection, testing and commissioning*.

Progress check 9.7

1   What is a throughput test?

2   Describe the half-split method for fault finding. Break a circuit into sections.

3   Describe the long lead method for measuring the continuity of a conductor.

THE FAULT RECTIFICATION PROCESS

LO3

Once the fault is diagnosed and its location identified, decisions have to be made about whether it can be repaired, how it will be repaired or what other action should be taken. Also, following any repair, a full set of inspections and tests must be carried out.

The rectification process

Rectify fault

If it is safe and appropriate to do so, the fault should be rectified. To do this you may have to:

- effect a repair – it may be possible to simply repair the faulty part, for example re-strip and re-terminate a cable or tighten a loose terminal screw. This does not mean that you should 'bodge' a repair by sticking a broken part together using insulation tape! Repair should only be carried out if the end result is a fully rectified fault and a safe circuit
- replace a part – this can be anything from a faulty switch to a printed circuit board. Always make sure that the replacement part is the right component for the job and that you have the information you need to correctly install and connect the component
- rewire – if it is not possible to re-terminate or joint a damaged cable, the faulty cable must be removed and replaced. This may mean damage to the building structure, so this type of repair must be discussed with the customer before it takes place. The damage will need to be made good and this may require the services of a decorator or other relevant trade.

Retest following fault rectification

Once a repair has been carried out the equipment or circuit needs to be put back into service. A number of tests must be completed before service is restored. It is important that the appropriate inspections and tests are carried out on the affected area. This is because further faults may have been caused by the fault diagnosis and repair process. For example, a length of metal trunking lid, removed then replaced, may have damaged the insulation on a cable, causing a second fault. The type of inspection and test certificates issued depend on the extent of the rectification work. There are three main types of test certificate, listed below.

- Electrical Installation Certificate – any work that involves the installation of a new (or replacement) circuit. The inspection and test only needs to be carried out on the affected circuit.
- Minor Works Certificate – any alteration to an existing circuit. As with the previous test, the inspection and test only needs to be carried out on the affected circuit.
- Periodic Inspection and Test Condition Report – issued for a routine test carried out on an existing installation. The customer may want you to carry out a periodic inspection and test once the fault is rectified to make sure that the rest of the installation is sound.

These tests are described in detail in *Chapter 8: Electrical installations: inspection, testing and commissioning*.

A functional test should be carried out on any equipment that you may have repaired as a result of fault diagnosis. Run the equipment or circuit and make sure it is working properly before handing it back to the customer. This may require the presence of the user because some equipment requires training to operate. Specialist equipment may have to be recommissioned and this may require a specialist to attend.

Only when a complete set of tests have been carried out and the electrician and equipment user are satisfied that the fault is cleared, and that the equipment or circuit is safe to use, should the fault be closed down and normal service resumed.

Restore installation to original condition

Once the circuit or equipment has been repaired and retested and is working normally, the installation must be restored to its original condition.

- Replace trunking and conduit lids.
- Replace covers generally, e.g. consumer unit and distribution boards and enclosures.
- Replace switches and other accessories and tighten screws.
- Tighten fixing screws.
- Reset time switches and thermostats and other automatic sensing or switching devices.
- Switch on circuit breakers and reset RCD.
- Switch on isolators.
- Refit lamps and ensure diffusers are in place.

Any damage to the fabric of the premises will need to be repaired. This may involve the simple replastering of small holes in the wall or ceiling, or it may require replacement of plasterboard, floorboards, or even wallpaper. If you do not feel skilled enough to effect a professional repair then a specialist should be called in. It is always good to warn a customer at the start of the work that this may be the case. You may have to negotiate the cost of making good with the customer.

Waste disposal

Any waste such as packaging and discarded faulty parts must be disposed of in a way that will not affect the environment or cause a health or safety hazard. The Control of Waste Regulations require that waste products are disposed of correctly and that specialist companies are employed to safely dispose of any hazardous materials.

Record and report

The inspection and test certificates and schedules of test results form the greater part of the documentation required on rectification of a fault. These must be distributed to the relevant parties and kept as a record of the work.

Your company will have its own documentation – a fault report, perhaps, that will detail not only the fault but the action you took to diagnose and rectify the problem. It may include a customer sign-off as a record that the fault has been cleared to their satisfaction.

It is important that records are kept, both as evidence that the work took place and the fault was cleared, but also as an information source for any future fault diagnosis that may occur in that premises.

Factors which can affect fault rectification

There are a number of issues around the actions to be taken after a fault is found. If it is a straightforward fault requiring a simple repair, then this should be carried out as soon as possible. Whether or not the electrician who found the fault is the person to carry out the repair work depends on the contract between the electrician and the customer.

There are times when repairing a fault is very difficult or expensive. There are a number of issues to consider.

- Is it cost-effective? In other words, would it be cheaper to replace the whole piece of equipment or even rewire the affected area?
- Is the repair a specialist job? Increasingly, equipment contains electronic circuitry which cannot simply be repaired with a pair of pliers and a screwdriver. It may have to be reset and recommissioned once repairs have taken place.
- Is it safe to carry out a repair? The location of the fault may be in a hazardous area and require specialist equipment or PPE to enter the area and carry out the work. For example, it may be near machinery or in a confined space.
- Is the equipment or installation still under guarantee or warranty? In this case, to replace or open any covers and remove any parts would void the warranty.
- Are there contractual issues? Is it part of your agreement with the customer to carry out repairs or only to test, inspect and diagnose a fault?
- Are replacement parts available or is the equipment or its components obsolete? If so, the equipment may have to be replaced with a newer version.
- Will the repair work have to be undertaken out of working hours? For example, there may be limited access during normal working hours, or it may not be possible to shut down power supplies until the evening or at weekends.

Whatever decision is made about repairing a fault, it cannot simply be left as it is. If immediate repair is not possible, any faulty equipment and circuits should be isolated so that they cannot be used. For faulty circuits, their fuses must be removed or circuit breakers locked in the 'off' position.

A temporary supply may have to be set up until the fault can be repaired and normal power restored. Barriers should be erected around isolated equipment and notices placed to warn that it must not be used. Any exposed cable ends must be terminated into a junction box or similar enclosure to prevent anyone coming into contact with live conductors.

Progress check 9.8

1 What are the three main types of fault rectification?
2 Why carry out a full inspection and test after diagnosing and repairing a fault?
3 State two issues to be considered before carrying out fault rectification.

Knowledge check

1 Which safety procedure document must be completed and authorised if the electrician has to work in a hazardous area?

a Fault record
b Permit to work
c Schedule of Requirements
d Minor Works Certificate

2 Which of the following does *not* pose an electric hazard when the supply is switched off?

a Uninterrupted power supply
b Generator
c Functional earth
d Photovoltaic panels

3 The first three steps of the fault diagnosis process are:

a analyse the evidence, check the supply and check protective devices
b identify symptoms, gather information and analyse the evidence
c switch off power, erect barriers and warning notices, and lock off the circuit breaker
d check protective devices, isolate and test, and rectify the fault

4 Information about an item of electrical equipment or a system such as a heating or alarm system can be found in:

a manufacturer's instructions
b building specification
c maintenance and test records
d fault report

5 The test instrument used to investigate a short circuit is a(n):

a low reading ohmmeter
b earth fault loop impedance tester
c phase rotation meter
d insulation resistance tester

6 The protective device that operates in the event of an open circuit is:

a RCD
b none
c fuse
d circuit breaker

7 The faulty operation of an RCD is called:

a intermittent-fault tripping
b earth-fault tripping
c nuisance-tripping
d hair-trigger tripping

8 A voltage value is read using a:

a voltage indicator
b voltmeter
c voltage surge detector
d voltage monitor

9 An ammeter must be connected in:

a parallel with the circuit
b series with the circuit
c series–parallel with the circuit
d synch with the circuit

10 Trying to establish the cause of a fault by putting in new components and switching equipment or a system back on is called:

a load and equipment check
b throughput testing
c self-diagnosis
d substitution

Understanding the fundamental principles and requirements of environmental technologies

This chapter covers:

■ solar heating and electrical production

■ electricity generation from wind and water

■ heat from ground, air and biomass sources

■ the principles of water conservation and reuse systems.

Introduction

Micro-generation and renewable technology is a new and exciting field with many opportunities for professionals, communities and individuals who want to be a part of it. This chapter introduces the most common technologies now being employed in the micro-generation and water conservation industries.

If you are training as a micro-generation installer, then this chapter will provide you with the fundamental working principles of each technology and the potential to install them (depending on location and regulatory requirements).

You will discover how much heat or energy may be gained from each type of system, and the benefits and savings available. You will also examine the legal and planning constraints on different types of systems and the factors which need to be considered before embarking on an installation.

This chapter will also prepare you for the specialist knowledge and competence units for the installation, commissioning, handover, inspection, service and maintenance of micro-renewable energy and water conservation technologies. The knowledge check questions will help you prepare for relevant tests at this level of study.

LO1

THE FUNDAMENTAL WORKING PRINCIPLES OF MICRO-RENEWABLE ENERGY AND WATER CONSERVATION TECHNOLOGIES

References are made in this chapter to **MCS** standards; these are standards of manufacture or installation as set out by the Micro-generation Certification Scheme, which is the industry-led quality assurance scheme. You will need to ensure micro-generation equipment and installers meet the requirements of the MCS. If you are training as an installer then you will need to be aware of their specialist training and certification process.

You will examine a range of micro-renewable technologies, from the very new, like solar photovoltaic electricity generation, to the old, such as water pressure (hydro) powered systems and wind power. You will learn:

- the basics of how each system works
- the important parts of each system
- how much power, electricity or heat could be generated.

The different technologies are organised according to the production of heat, power or a combination of both. Water conservation technology is also included as this is an energy- and resource-saving technology becoming increasingly popular in the domestic market. Many of the technologies have been developed from commercial-scale applications and are now becoming available for the growing domestic and micro-generation market.

Key term

MCS – the micro-generation industry's own standards scheme. Installers can become certified through the MCS and homeowners wishing to take advantage of government incentives such as the feed-in tariff or renewable heat incentive must use an MCS-certified installer and products (see the MCS website which is at: www.microgenerationcertification. org/)

Heat-producing technologies

Solar thermal

The technology of capturing energy from the sun and using it to heat water is relatively simple in principle. A closed loop system carries a volume of water which absorbs heat from the sun through a collector, then passes through a standard hot water storage cylinder, and in doing so heats up the water contained in the cylinder. As the liquid heats up the stored water, it loses its own heat, then returns back to the collector to regain heat energy from the sun and the cycle continues. The water heated by the closed loop system can then be used by the consumer for washing and heating in the same way as conventionally heated water. This system of transferring heat from one liquid or gas to another is known as a **heat exchange** and takes place in many renewable technology thermal systems.

The system usually supplements a conventional gas or oil heating system which provides additional heat to the storage cylinder when demand is high or when the solar system is not producing enough heat. Solar thermal systems may occasionally be designed to provide direct hot water for immediate use, instead of supplementing a conventional heat source. However, northern European climates do not offer enough consistent **irradiance** to make this type of design reliable or economically viable.

A good solar thermal system should be able to produce 400–600 kWh/year depending on the type and position of collector.

The system has a number of important components. They are:

- the collector
- the pipework and heat transfer fluid
- a storage tank/cylinder or heat exchange system linked to hot water/ heating
- a supplementary or existing heating system
- a cold water supply.

The collector

There are several types of solar heat collector. The two most common types are:

- flat plate collector
- evacuated tube collector.

In a domestic setting the collector is usually mounted on the roof and thus requires protection from frost and other extremes of climate.

Flat plate collectors have a continual loop of pipework mounted on a flat absorber panel, sealed over with a glass sheet. This design employs a **greenhouse effect** to trap heat from the sun and transfer it to the water in the pipes. The glass sheet also provides physical protection for the pipework underneath. To maximise heat gain, the absorber and exposed pipes should be matt black in colour and the absorber should be insulated underneath to prevent heat loss.

Key terms

Heat exchange – a system where heat is transferred from a warmer medium to a cooler medium. The media involved are usually fluids.

Irradiance – a measure of the sun's energy at a particular location, measured in watts per square metre (W/m^2).

Greenhouse effect – using glass or another material to trap and reflect sunlight inwards, thus raising the temperature within a structure.

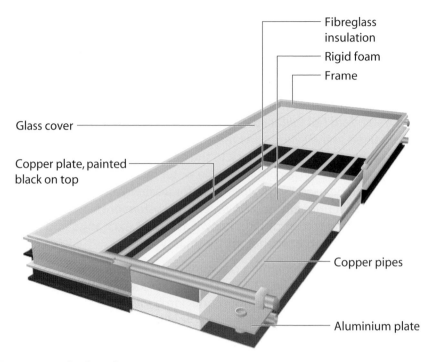

Figure 10.1: A flat plate collector

Evacuated tube systems employ a series of closed, vacuum cylinders, each containing a long flat collector plate with a single 'heat pipe' mounted on it. The 'heat pipe' contains water which boils and moves up the pipe where it discharges heat through a copper tip before condensing and returning back down the pipe and beginning the cycle again. The copper heat exchanger passes the heat on to a secondary circuit which then operates in a similar manner to the closed loop system described above.

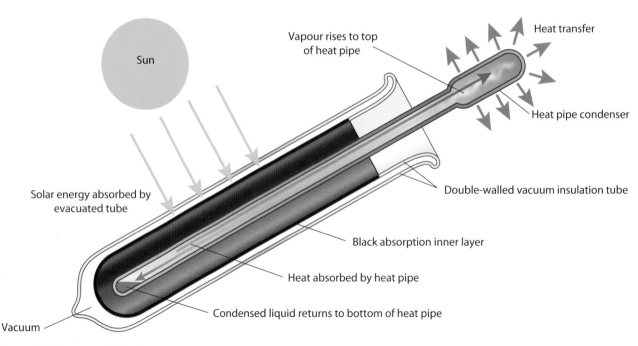

Figure 10.2: An evacuated tube collector

While any solar collector is mounted at an angle to optimise irradiance from the sun, evacuated tube collectors also use gravity to aid the evaporation/condensation cycle and thus they will not function if laid horizontally.

Pipework and heat transfer fluid

Pipework from the solar collector, through the hot water cylinder and back to the solar collector is referred to as the primary circuit. This consists of:

- the collector
- pipework carrying the fluid towards the storage cylinder (flow) and away from the cylinder back to the collector (return)
- a circulating pump
- drain and fill points
- associated pipe insulation and protection.

The heat transfer fluid is normally water and should have an antifreeze (glycol) content of up to 50 per cent. Glycol, an organic compound with antifreeze properties, was originally produced from ethylene and has a wide range of industrial applications. Solar-thermal glycol products are made specifically for use in solar installation applications and a corn-based glycol is now available with the same properties as traditional ethylene. Pipework should be insulated where exposed to frost, i.e. where pipes are penetrating the roof or other exposed locations. The drain point should be close to the lowest part of the pipework.

Storage tank/heat exchanger

Many homes have a hot water storage vessel in the form of a copper cylinder. In general heating and hot water applications this cylinder acts as a heat exchanger, in that it exchanges heat from a primary circuit fed by a boiler, fire or other heat source to a secondary circuit which forms the hot water supply. The hot water cylinder performs an identical function in a solar thermal system except that the primary heat supply circuit is replaced by the solar circuit and the conventional heat source should only be required as back-up. The cylinder should be fitted with high (top) and low (bottom) temperature sensors, relief valves and insulation (usually polyurethane foam).

Supplementary heating system

The solar heating system cannot be guaranteed to supply 100 per cent of the hot water demand at all times, therefore a supplementary system should also be in place. Since solar heating systems are often retro-fitted there will generally be an original heating system in place, usually a conventional gas or oil boiler, open fire or stove, or other heating appliance. This will normally become the supplementary system, used when there is low or no solar input or in times of exceptionally high demand.

Cold water supply

In principle, the cold water supply is not part of the solar thermal system but it must be considered as part of the hot water system as there must be a reliable supply of water to the heat exchanger. This is usually via a cold water storage tank placed in the loft or at height to allow the tank to be gravity fed.

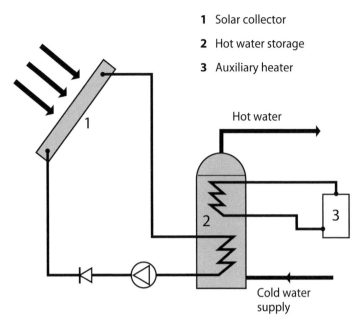

1 Solar collector

2 Hot water storage

3 Auxiliary heater

Figure 10.3: A solar thermal system

Ground source heat pump

It may be surprising to learn that the ground is often warmer than the air above it. This is because much of the heat energy from sunlight falling on the earth's surface is absorbed by the ground and released only very slowly. Ground temperatures at depths of up to about 2 metres rise and fall with seasonal temperatures but lag behind by about 1 month. At about 6 metres deep the ground temperature approaches a stable position at a temperature similar to the average air temperature above ground, the overall UK average air temperature being 10–14°C. After about 15 m deep the ground temperature begins to rise again, due to geothermal energy which is heat flowing outwards from the interior of the earth.

A ground source heat pump (GSHP) exchanges a wide volume of low temperature heat from the ground into higher temperature heat for dissipation within the home, using a collection system and a heat exchanger.

Heat is drawn from the ground into the collection system through liquid sealed within pipework which is buried at a suitable depth in the ground. The liquid, a water-glycol mix similar to that used in solar thermal applications, is pumped through the collection system and passed through a heat exchanger at ground level. The heat exchanger is connected to a domestic heat dissipation system, often underfloor heating, although radiators and space heating may also be designed in. The pipework is usually coiled and buried below frost level and at a pre-determined depth in a horizontal trench, usually at a minimum depth of 1.5 m. However, careful consideration must be given to the correct trench depth. A number of factors, including seasonal temperature and the thermal conductivity and general nature of the ground material, must be considered.

In areas where rock is close to the surface a trench system may not be suitable. In such cases a vertical 'borehole' design may be used. This design

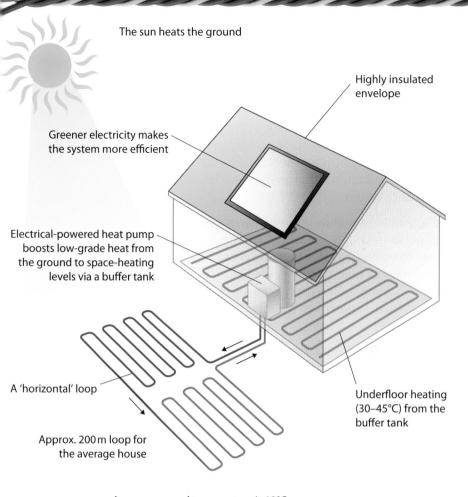

The sun heats the ground

Highly insulated envelope

Greener electricity makes the system more efficient

Electrical-powered heat pump boosts low-grade heat from the ground to space-heating levels via a buffer tank

A 'horizontal' loop

Underfloor heating (30–45°C) from the buffer tank

Approx. 200 m loop for the average house

Average ground temperature is 12°C

Figure 10.4: A ground source heat pump system

may also be employed where sufficient ground area is not available for adequate trench length, and is often used in compact or urban sites.

Ground source heat systems deliver a steady but relatively low level of heat and require a pump to push the fluid through the underground pipework. The pump will need to be operated electrically and so, for the ground source heat system to be viable, it must have a sufficiently high efficiency ratio, known as a coefficient of performance (CoP). The CoP compares the energy required to drive the system, i.e. to power the pump, with the heat energy gained from the system. A system with a high demand load, i.e. including radiators and space heating, will have a lower overall efficiency, whereas a system utilising only base heating requirements, i.e. an underfloor system, will have a much higher CoP. The input temperature also has an effect on the efficiency; the lower the overall 'uplift' (the difference in input and demand temperatures) the higher the overall efficiency. A CoP of 2.5 or above should be considered adequate, and up to 4 can be expected of well-designed systems, while below 2.5 the system may not be economic.

GSHPs can be considered carbon-neutral at point of use, i.e. no carbon is emitted in their use where they deliver the heat. However, the electricity used to pump the system must be considered.

Air source heat pump

Similar to ground source heat systems, air source systems extract heat from the air at low temperature and use heat exchanger technology to transfer the heat to the primary heating system of the house. An analogy often used is that of a refrigerator in reverse: the system takes heat from the surrounding air and pumps it into the heat exchanger. Heat pumps may be positioned to take advantage of warm exhaust air, e.g. by extracting heat from kitchen or bathroom ventilation outlets. The system should ideally be used in conjunction with conventional heating systems, most usefully an underfloor system.

Air source heat pumps resemble air conditioning units in size and shape, and perform largely the same function but in reverse. They require the space to position a large unit adjacent to an external wall on the outside and ductwork guiding the warm air into the house or to a heating system manifold. They should be positioned to allow a clear flow into the air intake with no obstructions and the minimum possibility of blockages, e.g. from foliage, vegetation, litter, etc.

Noise is an important consideration and units should be sited away from neighbouring properties and mounted on anti-vibration pads. The noise produced by a heat pump unit is comparable to that of a large refrigerator. Any commercially produced unit should meet the MCS-approved maximum noise limit of 42 dB. Heat pumps will also produce a small amount of condensate which should not be allowed to gather within or near the unit and should instead be directed away to a suitable drainage point.

Figure 10.5: A heat pump

Biomass

The term 'biomass' refers to the fuel source which has been derived from organically produced material, either grown specifically for this purpose or harvested as a waste product from agriculture or the timber processing industry. Commonly used biomass fuel can be produced from harvesting dedicated short rotation crops such as poplar or willow, which is then dried and processed into chips or pellets and distributed to the consumer. Timber waste produce or the products of timber recycling, often timber pallets, can also be processed into suitable fuel for biomass boilers. If raw timber logs are to be used as a fuel they should be allowed to season (dry out) for about 1 year. Agricultural waste such as chicken litter or animal manure and municipal waste can also be processed into biomass fuel, although this is a specialised application usually employed in larger community heat and power systems.

Burning biomass products is currently the most commonly used renewable energy technology in use in the UK. This is largely due to the tradition of home heating through open fires in habitable rooms or the use of a log burning stove in the kitchen. Many traditional houses still have redundant fireplaces in all the bedrooms. These would have been originally used for heating the room and would have burned wood or coal.

Biomass heat-generating systems have evolved from traditional log burning and, while many systems still use that type of technique, modern systems now usually burn processed wood pellets or wood chips in a specialised boiler directly linked to the domestic heating and hot water system. Specially designed modern boilers allow much more control of the combustion process and collection of heat than traditional stoves or an open fire. Fuel pellets are also designed to yield the maximum heat potential of the material.

While burning biomass releases carbon dioxide (CO_2), the gas released is equal to the CO_2 originally absorbed by the plant while growing. Therefore, in this way biomass can be considered to have little carbon impact; the only carbon emissions associated with the fuel production are due to transport and processing. Chips or pellets have more **embedded processing energy** than logs. However, for all biomass the processing energy is still small in comparison to other fuels.

Biomass fuel is usually delivered in bulk, either loose or in bags. Loose fuel such as pellets is delivered directly into a storage bin or 'hopper'. Hopper systems usually have an automatic feed system attached which delivers the fuel directly into the boiler through a screw drive system known as an **auger**. This type of hopper and boiler system should employ a fire break valve (or a sprinkler for larger or commercial installations) in the event of an accidental 'burn back', where the fuel is ignited outside the combustion chamber of the boiler.

Bagged pellets or woodchips are loaded into the boiler manually (hand-fired), usually on a daily basis. Pellet boilers should be generally maintenance free, apart from occasional ash removal and an annual service.

Many households also use traditional log-burning stoves which provide a cooking surface as well as heating and hot water through a back boiler. Stoves require more maintenance and are not as efficient as pellet-burning boilers but are cheaper and may also act as a focal point in a room. Maintenance issues with stoves include daily clearing of ash and flue cleaning, recommended annually. While most modern biomass boilers are of utilitarian design and are intended to be placed in a utility room or specially designed area, some products are available for installation in a living room, and incorporate a glass panel allowing the fire to be visible, thus retaining an 'open fire' feel to a room.

A biomass boiler should act as a stand-alone heat source and should not require supplementing with a conventional gas or oil-fired system, although it may well be supplemented with another renewable system for energy generation.

Key terms

Embedded processing energy – the energy required to manufacture or process a product. Usually expressed as W/kg or W/m^3, it allows a comparison between materials or products in terms of their energy use during manufacture.

Auger – helical screw device which carries material along an enclosed channel.

Chapter 10

Figure 10.6: A biomass boiler

Figure 10.7: Pellets and logs

Case study

A north Wales couple made the move to biomass heating for their three-bedroom cottage in 2009. A 20 kW pellet boiler was installed in an existing outhouse which also included storage capacity for the 250 bags of pellets required for one year's fuel supply. The outhouse was located about 30 m away from the cottage and so required additional insulated pipework in the ground. However, heat losses have proved to be minimal. The boiler system now meets all heating and hot water demands of the property, with an annual cost saving of about 70% on previous electric heating and hot water costs. Installation of the boiler and heating system was approximately £8,500 and was completed as part of the wider upgrading of the cottage.

Modern biomass boilers are manufactured to provide a wide range of outputs, typically for a single home in the region of 50–100 kW and for larger commercial applications up to several hundred kW.

Electricity-producing technologies

Solar photovoltaic

You are likely to have seen a recent increase in the use of solar panels in the UK in both new-build developments and those being retro-fitted to existing homes. This is largely due to the government incentive to pay micro-generators for surplus electricity generated and sold back to the supply company. The system of selling surplus power generated is known as a 'feed-in tariff' and is particularly applicable to solar photovoltaic technology.

Solar photovoltaic (or solar PV) uses semiconductor technology to generate electrical current from sunlight. A semiconductor is a special material which releases a small amount of electrical charge when light shines on it. Solar PV panels are made up from a network of connected PV cells which produce a large enough current to store or feed into a domestic supply. Solar PV technology does not require strong sunlight to function: panels will produce current as long as there is daylight. However, longer periods of sunshine unobstructed by cloud or shading will produce more current. For this reason panels should be positioned to avoid areas shaded by nearby buildings or trees.

Solar PV produces direct current (d.c.) while domestic appliances and all domestic systems are designed to work on alternating current (a.c.). A special component called an inverter must therefore be fitted to the d.c. power produced by the PV panels before it can be fed into the domestic supply. An inverter converts the direct current produced by the solar panels into the correct type of alternating current for use by common domestic appliances. To reduce power losses through the d.c. cabling, the inverter is usually placed as close as possible to the panels, usually within the roof structure for roof-mounted systems.

Solar PV, similar to solar thermal, works best when orientated south and tilted to maximise irradiance collection.

Figure 10.8: Roof-mounted solar photovoltaic panels

Activity 10.1

Use the Internet to research solar panel manufacturers. Find a range of solar panels (ten or more) and compare their efficiencies. Record each panel size in m² and output in watts. Divide output in watts by the panel area to find output in watts per m². You could extend this activity by comparing panel costs per watt produced.

On-grid and off-grid

An on-grid system links with the existing grid network and supplements the existing supply. As mentioned above, if the system is producing more power than is required, then this power can be fed back into the grid, or sold back under agreed arrangements known as a feed-in tariff. An off-grid system does not supplement an existing grid connection and relies solely on renewable technologies for power. Off-grid systems must therefore incorporate carefully considered methods of power storage, usually deep-cycle batteries which charge and discharge repeatedly.

Materials

The most common material in the manufacture of PV panels is monocrystalline silicon. Polycrystalline silicon is also used but is slightly less efficient. Currently, there is an emerging growth in the development of 'thin film' PV materials. These work on the same principle as semiconductors but are formed as a very thin, fabric-like material. Thin film materials have the potential to deliver much higher efficiency but are currently expensive to produce in the quantities needed for domestic applications. Thin film and crystalline technology are both used in building integrated photovoltaic (BIPV) applications. This is an emerging technology where solar cells are integrated into parts of the building fabric, for instance roof coverings and cladding or façade panels.

A standard-sized domestic solar PV installation in a suitable location will produce up to about 4 kW at peak production, potentially more depending on size and design. Individual panels have a wide range of individual power ratings, from 75 W upwards, increasing with the size of the panel. The number of panels and installation configuration will vary with each site. However, an optimised design for a suitable site will be able to generate enough electricity to feed back to the supplier at times of low domestic demand. Homeowners or supervisors of commercial installations can easily monitor output and thus sales back to the supplier.

Micro-wind

Similar to solar power, the wind is a large and untapped power resource. The UK is one of the windiest locations in Europe and receives about 40 per cent of the total wind energy available in Europe. Wind power and the use of small and large turbines is becoming a widely recognised, if often controversial, technology.

Homeowners may take advantage of the wind resource by using a small-scale turbine, mounted on a roof or on a mast (stand-alone) to generate electricity. The technology is rather basic: the wind spins the propellers of the turbine, connected to a generator, and current is generated and used or stored. Homeowners may take advantage of feed-in tariffs or the energy may be stored temporarily in batteries before being used on site. In practice, however, there are many more factors to consider, some of the most important being location and siting, wind speed, turbine sizing and maintenance.

Location

The turbine must be placed in such a position that the potential to harvest the wind power available is maximised. This means mounting the turbine at height as wind speed increases with altitude. For householders this will generally mean mounting the turbine to the roof or upper part of a gable wall. Clearly this will present an additional loading on the structure of the roof or wall and this must be considered when selecting the turbine and fixing to the structure. The structure must bear not only the weight of the turbine but also forces of torsion (twist) while the turbine is in operation.

Figure 10.9: The most common material used in solar PV panels is monocrystalline silicon

The turbine fixing must not compromise the weatherproofing ability of the building's exterior, nor interfere with services provision or flues. Consideration must also be given to neighbours with respect to noise, vibration and 'shadow flicker'. The latter is the moving shadow caused by the blades while in motion and is an inevitable occurrence which can only be managed through careful siting of the turbine.

To ensure maximum gain from wind movement the turbine must be placed within laminar air flow, i.e. in a steady wind stream, free from turbulence. Nearby buildings can cause air turbulence which can affect the turbine's performance and lead to reduced output and higher maintenance costs. Wind flow in urban areas will often be turbulent due to surrounding buildings.

The siting of wind turbines is often a matter of great concern and can easily cause acrimony between neighbours and within communities as they can be considered obtrusive and unsightly. Small roof-mounted micro-wind installations may take advantage of planning regulations under permitted development (see pages 493–494). However, large-scale wind turbine installations frequently generate hostility, often because the high wind pressure required is only available in remote and mountainous areas. Such sites are frequently regarded as a natural beauty and leisure resource, and the perceived despoiling of the territory with wind turbines may arouse strong feelings – hence there is usually considerable local opposition to proposals for wind farms. One solution to this has been the move towards off-shore wind farms. However, there remains opposition to these too, for reasons of visibility as well as concern for wildlife.

Wind speed

Electrical power generated by a turbine is a factor of the cube of the wind speed, so it is important to ensure that there will be adequate wind speed to drive the generator. A minimum of 6 metres per second would be required to make the turbine efficient so the average wind speed for the site should be established, either through the use of existing data for the site (available from a wide variety of sources) or by site monitoring for an extended period of time. A searchable database of mean wind speeds throughout the country is published by the Department of Energy and Climate Change. Alternatively, a homeowner could easily monitor and record wind speeds over a period of time using a simple anemometer. Readings should be taken as close as possible to the specific position of the turbine, including the actual height.

Figure 10.10: A digital anemometer measures wind speed

Turbine sizing

Homeowners and manufacturers face a dilemma with turbine sizing because more power is generated by larger blades. However, site constraints, including planning regulations, structural safety and other building constraints, mean that the swept area (turbine blade diameter) is often severely limited. As mentioned previously, shadow flicker is the intermittent shading effect that the blades produce while turning and any effects of this must be considered. Wind turbines suitable for domestic applications generally produce up to 1 kW under optimum conditions. If the power generated is not used then storage must be built into the system.

Micro-hydro

The potential energy stored in water has been utilised for centuries – consider water wheels used to drive mills. Water falling from height through a naturally occurring river or stream is directed onto a specially designed wheel which rotates under the weight of the falling water. The turning motion of the wheel is then used to drive other machinery. Micro-generation of power from a water source uses the same principle to turn turbine blades and generate electricity in the same way as a wind turbine does.

Hydro schemes are particularly site-specific. A suitable site must be close to a running water course, river or stream and it must also have enough 'head' or 'drop' (the vertical distance the water is falling through) to drive the turbine. Because of the specific nature of potential sites, and the historic use of running water for mills, etc., many sites have already been identified and have been used in the past or already use the water mechanically. In these cases it may be easy to replace or add electricity-generating equipment in the form of a suitable turbine and associated storage and distribution equipment.

For a new site it must first be established through a detailed survey whether the water supply provides enough kinetic energy to support a micro-generation scheme. Two factors must be determined. These are:

- the flow rate in m^3/s
- the head or drop in metres obtainable over the distance available.

Head is generally fixed and determined by the site topography, while flow rate will often vary seasonally and detailed checks should be made over the course of at least 1 year prior to commissioning a new scheme. Low head may be compensated for by a higher flow rate. However, it is better to have higher head. A head of less than 2.0 m is not usually viable for generation. Suitable flow rates vary with the head available and tables are available to determine generating power from a combination of head distances and flow rates. If it is established that the watercourse is viable, then the course must be diverted to form a weir from which the water can fall onto the turbine, or it must be channelled into a 'penstock' (a pressurised pipe which

Figure 10.11: A water mill

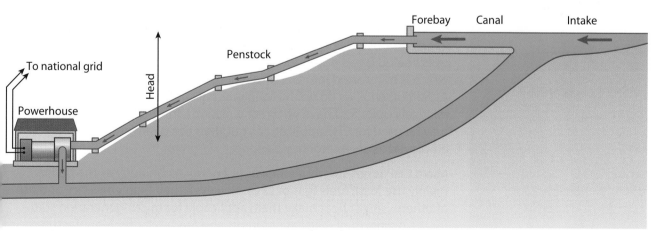

Figure 10.12: A micro-hydro system

discharges onto the turbine). Once channelled through the turbine, the water should be returned to rejoin the river slightly further down its course.

Micro-hydro provides completely clean renewable energy but can have high installation costs if no supporting structure is available and a system has to be constructed from scratch. It also has the advantage of usually supplying more power in the winter, when rainfall is heavier and hence flow rate is increased. This offers the combined advantage of providing more power at a time of year when energy demands are higher.

Case study

A derelict shell of a building beside a fast-flowing stream was all that was left of a one-time flax mill in rural Northern Ireland in 2010. As with many such sites the potential to generate power from water had already been identified and put to use in the past. The building's owner recognised this potential and restored the mill, also installing a 15 kW turbine. The site now generates power again. In its first year it generated over 50 000 kWh, easily enough to power the owner's home and provide a substantial contribution to the farm's energy requirements.

Co-generation technologies

Combined heat and power (CHP) technology produces both heat and electricity from the same source. The technology is well-established and is a product of large-scale systems where waste heat is captured from electricity-producing plants (often coal-fired plants) or other industrial sites and circulated for nearby community use. CHP technology is considered sustainable in principle because it recovers and uses energy which otherwise would be lost, regardless of the original power source.

Micro-combined heat and power (heat-led)

Micro CHP is a scaled-down version of industrial CHP and employs the same principle. A micro CHP system is essentially a boiler used for producing heat which is modified to also produce electricity. The term 'heat-led' refers to the fact that the boiler only produces electricity when it is producing heat, therefore the generation of electricity is led by heat demand.

The boiler produces heat in the same way as standard boilers but heat which would have been lost through exhaust emissions is captured for use to produce electricity through a Stirling engine. The Stirling engine uses heated gases to produce rotary motion in a crank and piston system which is transferred into electrical energy and fed into the domestic supply. As Stirling engines do not have an internal combustion cycle, they are relatively quiet and have no direct emissions. Internal combustion engines may be used for CHP applications but their use is rare and often only employed as an emergency back-up.

Micro CHP is not strictly a renewable-energy technology as the boilers mainly use fossil fuels (gas or LPG). However, their high efficiency and use of otherwise lost heat make them applicable for micro-generation status for the purposes of installation grants and feed-in tariffs. Manufacturers claim efficiencies of up to or more than 85 per cent, compared to about

Figure 10.13: A CHP boiler

75 per cent for standard boilers. Thus about 10 per cent of the lost heat is reclaimed in a CHP system.

Water conservation technologies

The conservation of water, while not producing energy or heat, is considered a renewable technology as it helps save water, which is not only a valuable natural resource but also costs money and energy to clean, store and distribute. The UK and northern Europe are naturally wet areas with high annual rainfall, hence water should be a common natural commodity, yet we are often in 'drought' conditions. This is because there are such high demands made on water storage and distribution, with a lot of drinking-quality water being used for non-drinking purposes, or simply wasted through bad management or inappropriate end use.

Rainwater harvesting

Most domestic rainwater management consists of directing the water from the roof into a drain and then either to a soakaway system, a rainwater drain or drainage combined with the foul sewerage system. Rainwater harvesting is concerned with collecting and storing rainwater for use in the home. Uses include most **non-potable** applications such as WC flushing, garden watering or laundry.

All cold water used in the home is of drinking quality but the majority of water used is for non-drinking purposes. Saving drinking-quality water results in lower water costs as well as conserving the wider water resource. Storing and using rainwater also reduces the total load of rainwater on the drainage system, which is particularly useful if rainwater is directed into a **combined sewerage system**. The management of rainwater through a combined system should be avoided where possible and removing rainwater from a combined system will ease the stresses placed on the foul water waste system in periods of heavy rainfall or flash storms. Wider rainwater management technology includes planning and design issues such as dry beds and swales which use landscape design to conduct water safely to a settlement area.

There are several variants on the basic rainwater harvesting system. Two main distinctions are gravity-fed and pumped systems.

- A gravity-fed system is the simplest: water is collected and stored above the point of use and distributed downwards, using gravity.
- In a pumped system, the water is collected and stored in a reservoir or storage tank and pumped to the various points of use. An indirect pumped system uses elements of both in that water is pumped from a larger storage tank to a header tank from where the system acts as a gravity-fed system.

Pumped systems tend to be installed in larger buildings with more space for storage, while gravity systems are usually smaller and better suited to residential use.

In any type of water harvesting system, particular care should be taken to avoid **backflow** or 'back-siphonage'. Backflow prevention is achieved by the correct use of valves and the inclusion of correctly designed air gaps in the water collection system.

> ### Key terms
>
> *Potable/non-potable* – potable means drinkable, or fit for human consumption.
>
> *Combined sewerage system* – a sewerage system which manages rainwater and foul waste water. This is regarded as inefficient as rainwater is not foul and does not require treatment in the way waste water does.
>
> *Backflow* – the flow of water in the reverse direction to that intended, particularly dangerous if the water is carrying any contaminants and if it may enter the supply system before being filtered.

Figure 10.14: A rainwater storage butt

Rainwater conservation is perhaps the easiest renewable technology to employ in a domestic setting. A simple gravity-fed system requires only basic fittings and can easily be achieved with the minimum of technical knowledge. Rainwater butts and diverting kits which fit onto the rainwater down pipe are widely available and are easily installed by anyone with basic DIY skills. The harvested rainwater is often used only for gardening purposes but with a little more plumbing the water can be diverted back into the house for use in non-potable applications. Ideally, the harvested rainwater should be reused without pumping as this will require energy input. As collection and reuse systems become more complex, more careful planning is required. Often, a simple gravity-fed system is the only realistic option for a retro-installation on a standard domestic property.

Grey water reuse

Grey water is defined as reuseable waste water generated by domestic activities such as washing, laundry and dish washing. It does not include water from the toilet or urinals which is referred to as foul water or occasionally 'black water'.

Grey water may be reused without treatment for watering plants or lawns, or it may be treated and reused in the home for laundry or toilet flushing.

There are a number of grey water treatment systems available commercially; they generally consist of a pre-treatment stage, then a filtration and settling system. Water can be taken off after pre-treatment for garden watering but should be treated to remove detergents, oil and any solids before being used for toilet flushing or laundry. Settlement boxes may also act as planters with vegetation growing on the top layer of organic material; further layers consist of fine and coarse sand, and gravel.

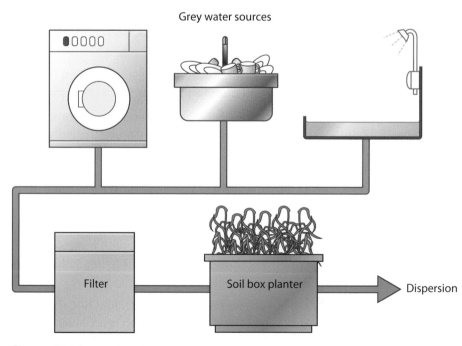

Figure 10.15: A grey water system

Plumbing to the treatment stage is relatively simple and only consists of a diverter from the outlets. Taking the filtered grey water back into domestic use requires more plumbing and will generally require storage and a pump. Storage and filtration may be underground to save external space. As with rainwater harvesting systems, particular care must be taken to avoid backflow where the water is collected.

While grey water is suitable for watering plants and lawns, it is not recommended for watering vegetables for human consumption. It is also found that while rainwater is very slightly acid, grey water tends to be slightly alkaline because of the detergent content of washing-up and laundry-waste water.

Progress check 10.1

1 What is the fundamental difference between solar thermal and solar photovoltaic systems?

2 What are the two types of solar thermal collector?

3 How might nearby tall buildings affect:
   - a solar photovoltaic installation
   - a wind turbine?

4 Describe the two main factors to be considered when reviewing a site for a micro-hydro system.

5 In addition to generating heat, what waste product does an air source heat pump produce, and how should this waste product be managed?

6 What component is used in a photovoltaic or wind installation to turn direct current into alternating current?

7 What is the minimum coefficient of performance for a ground source heat pump which is considered viable?

8 What is the meaning of 'heat-led' in describing a heat-led co-generation installation?

THE REQUIREMENTS OF BUILDING LOCATION AND FEATURES FOR THE POTENTIAL TO INSTALL MICRO-RENEWABLE ENERGY AND WATER CONSERVATION TECHNOLOGIES

LO**2**

This section discusses the features that a building or site needs in order to support a successful installation of a renewable technology. We will examine:

- the required features of a building to mount wind, solar PV or solar thermal technology
- the ground and building requirements to install heat pumps
- the site requirements for a successful micro-hydro system.

If you are examining a particular site with a view to installing a renewable system, or if you are becoming expert in a variety of technologies, then you need to be aware of the following information regarding site requirements.

Chapter
10

General requirements for a solar thermal or photovoltaic system

Similar building requirements apply to either of the two solar technologies. Most solar installations on a domestic scale are positioned on the roof. Therefore, it must be established that the roof is structurally safe and able to bear the increased load of the panels, supporting grid and pipework or cables. Checks should be made on rafter depth, roof structure in general and roof covering type prior to fitting.

The most effective position of a solar collector or panel is determined by a number of factors. Perhaps the most important of these is orientation (the direction in which the panels face). Other factors include shading from other buildings or trees, which may not always be apparent on an initial visit depending on the time of day, and angle of inclination, or 'tilt angle'.

Orientation

For any location in the northern hemisphere the sun's path through the sky is from east to west with the sun in the southern part of the sky at all times. The sun is at its strongest at midday when it is at its highest point in the sky. The sun's rising and setting point, and its height throughout the day, vary with location and time of year. Sun path charts and tables are widely available for a wide variety of locations and software is also available where building location and orientation can be entered and a sun path plotted for the building.

It follows therefore that solar thermal and photovoltaic panels, where positioned on a pitched roof, should be on the southern pitch. Most manufacturers recommend an orientation as close to due south as possible, with panels remaining effective up to 30° either side of due south. The ideal alignment of the roof is therefore along the east–west axis. If a standard double-pitched roof is aligned approximately north–south, then unfortunately a solar thermal installation will not be very effective and a solar PV installation would be inadvisable and unlikely to recoup its expense.

Inclination

This is the angle of tilt of the panels. As most domestic panels are to be fixed on the roof of the building then this is effectively the roof angle. The most effective angle of inclination is between 30° and 45° and most pitched roofs in the UK are within this range. If a roof has a more shallow pitch than the range indicated, but is not classified as a flat roof (<10°), it is not recommended to build the angle up artificially for structural and planning-permission reasons. For any pitched roof, panels should be installed in the same plane as the roof slope. This is a planning requirement as well as good structural practice. A system which is to be positioned on a flat roof must be fixed to a structural framework providing a suitable angle of inclination. Systems are available which can track the sun's path through the sky and adjust the orientation of the panels accordingly. However, this will be rare in a domestic setting and more likely to be used in a large commercial installation.

Shading

Nearby objects such as tall buildings, trees or landscape features may provide shading against the sunlight at certain times of the day. These may not be immediately apparent on an initial site visit, therefore care must be taken to establish the position of all such features and determine the position of any shadow they may cast during the day. Excessive shading will reduce the effectiveness of either type of panel so installations should be sited where any shading is at a minimum. In addition to reducing the amount of solar irradiance available to the cells or collectors, shading on PV panels can cause current 'bottlenecks' and lead to overheating in a PV **array**. Overheating will cause a drop in current supplied and could ultimately lead to permanent damage to the panel. Therefore, overshading presents more of a problem to PV installations but should nevertheless be avoided for both solar technologies.

In dense urban areas, overshading from nearby buildings will be highly likely. A detailed assessment of the site, with particular attention to potential for shading, should be made before deciding on any installation.

Figure 10.16: Partially shaded solar panels will be less effective

> **Key term**
>
> **Array** – a number of solar photovoltaic panels joined together.

Requirements for the potential to install a ground source heat pump system

A ground source heat pump requires either a large area of relatively shallow excavation, or a small area but with deep excavation. An approximate rule of thumb is to allow a collector area of about 2½ times the floor area required to be heated or 10 m of trench per kW heat pump rating.

A number of options for collector pipework are available: one of the most common is coiled polyurethane, commonly known as a 'slinky'. This is particularly useful where there is not enough space for a full horizontal array. The slinky may be installed horizontally or vertically. If only a small area is available for a GSHP system then a deep installation may be possible; in this case a borehole is drilled to a depth of 100 m or more. This option is more expensive than a shallow trench. However, economies of scale may apply if a number of boreholes are to be used. Boreholes are best suited to tight urban sites where there is often no open ground.

For a shallow trench system the land available must be easily accessible, easy to excavate and free from excessive services buried in the ground. Buried services will not necessarily impair the system's effectiveness but will have an effect on the ease of installation and subsequent cost. No maintenance should be required for the buried pipework, therefore the land should be freely available for its previous use after installation. Access must be allowed to the manifold connections, pump and heat exchanger for maintenance purposes.

A GSHP is not a stand-alone system because it requires an electricity supply to pump the collecting fluid around and to run the heat exchanger. Therefore, a mains connection or other reliable source of electricity should be available. In well-developed micro-generation systems the GSHP pump may run off another renewable technology such as solar or wind. However, storage back-up would be required.

Figure 10.17: A ground source heat collector – a 'slinky' is usually buried in the ground

Case study

A Derbyshire homeowner has invested in a retro-installed ground source system to provide heating for his stone-built cottage. The property is not supplied by gas or oil mains and the previous method of heating was by expensive electric storage heaters. After installing 'slinky' pipework in two 100 m-long trenches, and retro-insulating the house, the system provides more than enough heating and hot water for two people. Installation costs were high but energy bills were reduced immediately and the installation is on track to pay for itself within 10 years.

Requirements for the potential to install an air source heat pump system

Air source heat pumps (ASHPs) only require the space to accommodate the unit, and careful positioning to allow the correct flow of air to and away from the unit. Units should not be placed together nor should they be positioned in such a way that one unit's intake is liable to collect the output of another unit. They should not be placed in an internal space or in a loft, as there is unlikely to be adequate ventilation to provide fresh ambient air and the efficiency of the units will become greatly reduced as they try to extract heat from already cooled air.

Units should be positioned externally and close to the wall to reduce ductwork and to allow for access for maintenance. Allowance should be made to drain away a small amount of condensate; it should not be allowed to pool or spill onto a nearby surface, especially in cold conditions where it may freeze and cause a slip hazard. Pooling of condensate around the base of a unit may also corrode the unit structure.

A unit should have access to a free flow of air, so positioning in a corner or enclosed space will reduce the free ambient air available. While ASHPs make use of heat in the air and obviously perform better in warm, ambient air, positioning in relation to orientation (i.e. south-facing) does not have any significant effect on the overall output of the unit.

Some consideration should be given to noise. The units make a small amount of noise and so should not be positioned near to bedroom windows or where there is likely to be high reverberation or echoing effect.

Because of the relatively low additional pipework required, ASHPs work well being retro-fitted into existing buildings; they work best with underfloor heating systems or low-output radiators.

Requirements for the potential to install a biomass system

The main requirements for a biomass system, besides conventional plumbing and a hot water cylinder, are the need for a flue to safely emit gases and space for fuel storage.

The flue must meet Building Regulations, particularly Part J (the Building Regulations document covering heat-producing appliances) and may require planning permission, depending on location. Existing flues may be used but should be checked for suitability and may need to be fitted with a stainless steel liner. If the property is located in a smoke-control area then only approved appliances and fuels may be used. Installers will need to check with the local authority if the property is within a smoke-control area and, if so, whether the proposed appliance and fuel are approved for use. Manufacturers should provide details of such exemptions or limitations for their products.

Fuel for biomass boilers will generally be a form of processed wood such as chips or pellets, or it may be unprocessed logs. Both have a low calorific value against mass, meaning they are relatively bulky and require space for storage. Logs are the least efficient to store: they may be easily stacked but there will be a lot of air pockets between the logs which results in wasted storage space. Pellets are more space-efficient but still require storage and a means to move them into the boiler. Many biomass boiler systems include an auger drive to automatically feed the boiler. The pellets should therefore be stored in bulk close to the boiler, or loaded into a hopper connected to the delivery system.

If a separate building or extension is required to store the boiler and fuel, then installers should carefully check the rules on permitted development regarding floor area, size, shape and location of the proposed development. These are provided in detail from a number of sources, including Planning Portal, MCS and local authorities.

Requirements for the potential to install a micro-wind system

The main requirements for a successful micro-wind generation scheme are location and positioning to take advantage of available wind energy, and integrity of the supporting structure.

To be effective, a wind turbine requires a minimum average wind speed of 6 m/s. The site should therefore be located in an area meeting or exceeding this value. Wind speed maps and tables are widely available for the UK but, due to micro-climates, a long-term assessment is also advisable. An assessment should take place close to the exact proposed location, including at the proposed height as wind speed increases with altitude.

Wind direction in the UK is predominately from the south-west. While most turbines automatically rotate to the optimum angle to take advantage of the wind direction, it is not advisable to place a turbine where there is significant blocking of the wind from the prevalent direction due to buildings or land topography. Other factors concerning wind availability and flow should also be considered. **Laminar flow** is preferable and this is best achieved in open spaces; urban locations in general make poor sites for wind turbines as the wind tends to 'gust' and have a highly variable speed due to the surrounding building topography. Uneven gusting **turbulent flow** leads to poor overall performance and increased turbine maintenance issues. In cases where turbulent flow is unavoidable, the best solution is to raise the position of the turbine above the zone of turbulent flow. However, this may not always be possible, especially in urban sites.

Most micro-wind sites will be either close to the roof of a dwelling or mounted on a mast to achieve the maximum height advantage. In the case of building-mounted turbines, they will generally be mounted on a mast designed to project beyond the roof height and fixed to the wall at the

Key terms
Laminar flow – smooth, uninterrupted air flow.
Turbulent flow – the opposite of laminar flow – irregular, disorganised and uneven air flow caused by interruptions to laminar air flow.

Turbulent

Laminar

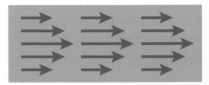

Figure 10.18: Differences between turbulent and laminar air

highest point which is structurally sufficient to bear the increased load. This will often be a gable wall close to the apex. The wall will have to withstand not only the increased load of the turbine and supporting structure but also the effects of torsion (twisting movement) when the turbine is in use. The fixings will need to be sufficient to withstand occasional or sustained gusts significantly higher than the yearly average and they also must not compromise the weatherproofing ability or the insulation properties of the wall. Through fixings are not recommended due to a cold bridging effect, and dampers may be required to counteract any vibration.

Micro-turbines on a mast located away from a building avoid all the structural issues associated with building-mounted turbines. Masts are specially designed for the purpose of supporting a turbine at height and the main consideration for such an installation is positioning to take maximum advantage of the wind resource. The mast should be sited away from obstructions, including large trees and buildings, and not in the wake of another turbine as the disturbed air is then in turbulent flow. Guys or cable stays should be clearly marked or fenced off and the mast should be secured to avoid vandalism or accidental access.

Requirements for the potential to install a micro-hydro system

Micro-hydro is perhaps the most site-specific of all the technologies discussed. Site and building requirements include a running watercourse with sufficient head (drop height) and flow rate (speed of water moving past). The geography of the site should also lend itself to construction related to channelling the water, housing the turbine, distributing or storing the power generated and returning the water to the watercourse. All of these factors need to be combined with the location such that the distribution of power generated is economically viable. A site may well meet all of the topographical requirements but be located so far away from any habitation that transferring the power is uneconomic.

Because of the historic use of water to produce mechanical power, many viable sites are already identified and in use. Examples are water mills for grinding corn, some of which are preserved for historical interest. These sites may be upgraded with electrical-producing turbines installed alongside mechanical energy-production equipment. Each of these sites will be different and will require a detailed assessment and design before installation.

Minimum useful head is 2 m for the smallest micro-generation system but the higher the head available the better. Up to 10 m is classed as 'low head' while over 10 m up to 100 m is classed as 'high head'. Higher constant flow

rate is also preferred although low head can be somewhat compensated for with increased flow rate. At a 10 m site, 0.16 m³/s flow rate will be required to produce 10 kW, which is close to the minimum amount to make a scheme viable.

A viable site may encompass some considerable distance to gain the required head; lateral distances from intake to turbine may be 1,000 m or more. This may raise issues of ownership, access and easement, as well as maintenance of the pipework.

Hydro max. power	Low-head hydropower sites			High-head hydropower sites		
	Gross head 2 m	Gross head 5 m	Gross head 10 m	Gross head 25 m	Gross head 50 m	Gross head 100 m
5 kW	0.414 m³/s	0.166 m³/s	0.083 m³/s	0.033 m³/s	0.017 m³/s	0.008 m³/s
10 kW	0.828 m³/s	0.331 m³/s	0.166 m³/s	0.066 m³/s	0.033 m³/s	0.017 m³/s
25 kW	2.070 m³/s	0.828 m³/s	0.414 m³/s	0.166 m³/s	0.083 m³/s	0.041 m³/s

Table 10.1: Minimum flow rates required for a range of (gross) heads. Source: Renewables First

Any micro-hydro scheme, either existing or refitted, should have the minimum possible impact on the natural environment, including vegetation and wildlife. This is particularly important when considering fish migration and breeding. A viable site must be able to retain a portion of the watercourse in its original flow, known as compensation flow, and the penstock intake must also be protected against wildlife ingress with a mesh or other type of guarding. This will also provide protection against debris, leaf litter, etc. It is important to note that due to potential effects on the natural environment and the abstraction of water from a natural course, any micro-hydro system is subject to a full planning-permission application and no permitted development rights apply to such schemes. This is discussed more fully in the following section on planning and permitted development.

Working practice 10.1

A potential client has approached you with regard to a micro-hydro site. The site is across a fast-flowing stream on hilly ground and on their land but it is not close to any mains electrical supply. There is an old building on the site which was once a mill but has been disused for over 80 years. Your initial site survey indicated a head height of about 7 m across a length of 250 m and a flow rate of about 0.15m³/s.

1 What size system might be suitable for this site?

2 How might output be increased?

3 Advise the client on the additional factors they will need to consider if they wish to make money by supplying electricity to the grid.

4 An earlier micro-hydro consultant who was keen to impress the client insisted that, because there was an old building already present, they would not have to apply for planning permission. How would you advise the client on the matter of planning permission?

You could extend this activity by researching micro-hydro sites in your area and investigating their output.

Requirements for the potential to install a micro-combined heat and power (heat-led) system

Building requirements for a micro-CHP system are similar to those for a standard gas or oil boiler. A micro-CHP boiler is comparable in size or only slightly larger than a standard boiler and requires:

- a fuel supply (usually LPG)
- electrical supply to control the boiler operation
- electrical distribution equipment to distribute the electricity generated
- plumbing pipework to distribute the hot water supplied.

Apart from electrical distribution, the equipment requirements are almost identical to a standard boiler. If you intend to take advantage of a feed-in tariff and export electricity back to the grid, then a grid connection and meter must also be provided.

No planning or permitted development is required and building regulations apply in the same way as with other heat-producing systems, namely Parts J and L.

Requirements for the potential to install a rainwater harvesting/grey water reuse system

Rainwater and grey water, while both water conservation technologies, do not necessarily need the same building requirements.

Rainwater is perhaps the easiest to manage. In its simplest terms, rainwater harvesting can be achieved with standard guttering and rainwater goods – a diverter fitted on a down pipe and a storage tank, usually in the form of a proprietary water butt. The stored rainwater can be used passively to water plants or piped to an irrigation system. A more developed rainwater system will require additional pipework and a pump to introduce the water into the house for specific tasks such as toilet flushing or laundry. In these cases the water would be supplied directly to these locations, not reintroduced into the storage tank. Further development of a rainwater harvesting system would place the diverter close to the roof with the storage also at roof level, and would employ gravity to distribute the water to where it is needed, thus disposing of the requirement for a pump. Such a system would be likely to be designed in at construction stage and would be difficult to retro-fit.

Grey water may also be used for irrigation without treatment. However, if you intend to reuse grey water within the home for laundry or flushing then it should be cleaned using a filtration system. Grey water filtration systems range from a fish tank-sized settlement system providing basic but adequate filtration and easily manufactured and installed with basic DIY, to commercially produced cleansing and storage systems requiring additional housing or buried in the ground to conserve space.

Combined grey water and rainwater collection and treatment systems are also available. However, larger systems such as these would be preferable for community buildings or projects greater than single domestic scale.

Activity 10.2

Use a rainfall map or search online to find the average rainfall for your area. Estimate the roof area of your house, college or work building and see if you can estimate how much rainfall could be harvested by a rainwater harvesting system.

You could extend this activity by looking at commercial premises, finding out the water supply price and calculating the potential savings on water supplied.

Progress check 10.2

1 What is the ideal orientation for a roof slope for a proposed solar thermal installation?

2 How is solar (photovoltaic *or* thermal) collection optimised on a flat roof?

3 How might a nearby large tree affect the design of a solar PV installation?

4 What alternative installation technique is available for a ground source heat pump scheme in a tight urban site with very little available space for installation of pipework?

5 State two locations where it would be inadvisable to place an air source heat pump.

6 Why does a biomass boiler system require additional space to that required to house only the boiler?

7 What is laminar air flow?

8 What is the minimum head value to make a micro-hydro scheme viable?

9 What factor may compensate for a low head value in a micro-hydro scheme?

10 What is the usual fuel source for a heat-led CHP boiler?

11 Describe a simple method of installing a basic rainwater harvesting system.

12 Roughly how much ground area is required by a ground source heat pump, in comparison with the building floor area it is required to heat?

REGULATIONS

LO3

This section will look at two of the most important regulatory requirements relating to micro-renewable technologies. If you are planning an installation it is important that you fully understand the legal requirements and obligations made on the installer and owner. Not all installations are automatically permitted under planning legislation and not understanding the correct status of a proposal from the outset could lead to costly mistakes, time lost and at worst criminal proceedings. Similarly with Building Regulations (an area which is often poorly understood by the non-professional), ill-informed installation work could be unsafe and potentially put lives at risk.

The two important types of regulatory requirements in the UK which need to be met by domestic installation of micro-renewable energy systems are planning legislation and the regulations arising out of the Building Act 1984, commonly known as Building Regulations.

Relevant planning legislation consists mainly of The Town and Country Planning Act 1980 (updated in 2010) and The Town and Country Planning (General Permitted Development) Order 2008 (updated in 2011).

Planning legislation and Building Regulations are separate areas of the design and construction process and are usually managed from different parts of a council administration and by entirely separate departments and staff.

Planning legislation examines the impact of a proposed development on the local physical and human environment and considers:

- privacy
- amenity
- impact on transport
- the natural environment
- noise
- rights to light
- local history
- conservation, amongst other factors.

Building control is concerned with:

- structural safety
- energy efficiency
- services' use and safety
- heating
- ventilation
- the factors involved in ensuring a building is safe, habitable and energy-efficient.

Planning permission and permitted development

Under planning legislation, any new buildings or change of use or extensions or alterations to existing buildings must be approved by the local authority planning department. This process consists of:

- the submission of a scheme with drawings and details concerning the scheme design and impact on the environment and surroundings
- the consideration of the scheme by the planning department
- consultation with neighbours and anyone who may be affected by the scheme
- an eventual decision by the local authority whether to grant or refuse the application.

However, homeowners may carry out certain small amendments to their property provided the work is within certain limits of size and scale, and the property fits specific descriptions. Development within these limits is known as 'permitted development' rights or 'lawful development' and is outlined in The Town and Country Planning (General Permitted Development) Order 2008. In the case of development falling within permitted development rights, the homeowner should make an application for a Lawful Development Certificate, not dissimilar to the planning process except that the householder describes the development and states how and why it is within the limits of permitted development as described in the Act. The planning department then checks to ensure that the proposal is correctly defined within the limits of the Act and then issues a Lawful Development Certificate.

Until recently, permitted development was generally used by householders building modest extensions, loft conversions, porches and garages, etc. However, recent interest in renewable energy and changes in the Order in effect since December 2011 now make provision for micro-renewable energy systems within lawful development.

Important exceptions to permitted development rights include:

- all listed buildings
- some buildings within the grounds of a listed building
- a Scheduled Monument
- properties within Sites of Special Scientific Interest (SSSIs)
- properties within a National Park
- properties within a Conservation Area or with an Article 4 directive.

The following is a brief summary of the limits applied to some micro-renewable technologies under current permitted development rights. However, any installer should check carefully with the local planning department regarding any specific restrictions relating to the property and discuss the proposal to ensure permitted development rights are correctly interpreted for the property. Thorough checks with the planning department should be made in advance of commencing work on the installation. If work has started and a scheme is subsequently found to be outside the remit of lawful development, then a full planning application will be required. An application may not necessarily be successful and any work carried out without permission will have to be reinstated and the site restored to its original condition.

While it is not a legal requirement to provide a Lawful Development Certificate, all schemes within the limits of lawful development should apply for one.

The definitive reference to what exactly is within permitted development is to be found in the Act itself, which is a freely available document. It can be viewed and downloaded from www.legislation.gov.uk/uksi/2011/2056/made.

Solar panels, both thermal and photovoltaic

- Roof-mounted panels should not be positioned higher than the main part of the roof, excluding the chimney, and should not project more than 200 mm beyond the plane of the roof slope.
- There is no limitation on the number of panels or proportion of roof coverage.
- Solar panels mounted on a flat roof require a supporting structure to provide an adequate angle of inclination. This structure should not be greater than 2 m higher than the existing flat roof and must be positioned a minimum of 2 m from the edge.

Stand-alone panels

- Only one stand-alone solar PV is permitted within the property boundary.
- Must be less than 4 m in height.
- Total surface area is not to exceed 9 m² and no dimension is to be more than 3 m long.

Micro-wind

- Any building-mounted wind turbine (including the blades) should not extend further than 3 m above the highest part of the roof excluding the chimney nor exceed an overall height of 15 m.
- Any part of the blade must be more than 5 m clear of the ground, and more than 5 m away from any boundary.

- The swept area of any building-mounted wind turbine blade must be no more than 3.8 m^2 squared.
- Only one micro-wind installation per building is permitted.
- In conservation areas a building-mounted turbine must not be positioned on any wall or roof of the main elevation, i.e. the side fronting the road.
- Further restrictions control stand-alone wind turbines and the installer is advised to consult planning regulations and the local authority before commencing any installation.

Air-source heat pumps, ground-source heat pumps

- Both are considered permitted development subject to property limitations, as discussed (listed buildings, etc.).
- Air source heat pumps are not allowed if the building already has a micro-wind installation.

Micro-hydro

There is no allowance under permitted development for micro-hydro systems. Any such system must be submitted for full planning consent prior to commencement.

Biomass boiler

No requirement for planning permission is necessary for a biomass boiler to be placed within an existing building. However, full planning may be required for a new building with a floor area of more than 10 m^2 to house a boiler.

Rainwater harvesting, grey water reuse

- Standard water butts will not require planning permission.
- Plant requirements are similar to those for existing buildings/new buildings above.
- It is of vital importance that the installer checks thoroughly any conditions relating to the property and consults with the local authority regarding planning permission and lawful development before commencing work on any installation.
- Useful guidance is available from a number of sources, including Planning Portal, the Communities and Local Government site, the local authority and the MCS Planning Standards.

Building Regulations

The Building Regulations enact the Building Act 1984 and for practical purposes are often regarded as the set of 14 Approved Documents (ADs) which offer guidance on how to meet the requirements of the Act. The Approved Documents are not the law and there may be other ways to meet the requirements of the Act. The local authority building control department will be able to offer advice for specialist applications.

The 14 Approved Documents are named A to P and cover the different parts of building construction, as shown in Table 10.2.

All of the Approved Documents are of importance. However, a number are more widely consulted than others and contain guidance more applicable to the installation of renewable technologies.

Approved document	Description
A	Structural safety
B	Fire safety
C	Resistance to contaminants and moisture
D	Toxic substances
E	Resistance to sound
F	Ventilation
G	Sanitation, hot-water safety and efficiency
H	Drainage
J	Heat-producing appliances
K	Protection from falling, collision and impact
L	Conservation of fuel and power
M	Access to – and use of – buildings
N	Glazing safety
P	Electrical safety

Table 10.2: The 14 Approved Documents, commonly referred to as the 'Building Regulations'

Part A: Structural safety – this document contains guidance on the structural safety of buildings up to six storeys, including important information on foundation types, wall thicknesses and roof design. It should be consulted in relation to roof- or wall-mounted apparatus and changes to building loadings or structural support, for example if cutting holes in joists to pass cables or pipework through.

Part L: Conservation of fuel and power – this is one of the largest and most widely consulted documents as it contains much of the guidance on insulation and thermal requirements of new buildings and alterations to existing buildings. It is the key document on energy use and efficiency and should be consulted in relation to any renewable energy or heat-producing technology. Often, the implementation of a renewable or energy-efficient technology may offset design requirements not met by conventional energy installations.

Part P: Electrical safety – guidance within this document must be referred to in any electrical work undertaken. This will particularly apply to solar PV, wind turbines and micro-hydro, etc. Any electricity-producing technology will count as **notifiable work**.

Guidance within other documents, particularly ventilation, drainage, sanitation and fire, is also of importance and a competent installer should have a good awareness of the relevant requirements of all the building regulations relating to their particular technology. Installers should also be aware that the Building Regulations Approved Documents are regularly updated and they should keep informed of any changes in the law and amendments to Buildings Regulations guidance.

> **Key term**
>
> **Notifiable work** – any work to a building which requires the local authority building control department to be notified so that it can be checked and deemed safe. This includes any significant structural work and all electrical work apart from like-for-like replacement.

Activity 10.3

Visit www.planningportal.gov.uk/buildingregulations/ and look at the approved documents. Download each PDF version and save it for your own reference. You will find it useful to have a basic working knowledge of the regulations, as well as having an electronic copy to refer to for design advice.

Progress check 10.3

1 A developer wishes to construct a solar photovoltaic stand-alone array consisting of 8 m² of panels in a 2×4 m configuration.
   - Will this scheme be permitted under the rules of lawful development?
   - If not, what adjustments might you recommend to ensure the design meets the conditions of permitted development?

2 A developer wishes to construct a micro-hydro installation and believes that it will not require a planning application as there is already an old derelict mill on the site. Is this assumption correct?

3 State the piece of legislation which gave rise to the Building Regulations.

4 Which Building Regulations Approved Document (AD) would an installer consult to aid with the electrical installation of a solar PV scheme?

5 Which Building Regulations Approved Document (AD) would an installer consult to aid with the hot water provision of a solar thermal scheme?

6 Is working to Building Regulations Approved Documents a legal requirement?

 LO4

ADVANTAGES AND DISADVANTAGES OF MICRO-RENEWABLE ENERGY AND WATER CONSERVATION TECHNOLOGIES

Each of the different renewable technologies below offer different advantages and disadvantages to the installer or consumer. The technologies are:

- solar thermal (hot water)
- solar photovoltaic
- ground source heat pump
- air source heat pump
- micro-wind
- biomass
- micro-hydro
- micro-combined heat and power (heat-led)
- rainwater harvesting
- grey water reuse.

For a consumer considering an installation, careful thought must be given to the balance between installation cost, effectiveness and output. Coupled with this, some of the technologies are easier to retro-fit than others, some are ideal for new build only and some, e.g. hydro, are particularly site-specific.

Planning and other regulatory matters must also be considered, for instance the structural and aesthetic impact on the host property and any possible effects on the wider area.

Government incentives or other financial drivers must be weighed against set-up costs and the payback time estimated. Allowance must also be made within this calculation for changing fuel costs in comparison with other energy sources, as well as changing incentive values. For example, feed-in tariffs vary from one technology to another, and reduce year-on-year as more micro-generation sites are added to the national network.

Read Table 10.3 on the next page before you attempt Activity 10.4 and the progress check.

Activity 10.4

Consider your house or college building and assess its suitability for an electricity-producing technology, either wind or solar photovoltaic. Summarise the advantages and disadvantages and apply these to your home or college building. Suggest the most appropriate installation and give reasons why.

You could extend this activity by estimating how much power might be generated by your chosen installation, and how much income could be made either from savings or by selling back to the grid. You will also need to research the latest price per unit for a feed-in tariff (FIT).

Progress check 10.4

1 Which heat-producing technology will provide continual low-level heat with minimal noise disruption?

2 What spatial factor is a disadvantage in planning for a biomass boiler?

3 Which energy-producing technology is often associated with remote, off-grid sites?

4 Is a micro-CHP boiler similar in cost to a conventional boiler?

5 Could a householder plan to source all the household's water supply, including drinking water, from a rainwater harvesting system?

Technology	Advantages	Disadvantages
Solar thermal Roof-mounted heat-generating technology	• Relatively low installation costs • No carbon emissions during use • Clean and efficient • High output under ideal conditions	• Unreliable • Limited capacity
Solar photovoltaic Roof-mounted electrical-generating technology	• Can generate electricity for resale • Proven technology • Free after installation • Little or no maintenance • Functions in low light • Infinite availability of energy source • No emissions after installation	• Expensive initial costs • Production and manufacture high in embedded energy • Substantial additional electrical installation required • Variable energy production depending on insolation (sunlight available) • Only generates power in daylight hours, thus requiring storage
Ground source heat pump	• Continual heating provided • Low cost after installation • Low maintenance requirement	• Low level of heat provided • Potential for high installation costs
Air source heat pump	• Continual heating provided • Low cost after installation • Low maintenance requirement • Often not subject to planning requirements	• Low level of heat provided • Some noise
Micro-wind	• Free after installation • Potential to export electricity for resale • Potential for high efficiency generation at selected sites • No emissions after installation • Suitable for remote areas where there may be no grid connection	• Potential for high set-up costs • Planning issues • Structural and siting issues • Variable and weather-dependent generation • Requires maintenance
Biomass	• High efficiency • Low-cost fuel • Usually not subject to planning requirements • Renewable fuel source • Low maintenance requirement	• Boilers are specialist devices and have high initial costs • Bulky fuel requiring significant space to store
Micro-hydro	• Potential for high-efficiency generation at selected sites • Low-cost power after set-up • Can provide power in remote settings • Once structure is complete it is long-lasting	• Limited siting potential • Potential for high set-up costs • Planning issues • Conservation issues • Potential for environmental impact • High structural and design input required • Variable and weather-dependent generation • High maintenance requirement
Micro-combined heat and power (heat-led)	• Highly efficient • Potential to export electricity for resale • Little additional plumbing or other equipment required • No planning issues or additional Building Control requirements above standard	• More expensive than standard boilers • Some additional electrical fittings required
Rainwater harvesting	• Free water for domestic use • Potential for high harvest in UK climate • Low additional plumbing requirement • Saves valuable water resources • Saves on water bills • Reduces pressure on existing drainage/sewerage system	• Not recommended for potable (drinking) use • Storage may be an issue on limited sites • Requires a pump if being introduced to the home
Grey water reuse	• Free water for non-potable use • Saves valuable water resources • Saves on water bills • Reduces pressure on existing drainage/sewerage system	• Not recommended for potable (drinking) use • Storage may be an issue on limited sites • Requires a pump if being introduced to the home • Requires equipment investment and installation to reuse in the home

Table 10.3: The advantages and disadvantages of renewable technologies

Knowledge check

1 The ideal orientation for photovoltaic or solar thermal panels in the UK is:

a south-east

b due south

c south-west

d west

2 Building Regulation Approved Document A relates to:

a ventilation

b electrical safety

c conservation of fuel and power

d structural safety

3 The Building Regulation Approved Documents most applicable to the installation of a CHP boiler would be:

a L (Conservation of fuel and power) and C (Resistance to contaminants and moisture)

b L (Conservation of fuel and power) and P (Electrical safety)

c J (Heat-producing appliances) and L (Conservation of fuel and power)

d J (Heat-producing appliances) and P (Electrical safety)

4 How many stand-alone solar photovoltaic installations are permitted within a domestic property, under lawful development rights?

a One

b Two

c Three

d As many as can be safely fitted within the area

5 In considering the design and installation of wind turbines, what is meant by 'laminar flow'?

a Cold northerly air flow

b Smooth, uninterrupted air flow

c The flow of electricity to the point of use

d The rotational speed of the turbine blades

6 Which component turns direct current into alternating current?

a A diverter

b A reverter

c A converter

d An inverter

7 Monocrystalline and polycrystalline silicon are the principal materials used in the manufacture of:

a wind turbine blades

b grey water sterilisation units

c photovoltaic panels

d biomass fuel pellets

8 Embedded energy means the amount of energy required to manufacture a particular material. Embedded energy is often measured in:

a man hours per kilogram

b watts per kilogram

c kilowatts per year

d metres per second

9 Glycol is an additive used in the collector systems of solar and heat pump systems. Its principal use is to:

a prevent freezing of the transfer fluid

b ensure fast and efficient flow of the transfer fluid

c aid the efficient transfer of energy within the heat exchanger

d aid in the locating of leaks in the pipework system

10 In a micro-hydro system, what is a penstock?

a Guarding to protect livestock from unsafe machinery

b Pressurised pipe delivering water to the turbine

c Metal mesh placed across the inlet to filter out debris

d Depressurised pipe returning water to the river

Career awareness in building services engineering

This chapter covers:

- **career planning**
- **goal and target setting**
- **CVs, application forms and covering letters**
- **interviews**
- **business plans**
- **becoming fully qualified.**

Introduction

In this highly competitive industry, planning for your future career has never been more important. Jobs are very rarely for life and you are expected to continuously develop skills if you are to compete in the job market. The industry is also changing and evolving and engineers are expected to have a broader range of skills. As a highly trained electrician you may have the opportunity to work in different countries as the skills market becomes more open. This chapter looks at equipping you with some basic principles and skills that will enable you to plan your career in a structured way.

CAREER PLANNING

Deciding what you want to do is a very important starting point in your career as there are many specialist areas to consider. A particular specialist area that interests you might take several years to train for and achieve. Making an informed choice depends on having all of the information before you make up your mind. The process of gathering information from as many different sources as possible can take time and needs to be researched and recorded carefully. When you have all of the information in front of you and readily available it can become much easier to make a job choice or to take a particular course of action.

Research the specialism that interests you

Research has become a lot easier with the advent of the Internet, but talking to people in the business is always the best starting point. Seek out as many different job titles in the industry as you can and create a file for each one. If you have carried out good research, you will find it easier to make decisions as you can read through, discuss and reflect on all the information you have gathered – organisation is the key.

Researching can also mean looking at your own needs and working out what exactly it is about a job that attracts you – of course it can just be a gut reaction or instinct!

A good starting point is to ask yourself a few questions.

What is it about this particular job that I think I like?

There may be elements of a particular job that you have discovered during work experience or from family and friends that attract you to it. The job may involve working in very interesting locations or with particular equipment or specialist companies. You might have a hobby that has links to the job which makes it even more appealing. Think very carefully and see if you can identify what it is that makes this job special.

Would this job match my personal needs and what are they?

This means practical things like money, location, family, but also aspirations and end goals – it may not be a good idea to start in a career if it cannot lead to where you want to be eventually.

What skills would I need to start out in this career?

This can be as simple as looking at the job advert. Adverts are generally written from the job specification. The job specification defines all the main requirements for the job. It will allow you to make an initial judgement about your level of skill and whether or not you match the job specification straight away without completing any additional work experience or training (although you can never have enough of these!). As well as being guided by the job advert you will need to carry out wide-reaching research by using the Internet and talking to people.

What skill gaps do I have and how can I get the skills I need?

From the advert you should be able to identify any skill gaps you have and if this job still really appeals to you. You may have to create a plan to get these extra skills. It may be that you can get the missing skills with some very specific work experience or voluntary work. This is another area for you to research.

Identifying your skills and the skill requirements for the job will also help you work out how long this will take to achieve.

Case study

Paul wanted to work in the electrical industry after he had completed a work experience week with an electrician. He particularly liked the idea of working with lighting as he had seen some really interesting systems that required quite a lot of design work and an artistic touch.

How could Paul research and decide if this was the career for him?

- He visisted a lighting design shop and spoke to the manager. He went armed with well-planned questions and found a lot of information about companies that worked in this area or with these particular technologies. He also made enquiries about further work experience.

- He spoke to someone in the industry and to some more experienced people – in doing so he found a number of other related jobs, including lighting design engineers working on film sets and stage lighting.

- He researched the official professional bodies related to his area of interest – the Institution of Lighting Engineers and the Institution of Engineering and Technology (IET). When you begin to research, you normally start with a few key terms or words that lead to others until you have exhausted a particular line of research. Your basic research can be taken a long way by speaking to people but all you will have is a collection of opinions – at some point you should find the official professional bodies that are involved.

Having identified a job you want to do and started the research, you now need to take some positive action to get you closer to that job. Thinking in a logical way and breaking down your activities, using SMART targets and setting goals will take time initially but ultimately it will speed up the actual process of getting a job.

- What are the logical steps to getting this perfect job?
- How can I break this down into a set of achievable goals?
- What are the first five things I can do to help myself?
- When can I get these actions done by?

Activity 11.1

1 Training to be an electrician might not be your only skill. You may wish to diversify, especially if you intend to set up a business. Use the Internet to research the qualifications required for being a:

- plumber
- heating and ventilation engineer
- refrigeration and air conditioning engineer
- gas installer.

2 Job monitoring – use the Internet to find 20 companies that you would like to work for. Research them further and reduce the list to your 10 best companies. Then complete a spreadsheet like the one below. You must monitor these companies on a weekly basis.

Company	Website	Job roles	Recruitment contact	Jobs available	Location	Action
1.						
2.						
3.						

Figure 11.1: Your tutor may be able to give useful advice about the career you have chosen

Goals

Setting some realistic goals may mean the difference between success and failure when trying to get a job. Goals can be split into different categories: short-, medium- and long-term.

Imagine you wish to be a lighting design engineer and you want to start planning how to do this. You can set yourself the following goals.

Short-term goals

- Find out the facts about the job.
- Check out job availability.
- Find out who the local companies are.
- Write a personal statement, CV and covering letter.

Medium-term goals

- Contact local companies.
- Speak to industry bodies.
- Send in your CV.
- Set up work experience or meetings with companies to get more information.
- Investigate outside your local area.

Long-term goals

- Begin training course or further education.
- Start work experience.

After thinking through these questions and making notes on your approach it may be that the next action is to make a CV tailored to the specific job advert you have seen.

Progress check 11.1

1   When researching a job, give an example of a short-, medium- and long-term target.
2   Describe two steps that will help you get the job you want.
3   How can you identify any skill gaps you may have for a potential job?

Target setting

To achieve each of these goals you will need sensible targets. Targets must be SMART.

Specific
Measurable
Achievable
Realistic
Time-constrained

Targets need to be broken down into *specific* points that can be focused on. You also need to know when you have completed a target, hence there needs to be some form of *measurement*. If a target is not *realistic* and *achievable* then it is not worth attempting and it must be *time-constrained*, i.e. it must have a planned end date. If there is an end date for a target you are much more likely to achieve it!

SMART target example

Investigate five potential employers in the area of lighting design and send a CV with covering letter to them all by 25 November this year.

Specific – you have identified a particular type of employer.
Measurable – you have five target employers (if you only find three, you have missed your target).
Achievable and realistic – this seems quite possible!
Time-constrained – you have decided that it is possible by 25 November.

Activity 11.2

1   Write five SMART targets related to getting into employment or progression within your current role.
2   Discuss your targets with a classmate or colleague to check they are SMART.

Progress check 11.2

1   Write down an example of a target that is specific.
2   Write down an example of a target that is measurable.
3   Write down an example of an achievable target and explain why it is achievable.

CV writing

Creating a perfect CV can be daunting and it can take time to make it look good and have the correct impact on a potential employer – but it is worth getting right. The job market is currently very competitive and making

the CV fit for purpose is one way of increasing your chances of getting employment. Most job adverts will receive many times more applications than there are jobs. It would not be unreasonable to have well over 100 applications for a single job. All of the applications and CVs received for this one job will be filtered out and sifted so only the most appropriate applications pass through to the next stage. This next stage could be a simple telephone interview or one of many formal interviews. Either way, your response to an application by sending a completed form, CV and covering letter are the first contact a potential employer will have with you, so it must be good. Some companies admit that they use the first 10 seconds of reading a CV or covering letter to sift an application. A potential employer needs to see if you are a possibility as quickly as they can if they have hundreds of applicants. This means that an application can be put in the reject pile for very simple errors in spelling, punctuation and grammar, or if the applicant has made an incorrect assumption.

CV research and preparation

A CV will be different for each job application – unless the adverts and companies are identical! Careful research into the company and role advertised must be done to make sure that what you say in your CV matches the skills and experience required. The advert will give the main points a successful applicant has to meet as a minimum standard. Your CV must show this evidence clearly, concisely and be highlighted or obvious to the reader.

General rules of a good CV

There are many different types of CV formats and different companies may expect different things. There is no such thing as a perfect CV as it is down to the personal preferences of the person reading it. However, to increase your chances with an application, keep it simple, keep it short and tailor it to the job. A typical CV should be no more than two pages – ideally one. You should not try and get everything into a CV by reducing the font size or using multiple column formats. Choose a font size and type that is easily read and can be scanned in/recognised by computer software (point size 10–12 and Arial or Times New Roman are suitable). With most CVs it is a case of taking information out rather than putting information in to make it more concise and easier for an employer to follow.

It is a good idea to not only spell check and grammar check but also to get one or more credible people to read your CV.

CV layout

The information you put into a CV depends on what you are applying for. Generally, there are some standard items that must be included.

Name and contact details – this information should be at the top of the CV as some companies scan this information into a database. Make sure you have a good email address as an inappropriate address may put off a potential employer or even worse not get through the company mail system. It would be a shame not to get employment because of something as simple as an email address issue! Some companies might ask you to include a photo and this might be a good place to put it. A photo could be requested for any interviews on secure sites such as MOD or airports.

Personal statement or personal profile – this is read in the first 10 seconds of a recruiter picking up your application and tells them exactly why they must give you the job over the next applicant. This statement has to be strong, truthful, concise and match the job specification in the advert. Remember, if you get to the next stage and an interview you will be expected to defend what you have written here.

Work experience and skills – this heading gives you the chance to show your strengths to the potential employer. Any work experience is good in an application but bullet-point the most appropriate work or training for the job first. This will ensure your key strengths are read as early as possible in your CV. Any achievements during work experience should also be highlighted here. Sifting lots of CVs can be as simple as ticking a box at this stage so it would be a shame to miss out due to a simple omission! Supporting letters from work experience also look very good and can be stored up as trump cards in your portfolio should they be required as proof. If this is your first job application it is important to include any additional skills you have gained. Skills that will show an employer you are dedicated or evidence of a hobby that demonstrates leadership or management are particularly valuable.

Employment – this is the history of your jobs and is in reverse order of when you held these posts. It is very important that dates are put against each role you held. Against each role you need to describe any achievements or targets met. If you do not have any work experience this is where you should write your academic qualifications. This does not necessarily mean every single grade, just the main achievements and certificates.

Additional information – any other information that you feel is important, has not been covered or has particular relevance to the application can be briefly described in this section. You may have a pastime that is relevant to the job or shows particular skills that an employer might like, such as leadership skills. You may be a weekend volunteer or run a club – all of which shows a potential employer that you are focused and self-motivated. You may compete at a high level in sport, or even coach. You may be a student representative, member of the student council or member of a school/college leadership team. All of these show an employer that you have skills that could apply to the personal specification in the job advert. You must make sure that they are relevant and tailored to the advert to increase your job prospects.

Pastime/hobby/skill	Possible relevance to employer that needs to come out in CV and interview
Run an Internet business	Computer skills, self-motivation, dedication, initiative, ambitious
Referee or run the line in weekend football league	Dedication, time-keeping, responsibility, can handle conflict, team player, decision-maker
Karate black belt	Dedication, self-motivated, not afraid of hard work, self-discipline
School/college leadership team Student representative	Dedication, self-motivated, self-discipline, initiative
Volunteer work	Self-motivated, maturity, time-keeping, hard-working, community spirit, initiative, focused
Help in class – peer teaching	Self-motivated, mature, hard-working, focused

Table 11.1: Spare-time pursuits and their relevance to job applications

CV examples

SARAH SMITH
sarah.230smith@email.co.uk 07771 8888888
100 Any street, Any town, N12 0TE

PERSONAL STATEMENT
I have always had an active interest in practical subjects, working with my uncle as an electrician's mate at the weekends for the past year. My naturally inquisitive nature means I always want to learn how things work and gain new skills. My ambition is to become a fully qualified electrician specialising in testing, inspection and fault finding as this is what I have enjoyed most during my work experience. My major strengths are working within a team and problem solving (see attached reference).

WORK EXPERIENCE
Electrician's mate (weekends) – September 2011 – present
Responsibilities:
• 1st fix installation work and 2nd fix – assist lead electrician as directed
• Stock ordering for all new installations

Cancer research charity shop volunteer – June 2011 – July 2011
Responsibilities:
• Customer reception – greeting customers with items to leave at the shop and sorting into saleable and non-saleable stock
• Sales – working with customers, dealing with money/credit card transactions

McDonald's weekend crew member – January 2011 – March 2011
Responsibilities:
• Customer sales – processing orders and money at till points
• Crew member – working in high intensity dynamic grill kitchen and general duties

EDUCATION

Subject	Qualification	Awarding body	Grade	Date
English	GCSE	Edexel	C	June 2012
Science (triple)	GCSE	OCR	BBB	June 2012
Maths	GCSE	Edexel	A	May 2011
ICT	OCR National	OCR	Distinction	June 2012
Construction	BTEC	Edexel	Distinction	June 2012

PERSONAL INTERESTS
Coaching junior team hockey and playing at County level for the past 3 years.

REFERENCES
Sir John James Humphreys – Founder and Chairman for Southgate Hockey Club
Julie Jones – Manager, McDonald's, Southgate
Bob Smith – Director, AJ'S Electrical Contractors

Figure 11.2: CV Example 1

Good words	Poor words/phrases
created	we followed
developed	we watched
managed	we helped
achieved	tried
led	quite liked it

Table 11.2: Examples of good words to use in your CV – and ones to avoid

Some common pitfalls with CVs

The most common issues with CVs are spelling and grammar mistakes. Spell check does not always pick up words which are spelt similarly but which have completely different meanings. A thorough read-through by an experienced person is a wise move.

CVs should be very positive documents so leave out anything negative, especially if it is about a previous job or manager. Bad mouthing a previous employer will not go down well with a potential new manager, so avoid it at all cost.

CV – Deepak Deepchand

Address: Any road, Plumpton, London, S14 0PU
Email address: deepak@email.co.uk
Telephone number: 0208 334 3344
Personal profile:
I am an attentive, hard-working individual. I have a keen interest in science and books, I also like to listen to music and play the piano (Grade 3). I am artistic, love photography and have taken A-level art a year early.
Educational details:
I attended Monopoly primary school from 1998–2005.
I attended Ashhurst school from 2005 – present, currently in the sixth form (year 12).
Examinations:
GCSEs taken and results:
English language – A; English literature – B; Chemistry – B; Physics – C; Biology – B; Maths – A*;
Art – A*; I.T. – Merit; R.S. – A; Spanish – C; Resistant Materials – D
AS levels taken and results:
Maths (fast track) – C
Work experience:
One week at Parkside Electrical Ltd, September 2011
Interests:
Science and Electronics – Junior member of the IET. I am a part-time football referee for a local Sunday league.
References:
Jonathan Mundain – Senior Electrician, Parkside Electrical Ltd

Figure 11.3: CV Example 2 – cut-down version, ready to be tailored

Take great care with the language you use in a CV. Avoid slang, abbreviations, jargon and clichés – keep the language simple and to the point. Some trade-specific phrases or terms can be used if they are used in the advertisement. The choice of verbs in your CV can change the way you come over dramatically. Use action words and phrases. Never use the word 'we' as this is all about what *you* have achieved.

Too much personal information can take the place of valuable space on a CV. You do not need to say that you have children or are married as this holds little relevance to the job. A date of birth is also irrelevant in a CV.

Portfolio

It is a good idea to build a portfolio of evidence that you can take to an interview. This portfolio could include references, certificates, awards and

academic achievements or evidence of personal hobbies (especially if they are at a very high level such as an instructor). Keep spare copies of your CV in your portfolio as you may be interviewed by several people and spare copies will make you look very prepared and organised.

Application forms

When applying for some jobs the process can involve completing an application form, by post or more increasingly online. These forms are very structured and great care must be taken to read them thoroughly before putting pen to paper or cursor to screen. Your initial response might change once you have read the form completely and it might be too late to change it. You must always consider the information in the advert when completing an application form, being careful to use the key words and clues in the advert.

Job Application Form

In order for your application to be processed, please complete all sections using BLOCK CAPITALS.

Position applied for:

Personal details

Name:

Address:

Postcode:

Telephone number:

Mobile number:

e-mail address:

Employment history

Please detail your current and previous job history and give an explanation for any gaps in employment.

Name & address of employer (Most recent first)	Job title & main responsibilities	Reason for leaving

Education

Please give details of the School/College/University you attended

School/College/University name and address	Dates attended

Figure 11.4: An example of a job application form. Further pages are likely to ask for details of your qualifications, other skills, and the names and contact details of your referees

If you are completing an online application, once you hit the submit button the next time you see the information could be when you are in an interview. Print off all screens of your application and keep copies in a file for that particular job. This will form the basis of your preparation if you get to the interview stage.

References

References are always a good idea as they can give the potential employer a further insight into the interview candidate. The best place to get a reference would be from an employer or previous training college or school. The reference must refer to the candidate, be dated and signed and give an indication of the candidate's work, attitude and time-keeping as a minimum. All of these points can and often are followed up by a phone call to the referee, so make sure they are willing to be contacted.

Covering letter

When someone who is recruiting receives an application, the first thing they will expect is a covering letter.

A Salmon
Blandfood Rd
Eckleston
BR1 1AS

Job reference: 123JG

Dear Mrs Jones,   28th April 2012

(Para 1 – introduction and acknowledgement of the job and advert)
My name is Andy Salmon and I am writing to you with reference to the recently advertised job on your website posted on 22nd April with a closing date of 22nd May.

(Para 2 – explanation for application)
I am currently in my final year of A levels and will be ready to start full-time employment from 1st July. Ever since I first worked with my Uncle and his electrical company I have wanted to do this as a full-time job.

(Para 3 – why you match the criteria)
I am applying for the role of electrician's mate based in Apsley Mills. I feel my strength as a team player and technical skills I have gained recently on work experience with an electrician will enable me to start straight away.

(Para 4 – strong conclusion)
I have enclosed my CV for you to consider. I would welcome the opportunity to present myself at an interview. I will call your secretary on Thursday to make sure my application has been received. I very much look forward to hearing from you next week.

Yours sincerely,

A Salmon

Figure 11.5: An example of a covering letter

One purpose of a covering letter is to introduce you as the applicant to the potential employer but ultimately it is to get you to interview. The covering letter is formal, polite and should set the scene for you. The content must include:

- who you are
- why you are applying
- what you are looking for
- how you match the needs of the role.

At the end of the letter make sure you thank the reader for taking the time to consider your application and don't forget to ask for an interview – if you don't ask, you won't get!

In total your covering letter should be no more than five small paragraphs and be specific for the job – never use a generic covering letter as these are easily spotted. The company has taken the trouble to open and read your letter so you should take the time to research and make the letter specific to them. You should also use the same language as that used in the advert, picking out the key words.

Activity 11.3

1 Write a CV for an advertised job in the specialist area you are interested in.
2 Write a covering letter for this job application using the information in the advert to make it specific.

Progress check 11.3

1 Name five important things to include in your CV.
2 Name five common mistakes in a CV.
3 What should be included in a covering letter?

The interview

Create an interview plan and checklist – leave nothing to chance. You can't control what will happen in the interview but you can make sure you have done everything possible up to the point when you walk into the interview.

An interview checklist should include everything that is within your control. It will make you calmer and allow you to get in the correct frame of mind to simply concentrate on what you have to do on the day. A checklist will make sure as many things as possible are planned, accounted for and organised. So what should be in an interview checklist? Here are some ideas.

Figure 11.6: Leave nothing to chance when you are preparing for an interview

Preparation planner for week before interview (interview is on 8 November)				
Planned activity	**By when**	**What is involved?**		**Completed**
Confirm location date, time, who is interviewing	1 Nov	Speak to company reception/secretary.		
Confirm local parking/restrictions	1 Nov	Speak to company reception/secretary.		
Pick the dress, trouser suit, suit, shirt, tie, shoes you will wear Try on, wash/iron if required	1 Nov	You may need to organise dry cleaning or go and buy new interview clothes or suit.		
Dry-run journey the same day but a week earlier	1 Nov	Drive/use the same method of transport as per interview day.		
Parking check	1 Nov	Work out the parking arrangements and money/drop-off points if you are getting a lift.		
Prepare your portfolio	3 Nov	Photocopy certificates, CV, supporting documents, references, awards, your application.		
Your pitch!	3 Nov	Refer to your application and CV and write a concise one-minute message about you and your strengths and how your skills match the job specification. You might end up in the lift with the managing director or the person interviewing you so it is worth having a well-rehearsed sales pitch about yourself.		
Create the right image	4 Nov	Work out your 'happy thought'. It is difficult to look unhappy if you are thinking happy – create an image in your head that you can call upon just before you walk into the interview. This will help with your nerves. Don't forget it is good to be nervous as nerves can make you sharper and more alert.		
Haircut	5 Nov	Invest in a good haircut – this will create the right image and make you feel more confident. Put off the pink streaks in the hair until after you have the job and know what is expected of you!		
Check your application	5 Nov	Read through all your application documents and CV to refresh yourself on exactly what you have said about yourself.		
Prepare answers	5 Nov	Look at the advert and any documents supplied to identify the skills and qualities they are looking for – find three examples for each point that prove you have what it takes. Write them down and practise how you will answer. Use friends or family to simulate the interview.		

The day

With all the preparation complete you should be able to start relaxing. You know what you are going to wear and are happy that all your clothes are clean, neat and appropriate. Your journey and parking are planned and you are as confident as you can be because you did a dry run exactly a week ago on the same day of the week. You have the correct change and also some spare just in case the machine swallows your money. Your alarm and back-

Chapter
11

Figure 11.7: Leave nothing to chance when you are preparing for an interview

up alarm have worked and got you up in time for a good breakfast. All you need to do is pick up the prepared interview pack and rehearse the possible questions and answers in your head and from your notes.

The last things you should do before going into the actual interview are to take a couple of minutes to deep breathe, maybe visit the toilet (nerves can play tricks on you!) and then start thinking about your pre-planned 'happy thought'. The happy thought should help settle any last-minute nerves that can't be helped. Even though you have planned all parts of the interview process that are within your control, nerves will still play a part, so a happy thought is essential just before you put your hand on the interview room door handle.

The interview questions

Interviewers are not there to trick you. They are there to find the right person for their company in a very restricted amount of time. Their job is to prise out evidence from an applicant that proves the person sitting in front of them is right for the job. This can be quite frustrating for the interviewer also.

Interviews can be quite formal with tables and chairs laid out like a board room meeting or the layout can be very casual like a table in a café. Either way you should treat them the same and prepare in the same way. The questions will still need to be answered and the information still needs to be put across so the interviewer is satisfied.

A formal type of interview that is run by a large number of big corporations is called the 'competency-based interview'. This interview takes the job specification and breaks it down into various 'competencies' that a successful candidate must meet. Each of these competencies is then questioned by the interviewer until they have all the evidence or they are confident that they can go no further. Some interviews appear to be going wrong but often it is simply time-constraints and frustration. The candidate may have all the experience and skills for the job but the interviewer can't get the information they need. This can be because the candidate does not understand the questions or they are rambling about an unrelated subject. The interviewer sometimes has to be quite strict and stern to get the interview back on course. If you feel an interview is starting to go wrong, it will generally be down to the interviewer not getting the information they need – don't forget, it's not personal!

The interview should begin with the interviewer setting the scene, making sure you are comfortable and are aware of the facilities. The next part of the interview can be where your CV is checked by a series of short questions asking you to explain certain parts and any gaps in employment or training you might have. The tricky part of the interview follows next where more in-depth searching questions are asked.

Sample questions

Here are a set of typical questions for you to consider. Try and imagine how you would answer them.

- Why are you here?
- Why do you want the job?
- Tell me about yourself.
- Why should I give you the job?
- What can you bring to the job that makes you special?
- How would a work colleague describe you?
- What three words best describe you?
- What are your strengths?
- What are your weaknesses?
- What do you think is the most important thing about the job you have applied for?
- What do you know about this company?
- What is the main strength of this company?
- What have you done to prepare for this interview?
- If you could change one thing about yourself, what would it be?

A different type of question will be asked when they are happy that all of your CV is in order and they understand a little about you as a person. These questions will deal with the specific skills required for the job. This line of questioning will be looking for evidence through personal experiences. These questions will be open questions designed to get you to talk about your skills. Some examples of these types of questions are given below.

- Give me a recent example of when you dealt with conflict.
- Give me a recent example of when you had to solve a difficult problem.
- Give me a recent example of when you had to work in a team to achieve a task.

Other follow-up questions often include the following.

- Why did you do that?
- What else did you do?
- What could you have done differently?
- What impact did that make?
- How do you know?
- How do you know you were successful?

These exploring questions are asked to make sure the interviewer has exhausted every possible avenue. The interviewer wants to know everything about the particular example you are discussing. To find as much information as possible, the interviewer will ask very open questions to avoid leading you into a particular area or subject. The interviewer generally does not want one-word answers. Simply, they want you to talk – but only about the areas they need to find out about. Be prepared for this type of questioning as it can seem intimidating and almost aggressive. It is not intended to be; it is just down to the interviewer, the limited time they have to get the information and the sheer volume of people they have to see. The easier you make it for them to get the information, the easier it will seem to you.

As a final part of the interview, you will be given the opportunity to ask questions. This is something you really need to research and think about before the day. You do not want to seem too smart by asking difficult questions but you can ask general things. It is also a good idea to ask questions as it shows you have prepared well. A key question is to find out when the results of the interview will be known. Other questions you could ask include the following.

- How well is the company doing? What are the company's growth plans?
- What development opportunities are there for a new employee?
- What is the interview decision-making process from now?
- When the decision is made, I would like some feedback – how do I go about getting this?
- What is the company policy on training?
- Who do you see as your main competitors?

Figure 11.8: Prepare some questions you can ask at the end of your interview

After the interview

Regardless of how you think the interview went, if you planned it well and carried out all possible actions, you have done your best. The important thing now is to stay focused on getting the job as your chances can be affected by your next actions. If you are told at the interview that they will let you know in one week, ask about the process for selection and who will make contact. You may be expected to contact them so it is good not to leave these things unsaid. If you have not heard anything after one week, you can either wait another day or contact the company for the result. You must also be mindful that if there are a lot of candidates going for the job, there may be a considerable number of people in exactly the same position as you – be patient.

If you were not the strongest candidate on the day and you are rejected, it does not need to stop there. You need to learn from the experience and improve for the next opportunity. Remember, most people do not get the first, second or even third job they apply for. You are within your rights to ask for feedback from the interviewer. They may choose not to give it to you but they can't be offended if you ask.

Once your interview is over, you will have used a lot of nervous energy but you still have work to do before you relax. Take down some notes on the interview questions and the answers you gave. This can help you with follow-on interviews and future opportunities. It may seem like the last thing you want to do but if you discuss this with a more experienced person they might be able to develop your interview techniques. Some people do this very well and can pass almost any interview simply by understanding how the process works.

Career progression

After a time working in the same job progression may be possible. You may have a change in circumstances or you may have identified a different role that really appeals to you. Planning for change can be approached in exactly the same way as getting your first job.

As a technician in building services engineering you have many routes available to you, either through extra training or sideway moves to different but related trades.

As a tradesperson you can progress from a basic labourer completing supervised tasks through to supervisory or managerial roles that carry much higher levels of responsibility and ultimately higher salaries. Within the electrical arm of building services there are also other more diverse routes that you can move into, such as specialist electricians on railways, high voltage or boats. As you progress through the various grades of electrician other roles are available to you through extra training or courses, such as assessor, estimator or project manager.

Activity 11.4

1 Use the Internet to investigate different specialist electrical roles that are available. Write a brief report to your training manager explaining the specialist area you would like to move to and the training courses/qualifications required.

2 Investigate the different levels of electrician using the Joint Industry Board (JIB) website.

3 Use the Internet to investigate the different qualifications available to become a fully qualified project manager.

4 Discuss with a colleague or classmate the questions you would find most difficult to answer in an interview. Write them down and work on your perfect answer.

Progress check 11.4

1 Name four things that should be in your interview checklist and put them in priority order for you.

2 List six questions you can expect to get asked at interview.

Of course progression could even mean it is time to investigate setting up your own company.

Business plans

If you are in a position to start your own business then a number of things will need to be considered. A first step when starting a business is the business plan. A carefully considered business plan is required to raise money with any bank. A business plan can be quite a complicated process but essentially it starts with making your mind up as to what exactly you are going to do – a business idea. The next step is to decide how you are going to achieve it.

Researching your business idea

For a start-up business to be a success you have to know your target market and if there is room in the market for you. There are lots of web applications that can help at the research stage, especially with market research (scavenger.net is one example). Here are typical questions you will need to answer before going much further and investing your own money in a business venture.

- Who are your competitors?
- What do they charge?
- Is there a growing need for your business?
- Is there anything that you do that would make your business unique?

As the business idea becomes firmer a formal plan will need to be created. Some people will just start a business but most go through some form of formal planning process.

The basic business plan

A business plan can help you to focus on exactly what it is you intend to do but its main purpose is to create a professional-looking document that can be taken to other businesses or banks so that they can lend money to finance what you intend to do. A business plan can vary according to what it is intended to be used for and will need to be changed and adapted as you research. A business plan will also need to be tailored to the bank you are approaching for finance or the partners you wish to join up with. Each bank will supply its own advice and format for the plan. Most business plans will contain a standard set of titles and content that you will need to complete and review as the business grows.

Executive summary

This section is like the personal statement in a CV. It will give all the main facts about what you are about and are trying to achieve. As a guideline, the executive summary should contain:

- a summary of what your business is about, what it does, what makes it unique
- your experience (to make any potential investors comfortable)
- how the business will be profitable
- how a lender can guarantee their investment.

Aims and objectives of the business

This section describes the mission of the business. Some people start a business solely as an investment. If this is the reason then it must be stated here as the business will need to start and get into profit as quickly as possible so it can be sold on. However, you might want to set up a business so that you can create a family business working with relatives and children. The aim of the business will have an impact on any potential investors and what they might want to put into your business.

A description of the business

In this section you need to describe exactly what your stated business is. This is often where you could write a company mission statement. An example of a mission statement for a small electrical company could be the following.

'A family-run electrical installation and 24-hour maintenance company specialising in domestic and small local commercial business in the South of England and home counties, using good-quality market-leading electrical products from local suppliers.'

Financial status

This section briefly describes your current business finance.

Following on from the finances required to start up, a forecast will be required. This will help the money lender or partner to make a judgement about the business they are about to invest in. It will also help the onward planning once the business has started.

Management

Although your company may only be one person initially a statement is still required to show how things will be managed. Any company, however small, will have lots of roles and functions that require thought and a degree of planning. As a sole trader (see page 518) you will have to do your own marketing and sales (work out how to get business in the most cost-effective way), complete administration, finances, stock control and research. Marketing and sales may require its own section so that a more detailed plan can be put down on paper showing how business will be found and grown for the new company.

SWOT analysis

A good way to show an investor you have thought about your business in detail is to include a SWOT analysis in your business plan.

Strengths
Weaknesses
Opportunities
Threats

The opportunities and threats are external to the company. Business opportunities might include changes in the law or planning rules that create a new demand. Threats might also arise due to changes in the law. The economy might create market difficulties that could threaten your business if they are not planned for in some way.

An easy way to start a SWOT analysis is by filling out a table like the one shown in Figure 11.9. The more honest you are, the better it will be for your business in the long term.

Strengths	Weaknesses
Highly qualified 5 years' specialist experience Finance secured for first year Market research completed and demand is there for the business Own a van	Difficult economic climate Small team of one Very specialist market Equipment is expensive Old van
Opportunities	**Threats**
Large company in area already successful Change in regulations means market demand will be growing Strong growth in local rental market	Large company in area already successful with market share Limited availability of low interest-rate finance for the equipment/new van required if company grows

Figure 11.9: A basic SWOT analysis for a testing company

Activity 11.5

Imagine you are turning your hobby into a business and complete a SWOT analysis.

Legal status

Any companies looking to invest money or resources will want to know the intended legal status of your company.

- Sole trader – if you are a start-up company you might be a sole trader.
- Partnership – if you intend to go into business with someone else you might need to register as a partnership.
- Limited company – if you are successful or have matured into a bigger business it may be the right time to register as a limited company.

Operational requirements

This section describes what you need to get the business going and to continue once you have started. You may need a new van or car, tools, test equipment, insurance, compliance certification, membership to official industry bodies and stock. This will give a realistic indication to any potential investor of what the borrowed money will be used for and also roughly how much is required to keep the business going during its start-up period.

BUILDING SERVICES ENGINEERING – BECOMING FULLY QUALIFIED

The building services industry – competent person schemes

The reputation of the building trade has always suffered because of the actions of a few rogue cowboy builders or tradespeople. In 2002 the government decided to take action and introduced 'the competent person self-certification scheme', otherwise known as competent person schemes (CPS). These schemes allowed businesses that were experienced and competent in their particular trade area to carry on trading but certify that their work met with all the latest requirements of the building standards.

The schemes also assist local authorities with the enforcement of building regulations. The consequences of not meeting these can be large fines and prosecution. Many prosecutions have been successful since the start of CPS, with some fines being as large as £15,000. Ultimately, if the level of work carried out is so poor that damage or injury occur, the consequence could be the loss of licence to trade or even prosecution leading to imprisonment.

There are several advantages to tradespeople, including:

- registered tradespeople no longer need to arrange site visits for inspectors to come and sign off each job
- clients' bills are reduced without the need for expensive extra inspection visits.

Most of the trade areas have a list of approved competence schemes. Currently there are around 18 approved schemes for building-related trades. Some trades, for instance electrical, have several approved schemes to choose from depending on the benefits of the scheme for the individual company.

Full legal name of scheme	Acronym	Web address (external links)
Ascertiva Group Limited	NICEIC	www.ascertiva.com
Association of Plumbing and Heating Contractors (Certification) Limited	APHC	www.aphc.co.uk
Benchmark Certification Limited	Benchmark	www.benchmark-cert.co.uk
BM Trada Certification Limited	BM Trada	www.bmtrada.com
British Institute of Non-Destructive Testing	BINDT	www.bindt.org
British Standards Institution	BSI	www.kitemark.com
Building Engineering Services Competence Assessment Limited	BESCA	www.besca.org.uk
Capita Gas Registration and Ancillary Services Limited	GSR	www.gassaferegister.co.uk
Cavity Insulation Guarantee Agency Limited	CIGA	www.ciga.co.uk
CERTASS Limited	CERTASS	www.certass.co.uk
ECA Certification Limited	ELECSA	www.elecsa.co.uk
Fensa Limited	FENSA	www.fensa.co.uk
HETAS limited	HETAS	www.hetas.co.uk
NAPIT Registration Limited	NAPIT	www.napit.org.uk
National Federation of Roofing Contractors Limited	NFRC	www.competentroofer.co.uk
Network VEKA Limited	Network VEKA	www.networkveka.co.uk
Oil Firing Technical Association Limited	OFTEC	www.oftec.org.uk
Stroma Certification Limited	STROMA	www.stroma.com

Table 11.3: Competence schemes for building services engineering

Competence schemes are authorised according to the type of building work and are voluntary. A trader has the option to join a scheme or continue organising work sign-off visits by local authority or approved inspectors. There may be quite a few schemes for each type of work and it is down to the individual trader to choose the scheme that best suits their organisation. Most members of these schemes are sole traders or small companies.

Consequences for the customer

Many jobs within the home need local authority approval. As well as the tradesperson, the customer also has a responsibility to fulfil. Failure by a customer to comply with building regulations could lead to them being fined up to £5,000.

Sub-standard work carried out by a sub-standard tradesperson employed by a customer looking for a cheap job will undoubtedly have further consequences. Sub-standard work will fail eventually and it is the level of failure that needs to be carefully considered. A bad window installation may simply leave a house cold but it could also fall out and injure someone If a passer-by is injured, the householder will be responsible. A sub-standard gas installation, however, has far-reaching consequences that could ultimately lead to serious injury or death. If a customer employs tradespeople that are not CPS registered and the work is found not to be to building regulation standards, the customer may well find that legally they have to pay for all corrective work. This expense could lead to a far greater outlay than if the job was carried out to the correct standard by registered tradespeople in the first instance.

When a property is sold and a solicitor requests a local authority search, any substantial work will be highlighted. At this point a solicitor can tell if the work was carried out by a CPS-registered tradesperson. This could lead to the customer having difficulty selling their property at a later date if the work has not been completed to the correct standard with the correct permission.

Work can still be signed off without a CPS-registered tradesperson but it will be the responsibility of the house owner to contact the local authority and arrange an inspector to come and sign the work off at the customer's expense.

The trades that are represented by the competency scheme, as detailed on www.gov.uk/building-regulations-competent-person-schemes, are broken down into:

- air pressure testing of buildings
- cavity wall insulation
- combustion appliances
- heating and hot water systems
- mechanical ventilation and air conditioning systems
- plumbing and water supply systems
- replacement windows, doors, roof windows or roof lights
- replacement of roof coverings on pitched and flat roofs and any necessary connected work (but not solar panels)
- micro-generation and renewable technologies (extra consumer protection is provided by the Micro-generation Certification Scheme and the Trustmark scheme)
- electrical installations.

Electrical installations authorised schemes include Benchmark, BESCA, BSI, ELECSA, NAPIT, NICEIC, OFTEC and STROMA. Electrical schemes can be split into electrical work in a dwelling (i.e. domestic house or flat), including lighting systems, or it can be further defined for those companies that do electrical work as extra work to their main job such as kitchen fitting, gas installations, bathroom fitting, security or fire alarm installations.

Each of the competent person schemes mentioned above cover a range of work. If the work is not carried out by a CPS member then the local authority needs to be notified and involved.

Examples of trade work that require CPS or local authority to be notified include:

- change or replacement windows and external doors
- replacement/change of roof covering on pitched/flat roofs
- replacement or installation of oil fuel tanks
- replacement or installation of boiler or heating system (all types of fuel)
- new bathrooms/kitchens if new electrical work/plumbing installed or altered
- new fixed air conditioning
- new electrical installations outdoors
- addition of radiators to existing heating system.

Examples of trade work that do not require local authority building control notification include:

- like-for-like replacement of baths, basins, sinks and toilets
- additional electrical power/light points
- alterations to existing electrical circuits except in kitchens, bathrooms and outdoors
- most minor repairs and replacements, except oil tanks, combustion appliances, electrical consumer units and double glazing.

Case study

A house owner had a new 'eco-friendly' house built, taking three years. The customer chose many new technologies including ground source heat pumps, grey water harvesting and wind turbines. The customer did not use a registered installer and decided to save money and contact the local authority later. After completion the owner invited the local authority to carry out a final certification/inspection visit. The installation did not meet all the required building regulation standards and the heating system had to be re-installed at great expense to the owner. The case ended up in court with the owner attempting to get back costs from the installer. The case took two further years to reach settlement.

Checking competence

Working in the building services industry will require proof of competence. On many construction sites access is denied until you can produce the correct paperwork or, increasingly, the correct competence card. Some construction sites actually have a 'no card, no job' policy. These cards are issued by the various schemes associated with the building services trades listed earlier.

The competency card is a good way of securing work on site but it also allows an employer to know exactly the skill sets and type of work they can expect from an employee. In summary, competency cards protect the individual, the customer and employer.

Construction Skills Certification Scheme (CSCS)

The CSCS card is becoming a requirement on all major construction sites and is valid for all trades. The idea of this card is to identify the owner of the card, give a skill level and identify the particular trade they are trained in. When the card is checked on site by the project manager, security or foreman, they are left in no doubt as to the competence of the individual.

The levels and the various trades that can be listed on the card are updated as the building services engineer progresses with training. The Gold card shows the owner is a skilled worker in their particular trade area. The minimum Red card standard proves the card holder will have completed the Health, Safety and Environment test. The CSCS card will give you the ability to work on any construction site in the UK.

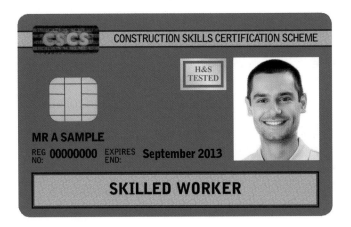

Figure 11.10: A CSCS card

Card colour	Qualification criteria
Red	Currently does not have NVQ at any level but is registered on a course and has taken the Health, Safety and Environment test
Green	Completed the Health, Safety and Environment test and a Level 1 construction NVQ
Blue	As above but with NVQ Level 2
Gold	As above but with NVQ Level 3
Black	This is for managerial, senior construction roles and is subject to level 4–7 NVQ qualifications

Table 11.4: The CSCS colour code

SKILLcard

For some tradespeople on site the NVQ route of education was not available or appropriate for their particular circumstances. A different card is available called the SKILLcard, issued by the same body, CSCS. This card covers occupations relating to heating and ventilation, air conditioning and also speciality trades associated with the building services industry, such as domestic heating and ductwork. This card scheme is also colour-coded and based on national qualifications and levels of experience but with no reference to NVQ levels. Although these cards are not as widely accepted they are still valid on some construction sites.

Electrotechnical Certification Scheme (ECS card)

There are different scheme cards for all the trade areas. For specialist areas like the electrotechnical industry it is very important to understand exactly what an operative is qualified to do before they are allowed to start work. This is to protect the operative as well as the customer, other tradespeople and employers.

The CSCS and ECS cards are combined into one for the electrotechnical trades. This means that when an electrician is asked for a CSCS card, the ECS card is actually the one relevant to the electrotechnical industry. ECS is the sole identity and competence card scheme for electrotechnicians in the UK and it is recognised and endorsed by the industry – the 'gold' standard.

To qualify for an ECS Installation Electrician card, applicants must:

- have completed an approved apprenticeship (successfully) *OR*
- hold the C&G 2360 part 1 and 2, or C&G 2351 or C&G 2330 Level 2 and 3 or an approved equivalent qualification *PLUS*
- hold an NVQ Level 3 in installation and commissioning *AND*
- hold a current, up-to-date health and safety certificate or recognised health and safety qualification such as the ECS Health and Safety test *AND*
- hold a formal BS7671 qualification in the current edition (BS7671:2008, 17th edition).

Figure 11.11: ECS Installation Electrician card

Knowledge check

1 What does CSCS stand for?

  a Certification Skills Construction Scheme

  b Construction Skills Certification Scheme

  c Certification Scheme for Construction Staff

  d Construction Staff Certification Scheme

2 What does ECS mean?

  a Electrician Certification Scheme

  b Electrical Construction Scheme

  c Electrotechnical Certification Scheme

  d Electrician and Construction Scheme

3 What does NAPIT stand for?

  a National Association for Professional Inspectors and Testers

  b National Association and Professional Institute for Technicians

  c National Association for Professional Inspectors and Technicians

  d National Association for Professional IT consultants

4 What is SWOT?

  a Strengths, weaknesses, opportunities and targets

  b Strengths, weaknesses, order and targets

  c Samples, weak points, order and targets

  d Strengths, weaknesses, opportunities and threats

5 What is a SMART target?

  a Stretching, measurable, achievable, realistic, time-constrained

  b Stretching, measurable, active, reachable, time-limited

  c Specific, measurable, achievable, realistic, time-constrained

  d Specific, measurable, achievable, realistic, topical

Index